STUDIES IN COMPUTATIONAL MATHEMATICS 12

Editors:

C.K. CHUI
Stanford University
Stanford, CA, USA

P. MONK
University of Delaware
Newark, DE, USA

L. WUYTACK
University of Antwerp
Antwerp, Belgium

ELSEVIER

Amsterdam – Boston – Heidelberg – London – New York – Oxford – Paris
San Diego – San Francisco – Singapore – Sydney – Tokyo

TOPICS IN MULTIVARIATE APPROXIMATION AND INTERPOLATION

TOPICS IN MULTIVARIATE APPROXIMATION AND INTERPOLATION

edited by

Kurt JETTER
Universität Hohenheim
Germany

Martin D. BUHMANN
Universität Giessen
Germany

Werner HAUSSMANN
Universität Duisburg-Essen
Germany

Robert SCHABACK
Universität Göttingen
Germany

JOACHIM STÖCKLER
Universität Dortmund
Germany

ELSEVIER

2006

Amsterdam – Boston – Heidelberg – London – New York – Oxford – Paris
San Diego – San Francisco – Singapore – Sydney – Tokyo

ELSEVIER B.V.
Radarweg 29
P.O. Box 211, 1000 AE
Amsterdam, The Netherlands

ELSEVIER Inc.
525 B Street
Suite 1900, San Diego
CA 92101-4495, USA

ELSEVIER Ltd.
The Boulevard
Langford Lane, Kidlington,
Oxford OX5 1GB, UK

ELSEVIER Ltd.
84 Theobalds Road
London WC1X 8RR
UK

First edition 2006

Library of Congress Cataloging in Publication Data
A catalog record is available from the Library of Congress.

British Library Cataloguing in Publication Data
A catalogue record is available from the British Library.

ISBN-10: 0-444-51844-4
ISBN-13: 978-0-444-51844-6
ISSN: 1570-579X

∞ The paper used in this publication meets the requirements of ANSI/NISO Z39.48-1992 (Permanence of Paper).
Printed in The Netherlands.

PREFACE

Multivariate Approximation and Interpolation has been an active research area in applied mathematics, for many years, and has had impact on various applications, in computer aided geometric design, in mathematical modeling, in computations with large scale data, in signal analysis and image processing, to mention a few. More recently, approximation theoretical ideas have shown to be useful even in the analysis of learning algorithms. It is the purpose of this book to give an overview of some - although selective - areas in this field in order to have a compact and up-to-date edition of issues in this stimulating topic. We hope that such a volume will be a good basis for graduate students and young researchers to dive into the subject, and a valuable resource of information for all researchers working in the field.

The eleven articles in this book are written by leading experts, who have been invited to communicate their experience and knowledge in a particular subject. The contributions are mainly written as surveys, with much background to start with, with a presentation of the main achievements from the past to the present, leading the reader finally to the forefront of research. The authors were also asked to provide an appropriate, although not comprehensive, list of references. We thank all the contributors for their support in this ambitious project, and for their immense efforts to make this collection of articles a highly valuable piece of work.

A short description to each chapter follows:

Durrmeyer Operators and Their Natural Quasi-Interpolants deals with a class of new polynomial reproducing quasi interpolants on simplices which were recently discovered by two of the authors. Their construction deviates from the usual approach using summability, and is based on new identities for Bernstein basis polynomials. The article not only provides a survey on the spectral analysis and the approximation properties of these operators, but in addition points to an interesting connection with hypergeometric series. In particular, a striking result on the property of a certain kernel function being pointwise completely monotonic is proved. The results are expected to provide a useful alternative for the construction of high order linear approximation schemes in function spaces of several variables.

The second chapter *Three Families of Nonlinear Subdivision Schemes* is written by Nira Dyn, who has been at the forefront of research in subdivision, for many years. The present article describes three more recent issues in the field, which deal with nonlinear schemes. First, control polygons with strong nonuniformity concerning the length of edges, are discussed. Next, local weighted essentially non oscillatory schemes are constructed which have the advantage to depend continuously on the data. And finally, subdivision schemes on manifolds are derived which are modifications of converging linear schemes, and which are analysed by their proximity to these.

The chapter *Parameterization for Curve Interpolation* by M. Floater and T. Surazhsky considers the approximation order for curve interpolation by parametric spline curves. The authors explain that, for the clamped cubic spline interpolant, the chord length parameterization gives full order of approximation as measured in the Hausdorff distance. Moreover, a bootstrapping method for improving the parameterization is proposed in order to obtain optimal approximation order for higher degree spline interpolants, such as the two-point quintic Hermite scheme of order 6. A short survey of degree-reduced schemes is also included.

In the chapter *Refinable Multivariate Spline Functions*, T. Goodman and D. Hardin present a very general view on what is probably the most important building block in wavelet analysis: refinable functions and especially those from spline spaces in one and more dimensions. Both gridded data and general triangulations are considered. With the former, the well-known box-splines and the so-called new multi-box-splines are linked. The latter are addressed in connection with continuous differentiable spline functions and with piecewise linear splines. The article is a very comprehensive review with several examples, where the numerical stability of the functions in the presented approaches is of special interest.

In the chapter *Adaptive Wavelets for Sparse Representations of Scattered Data* A. Kunoth considers the problem of scattered data fitting by a sparse wavelet representation. The presented schemes are based on least-squares approximation and wavelet thresholding. The considered methods are data-dependent and operate by adaptive refinement in a coarse-to-fine manner. In particular, the initial step of typical wavelet methods is avoided, where gridded data on a "finest" resolution level must be generated. The chapter also discusses the main ideas for solving large scattered data problems including the multilevel regularisation and the treatment of outliers in a concise way. With this chapter the author gives a very good survey on recent developments in this area.

The author of the chapter *Ready-to-Blossom Bases in Chebyshev Spaces* is a well-known expert especially in the theory of blossoming. In the present review paper, M.-L. Mazure gives a comprehensive survey on the concept of blossoming and the fundamental notion of extended Chebychev spaces. For the latter, characterisations are presented in many equivalent formulations, some of them known and reviewed here, some of them new. For the former, both existence and their properties are discussed and, for instance, the relationship between blossoms and Bernstein bases and the existence of Bézier points is explained. And, of course, special attention is given to blossoms in the EC (extended Chebyshev) spaces.

A comprehensive survey along with some new results on the structural analysis of subdivision surfaces near extraordinary vertices is offered in the chapter *Structural Analysis of Subdivision Surfaces – A Summary* by J. Peters and U. Reif. For "standard" surface subdivision schemes, whose subdivision matrix has a double subdominant eigenvalue, the issues of normal and C^1-continuity are discussed in detail. Here, the authors extend the known results to cases where the generating functions of the scheme may be linearly dependent. Moreover, a simplified test of injectivity for the so-called characteristic map is developed for subdivision schemes with certain symmetry properties. The Doo-Sabin scheme serves as an illustration of these new techniques. The C^2-regularity and corresponding constraints for the subdivision matrix are also discussed. The chapter closes with a detailed analysis of the limit curvature at extraordinary vertices, which is very useful for understanding the visual artifacts in specific subdivision surfaces.

Polynomial Interpolation in Several Variables: Lattices, Differences, and Ideals. In this chapter T. Sauer points out that when passing from one to several variables, the nature and structure of polynomial interpolation changes completely. The solvability of an interpolation problem with respect to a given finite dimensional space of multivariate polynomials does not only depend on the number of the nodes but significantly on their geometric position. This makes the theory of interpolation in several variables a highly difficult and non-trivial problem. The main reason is the loss of the Haar condition in domains different from univariate intervals or S^1. The author gives an excellent survey of some basic constructions of interpolation lattices which emerge from the geometric characterization due to Chung and Yao. Depending on the structure of the specific interpolation problem, there are different representations of the interpolation polynomial and of the error formulas, reflecting the underlying point geometry. In addition, the close relationship with algebraic concepts such as constructive ideal theory is pointed out.

A particularly elegant way of solving multivariate interpolation and approximation problems is provided by kernels or radial basis functions. These have plenty of applications, since they provide meshless methods for solving partial differential equations and are in the core of modern techniques for machine learning. The chapter *Computational Aspects of Radial Basis Function Approximation* by H. Wendland surveys recent progress in numerical methods connected to kernel techniques. Reduction of problem complexity and improvement of stability are the most important computational bottlenecks in this area. Both are treated comprehensively, in particular by multipole expansions, domain decompositions, partitions of unity, multilevel techniques, and regularization by smoothing.

Kernels and regularization are the link to the paper *Learning Theory: From Regression to Classification* by Q. Wu, Y. Ying, and D. X. Zhou which looks at recent developments in machine learning from the viewpoint of approximation theory. In particular, a regularization approach in reproducing kernel Hilbert spaces is used to analyze errors of regression and classification algorithms. This field connects multivariate approximation to optimization and stochastic processes, and it has quite a promising future due to its importance for the design of intelligent systems in engineering.

The final chapter *Coherent States from Nonunitary Representations* by G. Zimmermann provides an interesting and powerful alternative to periodic wavelets on the unit circle by employing Möbius transformations as generators for the building blocks in the analysis and in the synthesis process. The usual unitary representations of this group of transformations being not square integrable, however, the usual "wavelet" construction has to be modified. It is now built on a nonunitary representation and its contragredient counterpart. The chapter also deals with these aspects in a general, and abstract, way in order to provide the essential ingredients for this extension of wavelet-type expansion of functions in appropriate function spaces.

Many people have contributed to the production of the book. All the articles are peer-refereed and carefully edited. Our thanks go to the referees for their valuable support, guaranteeing top scientific standard of all chapters. During the editing procedure, we got help from Dr. Elena Berdysheva and Dr. Georg Zimmermann to whom we are grateful, in particular, for compiling the index. Last not least, we would like to thank the series editors, and the publisher for their kind cooperation.

Martin D. Buhmann
Werner Haußmann
Kurt Jetter
Robert Schaback
Joachim Stöckler

The editors

CONTENTS

Topics in Multivariate Approximation and Interpolation
K. Jetter et al., Editors

Durrmeyer Operators and Their Natural Quasi-Interpolants

Elena Berdysheva [a], **Kurt Jetter** [a,1], **Joachim Stöckler** [b]

[a]Universität Hohenheim, Institut für Angewandte Mathematik und Statistik, D-70593 Stuttgart, Germany
[b]Universität Dortmund, Institut für Angewandte Mathematik, D-44221 Dortmund, Germany

Dedicated to Professor Charles K. Chui on the occasion of his 65th birthday.

Abstract

This paper provides a survey on spectral analysis and approximation order of our quasi-interpolants of Durrmeyer type on simplices, together with various new aspects and achievements. The latter include Bernstein type inequalities which are proved using a striking property of appropriately modified Durrmeyer operators, namely, their kernel functions are pointwise completely monotonic.

Key words: Bernstein basis polynomial, Bernstein inequality, completely monotonic sequence, Durrmeyer operator, hypergeometric series, Jackson-Favard estimate, Jacobi polynomial, K-functional, Laplace type integral, Legendre differential operator, positive operator, quasi-interpolants, Voronovskaja theorem

2000 MSC: 41A10, 41A17, 41A36, 41A63, 33C45

Email addresses: `berdyshe@uni-hohenheim.de` (Elena Berdysheva),
`kjetter@uni-hohenheim.de` (Kurt Jetter),
`joachim.stoeckler@math.uni-dortmund.de` (Joachim Stöckler).

[1] Corresponding author

1. Introduction

The construction of quasi-interpolant operators through linear combinations of (Bernstein-)Durrmeyer operators has a long history in Approximation Theory. Durrmeyer operators have several desirable properties such as positivity and stability, and their analysis can be performed using their elegant spectral properties. Their approximation order is low, however, and for this reason quasi-interpolants with better approximation properties are necessary for more efficient approximation. In our aim at constructing good quasi-interpolants on triangulated domains, the natural first step is to consider a single triangle - or a simplex in higher dimensions.

A comprehensive description of our previous results in this direction is included as part of this article. We emphasize the close relation of quasi-interpolants to certain partial differential operators on the simplex, which are generalizations of the Legendre differential operator and its Jacobi-type analogue

$$P_r^{\alpha,\beta} := w_{\alpha,\beta}^{-1}(x)\frac{d^r}{dx^r}\left[w_{\alpha,\beta}(x)x^r(1-x)^r\frac{d^r}{dx^r}\right],$$

where $w_{\alpha,\beta}(x) = x^\alpha(1-x)^\beta$ and $\alpha, \beta > -1$ define the Jacobi weight for the standard interval $[0,1]$. In addition to this survey we also present new results which lead to a Bernstein estimate for the aforementioned differential operators (Section 6) and to direct estimates of the error of approximation of our quasi-interpolants by newly defined K-functionals on the simplex (Section 7). The key result in order to prove the Bernstein inequality is a beautiful property of the sequence of appropriately modified Durrmeyer operators: their kernels constitute a pointwise completely monotonic sequence (Theorem 2). Here we employ methods of Koornwinder and Askey for the Laplace integral of Jacobi-polynomials and the characterization of completely monotonic sequences by Hausdorff's theorem (Section 4).

The structure of the paper is as follows. In Section 2 we give the definition of the Durrmeyer operators (with Jacobi weights), and in Section 3 we review their spectral properties, see Theorem 1. Section 4 deals with the kernel function of the appropriately modified Durrmeyer operator, and provides the striking result of Theorem 2 showing the pointwise complete monotonicity of the associated kernels. We then give the definition of our quasi-interpolants in Section 5, together with the adequate partial differential operators of Jacobi type. Their spectral analysis leads to a valuable representation of the quasi-interpolants as a linear combination of Durrmeyer operators, in Theorem 9. Section 6 is devoted to the proof of the Bernstein inequalities, which are stated in Theorem 11 and Theorem 12. This is the second key section of the paper, which contains new and unpublished material. Its application in Section 7 follows along the lines of classical Approximation Theory and provides a valuable and elegant extension of several properties of the Durrmeyer operator to our quasi-interpolants: The estimate of Jackson-Favard type, the Voronovskaja type theorem (including its 'strong' version) and the so-called direct estimate in terms of the proper K-functional. Rather than giving complete references for each result within the text, we conclude in Section 8 with historical remarks in order to point out the development of the main results.

2. The Bernstein Basis Functions

The standard simplex in $\mathbb{R}^d$ is given by

$$\mathbf{S}^d := \{ (x_1,\ldots,x_d) \in \mathbb{R}^d \mid 0 \le x_1,\ldots,x_d \le 1 \,,\; x_1+\cdots+x_d \le 1 \} \,.$$

We shall use barycentric coordinates

$$\mathbf{x} = (x_0, x_1, \ldots, x_d) \,, \qquad x_0 := 1 - x_1 - \ldots - x_d \,,$$

in order to define the d-variate Bernstein basis polynomials. Namely, for given $\alpha = (\alpha_0, \alpha_1, \ldots, \alpha_d) \in \mathbb{N}_0^{d+1}$

$$B_\alpha(x_1,\ldots,x_d) := \binom{|\alpha|}{\alpha} \mathbf{x}^\alpha := \frac{|\alpha|!}{\alpha_0!\alpha_1!\cdots\alpha_d!}\, x_0^{\alpha_0} x_1^{\alpha_1} \cdots x_d^{\alpha_d} \,.$$

Here, we use standard multi-index notation. We also allow $\alpha \in \mathbb{Z}^d$ with $|\alpha| := \alpha_0 + \cdots + \alpha_d \in \mathbb{N}$. It is then convenient to put $B_\alpha \equiv 0$ if one α_i is negative.

For given $n \in \mathbb{N}$, the Bernstein basis polynomials $\{B_\alpha \mid \alpha \in \mathbb{N}_0^{d+1} \,,\; |\alpha| = n\}$ are a basis for $\mathbf{P}_n = \mathbf{P}_n^d$, the space of d-variate algebraic polynomials of (total) degree n. They are used for the definition of various polynomial operators. In this paper, we study quasi-interpolants based on the Bernstein-Durrmeyer operators with Jacobi weights. Here, the weight function is given by

$$\omega_\mu(x_1,\ldots,x_d) := \mathbf{x}^\mu = x_0^{\mu_0} x_1^{\mu_1} \cdots x_d^{\mu_d} \,,$$

where $\mu = (\mu_0, \mu_1, \ldots, \mu_d) \in \mathbb{R}^{d+1}$ with $\mu_i > -1$, $i = 0,\ldots,d$. Whence, $|\mu| := \mu_0 + \mu_1 + \cdots + \mu_d > -d-1$.

On the simplex, we use the (weighted) inner product

$$\langle f | g \rangle_\mu := \int_{\mathbf{S}^d} \omega_\mu \, f \, g \tag{1}$$

to define the (Bernstein-)Durrmeyer operator of degree n,

$$\mathbf{M}_{n,\mu} \;:\; f \mapsto \mathbf{M}_{n,\mu}(f) := \sum_{|\alpha|=n} \frac{\langle f | B_\alpha \rangle_\mu}{\langle \mathbf{1} | B_\alpha \rangle_\mu} \, B_\alpha \,. \tag{2}$$

Here, $\mathbf{1}$ denotes the function constant equal one. It is well-known that, for $\alpha = (\alpha_0, \alpha_1, \ldots, \alpha_d) \in \mathbb{N}_0^{d+1}$, with $|\alpha| = n$,

$$\begin{aligned} \langle \mathbf{1} | B_\alpha \rangle_\mu &= \binom{|\alpha|}{\alpha} \frac{\Gamma(\alpha_0+\mu_0+1)\Gamma(\alpha_1+\mu_1+1)\cdots\Gamma(\alpha_d+\mu_d+1)}{\Gamma(|\alpha|+|\mu|+d+1)} \\ &= \frac{\Gamma(n+1)}{\Gamma(n+|\mu|+d+1)} \prod_{i=0}^{d} \frac{\Gamma(\alpha_i+\mu_i+1)}{\Gamma(\alpha_i+1)} \,. \end{aligned} \tag{3}$$

For the unweighted case (where $\mu = (0, 0, \ldots, 0)$) this recovers the formula

$$\int_{\mathbf{S}^d} B_\alpha = \frac{n!}{(n+d)!} , \quad \alpha \in \mathbb{N}_0^{d+1} , \ |\alpha| = n .$$

3. Spectral Properties of the Durrmeyer Operators

The Durrmeyer operator is usually considered on the domain $L^p_\mu(\mathbf{S}^d)$, $1 \le p < \infty$, which is the weighted L^p space consisting of all measurable functions on S^d such that

$$\|f\|_{p,\mu} := \left(\int_{\mathbf{S}^d} \omega_\mu \, |f|^p \right)^{1/p}$$

is finite. For $p = \infty$, the space $C(\mathbf{S}^d)$ of continuous functions is considered instead. In this setting the following properties are more or less obvious.

- *The operator is positive:* $\mathbf{M}_{n,\mu}(f) \ge 0$ for every $f \ge 0$.
- *It reproduces constant functions:* $\mathbf{M}_{n,\mu}(p) = p$ for $p \in \mathbf{P}_0$.
- *It is contractive:* $\|\mathbf{M}_{n,\mu}(f)\|_{p,\mu} \le \|f\|_{p,\mu}$ for every $f \in L^p_\mu(\mathbf{S}^d)$.

The most striking and useful property refers to the Hilbert space setting,

$$\mathbf{H} := L^2_\mu(\mathbf{S}^d) .$$

This space can be written as the sum of spaces of orthogonal polynomials,

$$L^2_\mu(\mathbf{S}^d) = \sum_{m=0}^{\infty} \mathbf{E}_{m,\mu} , \quad \text{where}$$

$$\mathbf{E}_{0,\mu} := \mathbf{P}_0 \quad \text{and} \quad \mathbf{E}_{m,\mu} := \mathbf{P}_m \cap \mathbf{P}^{\perp}_{m-1} \quad \text{for} \quad m > 0 .$$

Here, orthogonality refers to the weighted inner product (1). It is clear that $\mathbf{M}_{n,\mu}$ is a bounded self-adjoint operator on $\mathbf{H}$. Its spectrum is given by the following result.

Theorem 1. *For all $n \in \mathbb{N}$, the spaces $\mathbf{E}_{m,\mu}$, $m \ge 0$, are eigenspaces of the Durrmeyer operator, and*

$$\mathbf{M}_{n,\mu}(p_m) = \gamma_{n,m,\mu} \, p_m \quad \textit{for} \quad p_m \in \mathbf{E}_{m,\mu} ,$$

where

$$\gamma_{n,m,\mu} := \frac{n!}{(n-m)!} \frac{\Gamma(n+d+|\mu|+1)}{\Gamma(n+d+|\mu|+m+1)} , \quad \textit{for} \quad n \ge m ,$$

while

$$\gamma_{n,m,\mu} = 0 \quad \textit{for} \quad n < m .$$

In particular,

$$\gamma_{n,0,\mu} = 1 , \quad 0 \le \gamma_{n,m,\mu} < 1 , \quad m > 0 ,$$

and

$$\lim_{n \to \infty} \gamma_{n,m,\mu} = 1 \quad \textit{for fixed } m . \tag{4}$$

Hence, for $f = \sum_{m=0}^{\infty} p_m$, with $p_m \in \mathbf{E}_{m,\mu}$, we find $\mathbf{M}_{n,\mu}(f) = \sum_{m=0}^{n} \gamma_{n,m,\mu}\, p_m$. In particular, the restrictions $\mathbf{M}_{n,\mu}|_{\mathbf{P}_k}$ act as isomorphisms on the spaces $\mathbf{P}_k$ as long as $k \le n$.

4. The Kernel Function

According to equations (2) and (3), the Durrmeyer operator of degree n can be written as

$$\{\mathbf{M}_{n,\mu}(f)\}(\mathbf{y}) = \int_{\mathbf{S}^d} \omega_\mu(\mathbf{x})\, f(\mathbf{x})\, K_{n,\mu}(\mathbf{x},\mathbf{y})\, d\mathbf{x}$$

with its kernel given by

$$K_{n,\mu}(\mathbf{x},\mathbf{y}) := \frac{\Gamma(n+|\mu|+d+1)}{\Gamma(n+1)} \sum_{|\alpha|=n} \Big(\prod_{i=0}^{d} \frac{\Gamma(\alpha_i+1)}{\Gamma(\alpha_i+\mu_i+1)}\Big)\, B_\alpha(\mathbf{x})\, B_\alpha(\mathbf{y})\,.$$

Putting

$$\underline{\mu} := \min_i \mu_i\,,$$

we are going to study properties of the modified kernel

$$\begin{aligned} T_{n,\mu}(\mathbf{x},\mathbf{y}) :&= \frac{\Gamma(n+\underline{\mu}+1)}{\Gamma(n+|\mu|+d+1)}\, K_{n,\mu}(\mathbf{x},\mathbf{y}) \\ &= \sum_{|\alpha|=n} \frac{\Gamma(n+\underline{\mu}+1)}{\prod_{i=0}^{d}\Gamma(\alpha_i+\mu_i+1)} \binom{n}{\alpha}\, \mathbf{x}^\alpha\, \mathbf{y}^\alpha\,, \qquad \mathbf{x},\mathbf{y}\in\mathbf{S}^d. \end{aligned} \tag{5}$$

This kernel is non-negative, but we are going to prove much more. Namely, under a slight restriction on the exponents of the Jacobi weight, the forward differences of the sequence $(T_{n,\mu}(\mathbf{x},\mathbf{y}))_{n\ge 0}$ alternate in sign. The result seems to be new even in the univariate setting.

Theorem 2. *Let $\mu = (\mu_0, \mu_1, \ldots, \mu_d)$ be such that $\mu_i \ge -1/2$, $i = 0, \ldots, d$. Then, for every $\mathbf{x}, \mathbf{y} \in \mathbf{S}^d$, the sequence $(T_{n,\mu}(\mathbf{x},\mathbf{y}))_{n\ge 0}$ is bounded and completely monotonic; i.e., the inequalities*

$$T_{n,\mu}^{(r)}(\mathbf{x},\mathbf{y}) := (-1)^r \Delta^r T_{n,\mu}(\mathbf{x},\mathbf{y}) = \sum_{\ell=0}^{r} (-1)^\ell \binom{r}{\ell} T_{n+\ell,\mu}(\mathbf{x},\mathbf{y}) \ge 0$$

hold true for all $r, n \ge 0$.

Here, $(\Delta\nu_n)_{n\ge 0} = (\nu_{n+1} - \nu_n)_{n\ge 0}$ denotes the forward difference of the sequence $(\nu_n)_{n\ge 0}$. For the proof of this result, we make use of the characterization of completely monotonic sequences due to Hausdorff.

Lemma 3. *(see [28], Chapter III, Theorem 4a) A real sequence $(\nu_n)_{n\ge 0}$ is completely monotonic if and only if there exists a non-decreasing bounded function g on $[0,1]$ such that*

$$\nu_n = \int_0^1 t^n \, dg(t), \qquad n \in \mathbb{N}_0 .$$

Here, the integral is to be understood as a Lebesgue-Stieltjes integral.

Remarks. The following facts will be useful for our discussion of complete monotonicity.

(a) *The sequences $(q^n)_{n\geq 0}$, with $0 \leq q \leq 1$, are completely monotonic.* This result is obvious.

(b) *The sum and the product of two completely monotonic sequences are completely monotonic.* The first statement is again trivial, and the second statement follows from

$$-\Delta(c_n d_n) = (-\Delta c_n) d_n + c_{n+1}(-\Delta d_n)$$

by induction.

(c) *For given $\mu_0 \geq \mu_1 > -1$ the sequence*

$$c_n = \frac{\Gamma(n+\mu_1+1)}{\Gamma(n+\mu_0+1)} , \qquad n \geq 0 ,$$

is completely monotonic. This follows from the formula

$$(-1)^k \Delta^k c_n = \frac{(\mu_0 - \mu_1)_k}{(n+1+\mu_0)_k} \, c_n , \quad k, n \geq 0 ,$$

with $(a)_0 := 1$ and $(a)_k := a(a+1)\cdots(a+k-1)$ for $k > 0$, the so-called shifted factorial or Pochhammer symbol.

(d) *The sequence of integrals $c_n = \int f_n \, dm$, $n \geq 0$, of a pointwise completely monotonic family $(f_n)_{n\geq 0}$ of functions which are integrable with respect to the non-negative measure dm, is completely monotonic.*

For the proof of Theorem 2 it is sufficient to consider the case

$$\mu_0 \geq \mu_1 \geq \cdots \geq \mu_d \geq -1/2 , \quad \text{whence } \underline{\mu} = \mu_d ,$$

since the kernel is invariant modulo a permutation of the variables. Under this assumption, we use induction on d, the number of variables. For $d = 1$, the statement is the special case $t = 1$, $(\alpha, \beta) = (\mu_0, \mu_1)$ of the following result.

Lemma 4. *Let $\alpha \geq \beta \geq -1/2$. Then, for every $x, y, t \in [0, 1]$, the sequence*

$$\begin{aligned} \nu_n(x, y; t) &= \nu_n^{(\alpha,\beta)}(x, y; t) \\ &= \sum_{k=0}^{n} \frac{\Gamma(n+\beta+1)}{\Gamma(k+\alpha+1)\Gamma(n-k+\beta+1)} \binom{n}{k} (xy)^k \left((1-x)(1-y)t\right)^{n-k} , \end{aligned}$$

$n \geq 0$, is bounded and completely monotonic.

Remark. The restriction on the Jacobi exponents is crucial. For example, a simple calculation shows that

$$-\Delta\nu_1^{(\beta,\beta)}\left(\frac{1}{2},\frac{1}{2};1\right) = \frac{1}{\Gamma(\beta+1)}\,\frac{2\beta+1}{8(\beta+1)},$$

which is negative for $-\frac{1}{2} > \beta > -1$.

Taken the result of Lemma 4 for granted, the *proof of Theorem 2* is finished by the following induction step.

For $d \geq 2$, we write the kernel $T_{n,\mu} = T^d_{n,\mu}$ in equation (5) in terms of kernels of fewer variables (For clarity, we mark the number of variables as a superscript). We put $\mathbf{x} = (x_0, \mathbf{x}^*)$ with $\mathbf{x}^* = (x_1, \ldots, x_d)$ and $x_0 = (1 - x_1 - \cdots - x_d)$, and $\mathbf{y} = (y_0, \mathbf{y}^*)$ with analogous notation. Also, $\alpha = (\alpha_0, \alpha^*)$ with $\alpha^* = (\alpha_1, \ldots, \alpha_d) \in \mathbb{N}_0^d$, and $\mu = (\mu_0, \mu^*)$ with $\mu^* = (\mu_1, \ldots, \mu_d) \in \mathbb{R}^d$.

If $x_0 = 1$, then $x_1 = \cdots = x_d = 0$ and

$$T^d_{n,\mu}(\mathbf{x},\mathbf{y}) = \frac{1}{\prod_{i=1}^d \Gamma(\mu_i+1)}\,\frac{\Gamma(n+\mu_d+1)}{\Gamma(n+\mu_0+1)}\,y_0^n\,, \qquad n \geq 0\,,$$

which is completely monotonic by an application of cases (a)-(c) of the remarks above. The same argument applies if $y_0 = 1$. So we may assume henceforth that

$$0 \leq x_0, y_0 < 1\,.$$

Here, both $\tilde{\mathbf{x}} := \frac{1}{1-x_0}\mathbf{x}^*$ and $\tilde{\mathbf{y}} := \frac{1}{1-y_0}\mathbf{y}^*$ are elements of $\mathbf{S}^{d-1}$. By simple computations and by letting $k := \alpha_0$, we obtain

$$\begin{aligned} T^d_{n,\mu}(\mathbf{x},\mathbf{y}) &= \sum_{k=0}^{n} \frac{\Gamma(n+\mu_d+1)}{\Gamma(k+\mu_0+1)}\binom{n}{k}(x_0y_0)^k\left((1-x_0)(1-y_0)\right)^{n-k} \\ &\qquad \sum_{|\alpha^*|=n-k} \frac{1}{\prod_{i=1}^d \Gamma(\alpha_i+\mu_i+1)}\binom{n-k}{\alpha^*}(\tilde{\mathbf{x}})^{\alpha^*}(\tilde{\mathbf{y}})^{\alpha^*} \\ &= \sum_{k=0}^{n} a_{k,n}(x_0,y_0)\,T^{d-1}_{n-k,\mu^*}(\tilde{\mathbf{x}},\tilde{\mathbf{y}})\,, \end{aligned} \tag{6}$$

with

$$a_{k,n}(x_0,y_0) := \frac{\Gamma(n+\mu_d+1)}{\Gamma(k+\mu_0+1)\Gamma(n-k+\mu_d+1)}\binom{n}{k}(x_0y_0)^k\left((1-x_0)(1-y_0)\right)^{n-k}.$$

Using the induction hypothesis we find - according to Lemma 3 - for each pair $(\tilde{\mathbf{x}},\tilde{\mathbf{y}})$ a bounded, nondecreasing function g^* such that

$$\nu_n^* := T^{d-1}_{n,\mu^*}(\tilde{\mathbf{x}},\tilde{\mathbf{y}}) = \int_0^1 t^n\,dg^*(t)\,, \qquad n \geq 0\,.$$

Inserting this in equation (6),

$$T^d_{n,\mu}(\mathbf{x},\mathbf{y}) = \int_0^1 \sum_{k=0}^{n} a_{k,n}(x_0,y_0)\,t^{n-k}\,dg^*(t)\,.$$

Here, the integrand is given by the sequence $\nu_n(x_0, y_0; t)$ considered in Lemma 4, by putting $(\alpha, \beta) = (\mu_0, \mu_d)$, and the induction step is completed by applying item (d) of the remarks above. This finishes the proof of Theorem 2.

For the *proof of Lemma 4*, it is sufficient to assume $\alpha > \beta > -\frac{1}{2}$, since the limit case $\alpha = \beta$ or $\beta = -\frac{1}{2}$ then follows by continuity. Here we make use of Koornwinder's integral representation (of Laplace type) for the normalized Jacobi polynomials which are given by a hypergeometric series as follows:

$$\begin{aligned} R_n^{(\alpha,\beta)}(x) := \frac{P_n^{(\alpha,\beta)}(x)}{P_n^{(\alpha,\beta)}(1)} &:= F_1^2\big(-n, n+\alpha+\beta+1; \alpha+1; \frac{1-x}{2}\big) \\ &= \big(\frac{x+1}{2}\big)^n F_1^2\big(-n, -n-\beta; \alpha+1; \frac{x-1}{x+1}\big) \end{aligned} \tag{7}$$

for $\alpha, \beta > -1$. The latter identity follows from Euler's linear transformation,

$$F_1^2(a, b; c; z) = (1-z)^{-a} F_1^2\big(a, c-b; c; \frac{z}{z-1}\big) ,$$

(cf. [22], Section 2.4). Koornwinder's result (see [21], Section 3) reads as follows; for an easy analytic proof we refer to Askey [2].

Lemma 5. *For* $\alpha > \beta > -\frac{1}{2}$ *we have*

$$\begin{aligned} &R_n^{(\alpha,\beta)}(x) \\ &= \frac{2\Gamma(\alpha+1)}{\Gamma(\beta+\frac{1}{2})\Gamma(\frac{1}{2})\Gamma(\alpha-\beta)} \int_{u=0}^{1} \int_{\varphi=0}^{\pi} \Big[\frac{x+1}{2} + \frac{x-1}{2}u^2 + u\sqrt{x^2-1}\cos\varphi\Big]^n \\ &\qquad u^{2\beta+1}(1-u^2)^{\alpha-\beta-1}\big(\sin\varphi\big)^{2\beta}\, d\varphi\, du . \end{aligned}$$

With this result at hand, the proof is finished by a straightforward, but lengthy computation. Using (7), we find

$$\begin{aligned} \nu_n(x, y; t) &= \frac{[(1-x)(1-y)t]^n}{\Gamma(\alpha+1)} F_1^2\big(-n, -n-\beta; \alpha+1; \frac{xy}{(1-x)(1-y)t}\big) \\ &= \frac{1}{\Gamma(\alpha+1)}\big[(1-x)(1-y)t - xy\big]^n R_n^{(\alpha,\beta)}\Big(\frac{(1-x)(1-y)t + xy}{(1-x)(1-y)t - xy}\Big) , \end{aligned}$$

and Koornwinder's integral gives

$$\nu_n(x, y; t) = \int_{u=0}^{1} \int_{\varphi=0}^{\pi} \big[\Phi(x, y, t; u, \varphi)\big]^n\, dm_{\alpha,\beta}(u, \varphi) . \tag{8}$$

Here,

$$\Phi(x, y, t; u, \varphi) = (1-x)(1-y)t + xyu^2 + 2\varepsilon\sqrt{(1-x)(1-y)txy}\; u\, \cos\varphi \tag{9}$$

with $\varepsilon = \operatorname{sign}((1-x)(1-y)t - xy)$ and

$$dm_{\alpha,\beta}(u, \varphi) = \frac{2}{\Gamma(\beta+\frac{1}{2})\Gamma(\frac{1}{2})\Gamma(\alpha-\beta)}\, u^{2\beta+1}(1-u^2)^{\alpha-\beta-1}\big(\sin\varphi\big)^{2\beta}\, d\varphi\, du ,$$

which is a positive measure. From (9) we see that

$$
\begin{aligned}
0 &\le \left(\sqrt{(1-x)(1-y)t} - \sqrt{xyu} \right)^2 \\
&\le \Phi(x,y,t;u,\varphi) \le \left(\sqrt{(1-x)(1-y)} + \sqrt{xy} \right)^2 \\
&\le \left(\frac{1-x+1-y}{2} + \frac{x+y}{2} \right)^2 = 1
\end{aligned}
$$

for $0 \le x,y,u,t \le 1$. Whence - by remark (d) above - the sequence in (8) is completely monotonic. This finishes the proof of Lemma 4, and settles the proof of Theorem 2.

Remark. For ultraspherical polynomials, the representation formula of Lemma 5 has the following limit as $\alpha \to \beta = \lambda - \frac{1}{2}$,

$$
\frac{C_n^{(\lambda)}(x)}{C_n^{(\lambda)}(1)} = \frac{\Gamma(\lambda+\frac{1}{2})}{\Gamma(\frac{1}{2})\Gamma(\lambda)} \int_0^\pi \left[x + \sqrt{x^2-1}\, \cos\varphi \right]^n \left(\sin\varphi \right)^{2\lambda-1} d\varphi\,, \quad \lambda > 0\,.
$$

In particular, for the Legendre polynomials P_n normalized by $P_n(1) = 1$ (case $\lambda = \frac{1}{2}$), we recover the Laplace integral

$$
P_n(x) = \frac{1}{\pi} \int_0^\pi \left[x + \sqrt{x^2-1}\, \cos\varphi \right]^n d\varphi\,.
$$

For details, see again [2] and [21].

5. The Quasi-Interpolants

The following second order differential operator $\mathbf{U}_\mu$ plays a prominent role in our analysis,

$$
\begin{aligned}
-\mathbf{U}_\mu :=& \sum_{i=1}^{d} \left(\omega_\mu(\mathbf{x})\right)^{-1} \frac{\partial}{\partial x_i} \left\{ \omega_\mu(\mathbf{x})\, x_0 x_i \frac{\partial}{\partial x_i} \right\} \\
&+ \sum_{1 \le i < j \le d} \left(\omega_\mu(\mathbf{x})\right)^{-1} \left(\frac{\partial}{\partial x_j} - \frac{\partial}{\partial x_i} \right) \left\{ \omega_\mu(\mathbf{x})\, x_i x_j \left(\frac{\partial}{\partial x_j} - \frac{\partial}{\partial x_i} \right) \right\}.
\end{aligned}
\tag{10}
$$

Here, as before, $\mathbf{x} = (x_0, x_1, \ldots, x_d)$ with $x_0 = 1 - x_1 - \cdots - x_d$. In the definition of the operator, we take the negative sign in order to have a positive spectrum.

Lemma 6. *The differential operator* $\mathbf{U}_\mu$ *is densely defined on the Hilbert space* $\mathbf{H}$, *and symmetric. We have*

$$
\mathbf{U}_\mu(p_m) = m(m+d+|\mu|)\, p_m\,, \quad m \ge 0\,, \quad p_m \in \mathbf{E}_{m,\mu}\,,
$$

i.e., the spaces $\mathbf{E}_{m,\mu}$ *are also eigenspaces of* $\mathbf{U}_\mu$. *In particular,*

$$
\mathbf{U}_\mu(\mathbf{M}_{n,\mu}(f)) = \mathbf{M}_{n,\mu}(\mathbf{U}_\mu(f))
$$

for $f \in C^2(\mathbf{S}^d)$.

We put $\mathbf{U}_{0,\mu} := \mathbf{I}$ and define the differential operators $\mathbf{U}_{\ell,\mu}$ of order 2ℓ by

$$\mathbf{U}_{\ell+1,\mu} := \frac{1}{(\ell+1)^2}\Big(\,\mathbf{U}_\mu - \ell(\ell+d+|\mu|)\mathbf{I}\,\Big)\,\mathbf{U}_{\ell,\mu}\,, \qquad \ell \geq 0\,. \tag{11}$$

Thus, $\mathbf{U}_\mu = \mathbf{U}_{1,\mu}$. The operators are again densely defined and symmetric, with spectral properties as follows.

$$\mathbf{U}_{\ell,\mu}(p_m) = \frac{\sigma_{\ell,m,\mu}}{\ell!}\,p_m\,, \quad m \geq 0\,, \quad p_m \in \mathbf{E}_{m,\mu}\,,$$

with

$$\sigma_{0,m,\mu} = 1 \quad \text{and}$$
$$\sigma_{\ell+1,m,\mu} = \frac{m(m+d+|\mu|) - \ell(\ell+d+|\mu|)}{\ell+1}\,\sigma_{\ell,m,\mu}\,, \qquad \ell \geq 0\,.$$

In particular, $\sigma_{\ell,m,\mu} = 0$ for $\ell > m$.

The relation to the spectrum of the Durrmeyer operators is explained by the following identity.

$$\gamma_{n,m,\mu}\sum_{\ell=0}^{m}\frac{\sigma_{\ell,m,\mu}}{n(n-1)\cdots(n-\ell+1)} = 1\,, \qquad 0 \leq m \leq n\,. \tag{12}$$

From this,

$$\sum_{\ell=0}^{r}\frac{1}{\binom{n}{\ell}}\mathbf{U}_{\ell,\mu}(\mathbf{M}_{n,\mu}p) = p\,, \qquad p \in \mathbf{P}_r\,, \qquad 0 \leq r \leq n\,,$$

and in particular

$$\left(\sum_{\ell=0}^{n}\frac{1}{\binom{n}{\ell}}\mathbf{U}_{\ell,\mu}(\mathbf{M}_{n,\mu})\right)\Bigg|_{\mathbf{P}_n} = \mathbf{I}\Big|_{\mathbf{P}_n}\,, \qquad n \in \mathbb{N}\,.$$

These properties show that the following natural definition of (Bernstein-)Durrmeyer quasi-interpolant operators of order (r,n) arises,

$$\mathbf{M}_{n,\mu}^{(r)}(f) := \sum_{\ell=0}^{r}\frac{1}{\binom{n}{\ell}}\mathbf{U}_{\ell,\mu}(\mathbf{M}_{n,\mu}(f))\,, \quad 0 \leq r \leq n\,. \tag{13}$$

These are polynomial operators with range $\mathbf{P}_n$, and they show the reproduction property

$$\mathbf{M}_{n,\mu}^{(r)}(p) = p\,, \qquad p \in \mathbf{P}_r\,.$$

In particular, $\mathbf{M}_{n,\mu}^{(0)} = \mathbf{M}_{n,\mu}$ and $\mathbf{M}_{n,\mu}^{(n)}|_{\mathbf{P}_n} = \mathbf{I}|_{\mathbf{P}_n}$. The spectral properties of these operators are given by

Theorem 7. *For all $n,m,r \in \mathbb{N}_0$, $0 \leq r \leq n$, the spaces $\mathbf{E}_{m,\mu}$ are eigenspaces of the operator $\mathbf{M}_{n,\mu}^{(r)}$. Namely, for $p_m \in \mathbf{E}_{m,\mu}$ we have*

$$\mathbf{M}_{n,\mu}^{(r)}(p_m) = \lambda_{n,m,\mu}^{(r)}\,p_m$$

with the eigenvalues

$$\lambda^{(r)}_{n,m,\mu} = \gamma_{n,m,\mu} \sum_{\ell=0}^{r} \frac{1}{\binom{n}{\ell}} \frac{\sigma_{\ell,m,\mu}}{\ell!} .$$

In particular, $\lambda^{(r)}_{n,m,\mu} = 1$ *for* $m \le r \le n$, *while* $\lambda^{(r)}_{n,m,\mu} = 0$ *for* $m > n$. *Also,*

$$\lambda^{(r)}_{n,m,\mu} = 1 - \gamma_{n,m,\mu} \sum_{\ell=r+1}^{m} \frac{1}{\binom{n}{\ell}} \frac{\sigma_{\ell,m,\mu}}{\ell!} , \qquad r < m \le n ,$$

and, for fixed m,

$$\lim_{n\to\infty} \lambda^{(r)}_{n,m,\mu} = 1 .$$

The latter properties follow from (12) and (4). Some further identities will be used in the sequel. The basic lemma here is

Lemma 8. *For* $n, m, r \in \mathbb{N}_0$, $0 \le r \le n$, *the following difference equation holds true:*

$$\lambda^{(r)}_{n,m,\mu} - \lambda^{(r)}_{n-1,m,\mu} = \frac{r+1}{n+d+|\mu|} \frac{1}{\binom{n}{r+1}} \frac{\sigma_{r+1,m,\mu}}{(r+1)!} \gamma_{n,m,\mu} .$$

From this we find two important identities. The first one is

$$\mathbf{M}^{(r)}_{n,\mu} - \mathbf{M}^{(r)}_{n-1,\mu} = \frac{r+1}{n+d+|\mu|} \frac{1}{\binom{n}{r+1}} \mathbf{U}_{r+1,\mu}(\mathbf{M}_{n,\mu}) . \tag{14}$$

This is a difference equation with respect to the index n. The second identity follows from this,

$$\begin{aligned} \mathbf{M}^{(r)}_{n,\mu} &= \mathbf{M}^{(r-1)}_{n,\mu} + \frac{1}{\binom{n}{r}} \mathbf{U}_{r,\mu}(\mathbf{M}_{n,\mu}) \\ &= \frac{1}{r} \left\{ (n+d+|\mu|+r)\, \mathbf{M}^{(r-1)}_{n,\mu} - (n+d+|\mu|)\, \mathbf{M}^{(r-1)}_{n-1,\mu} \right\} . \end{aligned}$$

This is a recurrence relation with respect to the superscript r showing that higher order quasi-interpolants are built from lower order ones through successive 'linear interpolation'. It can be used in order to show the following identity which writes the quasi-interpolants as a linear combination of Durrmeyer operators, viz. using induction with respect to r we find

Theorem 9. *For* $0 \le r \le n$ *the following identity holds true:*

$$\mathbf{M}^{(r)}_{n,\mu} = \sum_{\ell=0}^{r} (-1)^{\ell} \binom{r}{\ell} \binom{n+d+|\mu|+r-\ell}{r} \mathbf{M}_{n-\ell,\mu} .$$

6. The Bernstein Inequality

In order to find an upper bound for the norm of the operator $\mathbf{U}_{r,\mu}(\mathbf{M}_{n,\mu})$, we make use of the identity

$$\frac{1}{\binom{n}{r}}\mathbf{U}_{r,\mu}(\mathbf{M}_{n,\mu}) = \frac{n+d+|\mu|}{r}\sum_{\ell=0}^{r}(-1)^{\ell}\binom{r}{\ell}\binom{n+d+|\mu|+r-\ell-1}{r-1}\mathbf{M}_{n-\ell,\mu}\,, \tag{15}$$

which follows from equation (14) and Theorem 9 by straightforward calculation. We relate this to the modified operators

$$\{\mathbf{T}_{n,\mu}(f)\}(\mathbf{y}) = \int_{\mathbf{S}^d} \omega_\mu(\mathbf{x})\, f(\mathbf{x})\, T_{n,\mu}(\mathbf{x},\mathbf{y})\, d\mathbf{x}\,,$$

with the kernel $T_{n,\mu}$ defined in (5), and to the 'difference' operators

$$\mathbf{T}^{(r)}_{n,\mu} := (-1)^r\Delta^r\mathbf{T}_{n,\mu} = \sum_{\ell=0}^{r}(-1)^{\ell}\binom{r}{\ell}\mathbf{T}_{n+\ell,\mu}\,, \quad r\geq 0\,. \tag{16}$$

By equation (5)

$$\mathbf{T}_{n,\mu} = \frac{\Gamma(n+\underline{\mu}+1)}{\Gamma(n+d+|\mu|+1)}\mathbf{M}_{n,\mu}\,,$$

with $\underline{\mu} := \min_i \mu_i$, and according to Theorem 2, the operators $\mathbf{T}^{(r)}_{n,\mu}$ are positive.

Lemma 10. *For* $0 \leq r \leq n$ *we have*

$$\frac{1}{\binom{n}{r}}\mathbf{U}_{r,\mu}(\mathbf{M}_{n,\mu}) = \frac{\Gamma(n+d+|\mu|+1)}{\Gamma(n+\underline{\mu}+1)}\sum_{\ell=0}^{r}(-1)^{\ell}\binom{n+\underline{\mu}}{\ell}\binom{d+|\mu|-\underline{\mu}+r-1}{r-\ell}\mathbf{T}^{(\ell)}_{n-\ell,\mu}\,.$$

Proof. We evaluate the right-hand side by inserting

$$\mathbf{T}^{(\ell)}_{n-\ell,\mu} = \sum_{\lambda=0}^{\ell}(-1)^{\ell-\lambda}\binom{\ell}{\lambda}\frac{\Gamma(n-\lambda+\underline{\mu}+1)}{\Gamma(n-\lambda+d+|\mu|+1)}\mathbf{M}_{n-\lambda,\mu}\,,$$

which follows from (16). The result is

$$\frac{\Gamma(n+d+|\mu|+1)}{\Gamma(n+\underline{\mu}+1)}\sum_{\ell=0}^{r}(-1)^{\ell}\binom{n+\underline{\mu}}{\ell}\binom{d+|\mu|-\underline{\mu}+r-1}{r-\ell}\mathbf{T}^{(\ell)}_{n-\ell,\mu}$$
$$= \frac{n+d+|\mu|}{r}\sum_{\lambda=0}^{r}(-1)^{\lambda}c^{(r)}_{\lambda}\,\mathbf{M}_{n-\lambda,\mu}$$

with

$$\begin{aligned} c_\lambda^{(r)} &= \frac{r}{n+d+|\mu|}\binom{n+d+|\mu|}{\lambda}\sum_{\ell=\lambda}^{r}\binom{n-\lambda+\underline{\mu}}{\ell-\lambda}\binom{d+|\mu|-\underline{\mu}+r-1}{r-\ell} \\ &= \frac{r}{n+d+|\mu|}\binom{n+d+|\mu|}{\lambda}\sum_{\ell=0}^{r-\lambda}\binom{n-\lambda+\underline{\mu}}{\ell}\binom{d+|\mu|-\underline{\mu}+r-1}{r-\lambda-\ell} \\ &= \frac{r}{n+d+|\mu|}\binom{n+d+|\mu|}{\lambda}\binom{n-\lambda+d+|\mu|+r-1}{r-\lambda} \\ &= \binom{r}{\lambda}\binom{n+d+|\mu|+r-\lambda-1}{r-1}. \end{aligned}$$

This proves the lemma via identity (15). □

We are now prepared for our main result in this section.

Theorem 11. *Let $\mu = (\mu_0, \mu_1, \ldots, \mu_d) \in \mathbb{R}^{d+1}$ with $\underline{\mu} := \min \mu_i \geq -\frac{1}{2}$. Then, for $1 \leq p \leq \infty$ and $n, r \in \mathbb{N}_0$, $0 \leq r \leq n$, we have*

$$\left\|\mathbf{U}_{r,\mu}(\mathbf{M}_{n,\mu}(f))\right\|_{p,\mu} \leq 2^r \binom{n}{r}\binom{d+|\mu|-\underline{\mu}+r-1}{r}\|f\|_{p,\mu}.$$

Proof. Following the standard approach in the literature, we look at the dual pairing $\left(L^p_\mu(\mathbf{S}^d), L^q_\mu(\mathbf{S}^d)\right)$, with $\frac{1}{p}+\frac{1}{q}=1$, and the corresponding bilinear form

$$\langle\, f \mid g\,\rangle = \int_{\mathbf{S}^d} \omega_\mu(\mathbf{x})\, f(\mathbf{x})\, g(\mathbf{x})\, d\mathbf{x}\,.$$

Since the operators $\mathbf{U}_{r,\mu}$ and $\mathbf{M}_{n,\mu}$ commute, and are symmetric, the result for $p=\infty$ implies the result for $p=1$, and the Riesz-Thorin interpolation theorem then provides the full result for all p. Thus, it is sufficient to consider $p=\infty$.

Since the operators $\mathbf{T}^{(\ell)}_{n-\ell,\mu}$ are positive,

$$\|\mathbf{T}^{(\ell)}_{n-\ell,\mu}\|_{C(\mathbf{S}^d)\to C(\mathbf{S}^d)} = \|\mathbf{T}^{(\ell)}_{n-\ell,\mu}(\mathbf{1})\|_\infty\,. \tag{17}$$

For $\ell = 0$, we find

$$\mathbf{T}^{(0)}_{n,\mu}(\mathbf{1}) = \frac{\Gamma(n+\underline{\mu}+1)}{\Gamma(n+d+|\mu|+1)}\mathbf{M}_{n,\mu}(\mathbf{1}) = \frac{\Gamma(n+\underline{\mu}+1)}{\Gamma(n+d+|\mu|+1)}\mathbf{1}\,,$$

while for $\ell > 0$,

$$\begin{aligned} \mathbf{T}^{(\ell)}_{n-\ell,\mu}(\mathbf{1}) &= \sum_{\lambda=0}^{\ell}(-1)^{\ell-\lambda}\binom{\ell}{\lambda}\mathbf{T}^{(0)}_{n-\lambda,\mu}(\mathbf{1}) \\ &= \sum_{\lambda=0}^{\ell}(-1)^{\ell-\lambda}\binom{\ell}{\lambda}\frac{\Gamma(n-\lambda+\underline{\mu}+1)}{\Gamma(n-\lambda+d+|\mu|+1)}\mathbf{1}\,. \end{aligned}$$

Now

$$\sum_{\lambda=0}^{\ell}(-1)^{\ell-\lambda}\binom{\ell}{\lambda}\frac{\Gamma(n-\lambda+\underline{\mu}+1)}{\Gamma(n-\lambda+d+|\mu|+1)}$$

$$=\frac{\ell!\,\Gamma(n-\ell+\underline{\mu}+1)}{\Gamma(n+d+|\mu|+1)}\sum_{\lambda=0}^{\ell}(-1)^{\ell-\lambda}\binom{n-\lambda+\underline{\mu}}{\ell-\lambda}\binom{n+d+|\mu|}{\lambda}$$

$$=\frac{\ell!\,\Gamma(n-\ell+\underline{\mu}+1)}{\Gamma(n+d+|\mu|+1)}\sum_{\lambda=0}^{\ell}\binom{-n+\ell-\underline{\mu}-1}{\ell-\lambda}\binom{n+d+|\mu|}{\lambda}$$

$$=\frac{\ell!\,\Gamma(n-\ell+\underline{\mu}+1)}{\Gamma(n+d+|\mu|+1)}\binom{d+|\mu|-\underline{\mu}+\ell-1}{\ell},$$

whence the operator norm in (17) is given by

$$\|\mathbf{T}^{(\ell)}_{n-\ell,\mu}\|_{C(\mathbf{S}^d)\to C(\mathbf{S}^d)}=\frac{\ell!\,\Gamma(n-\ell+\underline{\mu}+1)}{\Gamma(n+d+|\mu|+1)}\binom{d+|\mu|-\underline{\mu}+\ell-1}{\ell},\quad 0\le\ell\le n. \tag{18}$$

The proof is finished by an application of Lemma 10. We have

$$\frac{1}{\binom{n}{r}}\left\|\mathbf{U}_{r,\mu}(\mathbf{M}_{n,\mu})\right\|_{C(\mathbf{S}^d)\to C(\mathbf{S}^d)}$$

$$\le\frac{\Gamma(n+d+|\mu|+1)}{\Gamma(n+\underline{\mu}+1)}\sum_{\ell=0}^{r}\binom{n+\underline{\mu}}{\ell}\binom{d+|\mu|-\underline{\mu}+r-1}{r-\ell}\|\mathbf{T}^{(\ell)}_{n-\ell,\mu}\|_{C(\mathbf{S}^d)\to C(\mathbf{S}^d)}$$

and the bound, by (18), takes the form

$$\sum_{\ell=0}^{r}\binom{n+\underline{\mu}}{\ell}\binom{d+|\mu|-\underline{\mu}+r-1}{r-\ell}\frac{\ell!\,\Gamma(n-\ell+\underline{\mu}+1)}{\Gamma(n+\underline{\mu}+1)}\binom{d+|\mu|-\underline{\mu}+\ell-1}{\ell}$$

$$=\sum_{\ell=0}^{r}\binom{d+|\mu|-\underline{\mu}+r-1}{r-\ell}\binom{d+|\mu|-\underline{\mu}+\ell-1}{\ell}$$

$$=\binom{d+|\mu|-\underline{\mu}+r-1}{r}\sum_{\ell=0}^{r}\binom{r}{\ell}$$

$$=2^r\binom{d+|\mu|-\underline{\mu}+r-1}{r}.$$

This proves the theorem and prepares us to state a few implications thereof. □

Theorem 12. *For given $n,r,\rho\in\mathbb{N}_0$, $0\le r,\rho\le n$, and $1\le p\le\infty$, we have*

$$\left\|\mathbf{U}_{\rho,\mu}(\mathbf{M}^{(r)}_{n,\mu}(f))\right\|_{p,\mu}\le c_{r,\rho}\binom{n}{\rho}\left\|f\right\|_{p,\mu},\quad f\in L^p_\mu(\mathbf{S}^d),$$

where the constant $c_{r,\rho}=c_{r,\rho,\mu,d}$ depends only on r,ρ,μ and d. Here, $\mu\in\mathbb{R}^{d+1}$ with $\underline{\mu}\ge-\frac{1}{2}$.

Proof. From equation (11) we find a representation

$$\mathbf{U}_{\rho,\mu}\mathbf{U}_{\ell,\mu} = \sum_{k=0}^{\rho+\ell} \alpha_{\ell,\rho}^{(k)}\, \mathbf{U}_{k,\mu}\ .$$

We insert this in the definition (13) of the quasi-interpolant to give

$$\mathbf{U}_{\rho,\mu}(\mathbf{M}_{n,\mu}^{(r)}(f)) = \sum_{\ell=0}^{r} \frac{1}{\binom{n}{\ell}} \sum_{k=0}^{\rho+\ell} \alpha_{\ell,\rho}^{(k)}\, \mathbf{U}_{k,\mu}\left(\mathbf{M}_{n,\mu}(f)\right)\ .$$

An application of Theorem 11 finishes the proof. □

The special case $\rho = 0$ is of particular importance, since this shows that our quasi-interpolants are uniformly bounded.

Theorem 13. *For given* $n, r \in \mathbb{N}_0$, $0 \le r \le n$, $\mu \in \mathbb{R}^{d+1}$ *with* $\underline{\mu} \ge -\frac{1}{2}$ *and* $1 \le p \le \infty$, *we have*

$$\left\|\mathbf{M}_{n,\mu}^{(r)}(f)\right\|_{p,\mu} \le c_{r,0}\, \left\|f\right\|_{p,\mu}\,, \quad f \in L_{\mu}^{p}(\mathbf{S}^d)\,, \tag{19}$$

where

$$c_{r,0} := \sum_{\ell=0}^{r} 2^{\ell} \binom{d+|\mu|-\underline{\mu}+\ell-1}{\ell}\,. \tag{20}$$

Another way to get this result is via an extension of Lemma 10 by writing the quasi-interpolants as linear combinations of the positive operators $\mathbf{T}_{n-\ell,\mu}^{(\ell)}$. We state this important identity as a separate result.

Theorem 14. *For* $0 \le r \le n$ *we have*

$$\mathbf{M}_{n,\mu}^{(r)} = \frac{\Gamma(n+d+|\mu|+1)}{\Gamma(n+\underline{\mu}+1)} \sum_{\ell=0}^{r} (-1)^{\ell} \binom{n+\underline{\mu}}{\ell} \binom{d+|\mu|-\underline{\mu}+r}{r-\ell} \mathbf{T}_{n-\ell,\mu}^{(\ell)}\,.$$

This follows from Lemma 10 via (13), and provides the following expression for the constant in (20).

$$\begin{aligned} c_{r,0} &= \frac{\Gamma(n+d+|\mu|+1)}{\Gamma(n+\underline{\mu}+1)} \sum_{\ell=0}^{r} \binom{n+\underline{\mu}}{\ell} \binom{d+|\mu|-\underline{\mu}+r}{r-\ell} \mathbf{T}_{n-\ell,\mu}^{(\ell)}(1) \\ &= \binom{d+|\mu|-\underline{\mu}+r-1}{r} \sum_{\ell=0}^{r} \binom{r}{\ell} \frac{d+|\mu|-\underline{\mu}+r}{d+|\mu|-\underline{\mu}+\ell}\,. \end{aligned}$$

7. Convergence and Direct Results

The results of the previous section allow us to prove convergence. The first statement is a direct consequence of the uniform bound (19) for the quasi-interpolants. Namely,

$$\lim_{n\to\infty} \|f - \mathbf{M}_{n,\mu}^{(r)}(f)\|_{p,\mu} \to 0\,, \quad n \to \infty\,,$$

for $f \in L^p_\mu(\mathbf{S}^d)$ or $f \in C(\mathbf{S}^d)$, respectively, since the convergence holds for polynomials, due to Theorem 7. Inserting this in (14) leads to the following expansion, via a telescoping argument.

Lemma 15. *For $n, r \in \mathbb{N}_0$, $0 \le r \le n$, and $\underline{\mu} \ge -\frac{1}{2}$ we have*

$$f - \mathbf{M}_{n,\mu}^{(r)}(f) = \sum_{\ell=n+1}^{\infty} \frac{r+1}{\ell + d + |\mu|} \frac{1}{\binom{\ell}{r+1}} \mathbf{U}_{r+1,\mu}(\mathbf{M}_{\ell,\mu}(f)) \tag{21}$$

for $f \in L^p_\mu(\mathbf{S}^d)$ or $f \in C(\mathbf{S}^d)$, respectively, with convergence in the norm.

Remark. For smooth functions $f \in C^{2r+2}(\mathbf{S}^d)$ the error expansion (21) holds true for general μ with the usual restriction $\underline{\mu} > -1$, see [4], Theorem 5, where a proof is given with no recourse to the Bernstein inequality.

This error expansion is the basis for several quantitative convergence results. The arguments are more or less standard. We start with the Jackson-Favard type estimate.

Theorem 16. *Let $n, r \in \mathbb{N}_0$, $0 \le r \le n$, and $\mu \in \mathbb{R}^{d+1}$ with $\underline{\mu} > -1$. Then, for $f \in C^{2r+2}(\mathbf{S}^d)$,*

$$\left\| f - \mathbf{M}_{n,\mu}^{(r)}(f) \right\|_{p,\mu} \le \frac{C_{n,r,d,\mu}}{\binom{n}{r+1}} \left\| \mathbf{U}_{r+1,\mu}(f) \right\|_{p,\mu}$$

with

$$C_{n,r,d,\mu} := \sum_{\ell=n+1}^{\infty} \frac{r+1}{\ell + d + |\mu|} \frac{\binom{n}{r+1}}{\binom{\ell}{r+1}} \to 1\,, \quad n \to \infty\,. \tag{22}$$

Proof. Under the smoothness assumption for f the differential operator commutes with the Durrmeyer operator, and we can estimate the series in (21) by using

$$\left\| \mathbf{U}_{r+1,\mu}(\mathbf{M}_{\ell,\mu}(f)) \right\|_{p,\mu} = \left\| \mathbf{M}_{\ell,\mu}(\mathbf{U}_{r+1,\mu}(f)) \right\|_{p,\mu} \le \left\| \mathbf{U}_{r+1,\mu}(f) \right\|_{p,\mu}\,.$$

□

The second quantitative result is of Voronovskaja type.

Theorem 17. *Let $r \in \mathbb{N}$ and $\mu \in \mathbb{R}^{d+1}$ with $\underline{\mu} > -1$. Then, for $f \in C^{2r+2}(\mathbf{S}^d)$,*

$$\begin{aligned} \lim_{n\to\infty} \binom{n}{r+1} \{f - \mathbf{M}_{n,\mu}^{(r)}(f)\}(\mathbf{x}) &= \lim_{n\to\infty} C_{n,r,d,\mu} \left\{\mathbf{U}_{r+1,\mu}(f)\right\}(\mathbf{x}) \\ &= \left\{\mathbf{U}_{r+1,\mu}(f)\right\}(\mathbf{x})\,, \end{aligned}$$

where the convergence is uniform in $\mathbf{S}^d$.

Proof. Using Lemma 15 and (22), we find

$$\begin{aligned}\Big\{\binom{n}{r+1}\{f-\mathbf{M}_{n,\mu}^{(r)}(f)\}-C_{n,r,d,\mu}\,\mathbf{U}_{r+1,\mu}(f)\Big\}\\ =\sum_{\ell=n+1}^{\infty}\frac{r+1}{\ell+d+|\mu|}\frac{\binom{n}{r+1}}{\binom{\ell}{r+1}}\{\mathbf{U}_{r+1,\mu}(\mathbf{M}_{\ell,\mu}(f))-\mathbf{U}_{r+1,\mu}(f)\}\,.\end{aligned}\tag{23}$$

The result follows from

$$\lim_{\ell\to\infty}\{\mathbf{U}_{r+1,\mu}(\mathbf{M}_{\ell,\mu}(f))-\mathbf{U}_{r+1,\mu}(f)\}=\lim_{\ell\to\infty}\{\mathbf{M}_{\ell,\mu}(\mathbf{U}_{r+1,\mu}(f))-\mathbf{U}_{r+1,\mu}(f)\}=0$$

in the uniform norm. □

Our third quantitative convergence result is of 'strong' Voronovskaja type. It follows from another application of Lemma 15 (case $r=0$) on the right-hand side of (23).

Theorem 18. *Let $n,r\in\mathbb{N}_0$, $0\le r\le n$, and $\mu\in\mathbb{R}^{d+1}$ with $\underline{\mu}>-1$. Then, for $f\in C^{2r+2}(\mathbf{S}^d)$,*

$$\begin{aligned}&\binom{n}{r+1}\{f-\mathbf{M}_{n,\mu}^{(r)}(f)\}-C_{n,r,d,\mu}\mathbf{U}_{r+1,\mu}(f)\\ &=-\sum_{\ell_0=n+1}^{\infty}\frac{r+1}{\ell_0+d+|\mu|}\frac{\binom{n}{r+1}}{\binom{\ell_0}{r+1}}\sum_{\ell_1=\ell_0+1}^{\infty}\frac{1}{\ell_1(\ell_1+d+|\mu|)}\mathbf{U}_{r+1,\mu}\{\mathbf{U}_{1,\mu}(\mathbf{M}_{\ell_1,\mu}(f))\}\,,\end{aligned}$$

with convergence in the uniform norm. In particular, for $f\in C^{2r+4}(\mathbf{S}^d)$,

$$\binom{n}{r+1}\{f-\mathbf{M}_{n,\mu}^{(r)}(f)\}-C_{n,r,d,\mu}\mathbf{U}_{r+1,\mu}(f)=O(n^{-1})\quad as\quad n\to\infty\,.$$

The idea can be iterated in order to find an error expansion in powers of n^{-1}. We omit the details but rather give a final result in terms of the K-functional

$$K(f,t):=K_{\ell,\mu,p}(f,t):=\inf_{g\in C^{2\ell}(\mathbf{S}^d)}\Big\{\,\|f-g\|_{p,\mu}+t\,\|\mathbf{U}_{\ell,\mu}(g)\|_{p,\mu}\,\Big\}\,,\quad \ell>0\,.$$

Theorem 19. *Let $n,r\in\mathbb{N}_0$, $0\le r\le n$, and $\mu\in\mathbb{R}^{d+1}$ with $\underline{\mu}\ge-\frac{1}{2}$. Then, for $f\in L^p_\mu(\mathbf{S}^d)$ or $f\in C(\mathbf{S}^d)$, respectively,*

$$\|f-\mathbf{M}_{n,\mu}^{(r)}(f)\|_{p,\mu}\le c_r\,K_{r+1,\mu,p}\Big(f,\binom{n}{r+1}^{-1}\Big)\,.$$

The *proof* is again standard. Since

$$f - \mathbf{M}_{n,\mu}^{(r)}(f) = (f - g) - \mathbf{M}_{n,\mu}^{(r)}(f - g) + (g - \mathbf{M}_{n,\mu}^{(r)}(g)) ,$$

we can apply Theorems 13 and 16 to find

$$\|f - \mathbf{M}_{n,\mu}^{(r)}(f)\|_{p,\mu} \le (1 + c_{r,0})\|f - g\|_{p,\mu} + \frac{C_{n,r,d,\mu}}{\binom{n}{r+1}}\|\mathbf{U}_{r+1,\mu}(g)\|_{p,\mu}$$

for arbitrary $g \in C^{2r+2}(\mathbf{S}^d)$. Taking the infimum with respect to g proves the result with constant

$$c_r := \max(\ 1 + c_{r,0}\ ,\ \sup_n\ C_{n,r,d,\mu}\ ) .$$

8. Additional Notes

The Durrmeyer operators (2) were introduced by Durrmeyer in his thesis [15]. The study of their approximation properties was initiated by Derriennic in several papers, and later studied by many authors. The spectral properties of Theorem 1 appear in [10], [11] and [12] for the unweighted case, in [6] and [7] for the weighted univariate case, and in [14] for the weighted multivariate case. See also Chapter 5.2 in Paltanea's recent book [23], and the references given there.

The statement of Theorem 2 is original, as is its application in Section 6. For the properties and formulas for hypergeometric functions, we refer to standard tablework, such as [1] and [22]. For the Laplace type integrals, a direct approach can be taken from Koornwinder [21] and, in particular, from Askey's [2] elegant work related to this. It is also worthwhile to consult Szegö's [27] chapter on Jacobi polynomials.

The differential operator $\mathbf{U}_\mu = \mathbf{U}_{1,\mu}$ in (10) and its powers play a prominent role in the study of direct and inverse theorems for the Durrmeyer operator; see again the papers by Derriennic, Berens and Xu, and Ditzian, where also the spectral properties of Lemma 6 can be found. The higher order case $\mathbf{U}_{\ell,\mu}$ was first investigated in [4], with different notation. In the univariate (unweighted) case, these operators are called Legendre differential operators in Heilmann's Habilitationsschrift [16]. In the present paper, we have chosen the recursive definition (11) leading to a product representation for $\mathbf{U}_{\ell,\mu}$ which was communicated to us by Michael Felten.

The quasi-interpolants (13) were introduced in [17], for the unweighted case. The weighted case was considered in [18] and [4], where also the statements of Theorem 7 and Lemma 8 can be found. The expression of the quasi-interpolants in terms of Durrmeyer operators in Theorem 9, however, is new. This latter result embeds our operators into the class of quasi-interpolants constructed as linear combinations of Durrmeyer operators. However, this approach is usually less direct and less explicit than ours. We refer to work of Derriennic [13] again, and to Sablonnière [24], [25] and Heilmann [16]. Sablonnière's recent paper [26] presents a good account on these constructions.

The Bernstein inequality, Theorem 11, is again original. It confirms our conjecture posed in [3] where a different proof was given for the special case $d = 1$ and $\mu = (0,0)$. We expect that the representation of the quasi-interpolants as a linear combination of the positive operators $\mathbf{T}_{n-\ell,\mu}^{(\ell)}$ as given in Theorem 14 will have far-reaching applications. Last not least, the case $\mathbf{M}_{n,\mu}^{(n)}$ is expected to lead to a new representation of polynomial reproducing kernels of full order on the simplex $\mathbf{S}^d$, with connections to addition theorems for orthogonal polynomials.

The direct results in Section 7 are more or less immediate consequences of the Bernstein inequality. We refer to our paper [4], and again to earlier work by Derriennic [13], Berens et al. [5]-[7] and Ditzian [14]. However, we still do not settle the natural question of 'converse' or even 'strong converse' theorems for our quasi-interpolants. Concerning this, the paper of Chen, Ditzian and Ivanov [8], the refined techniques of Knoop and Zhou in [19], [20] and Zhou's Habilitationsschrift [29] might be helpful.

The initial motivation for our studies came from the article by Chui et al. [9], in which univariate quasi-interpolants on irregular partitions of a bounded interval $I \subset \mathbb{R}$ were constructed as linear combinations of B-splines. In their approach, the quasi-interpolants $M_n^{(r)}$ (for the unweighted case) are the starting point for an inductive method of knot insertion. The quasi-interpolants in [9] give rise to the definition of "approximate duals" of B-splines, which are the centerpiece for their construction of nonstationary wavelet frames.

References

1. Andrews, G. E., Askey, R. and Roy, R., *Special Functions*, Cambridge University Press, Cambridge, 2000.
2. Askey, R., Jacobi polynomials, I. New proofs of Koornwinder's Laplace type integral representation and Bateman's bilinear sum, *SIAM J. Math. Anal.* **5** (1974), 119–124.
3. Berdysheva, E., Jetter, K. and Stöckler, J., Bernstein-Durrmeyer type quasi-interpolants on intervals, in: *Approximation Theory* (D. K. Dimitrov, G. Nikolov, R. Uluchev, Eds.), Marin Drinov Academic Publishing House, Sofia, 2004, pp. 32–42.
4. Berdysheva, E., Jetter, K. and Stöckler, J., New polynomial preserving operators on the simplex: direct results, *J. Approx. Theory* **131** (2004), 59–73.
5. Berens, H., Schmid, H. J. and Xu, Y., Bernstein-Durrmeyer polynomials on a simplex, *J. Approx. Theory* **68** (1992), 247–261.
6. Berens, H. and Xu, Y., On Bernstein-Durrmeyer polynomials with Jacobi-weights, in: *Approximation Theory and Functional Analysis* (C. K. Chui, Ed.), Academic Press, Boston, 1991, pp. 25–46.

7. Berens, H. and Xu, Y., On Bernstein-Durrmeyer polynomials with Jacobi-weights: the cases $p = 1$ and $p = \infty$, *Israel Math. Conf. Proc.* **4** (1991), 51–62.
8. Chen, W., Ditzian, Z. and Ivanov, K., Strong converse inequality for the Bernstein-Durrmeyer operator, *J. Approx. Theory* **75** (1993), 25–43.
9. Chui, C. K., He, W. and Stöckler, J., Nonstationary tight wavelet frames, I: bounded intervals, *Applied Comp. Harmonic Analysis* **17** (2004), 141–197.
10. Derriennic, M.-M., Sur l'approximation de fonctions intégrables sur $[0, 1]$ par des polynômes de Bernstein modifiés, *J. Approx. Theory* **31** (1981), 325–343.
11. Derriennic, M.-M., Polynômes de Bernstein modifiés sur un simplex T de R^l, problème des moments, in: *Proc. Polynômes Orthogonaux et Applications*, Bar-le-Duc, LNM 1171, Springer-Verlag, 1984, pp. 296–301.
12. Derriennic, M.-M., On multivariate approximation by Bernstein-type polynomials, *J. Approx. Theory* **45** (1985), 155–166.
13. Derriennic, M.-M., Linear combinations of derivatives of Bernstein-type polynomials on a simplex, *Coll. Math. Soc. Janos Bolyai* **58** (1990), 197–220.
14. Ditzian, Z., Multidimensional Jacobi-type Bernstein-Durrmeyer operators, *Acta Sci. Math. (Szeged)* **60** (1995), 225–243.
15. Durrmeyer, J.-L., Une formule d'inversion de la transformée de Laplace: Applications à la théorie des moments, Thèse de 3e cycle, Faculté des Sciences de l'Université de Paris, 1967.
16. Heilmann, M., *Erhöhung der Konvergenzgeschwindigkeit bei der Approximation von Funktionen mit Hilfe von Linearkombinationen spezieller positiver linearer Operatoren*, Habilitationsschrift, Universität Dortmund, 1992.
17. Jetter, K. and Stöckler, J., An identity for multivariate Bernstein polynomials, *Computer Aided Geometric Design* **20** (2003), 563–577.
18. Jetter, K. and Stöckler, J., New polynomial preserving operators on the simplex, Ergebnisberichte Angewandte Mathematik, Nr. 242, Universität Dortmund (November 2003).
19. Knoop, H.-B. and Zhou, X. L., The lower estimate for linear positive operators, I, *Constr. Approximation* **11** (1995), 53–66.
20. Knoop, H.-B. and Zhou, X. L., The lower estimate for linear positive operators, II, *Results Math.* **25** (1994), 315–330.
21. Koornwinder, T., Jacobi polynomials, II. An analytic proof of the product formula, *SIAM J. Math. Anal.* **5** (1974), 125–137.
22. Magnus, W., Oberhettinger, F. and Soni, R. P., *Formulas and Theorems for the Special Functions of Mathematical Physics*, 3rd ed., Springer-Verlag, Berlin, 1966.
23. Paltanea, R. , *Approximation Theory Using Positive Linear Operators*, Birkhäuser-Verlag, Boston, 2004.
24. Sablonnière, P., Bernstein-type quasi-interpolants, in: *Curves and Surfaces* (P.-J. Laurent, A. Le Méhauté, L. L. Schumaker, Eds.), Academic Press, Boston, 1991, pp. 421-426.
25. Sablonnière, P., Representation of quasi-interpolants as differential operators and applications, in: *New Developments in Approximation Theory*

(M. D. Buhmann, M. Felten, D. H. Mache and M. W. Müller, Eds.), Birkhäuser-Verlag, Basel, 1999, pp. 233–253.

26. Sablonnière, P., Recent progress on univariate and multivariate polynomial and spline quasi-interpolants, Prépublication 04-20, INSA Rennes, March 2004.
27. Szegö, G., *Orthogonal Polynomials*, 4th ed., Amer. Math. Soc., Providence, RI, 1975.
28. Widder, D. V., *The Laplace Transform*, 8th printing, Princeton University Press, Princeton, 1972.
29. Zhou, X., *Approximationsordnung und Regularität von Differentialoperatoren*, Habilitationsschrift, Gerhard-Mercator-Universität Duisburg, 1996.

Topics in Multivariate Approximation and Interpolation
K. Jetter et al., Editors

Three Families of Nonlinear Subdivision Schemes

Nira Dyn

School of Mathematical Sciences, Tel-Aviv University, Israel

Abstract

Three families of nonlinear subdivision schemes, derived from linear schemes, are reviewed. The nonlinearity is introduced into the linear schemes by adapting the schemes to the data. The first family, derived from the four-point interpolatory linear subdivision scheme, consists of geometrically controlled schemes, which are either shape preserving or artifact-free. The second family of schemes is designed for the functional setting, to be used in constructions of multiscale representations of piecewise smooth functions. The schemes are extensions of the Dubuc-Deslauriers $2N$-point interpolatory schemes, with the classical local interpolation replaced by ENO or WENO local interpolation. The third family consists of subdivision schemes on smooth manifolds. These schemes are derived from converging linear schemes, represented in terms of repeated binary averages. The analysis of the nonlinear schemes is done either by proximity to the linear schemes from which they are derived, or by methods adapted from methods for linear schemes.

Key words: linear and nonlinear subdivision scheme, adaptive tension parameter, convexity preserving scheme, ENO interpolation, data dependent scheme, repeated binary averages, refinement on a manifold, geodesics, projection onto a manifold

2000 MSC: 65D05, 65D07, 65D10, 65D15, 65D17, 65U05, 65U07

Email address: niradyn@post.tau.ac.il (Nira Dyn).

1. Introduction

Linear subdivision schemes are an important tool for the design and generation of curves and surfaces in geometric modelling. A further motivation for the study of subdivision schemes is their close relation to multiresolution analysis and wavelets (see e.g. [4]).

Linear subdivision schemes have been investigated for the last 20 years. Tools for analyzing their convergence and smoothness are now available. Yet, linear schemes have many limitations being independent of the data upon which they operate.

In this paper we review three families of nonlinear subdivision schemes, handling three different settings in which linear schemes fail. The construction of the nonlinear schemes is closely related to linear schemes and so is their analysis.

The first setting is a geometric setting of control polygons in 2D or 3D. The performance of linear subdivision schemes on initial control polygons with edges of comparable lengths is known to be satisfactory. But curves generated by linear schemes from control polygons with edges of significantly different lengths tend to have artifacts, such as self intersections and inflection points which do not correspond to the shape of the initial control polygon. We present a construction of a nonlinear scheme which alleviates the artifacts. The construction is based on the linear 4-point interpolatory scheme with a tension parameter. The nonlinearity is introduced by choosing the tension parameter adaptively for each inserted point, according to the geometry of the relevant local control polygon, which consists of the four points involved in the definition of the inserted point. The idea of adaptive tension parameter is used also for obtaining shape preserving schemes in 2D, such as convexity-preserving schemes. These two types of geometrically controlled 4-point schemes are reviewed in Section 2, following the paper [11]. A different construction of a nonlinear four-point scheme, which is circle-preserving, is studied in [16].

The second setting is a functional setting with data sampled from a piecewise smooth function [2]. The nonlinear subdivision schemes are data dependent, and are extensions of the Dubuc-Deslauries 2N-point interpolatory schemes. The nonlinearity is introduced by using the ENO (Essentially Non Oscillatory) idea of choosing a stencil for local interpolation, among several possible ones, according to the data. The choice of the stencil for estimating an inserted point aims at stencils which consist of points all from the same smoothness region. In order to obtain "stable schemes", namely schemes which depend continuously on the data, the WENO (Weighted ENO) idea is used instead of the ENO idea. Section 3 reviews mainly the special case corresponding to the 4-point scheme ($N = 2$).

The third setting is that of curves on smooth manifolds. Converging linear schemes are so changed to generate control points on the manifold at each refinement level. The analysis of the resulting nonlinear schemes is done by their proximity to the linear schemes from which they are derived, and from properties of the linear schemes obtained by known methods (see e.g. [6]). We discuss two general constructions of nonlinear schemes from linear schemes. The material is taken mainly from [17] and partially from [18], and is presented in Section 4. Similar constructions and results are presented in [17] and [18] for subdivision schemes in certain matrix groups and in certain Lie groups. This material is not reviewed here.

There are not many works in the literature on nonlinear subdivision schemes on manifolds. Analogous schemes to quadratic B-spline schemes on manifolds are studied in [12], [13]. A general approach to the construction of nonlinear schemes for manifold data from linear schemes is developed in [15]. The construction is different from those presented in Section 4. The schemes in [15] are used for multiscale representations of manifold-valued data.

The idea of analyzing subdivision schemes, related to linear ones, by proximity is not new. In [7] non-stationary linear schemes are analyzed by their proximity to stationary linear schemes. Here we mention two other types of nonlinear schemes that were analyzed in relation to linear ones. The first type is that of the "median interpolating" subdivision schemes and their extensions [19], [20], [14], where the nonlinearity is rather weak. The second type is that of the "normal curves" [5].

2. Geometrically Controlled 4-point Interpolatory Schemes

In this section we present nonlinear versions of the linear 4-point interpolatory scheme [8], which adapt the tension parameter to the geometry of the control points. It is well known that the linear 4-point scheme with the refinement rules

$$P_{2j}^{k+1} = P_j^k, \quad P_{2j+1}^{k+1} = -w(P_{j-1}^k + P_{j+2}^k) + (\frac{1}{2} + w)(P_j^k + P_{j+1}^k) \;, \qquad (1)$$

where w is a fixed tension parameter, generates a good curve from initial control points $\{P_j^0\}$, if the edges of the control polygon, $e_j^0 = P_{j+1}^0 - P_j^0$, have comparable length. The generated curve is C^1 for $w \in (0, w^*)$, where the value of w^* is not known, but it is known that $w^* < \frac{1}{8}$ [8].

In case of a control polygon with edges of significantly different lengths, the curve generated by (1) has artifacts such as self-intersections and inflection points which are not seen in the initial control polygon (see Fig. 1).

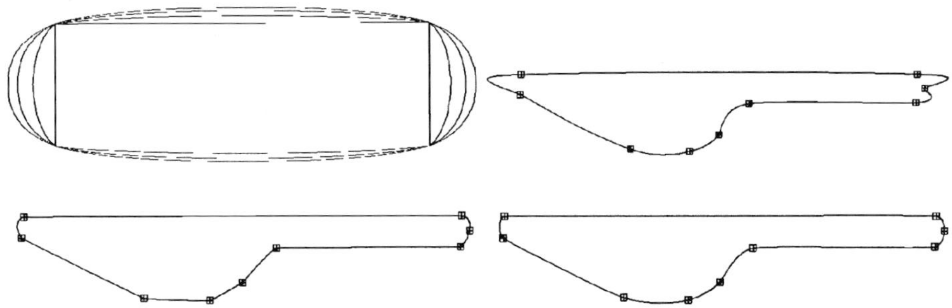

Fig. 1. Curves generated by the linear 4-point scheme: (upper left) the effect of different tension parameters, (upper right) artifacts in the curve generated with $w = \frac{1}{16}$, (lower left) artifact-free but visually non-smooth curve generated with $w = 0.01$. (Lower right) artifact-free and visually smooth curve generated in a nonlinear way with adaptive tension parameters.

To overcome this problem we developed in [11] nonlinear 4-point interpolatory schemes of the form (1) but with w chosen adaptively according to the geometry of the control polygon of the 4 points in (1).

Writing the insertion rule for P_{2j+1}^{k+1} in (1) in terms of the edges $e_j^k = P_{j+1}^k - P_j^k$, and relating the inserted point P_{2j+1}^{k+1} to the edge e_j^k, we get

$$P_{e_j^k} = M_{e_j^k} + w_{e_j^k}(e_{j-1}^k - e_{j+1}^k) \tag{2}$$

with $M_{e_j^k}$ the midpoint of e_j^k and $w_{e_j^k}$ the adaptive tension parameter for the refinement of e_j^k. Defining $d_{e_j^k} = w_{e_j^k}(e_{j-1}^k - e_{j+1}^k)$ as the displacement from $M_{e_j^k}$, we control its size by choosing $w_{e_j^k}$ according to a chosen geometrical criterion.

Here we review two such criteria. More criteria can be found in [11].

2.1. Displacement-safe schemes

In this family of schemes, $w_{e_j^k}$ is chosen so that

$$\|d_{e_j^k}\| < \frac{1}{2}\|e_j^k\| \quad \text{for} \quad \|e_j^k\| > 0\,. \tag{3}$$

This choice guarantees that the inserted control point $P_{e_j^k}$ is different from the boundary points of the edge e_j^k and that the length of each of the two edges replacing e_j^k is less than the length of e_j^k, namely

$$\|e_{2j}^{k+1}\| = \|P_{e_j^k} - P_j^k\| < \|e_j^k\|$$
$$\|e_{2j+1}^{k+1}\| = \|P_{j+1}^k - P_{e_j^k}\| < \|e_j^k\|$$

There are many ways to impose (3). We restrict the adaptive tension parameter $w_{e_j^k}$ to the interval $(0, \frac{1}{16}]$, so that a tension close to 1/16 is assigned to "regular stencils" e.g. stencils of four points with three edges of almost equal length. The less regular the stencil, the closer to zero is the tension parameter assigned to it. These heuristics are based on these observations about the linear 4-point scheme:

- This scheme generates "good" curves when applied to control polygons with edges of comparable length.
- This scheme generates curves which become smoother, the closer the tension parameter is to 1/16.
- For initial control polygons with edges of significantly different lengths, this scheme generates curves which preserve the shape of the initial control polygons, only for very small values of the tension parameters. (Recall that the control polygon itself corresponds to the generated curve with zero tension parameter.)

A "natural" choice of an adaptive tension parameter obeying (3) is

$$w_{e_j^k} = \min\left\{\frac{1}{16}, c\,\frac{\|e_j^k\|}{\|e_{j-1}^k - e_{j+1}^k\|}\right\}, \quad \text{with a fixed } c \in (0, \frac{1}{2}) . \tag{4}$$

Furthermore, we restrict c to the interval $[\frac{1}{8}, \frac{1}{2})$ to guarantee that $w_{e_j^k} = \frac{1}{16}$ for stencils with $\|e_{j-1}^k\| = \|e_j^k\| = \|e_{j+1}^k\|$. Indeed in this case, $\|e_{j-1}^k - e_{j+1}^k\| = 2\sin\frac{\theta}{2}\|e_j^k\|$, with θ, $0 < \theta \leq \pi$, the angle between the two vectors e_{j-1}^k, e_{j+1}^k. Thus $\|e_j^k\|/\|e_{j-1}^k - e_{j+1}^k\| = (2\sin\frac{\theta}{2})^{-1} \geq \frac{1}{2}$, and if $c \geq \frac{1}{8}$ then the minimum in (4) is $\frac{1}{16}$. The choice (4) defines irregular stencils (corresponding to small $w_{e_j^k}$) as those with $\|e_j^k\|$ much smaller than at least one of $\|e_{j-1}^k\|, \|e_{j+1}^k\|$, and such that when these two edges are of comparable length, the angle between them is not close to zero.

Moreover, for $\|e_j^k\| = 0$ we take $w_{e_j^k} = 0$, and then $P_{2j+1}^{k+1} = P_j^k = P_{j+1}^k$. Thus two repeated control points in the initial control polygon generate a corner in the limit curve, while due to (3), new repeated control points are not generated during the refinement process.

This feature of the displacement-safe scheme allows to generate piecewise smooth curves from initial control polygons with some repeated consecutive control points.

The convergence of the displacement-safe scheme, namely the convergence of the sequence of polygonal lines through the control points at each refinement level, is a consequence of the following result on the 4-point scheme with variable tension parameters, applied to scalar data [10].

Theorem 1. *The 4-point scheme with variable tension parameters has the refinement rules*

$$f_{2j}^{k+1} = f_j^k , \; f_{2j+1}^{k+1} = (\frac{1}{2} + w_j^k)(f_j^k + f_{j+1}^k) - w_j^k(f_{j-1}^k + f_{j+2}^k) . \tag{5}$$

It converges to continuous limit functions, if there is an $\varepsilon > 0$ such that $w_j^k \in [0, \frac{1}{8} - \varepsilon]$ for all (j,k). The scheme generates C^1 limit functions, if for all (j,k), $w_j^k \in [\varepsilon, \frac{1}{8} - \varepsilon]$, for some $\varepsilon > 0$.

The choice (4) of the adaptive tension parameters guarantees that for all (j,k), $w_{e_j^k} \in [0, \frac{1}{16}]$ and therefore by Theorem 1 applied to the components of the curve, the displacement-safe scheme defined by (4) converges and generates continuous curves. Yet we cannot conclude from Theorem 1 that the limit curves generated by this scheme are C^1, since the tension parameters used, during the subdivision process, are not bounded away from zero.

Nevertheless, our many simulations indicate that the curves generated by the displacement-safe scheme, based on (4), are C^1 (see Fig. 4).

If indeed these curves are C^1, then the displacement safe scheme based on (4) does not generate corners in the limit, and the resulting curves are piecewise smooth with corners only at repeated control points of the initial control polygon.

2.2. Convexity-preserving schemes in the plane

A shape property of planar control polygons, which is important to have in the curves generated by subdivision, is convexity. This can be achieved by a proper choice of the adaptive tension parameters.

To present this choice, we first introduce some local geometrical notions.

An edge such that its two neighboring edges are in the same half-plane, determined by the line through the edge, is termed a "convex edge".

An edge which is on the same line as one of its neighboring edges is termed a "straight edge".

A line through a control point, such that the two edges meeting at the control point are on the same side of the line, is termed a "convex tangent". A "straight tangent" at a control point is a line through one of the edges emanating from the point.

A polygon consisting of convex and straight edges is termed a "convex polygon". It is a "strictly convex polygon" if all its edges are convex. In Fig. 2 three examples of strictly convex polygons are given.

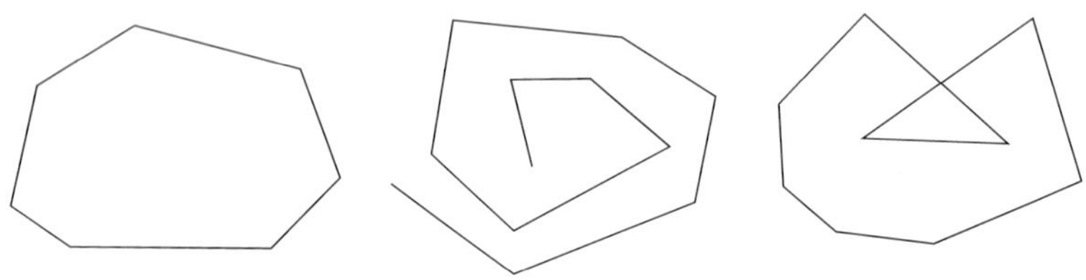

Fig. 2. Convex polygons: (left) closed, (middle) open, (right) self-intersecting.

The algorithm we present refines convex (strictly convex) control polygons into convex (strictly convex) control polygons. The construction of the inserted point $P_{e_i^k}$ is a geometric construction, which determines the displacement vector $w_{e_j^k}(e_{j-1}^k - e_{j+1}^k)$ rather than $w_{e_j^k}$ directly as in (4).

As a first step in the construction, at each control point from which at least one convex edge emanates, a convex tangent is constructed. At all other control points a straight tangent is constructed, coinciding with one of the straight edges meeting at the control point. We denote the tangent at P_j^k by t_j^k.

In case of a straight edge e_j^k, $P_{e_j^k} = M_{e_j^k}$. For a convex edge e_j^k, the construction of $P_{e_j^k}$ is illustrated schematically in Fig. 3. In this case, the tangents t_j^k and t_{j+1}^k together with e_j^k determine a triangle, $T_{e_j^k}$. By construction, the triangle $T_{e_j^k}$ and the edges e_{j-1}^k, e_{j+1}^k are on two different sides of the line through e_j^k.

The line from $M_{e_j^k}$ along the direction $e_{j-1}^k - e_{j+1}^k$ has a segment, $I_{e_j^k}$, inside $T_{e_j^k}$. Let $L_{e_j^k}$ denote the length of $I_{e_j^k}$. The point $P_{e_j^k}$ is chosen on $I_{e_j^k}$ so that

$$||P_{e_j^k} - M_{e_j^k}|| = \min\left\{\frac{1}{16}||e_{j-1}^k - e_{j+1}^k||,\ CL_{e_j^k}\right\},$$

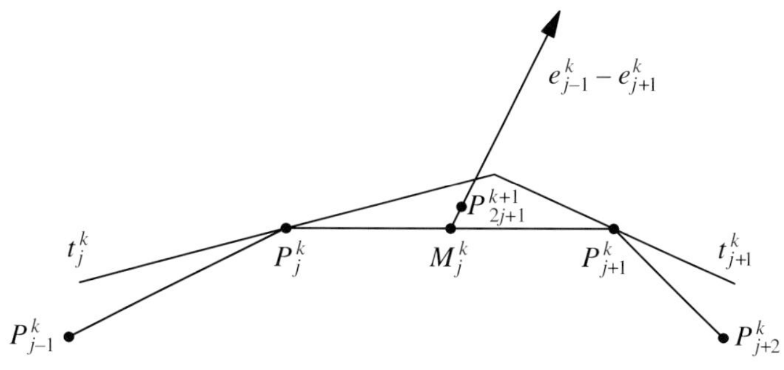

Fig. 3. Construction of an inserted point for a convex edge.

with a fixed C, satisfying $\frac{1}{2} < C < 1$. This choice guarantees that $0 < w_{e^k_j} \leq \frac{1}{16}$ and that the refined control polygon $\{P^{k+1}_j\}$ with

$$P^{k+1}_{2j} = P^k_j \ , \ P^{k+1}_{2j+1} = P_{e^k_j} \ ,$$

is convex (strictly convex) if the control polygon $\{P^k_j\}$ is (see [11]).

This construction of refined control polygons when repeated generates a sequence of convex (strictly convex) polygons from an initial convex (strictly convex) polygon. It is proved in [11] that the limit of this sequence is a continuous convex (strictly convex) curve. Moreover, it is proved that the curve between two consecutive initial control points is either a line segment when the edge connecting these two points in the initial control polygon is straight, or otherwise a strictly convex curve.

Note that the subdivision scheme is interpolatory and that P^{k+1}_{2j+1} depends on the points $P^k_{j-1}, P^k_j, P^k_{j+1}, P^k_{j+2}$ as in the linear 4-point scheme.

The convex tangents in this construction can be chosen in different ways. A natural choice of such a tangent is

$$t^k_j = P^k_{j+1} - P^k_{j-1} = e^k_j + e^k_{j-1} \ .$$

This choice was tested in many numerical experiments, and was found superior to other choices.

In Fig. 4, the performance of the convexity-preserving scheme is compared on several examples with that of the displacement-safe scheme of subsection 2.1 and with that of the linear 4-point scheme.

The convexity-preserving scheme is extended in [11] to a co-convexity preserving scheme for general planar polygons. This material is not reviewed here.

3. Quasilinear ENO-based 4-point Interpolatory Schemes

This family of schemes (ENO stands for Essentially Non Oscillatory) was designed to refine scalar data sampled from piecewise smooth functions. In such a scheme

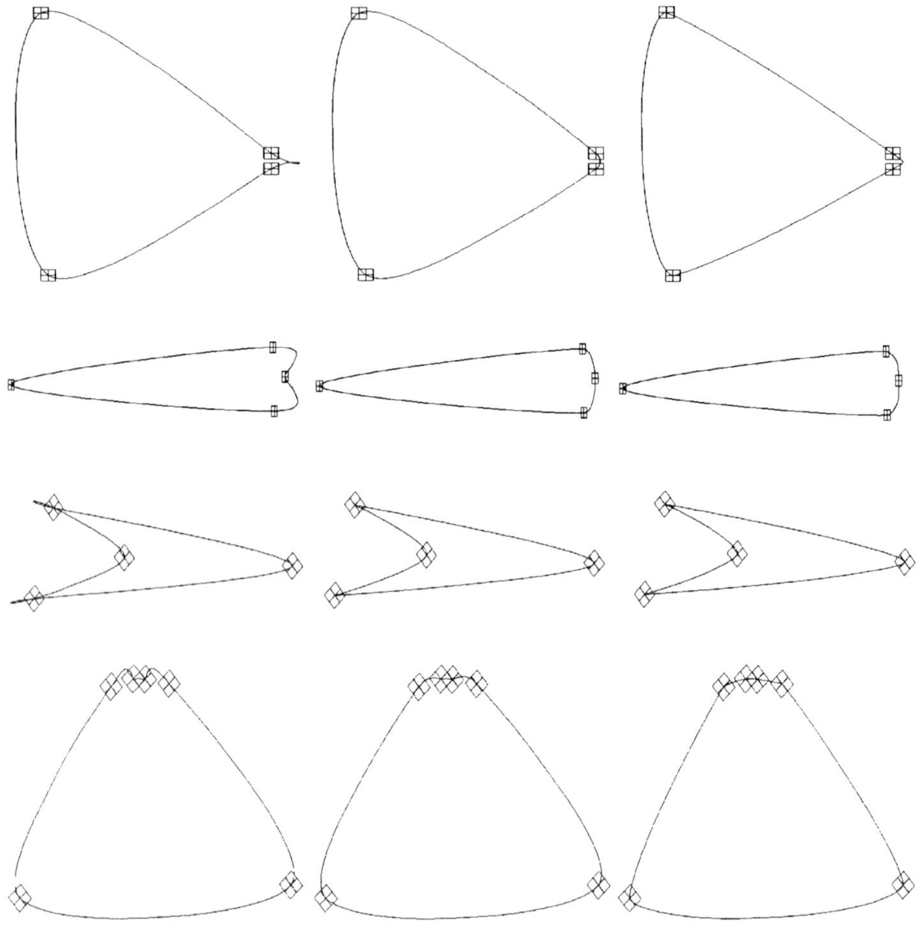

Fig. 4. Examples: (left column) the linear 4-point scheme with $w = 1/16$, (middle column) displacement-safe scheme of (4) with $C = 0.2$, (right column) convexity preserving scheme with $C = 0.9$.

the insertion rules during the refinement process, aim at estimating the function's values in terms of information taken from the same smoothness region. This idea can lead to schemes with quality of approximation similar to that of the linear 4-point scheme with $w = \frac{1}{16}$, when applied to data sampled from smooth functions. In the latter case the values f_j^{k+1} attached to the points $2^{-k-1}j, j \in \mathbb{Z}$, are, in fact, estimated from the values at refinement level k, by

$$f_{2j}^{k+1} = f_j^k, \quad f_{2j+1}^{k+1} = Q_{k,j}^c(2^{-(k+1)}(2j+1)), \quad j \in \mathbb{Z} ,$$

with $Q_{k,j}^c$, a cubic polynomial satisfying

$$Q_{k,j}^c(2^{-k}(j+i)) = f_{j+i}^k , \quad i = -1, 0, 1, 2 . \tag{6}$$

By the ENO approach, we can use two other interpolating polynomials. One is a cubic polynomial interpolating the data in the two endpoints of the interval

$2^{-k}(j, j+1)$, and in two points to its left. This is called a left stencil, and the interpolating cubic polynomial, $Q^{\ell}_{k,j}$ satisfies

$$Q^{\ell}_{k,j}(2^{-k}((j+i)) = f^k_{j+i}\ , \quad i = -2, -1, 0, 1\ . \tag{7}$$

The other polynomial is the cubic polynomial satisfying

$$Q^{r}_{k,j}(2^{-k}(j+i)) = f^k_{j+i}\ , \quad i = 0, 1, 2, 3\ . \tag{8}$$

The polynomial in (6) is based on a central stencil and that in (8) on a right stencil. The three different estimations for f^{k+1}_{2j+1} resulting from (7),(6) and (8) evaluated at $2^{-(k+1)}(2j+1)$, are

$$f^{k+1,\ell}_{2j+1} = \frac{1}{16}f^k_{j-2} - \frac{5}{16}f^k_{j-1} + \frac{15}{16}f^k_j + \frac{5}{16}f^k_{j+1}$$

$$f^{k+1,c}_{2j+1} = -\frac{1}{16}f^k_{j-1} + \frac{9}{16}f^k_j + \frac{9}{16}f^k_{j+1} - \frac{1}{16}f^k_{j+2}$$

$$f^{k+1,r}_{2j+1} = \frac{5}{16}f^k_j + \frac{15}{16}f^k_{j+1} - \frac{5}{16}f^k_{j+2} + \frac{1}{16}f^k_{j+3}$$

The selection of the stencil for the estimation of f^{k+1}_{2j+1} is data dependent. It follows the idea of Harten, Enquist, Osher and Chakravartly [9], that in the vicinity of a singularity in the smoothness, data taken from the same smoothness region of the function yields better estimates, and that such data is less oscillatory than data taken from the two sides of the singularity point. An example of a selection mechanism of a stencil is the choice of the "least oscillatory" interpolating polynomial, namely among $Q^{\ell}_{k,j}, Q^{c}_{k,j}, Q^{r}_{k,j}$ the one with least L_2-norm over the interval $2^{-k}(j, j+1)$.

The resulting ENO-based interpolatory subdivision scheme, converges for any initial data to a Hölder continuous limit function with exponent at least 0.66, and has the important property that if the initial data is sampled from a cubic polynomial, the limit function is that cubic polynomial.

The convergence of the ENO-based scheme, and its property of reproduction of cubic polynomials, guarantee that the limit function generated from the data $f(ih)$, $i \in \mathbb{Z}$, approximates, when scaled properly, the function f, at the rate $0(h^4)$, provided that f is smooth.

Since the selection mechanism does not guarantee that the selected stencil is indeed contained in the same smoothness region, it is not clear how to extend the above approximation result to piecewise smooth functions. For an analysis of a related approximation operator based on the ENO idea, see [1].

The application of a subdivision scheme in a multiscale decomposition and reconstruction with thresholding (e.g. for compression), requires the scheme to be stable, in the sense that small changes in the data result in small changes in the limit function. Converging linear schemes are stable, but the ENO-based 4-point scheme is not, because the selection mechanism is not continuous in the data.

To obtain a stable scheme, the ENO selection mechanism is replaced by the WENO selection mechanism (Weighted ENO), which is continuous in the data on the one hand, and retains the ENO idea on the other hand. In this scheme

$$f_{2j}^{k+1} = f_j^k \ , \ f_{2j+1}^{k+1} = Q_{k,j}^w(2^{-(k+1)}(2j+1)) \ ,$$

where $Q_{k,j}^w$ is a convex combination of $Q_{k,j}^r, Q_{k,j}^c, Q_{k,j}^\ell$ with coefficients $\alpha_0, \alpha_1, \alpha_2$ respectively.

There are various possible choices of $\alpha_0, \alpha_1, \alpha_2$ as continuous functions of the data $f_{j-2}^k, f_{j-1}^k, \ldots, f_{j+3}^k$. The idea of WENO is developed in [3], where an explicit choice is suggested. In this choice, the weights attached to stencils containing a singularity are small.

With the above suggested selection mechanism of the weights, the resulting subdivision scheme converges to Hölder continuous limit functions with exponent bounded below by 0.66. The scheme is C^s-stable for $s \leq 0.66$, namely for $u, v \in \ell_\infty(\mathbb{Z})$, and with $S^\infty(u)$ denoting the limit function generated by the scheme from the initial data u,

$$||S^\infty(u) - S^\infty(v)||_{C^s} \leq C||u - v||_{\ell_\infty},$$

where the constant C depends in a non-decreasing way on $\max\{||v||_{\ell_\infty}, ||u||_{\ell_\infty}\}$, and where

$$||f||_{C^s} = \sup_{x,y} |\frac{f(x) - f(y)}{(x-y)^s}|.$$

The method of analysis of convergence and smoothness in [2] is an extension of the method in the linear case, which is based on the scheme for the differences [6]. The reproduction of low degree polynomials and the locality of the investigated schemes, are central to the analysis. In the stability analysis, the existence of a restriction operator from a fine level back to the coarser level is used. The theory is developed for the class of quasi-linear schemes, defined in terms of a data dependent subdivision operator S, which is applied repeatedly. S associates with each $v \in \ell_\infty(\mathbb{Z})$ a linear operator $S(v) : \ell_\infty(\mathbb{Z}) \to \ell_\infty(\mathbb{Z})$ of the form

$$(S(v)w)_k = \sum_\ell a_{k,\ell}(v) w_\ell$$

with $a_{k,\ell}(v) = 0$ if $|k - 2\ell| > M$, for some $M > 0$ independent of v. The values generated by the quasi-linear subdivision scheme at refinement level k are

$$f^k = S(f^{k-1}) \cdots S(f^1) S(f^0) f^0 \ , \quad k = 1, 2, \ldots$$

The general theory for quasi-linear schemes, developed in [2], is applied there to the special case of $2N$-point ENO-based and WENO-based schemes.

4. Curve Subdivision Schemes on Manifolds

Linear subdivision schemes for curves refine control polygons in 3D or 2D. Starting from an initial control polygon $\mathcal{P}^0 = \{P_i^0\}$, which is the polygonal line through the control points $\{P_i^0\}$, and refining repeatedly, a linear subdivision scheme S generates a sequence of control polygons

$$S^\ell \mathcal{P}^0 \ , \quad \ell = 1, 2, \ldots$$

To design subdivision schemes for curves on a manifold, we require that the control points generated at each refinement level are on the manifold, and that the limit of the sequence of corresponding control polygons is on the manifold. Such schemes are nonlinear. Here we discuss two constructions of subdivision schemes on manifolds from converging linear schemes. Both constructions rely on the observation that any converging linear scheme can be calculated by repeated binary averages.

4.1. Converging linear schemes by repeated binary averages

A linear scheme for curves, S, is defined by two refinement rules of the form,

$$P_j^{k+1} = \sum_i a_{j-2i} P_i^k \ , \quad j = 0 \text{ or } 1(\text{mod } 2) \ . \tag{9}$$

Any converging linear scheme is affine invariant, namely $\sum_i a_{j-2i} = 1$ (see e.g.[6]). As is shown in [17], for a converging linear scheme, each of the refinement rules in (9) is expressible, in a non-unique way, by repeated binary averages. A reasonable choice is a symmetric representation relative to the topological relations in the control polygon.

For example the 4-point scheme (1) can be rewritten as

$$P_{2j}^{k+1} = P_j^k \ , \quad P_{2j+1}^{k+1} = Av_{\frac{1}{2}}(Av_{(-2w)}(P_j^k, P_{j-1}^k) \ , \ Av_{(-2w)}(P_{j+1}^k, P_{j+2}^k)) \ ,$$

where $Av_\alpha(P, Q) = (1-\alpha)P + \alpha Q \ , \quad \alpha \in \mathbb{R}, \ P, Q \in \mathbb{R}^n$. Refinement rules represented in this way are termed hereafter "refinement rules in terms of repeated binary averages".

Among the linear schemes there is a class of "factorizable schemes" for which the symbol $a(z) = \sum_i a_i z^i$, can be written as a product of linear real factors. For such a scheme, the control polygon obtained by one refinement step of the form (9), can be achieved by several simple global steps, uniquely determined by the factors of the symbol.

To be more specific, let us consider a symbol of the form

$$a(z) = z^{-\nu}(1+z)\frac{1+x_1 z}{1+x_1} \cdots \frac{1+x_m z}{1+x_m} \ , \tag{10}$$

with ν a positive integer. Note that this symbol corresponds to an affine invariant scheme since $a(1) = 2$, and $a(-1) = 0$, and that the symbol of any affine invariant scheme has the factor $1+z$, and satisfy $a(1) = 2$. Thus the form of the symbol in (10) is general for converging factorizable schemes.

Let $\{P_i^k\}$ denote the control points at refinement level k. The first step in calculating the control points at level $k+1$ corresponds to the factor $1+z$, and consists of splitting,

$$P_{2i}^{k+1,0} = P_{2i+1}^{k+1,0} = P_i^k \ . \tag{11}$$

This step is followed by m averaging steps corresponding to the factors $\frac{1+x_j z}{1+x_j}$:

$$P_i^{k+1,j} = \frac{1}{1+x_j}(P_i^{k+1,j-1} + x_j P_{i-1}^{k+1,j-1}) \ , \quad j = 1, \ldots, m \ . \tag{12}$$

Due to the factor $z^{-\nu}$, the control points at level $k+1$ are $P_i^{k+1} = P_{i+\nu}^{k+1,m}$. We term this procedure "global refinement procedure by repeated averaging".

Note that the symbol of a symmetric scheme ($a_{-j} = a_j$) has even m, $\nu = \frac{m}{2}$ and $x_j^{-1} = x_{m+1-j}$, $j = 1, \ldots, \frac{m}{2}$.

A very well known family of factorizable schemes are the B-spline schemes. The symbol of a scheme generating spline curves of order m+1 (of degree m) is

$$a(z) = (1+z)^{m+1}/2^m .$$

The 4-point scheme (1) is a symmetric scheme. With $w = \frac{1}{16}$ it is also factorizable. Its symbol has the form

$$a(z) = z^{-3}(1+z)\left(\frac{1+z}{2}\right)^3 \frac{1-(2-\sqrt{3})z}{\sqrt{3}-1} \cdot \frac{1-(2+\sqrt{3})z}{-(\sqrt{3}+1)} .$$

4.2. Construction of subdivision schemes on manifolds

The construction of nonlinear schemes on manifolds starts from a converging linear scheme, S, given either by local refinement rules in terms of repeated binary averages, or given by a global refinement procedure in terms of repeated binary averages. The second representation is prefered, if it exists.

The first construction of a subdivision T on a manifold M, "analogous to S", replaces every binary average in the representation of S, by a corresponding geodesic average on M. Thus $Av_\alpha(P,Q)$ is replaced by $gAv_\alpha(P,Q)$, where $gAv_\alpha(P,Q) = c(\alpha\tau)$, with $c(t)$ the geodesic curve on M from P to Q, satisfying $c(0) = P$ and $c(\tau) = Q$. The resulting subdivision scheme is termed geodesic subdivision scheme.

The second construction uses a smooth projection mapping onto M, and replaces every binary average by its projection onto M. The resulting nonlinear scheme is termed a projection subdivision scheme. One possible choice of the projection mapping is the orthogonal projection onto the manifold.

Example 2. In this example the linear scheme is the Chaikin algorithm ,

$$P_{2j}^{k+1} = Av_{\frac{1}{4}}(P_j^k, P_{j+1}^k) , \quad P_{2j+1}^{k+1} = Av_{\frac{3}{4}}(P_j^k, P_{j+1}^k) , \tag{13}$$

with the symbol $a(z) = (1+z)^3/4$.

Chaikin algorithm, calculated by a global procedure in terms of repeated binary averages:

$$P_{2i}^{k+1,0} = P_{2i+1}^{k+1,0} = P_i^k , \quad P_i^{k+1,j} = \frac{1}{2}(P_i^{k+1,j-1} + P_{i-1}^{k+1,j-1}) , \ j = 1,2 . \tag{14}$$

Chaikin geodesic scheme, derived from (13):

$$P_{2j}^{k+1} = gAv_{\frac{1}{4}}(P_j^k, P_{j+1}^k) , \quad P_{2j+1}^{k+1} = gAv_{\frac{3}{4}}(P_j^k, P_{j+1}^k) .$$

Chaikin geodesic scheme, derived from (14):

$$P_{2i}^{k+1,0} = P_{2i+1}^{k+1,0} = P_i^k , \quad P_i^{k+1,j} = gAv_{\frac{1}{2}}(P_i^{k+1,j-1}, P_{i-1}^{k+1,j-1}) , \ j = 1,2 .$$

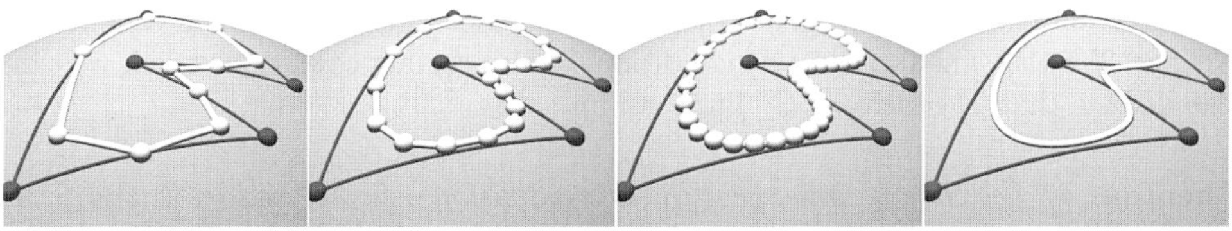

Fig. 5. Geodesic B-Spline subdivision of degree three. From left to right: Tp, T^2p,T^3p, $T^\infty p$.

Chaikin projection scheme derived from (13):

$$P_{2j}^{k+1} = G(Av_{\frac{1}{4}}(P_j^k, P_{j+1}^k)) , \quad P_{2j+1}^{k+1} = G(Av_{\frac{3}{4}}(P_j^k, P_{j+1}^k)) .$$

Chaikin projection scheme derive from (14):

$$P_{2i}^{k+1,0} = P_{2i+1}^{k+1,0} = P_i^k , \quad P_i^{k+1,j} = G(Av_{\frac{1}{2}}(P_i^{k+1,j-1}, P_{i-1}^{k+1,j-1}) , \ j = 1, 2 .$$

In the above G is a specific projection mapping on the manifold M.

Figure 5 displays a curve on a sphere, created by a geodesic analogous scheme to a third degree B-spline scheme, from a finite number of initial control points on the sphere.

4.3. Analysis of convergence and smoothness

The analysis of convergence and smoothness of the geodesic and the projection schemes is based on their proximity to the linear scheme from which they are derived, and on the smoothness properties of this linear scheme. This method of proof works for at most C^2 smoothness. To formulate the proximity conditions we introduce some notation.

For a control polygon $\mathcal{P} = \{P_i\}$, we define $\Delta^0\mathcal{P} = \mathcal{P}$, $\Delta\mathcal{P} = \{P_{i+1} - P_i\}$, and

$$\Delta^{\ell+1}\mathcal{P} = \Delta(\Delta^\ell\mathcal{P}), \ d_\ell(\mathcal{P}) = \max_i \|(\Delta^\ell\mathcal{P})_i\|, \ \ell = 0, 1, ...$$

The difference between two control polygons $\mathcal{P} = \{P_i\}$, $\mathcal{Q} = \{Q_i\}$, is defined as $\mathcal{P} - \mathcal{Q} = \{P_i - Q_i\}$.

With this notation the two proximity relations of interest to us are the following. Here, C is a generic constant.

Definition 3.

(i) Two schemes S and T are in 0-proximity if

$$d_0(S\mathcal{P} - T\mathcal{P}) \leq Cd_1(\mathcal{P})^2$$

for all control polygons $\mathcal{P}$ with $d_1(\mathcal{P})$ small enough.

(ii) Two schemes S and T are in 1-proximity if

$$d_1(S\mathcal{P} - T\mathcal{P}) \leq C[d_1(\mathcal{P})d_2(\mathcal{P}) - d_1(\mathcal{P})^3] .$$

for all control polygons $\mathcal{P}$ with $d_1(\mathcal{P})$ small enough.

We can deduce the convergence of T from the convergence of a linear scheme S, if S and T satisfy the 0-proximity condition. Under certain conditions on S, we can also deduce that T generates C^1 limit curves.

Theorem 4. *Let S be a convergent linear subdivision scheme, and let T and S be in 0-proximity. Then the sequence of control polygons $\{T^\ell \mathcal{P}\}_\ell$ for $\mathcal{P}$, with $d_1(\mathcal{P})$ small enough, converges in the sup-norm to a continuous curve.*
Moreover, if S generates C^1 limit curves and satisfies, for all $\mathcal{P}$, that $d_1(S^L\mathcal{P}) \leq \mu d_1(\mathcal{P})$ for some $L \in \mathbb{Z}_+$ and $\mu \leq (\frac{1}{2})^{\frac{L}{2}}$, then the limit of $\{T^\ell \mathcal{P}\}$ is a C^1 curve whenever such a limit exists.

Results on C^2 smoothness of T require also the 1-proximity condition [18].

Theorem 5. *Let S and T be in 0-proximity and in 1-proximity. If S is a linear scheme satisfying for all control polygons $\mathcal{P}$,*

$$d_j(2^{L(j-1)}S^L\mathcal{P}) \leq \mu_j d_j(\mathcal{P}), \quad j = 1,2,3 \ , \tag{15}$$

for some $L \in \mathbb{Z}_+$, and with μ_i, $i = 1,2,3$, satisfying

$$\mu_i < \mu_i^* \leq 1 \ , \quad , \ i = 1,2,3 \ , \tag{16}$$

$$\mu_3^* = 1, \quad \mu_1^{*2} \leq \frac{\mu_2^*}{2^L}, \quad \mu_1^{*3} \leq \frac{1}{2^{2L}}, \quad \mu_1^*\mu_2^* \leq \frac{1}{2^{2L}}, \tag{17}$$

then the curves generated by T are C^2.

It should be noted that a linear scheme S, generates C^2 limit curves if it satisfies equation(15) with $0 < \mu_i < 1$ for $i = 1,2,3$ (see e.g. [6]). Yet, the proof in [18] that the proximity relations between S and T imply that T also generates C^2 curves, is valid only for a restricted class of linear schemes satisfying equation (15) under the restrictions of (16) and (17).

The following theorem gives conditions for a nonlinear scheme T, generating curves on a manifold, and a converging linear scheme S, to satisfy the proximity conditions of Definition 3.

Theorem 6. *Let T be a geodesic or a projection subdivision scheme on a smooth manifold M, derived from a converging linear scheme S. Then S and T satisfy the two proximity conditions of Definition 3. In case M is a compact manifold or a surface with bounded normal curvatures, there exists a global $\delta > 0$, such that the proximity conditions hold for all $\mathcal{P}$ satisfying $d_1(\mathcal{P}) < \delta$.*

Example 7. It is well known that a factorizable linear scheme S, generating C^2 curves, has at least three of the $\{x_i\}$ in (10) equal to 1. Since for B-spline schemes $L = 1$ and $\mu_i = \frac{1}{2}$ in (15), it follows from the last two theorems that the geodesic and the projection analogous schemes of the B-spline schemes of degree ≥ 3 generate C^2 curves.

The linear 4-point scheme with $w = \frac{1}{16}$ is only C^1 and not C^2, although its symbol has the factor $(1+z)^4$. Therefore its nonlinear analogous schemes are not C^2, as the conditions of Theorem 5 are not satisfied.

References

1. Arandiga, P., Cohen, A., Donat, R. and Dyn, N., Interpolation of piecewise smooth functions, *Applied Comp. Harmonic Analysis*, to appear.
2. Cohen, A., Dyn, N. and Matei, B., Quasilinear subdivision schemes with applications to ENO interpolation, *Applied Comp. Harmonic Analysis* **15** (2003), 89–116.
3. Chang, T., Liu, X.-D. and Osher, S., Weighted essentially non-oscillatory schemes, *J. Comput. Phys.* **115** (1994), 200–212.
4. Daubechies, I., *Ten Lectures on Wavelets*, SIAM, Philadelphia, 1992.
5. Daubechies, I., Runborg, O. and Sweldens, W., Normal multiresolution approximation of curves, *Constr. Approximation* **20** (2004), 399–463.
6. Dyn, N., Subdivision schemes in Computer-Aided Geometric Design, in: *Advances in Numerical Analysis - Vol. II, Wavelets, Subdivision Algorithms and Radial Basis Functions* (W. Light, Ed.), Clarendon Press, Oxford, 1992, pp. 36–104.
7. Dyn, N. and Levin, D., Analysis of asymptotically equivalent binary subdivision schemes, *J. Math. Analysis and Applications* **193** (1995), 594–621.
8. Dyn, N., Gregory, J. and Levin, D., A 4-point interpolatory subdivision scheme for curve design, *Computer Aided Geometric Design* **4** (1987), 257–268.
9. Harten, A., Enquist, B., Osher, S. and Chakravartly, S., Uniformly high order accurate essentially non-oscillatory schemes III, *J. Comput. Phys.* **71** (1987), 231–303.
10. Levin, D., Using Laurent polynomial representation for the analysis of the non-uniform binary subdivision schemes, *Advances in Computational Mathematics* **11** (1999), 41–54.
11. Marinov, M., Dyn, N. and Levin, D., Geometrically controlled 4-point interpolatory schemes, in: *Advances in Multiresolution for Geometric Modelling* (N. A. Dodgson, M. S. Floater and M. A. Sabin, Eds.), Springer-Verlag, New York, 2005.
12. Noaks, L., Nonlinear corner cutting, *Advances in Computational Mathematics* **8** (1998), 165–177.
13. Noaks, L., Accelerations of Riemannian quadrations, *Proc. Amer. Math. Soc.* **127** (1999), 1827–1836.
14. Oswald, P., Smoothness of nonlinear subdivision schemes, in: *Curve and Surface Fitting*, Proc. Saint-Malo 2002, (A. Cohen, J.-L. Merrien and L. L. Schumaker, Eds.), Nashboro Press, Brentwood, 2003, pp. 323–332.

15. Rahman, I. U., Drori, I., Stodden, V., Donoho, D. L. and Schroeder, P., Multiscale representations for manifold-valued data, preprint.
16. Sabin, M., A circle-preserving variant of the four-point subdivision scheme, preprint.
17. Wallner, J. and Dyn, N., Convergence and C^1 analysis of subdivision schemes on manifolds by proximity, *Computer Aided Geometric Design*, to appear.
18. Wallner, J., Smoothness analysis of subdivision schemes by proximity, preprint.
19. Xie, G. and Yu, T. P.-Y., On a linearization principle for nonlinear p-mean subdivision schemes, in: *Advances in Constructive Approximation* (M. Neamtu and E. B. Saff, Eds.), Nashboro Press, Brentwood, 2004, pp. 519–533.
20. Xie, G. and Yu, T. P.-Y., Smoothness analysis of nonlinear subdivision schemes of homogeneous and affine invariant type, *Constr. Approximation*, to appear.

Topics in Multivariate Approximation and Interpolation
K. Jetter et al., Editors

Parameterization for Curve Interpolation

Michael S. Floater [1] and Tatiana Surazhsky

Centre of Mathematics for Applications, Department of Informatics, University of Oslo, P. O. Box 1035, 0316 Oslo, Norway

Abstract

A common task in geometric modelling is to interpolate a sequence of points or derivatives, sampled from a curve, with a parametric polynomial or spline curve. To do this we must first choose parameter values corresponding to the interpolation points. The important issue of how this choice affects the accuracy of the approximation is the focus of this paper. The underlying principle is that full approximation order can be achieved if the parameterization is an accurate enough approximation to arc length. The theory is illustrated with numerical examples.

Key words: interpolation, curve, arc length, approximation order

2000 MSC: 65D05, 65D10

1. Introduction

One of the most basic tasks in geometric modelling is to fit a smooth curve through a sequence of points $\mathbf{p}_0, \ldots \mathbf{p}_n$ in $\mathbb{R}^d$, with $d \geq 2$. The usual approach is to fit a parametric curve. We choose parameter values $t_0 < \cdots < t_n$ and find a parametric curve $\mathbf{c} : [t_0, t_n] \to \mathbb{R}^d$ such that

$$\mathbf{c}(t_i) = \mathbf{p}_i, \qquad i = 0, 1, \ldots, n. \tag{1}$$

Email addresses: `michaelf@ifi.uio.no` (Michael S. Floater), `tatiana@surazhsky.com` (Tatiana Surazhsky).

[1] Corresponding author

Popular examples of $\mathbf{c}$ are polynomial curves (of degree at most n) for small n, and spline curves for larger n.

In this paper we wish to focus on the role of the parameter values $t_0, \ldots, t_n$. Empirical evidence has shown that the choice of parameterization has a significant effect on the visual appearance of the interpolant $\mathbf{c}$. We will distinguish between two different applications:

- *Design.* If a designer wants to use interpolation to design a curve, then we can simply view parameterization as a *design tool.* In addition to choosing the points $\mathbf{p}_i$, the designer is also free to choose the parameter values t_i in order to increase the degrees of freedom at his disposal.
- *Approximation.* If the points $\mathbf{p}_i$ have been sampled from a smooth curve, e.g.
  - a height contour,
  - a curve of intersection between two surfaces,
  - or a physical curve,

  then the curve $\mathbf{c}$ will be an approximation to the curve the points $\mathbf{p}_i$ were sampled from. The important issue is now the *approximation error.* An interpolation method with higher approximation order will lead to faster convergence to the sampled curve as the sampling density increases.

While a lot of effort has been put into the effect of parameterization from the view point of *design*, little was known until recently about its effect on *approximation order.* The purpose of this paper is to summarize the recent results of [7,8] on approximation order and to illustrate the theory with numerical examples. Some comparisons with 'geometric interpolation' are made later in the paper.

2. Parameterization Methods

In order to discuss different parameterization methods, we will use the example of C^2 cubic spline interpolation. Suppose $\boldsymbol{\sigma} : [t_0, t_n] \to \mathbb{R}^d$ is the C^2 cubic spline curve such that

$$\boldsymbol{\sigma}(t_i) = \mathbf{p}_i, \qquad i = 0, 1, \ldots, n,$$

and

$$\boldsymbol{\sigma}'(t_i) = \mathbf{m}_i, \qquad i = 0, n, \tag{2}$$

for some chosen vectors $\mathbf{m}_0$, $\mathbf{m}_n$. By 'cubic spline curve' we understand that $\boldsymbol{\sigma}$ is a cubic polynomial curve in each interval $[t_i, t_{i+1}]$, and that $\boldsymbol{\sigma}$ has C^2 continuity at the break points $t_1, \ldots, t_{n-1}$. As is well-known [1,2,6], $\boldsymbol{\sigma}$ is unique. The end conditions (2) are sometimes known as 'clamped' end conditions, and this kind of spline interpolation is sometimes called 'complete' spline interpolation.

The first thing to note about parameterization is that applying a linear transformation to $t_0, \ldots, t_n$ does not change the resulting spline curve. Provided we treat the end conditions (2) correctly, we simply get a reparameterization of $\boldsymbol{\sigma}$. To be precise, if we define

$$\hat{t}_i := \lambda t_i + \mu, \qquad i = 0, 1, \ldots, n,$$

for $\lambda > 0$ and we define the curve $\hat{\boldsymbol{\sigma}} : [\hat{t}_0, \hat{t}_n] \to \mathbb{R}^d$ by

$$\hat{\boldsymbol{\sigma}}(t) := \boldsymbol{\sigma}((t-\mu)/\lambda),$$

then $\hat{\boldsymbol{\sigma}}$ is clearly a C^2 cubic spline curve over the partition $\hat{t}_0 < \hat{t}_1 < \cdots < \hat{t}_n$, satisfying the conditions

$$\hat{\boldsymbol{\sigma}}(\hat{t}_i) = \mathbf{p}_i, \qquad i = 0, 1, \ldots, n,$$

and

$$\hat{\boldsymbol{\sigma}}'(\hat{t}_i) = \mathbf{m}_i/\lambda, \qquad i = 0, n.$$

For this reason, we may as well set $t_0 = 0$ and we typically specify a parameterization by recursively setting

$$t_{i+1} := t_i + d_i$$

for some chosen interval lengths $d_0, d_1, \ldots, d_{n-1} > 0$. Multiplying the interval lengths d_i by a common factor λ will not change the intrinsic geometry of the spline curve $\boldsymbol{\sigma}$ as long as we divide the vectors $\mathbf{m}_0$ and $\mathbf{m}_n$ by the same factor λ.

The simplest choice is the *uniform parameterization* defined by $d_i = 1$, where the values t_i are uniformly spaced. But as early as 1967, Ahlberg, Nilson, and Walsh [1] proposed using the *chordal parameterization* in which

$$d_i := |\mathbf{p}_{i+1} - \mathbf{p}_i|, \tag{3}$$

and $|\cdot|$ denotes the Euclidean norm in $\mathbb{R}^d$. The motivation behind this was that the distance between two points on a curve is a reasonable approximation to the length of the associated curve segment. Thus the hope was that the 'speed' $|\boldsymbol{\sigma}'(t)|$ of the spline curve might be close to unity at all $t \in [t_0, t_n]$.

Epstein [5] showed that when using periodic boundary conditions, a chordal C^2 cubic spline interpolant is always regular, i.e., $\boldsymbol{\sigma}'(t) \neq \mathbf{0}$ at every $t \in [t_0, t_n]$.

Later, it was realized that the uniform and chordal parameterizations are the special cases $\mu = 0$ and $\mu = 1$ of the more general parameterization

$$d_i := |\mathbf{p}_{i+1} - \mathbf{p}_i|^\mu,$$

with $0 \le \mu \le 1$ acting as a kind of blending parameter. The choice $\mu = 1/2$ was termed by Lee [11] the *centripetal parameterization*, and suggested that it sometimes leads to a spline curve which is closer to the polygon passing through the data points $\mathbf{p}_i$.

Later still, Foley and Nielson [9] observed that the chordal parameterization lacks affine invariance. If a non-uniform scaling is applied to the data points $\mathbf{p}_i$ and the new chord lengths are computed, then the resulting spline curve will not in general be a scaling of the original. This suggested replacing the Euclidean distance in (3) by a metric which is invariant to affine transformations of the data points. Nielson had earlier found such an affine-invariant metric, and they defined d_i with the aid of both the affine-invariant distance between $\mathbf{p}_i$ and $\mathbf{p}_{i+1}$ and the angles of the polygon passing through $\mathbf{p}_{i-1}, \mathbf{p}_i, \mathbf{p}_{i+1}, \mathbf{p}_{i+2}$. Thus, unlike the previous methods, every value of d_i depends on all the data points $\mathbf{p}_0, \ldots, \mathbf{p}_n$.

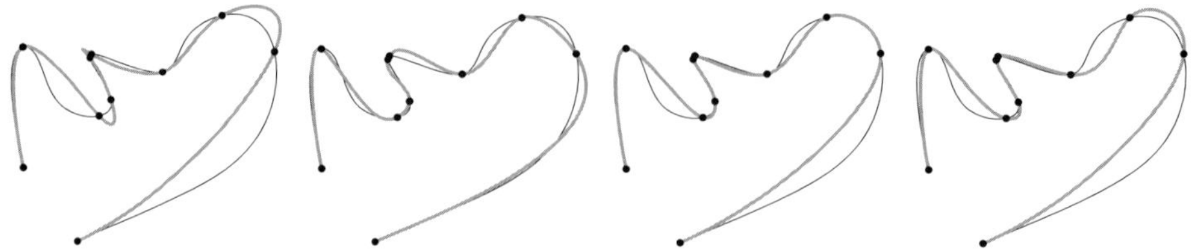

(a) uniform (b) chordal (c) centripetal (d) Foley-Nielson

Fig. 1. Choice of knot-vector for cubic spline interpolation.

Our goal in this paper is to study the effect of parameterization on *approximation* and with this in mind we have sampled points $\mathbf{p}_i$ from a fixed smooth curve, and computed the four spline curves $\boldsymbol{\sigma}$ resulting from the uniform, chordal, centripetal, and Foley-Nielson knot-vectors. Figure 1 shows the fixed curve in black and the four spline curves shown in grey. We chose the end vectors $\mathbf{m}_0$, $\mathbf{m}_n$ to have the same direction as the given curve and to have the magnitude

$$|\mathbf{m}_0| := |\mathbf{p}_1 - \mathbf{p}_0|/d_0, \qquad |\mathbf{m}_n| := |\mathbf{p}_n - \mathbf{p}_{n-1}|/d_{n-1},$$

the motivation being that $\boldsymbol{\sigma}'(t_0) \approx (\mathbf{p}_1 - \mathbf{p}_0)/d_0$.

It is quite evident from the figure that the chordal curve appears to be a particularly good approximation to the original curve. As we will see, this is not coincidence, and due to the fact that the length of a chord is a relatively good approximation to the length of an arc.

3. Approximation Error

The example of Figure 1 clearly shows that parameterization affects approximation error. In order to study approximation, we will assume that the points $\mathbf{p}_i$ are samples, $\mathbf{p}_i = \mathbf{f}(s_i)$, from a parametric curve $\mathbf{f} : [a,b] \to \mathbb{R}^d$, $d \geq 2$, with $a \leq s_0 < \cdots < s_n \leq b$. We will assume that $\mathbf{f}$ is parameterized with respect to arc length, by which we mean that it is (at least) continuously differentiable and that $|\mathbf{f}'(s)| = 1$ for all $s \in [a,b]$. Then the interpolation problem becomes

$$\mathbf{c}(t_i) = \mathbf{f}(s_i), \qquad i = 0, 1, \ldots, n, \tag{4}$$

and the error is the distance between the curve piece $\mathbf{f}|_{[s_0,s_n]}$ and the curve $\mathbf{c}$, see Figure 2. One well-known way of measuring the distance between two such curves is the Hausdorff distance,

$$\mathrm{dist}_H(\mathbf{f}|_{[s_0,s_n]}, \mathbf{c}) = \max\{\max_{t_0 \leq t \leq t_n} \min_{s_0 \leq s \leq s_n} |\mathbf{f}(s) - \mathbf{c}(t)|, \max_{s_0 \leq s \leq s_n} \min_{t_0 \leq t \leq t_n} |\mathbf{f}(s) - \mathbf{c}(t)|\}.$$

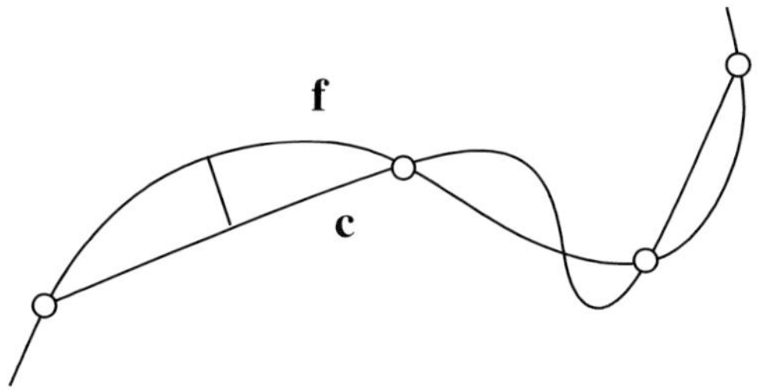

Fig. 2. Distance between two curves.

4. Approximation Order

It seems difficult to derive bounds on the error between **f** and $\boldsymbol{\sigma}$. We will instead study the approximation *order*. A higher approximation order implies faster convergence of the interpolating curves to the given curve as the sampling density increases. Consider for simplicity polynomial interpolation, i.e., let **c** in (4) be the Lagrange polynomial interpolant $\mathbf{p} : [t_0, t_n] \to \mathbb{R}^d$ of degree at most n, and let h be the length of the curve piece $\mathbf{f}|_{[s_0,s_n]}$, i.e.,

$$h = L(\mathbf{f}|_{[s_0,s_n]}) = \int_{s_0}^{s_n} |\mathbf{f}'(s)|\, ds = s_n - s_0.$$

The following table gives numerical estimates of k such that $\mathrm{dist}_H(\mathbf{f}|_{[s_0,s_n]}, \mathbf{p}) = O(h^k)$ as $h \to 0$ for smooth curves **f**:

Degree	1	2	3	4	5	$\cdots$
Uniform	2	2	2	2	2	$\cdots$
Chordal	2	3	4	4	4	$\cdots$

The table suggests that the uniform parameterization only provides an $O(h^2)$ approximation when applied to interpolation of any degree $n \geq 1$, while the chordal parameterization provides full approximation order $O(h^{n+1})$ for $n = 1, 2, 3$, and then stays at $O(h^4)$ for higher degrees $n > 3$. Arguably the most interesting of these numerical results is that of cubic interpolation ($n = 3$) with the chordal parameterization sketched in Figure 3. The following theorem is a special case of a result proved in [7].

Fig. 3. Cubic interpolation.

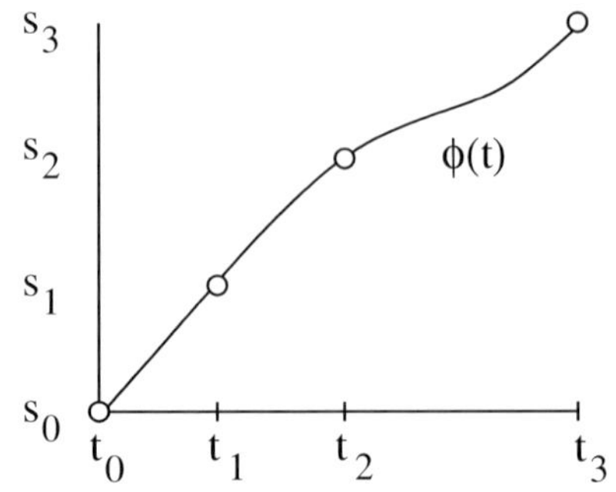

Fig. 4. Reparameterization ϕ.

Theorem 1. *Suppose* $\mathbf{f} \in C^4[a,b]$ *and for each sample* $s_0 < s_1 < s_2 < s_3$ *in* $[a,b]$*, let* $t_0 = 0$ *and* $t_{i+1} - t_i = |\mathbf{f}(s_{i+1}) - \mathbf{f}(s_i)|$ *for* $i = 0,1,2$*, and let* $\mathbf{p} : [t_0,t_3] \to \mathbb{R}^d$ *be the cubic polynomial such that* $\mathbf{p}(t_i) = \mathbf{f}(s_i)$*, for* $i = 0,1,2,3$*. Then* $\mathrm{dist}_H(\mathbf{f}|_{[s_0,s_3]}, \mathbf{p}) = O(h^4)$ *as* $h \to 0$*, where* $h = s_3 - s_0$*.*

The basic steps in the proof are:

(i) Show that

$$(t_{i+1} - t_i) - (s_{i+1} - s_i) = O\big((s_{i+1} - s_i)^3\big). \tag{5}$$

(ii) Use (5) to show that if $\phi : [t_0,t_3] \to \mathbb{R}$ is the cubic polynomial such that $\phi(t_i) = s_i$, $i = 0,1,2,3$, then

$$\|\mathbf{f} \circ \phi - \mathbf{p}\|_\infty = O(h^4),$$

where $\|\mathbf{q}\|_\infty := \max_{t_0 \le t \le t_3} |\mathbf{q}(t)|$, see Figure 4.

Consider now some of the details. Step (i) is to establish (5), that the error between the length of a chord and the length of an arc is of third order. This follows immediately from the following explicit error bound.

Lemma 2. *If* $\mathbf{f} \in C^2[a,b]$ *then*

$$0 \le (s_{i+1} - s_i) - |\mathbf{f}(s_{i+1}) - \mathbf{f}(s_i)| \le \frac{1}{12}(s_{i+1} - s_i)^3 \max_{s_i \le s \le s_{i+1}} |\mathbf{f}''(s)|.$$

The proof follows from the use of Taylor series and the identities $\mathbf{f}'(s) \cdot \mathbf{f}'(s) = 1$ and $\mathbf{f}'(s) \cdot \mathbf{f}''(s) = 0$ [7].

In Step (ii) we use the reparameterization ϕ. If $\mathbf{g}(t) := \mathbf{f}(\phi(t))$, then

$$\mathbf{p}(t_i) = \mathbf{g}(t_i), \qquad i = 0,1,2,3,$$

and so

$$\|\mathbf{g} - \mathbf{p}\|_\infty \le (t_3 - t_0)^4 \|\mathbf{g}^{(4)}\|_\infty / 4! \le h^4 \|\mathbf{g}^{(4)}\|_\infty / 4!.$$

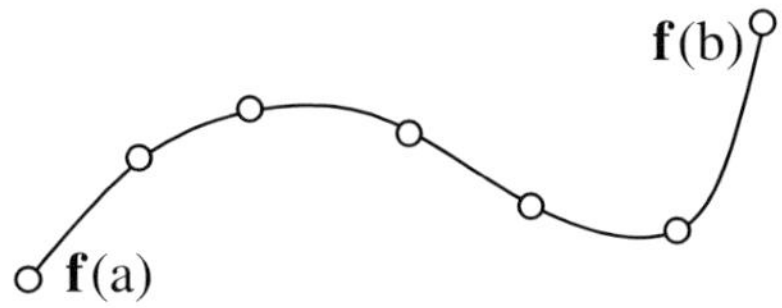

Fig. 5. Spline interpolation to **f**.

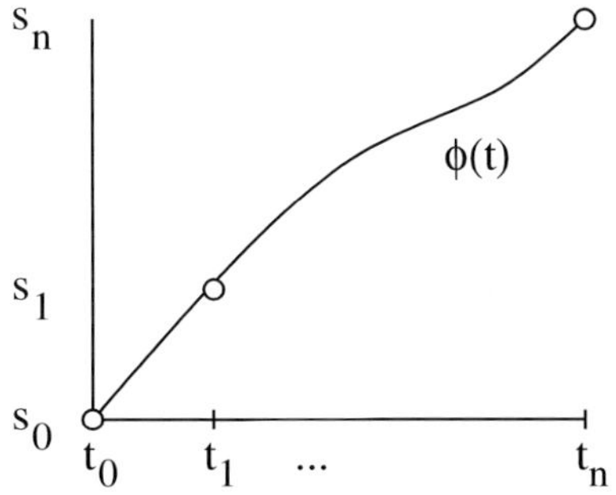

Fig. 6. Reparameterization ϕ.

So it is enough to show that $\phi' > 0$ for small enough h (so that $\mathbf{g}$ is well-defined) and that $\mathbf{g}^{(4)}$ is bounded as $h \to 0$. By the chain rule,

$$\mathbf{g}^{(4)} = (\phi')^4\mathbf{f}^{(4)} + 6(\phi')^2\phi''\mathbf{f}''' + (3(\phi'')^2 + 4\phi'\phi''')\mathbf{f}'' + \phi^{(4)}\mathbf{f}',$$

and so $\mathbf{g}^{(4)}$ is indeed bounded if ϕ', ϕ'', ϕ''' are bounded as $h \to 0$. To see this, observe that (5) implies

$$[t_i, t_{i+1}]\phi - 1 = \frac{s_{i+1} - s_i}{t_{i+1} - t_i} - 1 = O\left(\frac{(s_{i+1} - s_i)^3}{t_{i+1} - t_i}\right) = O\big((s_{i+1} - s_i)^2\big), \qquad i = 0, 1, 2.$$

Therefore,

$$[t_i, t_{i+1}, t_{i+2}]\phi = O(s_{i+1} - s_i), \quad i = 0, 1, \qquad [t_0, t_1, t_2, t_3]\phi = O(1).$$

Thus all divided differences of ϕ are bounded and by expressing ϕ in its Newton form and differentiating, it follows that all derivatives of ϕ are indeed bounded, and that $\phi' > 0$ for small enough h.

5. Extension to Cubic Splines

The error analysis of chordal cubic polynomial interpolation extends to complete C^2 cubic spline interpolation (see Figure 5 and Figure 6). Again, chord lengths provide full fourth order approximation. The following was proved in [8]. We continue to assume that $|\mathbf{f}'(s)| = 1$ for all $s \in [a, b]$.

Theorem 3. *Suppose* $\mathbf{f} \in C^4[a,b]$ *and for each sample* $a = s_0 < \cdots < s_n = b$, *let* $t_0 = 0$ *and* $t_{i+1} - t_i = |\mathbf{f}(s_{i+1}) - \mathbf{f}(s_i)|$, $0 \le i < n$, *and let* $\boldsymbol{\sigma} : [t_0, t_n] \to \mathbb{R}^d$ *be the* C^2 *cubic spline curve such that*

$$\boldsymbol{\sigma}(t_i) = \mathbf{f}(s_i), \quad i = 0, 1, \ldots, n,$$
$$\boldsymbol{\sigma}'(t_i) = \mathbf{f}'(s_i), \quad i = 0, n.$$

Then $\mathrm{dist}_H(\mathbf{f}, \boldsymbol{\sigma}) = O(h^4)$ *as* $h \to 0$, *where* $h = \max_{0 \le i < n}(s_{i+1} - s_i)$.

The basic ideas in the proof are similar to the polynomial case. But we now let $\phi : [t_0, t_n] \to \mathbb{R}$ be the C^2 cubic spline function satisfying

$$\phi(t_i) = s_i, \quad i = 0, 1, \ldots, n,$$
$$\phi'(t_i) = 1, \quad i = 0, n,$$

and use (5) to show that $\|\mathbf{f} \circ \phi - \boldsymbol{\sigma}\|_\infty = O(h^4)$, where $\|\mathbf{q}\|_\infty := \max_{t_0 \le t \le t_n} |\mathbf{q}(t)|$.

6. Parameter Improvement for Higher Degree Interpolation

Chord lengths do not give full approximation order $O(h^{n+1})$ for polynomial interpolation of degree $n > 3$. A solution is to use a parameterization which more accurately approximates the arc length of $\mathbf{f}$. A first improvement can be made by using the length of the cubic polynomial $\mathbf{p} : [t_0, t_3] \to \mathbb{R}^d$ in Theorem 1. It was shown in [7] that if $\mathbf{f} \in C^4[a,b]$ and $|\mathbf{f}'(s)| = 1$, for $s \in [a,b]$, then

$$L(\mathbf{p}|_{[t_0,t_1]}) - (s_1 - s_0) = O\Big((s_1 - s_0)^3 (s_2 - s_0)(s_3 - s_0)\Big),$$
$$L(\mathbf{p}|_{[t_1,t_2]}) - (s_2 - s_1) = O\Big((s_2 - s_0)(s_2 - s_1)^3 (s_3 - s_1)\Big),$$
$$L(\mathbf{p}|_{[t_2,t_3]}) - (s_3 - s_2) = O\Big((s_3 - s_0)(s_3 - s_1)(s_3 - s_2)^3\Big),$$

Thus, for example, the length of the cubic piece $\mathbf{p}|_{[t_1,t_2]}$ is a better approximation to the length of $\mathbf{f}|_{[s_1,s_2]}$ than the length of the chord $|\mathbf{f}(s_2) - \mathbf{f}(s_1)|$. Indeed the order of approximation has risen by two. Suppose now that $n = 4$ or $n = 5$ in (1). We start by letting $t_0, \ldots, t_n$ be chordal parameter values. Then we improve the parameterization as follows. For each $i = 0, \ldots, n-1$, we choose any sequence of four points $\mathbf{p}_j, \ldots, \mathbf{p}_{j+3}$ which includes $\mathbf{p}_i$ and $\mathbf{p}_{i+1}$ and we let $\mathbf{p} : [t_j, t_{j+3}] \to \mathbb{R}^d$ be the cubic interpolant $\mathbf{p}(t_k) = \mathbf{p}_k$, $j \le k \le j+3$. Then we set

$$\hat{t}_{i+1} - \hat{t}_i := L(\mathbf{p}|_{[t_i,t_{i+1}]}).$$

For most i there is a choice of which cubic to use. Nevertheless, it can be shown, using a similar approach to the proof of Theorem 1, that if $\mathbf{q} : [\hat{t}_0, \hat{t}_n] \to \mathbb{R}^d$ is the interpolant of degree at most n with $\mathbf{q}(\hat{t}_i) = \mathbf{p}_i$, and $\mathbf{f}$ is in C^{n+1}, then

$$\mathrm{dist}_H(\mathbf{f}|_{[s_0,s_n]}, \mathbf{q}) = O(h^{n+1}), \qquad n = 4, 5.$$

Continuing this idea, one can write a recursive algorithm which generates a parameterization for any n which supports polynomial interpolation of degree n, see [7]. We can even view the uniform parameterization (or any other parameterization) as the start point for the iteration, because the length of the chord between two points is also the length of the linear interpolant to those points, with respect to any parameterization. Schematically, the parameter improvement looks as follows:

$$\begin{array}{l}
\text{Uniform} \quad \rightarrow n = 1 \rightarrow O(h^2) \\
\qquad\qquad\quad \swarrow \\
\text{Chordal} \quad \rightarrow n = 3 \rightarrow O(h^4) \\
\qquad\qquad\quad \swarrow \\
\text{Improved I} \rightarrow n = 5 \rightarrow O(h^6) \\
\qquad\qquad\quad \swarrow \\
\text{Improved II} \rightarrow n = 7 \rightarrow O(h^8)
\end{array}$$

7. Hermite Interpolation

The theory of parameterization for polynomial interpolation extends to Hermite interpolation as long as we match arc length derivatives at every point.

7.1. Two-point cubic Hermite

Suppose we want to fit the Hermite cubic polynomial $\mathbf{p} : [t_0, t_1] \rightarrow \mathbb{R}^d$ to $\mathbf{f}$, i.e., such that

$$\mathbf{p}^{(k)}(t_i) = \mathbf{f}^{(k)}(s_i), \qquad i = 0, 1, \qquad k = 0, 1, \tag{6}$$

see Figure 7. If we use chordal parameter values, i.e., $t_1 - t_0 = |\mathbf{f}(s_1) - \mathbf{f}(s_0)|$, then, noting that $|\mathbf{f}'(s_i)| = 1$, if $\mathbf{f} \in C^4[a, b]$ then

$$\text{dist}_H(\mathbf{f}|_{[s_0,s_1]}, \mathbf{p}) = O(h^4) \quad \text{as } h \rightarrow 0.$$

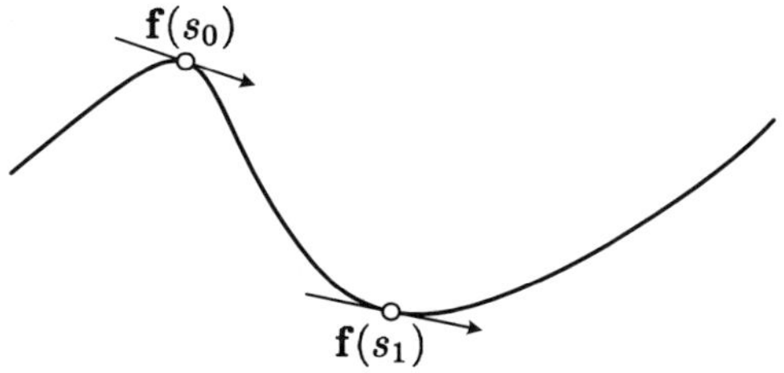

Fig. 7. Two-point cubic Hermite interpolation.

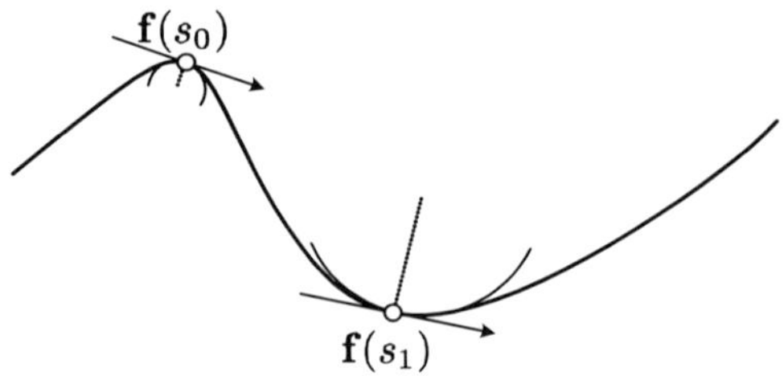

Fig. 8. Two-point quintic Hermite interpolation.

7.2. Two-point quintic Hermite

How do we choose a suitable parameterization $\hat{t}_0 < \hat{t}_1$ when fitting a quintic polynomial $\mathbf{q} : [\hat{t}_0, \hat{t}_1] \to \mathbb{R}^d$ such that

$$\mathbf{q}^{(k)}(\hat{t}_i) = \mathbf{f}^{(k)}(s_i), \qquad i = 0, 1, \qquad k = 0, 1, 2, \tag{7}$$

where $|\mathbf{f}'(s_i)| = 1$, and $\mathbf{f}'(s_i) \cdot \mathbf{f}''(s_i) = 0$ (see Figure 8)? One way of getting a sixth order approximation is to use the improved parameterization

$$\hat{t}_1 - \hat{t}_0 = (t_1 - t_0)(|\mathbf{p}'(\xi_-)| + |\mathbf{p}'(\xi_+)|)/2 \approx L(\mathbf{p}), \tag{8}$$

where $\xi_\pm = (t_0 + t_1)/2 \pm (t_1 - t_0)/(2\sqrt{3})$. Here $\mathbf{p}$ is the Hermite cubic interpolant in (6) using chordal values t_0, t_1, and we have used 2-point Gauss quadrature to estimate the integral

$$L(\mathbf{p}) = \int_{t_0}^{t_1} |\mathbf{p}'(t)| \, dt.$$

If $\mathbf{f} \in C^6[a, b]$, then

$$\mathrm{dist}_H(\mathbf{f}|_{[s_0,s_1]}, \mathbf{q}) = O(h^6) \quad \text{as } h \to 0.$$

Is the parameter improvement robust? Yes, in the sense that the new parameter intervals are always longer than the initial (chordal) parameter intervals. To see this, observe that since 2-point Gauss quadrature has quadratic (cubic) precision and positive weights,

$$\begin{aligned} \hat{t}_1 - \hat{t}_0 &= (t_1 - t_0)(|\mathbf{p}'(\xi_-)| + |\mathbf{p}'(\xi_+)|)/2 \\ &\geq |(t_1 - t_0)(\mathbf{p}'(\xi_-) + \mathbf{p}'(\xi_+))/2| \\ &= \left| \int_{t_0}^{t_1} \mathbf{p}'(t) \, dt \right| = |\mathbf{p}(t_1) - \mathbf{p}(t_0)| = |\mathbf{f}(s_1) - \mathbf{f}(s_0)| = t_1 - t_0. \end{aligned} \tag{9}$$

Thus the new parameter intervals certainly do not shrink to zero. The above inequality is very natural because we know that since the straight line between two points is the shortest path between them,

$$L(\mathbf{f}|_{[s_0,s_1]}) \geq t_1 - t_0.$$

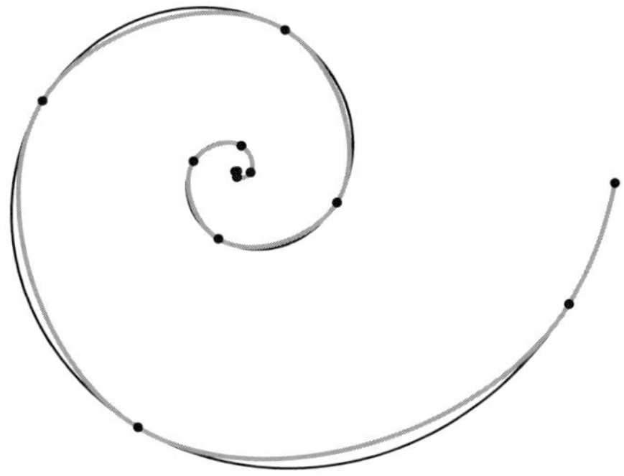

Fig. 9. Chordal Hermite cubic, C^1, $O(h^4)$.

We could instead use Simpson's rule to get sixth-order accuracy:

$$\begin{aligned}\hat{t}_1 - \hat{t}_0 &= (t_1 - t_0)(|\mathbf{p}'(t_0)| + 4|\mathbf{p}'((t_0 + t_1)/2)| + |\mathbf{p}'(t_1)|)/6 \\ &= (t_1 - t_0)(1 + 2|\mathbf{p}'((t_0 + t_1)/2)|)/3 \\ &\approx L(\mathbf{p}).\end{aligned}$$

Again, $\hat{t}_1 - \hat{t}_0 \geq t_1 - t_0$. This same property holds for any quadrature method with positive weights and degree of precision at least two.

7.3. Examples

Figure 9 shows the C^1 cubic spline built from cubic 2-point Hermite interpolation, based on chordal parameter values. In this figure and in all subsequent similar figures, the black curve is the original and the grey one the approximation. Figure 10 shows the C^2 quintic splines built from quintic 2-point Hermite interpolation, based on chordal parameter values, and the improved parameter values. Observe that the error in Figure 10(b) is noticeably smaller than that in Figure 10(a). Figure 11 shows the chordal cubic and the improved quintic interpolants for a different data set.

Recall that from (9), the length $\hat{t}_1 - \hat{t}_0$ of the improved parameter interval is greater or equal to the original chord length $t_1 - t_0$ and it appears that lengthening the parameter interval has the effect of lengthening the quintic curve in Figure 10. We decided to explore this behaviour further and scaled the length of the chord by various factors: 1/2, 1, 2, and 5. The resulting curves are shown in Figure 12 which provides a striking illustration of what a dramatic effect parameterization can have on Hermite interpolation when the derivatives are fixed (not scaled as in Section 2). Clearly a shorter parameter interval leads to a tighter curve.

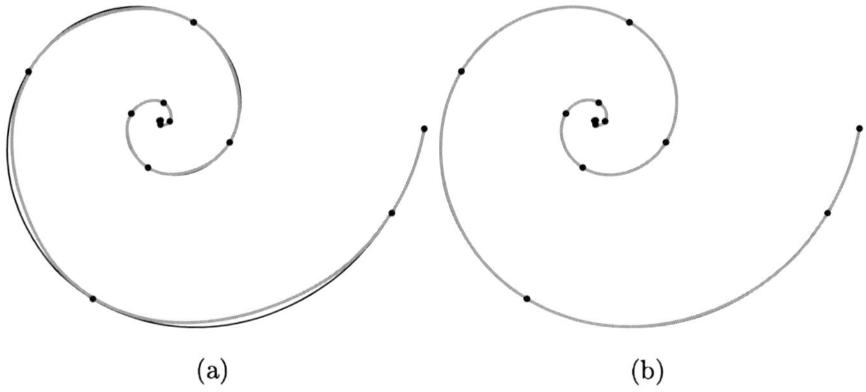

Fig. 10. Chordal quintic Hermite, C^2, $O(h^4)$, (a) and improved quintic Hermite, C^2, $O(h^6)$, (b).

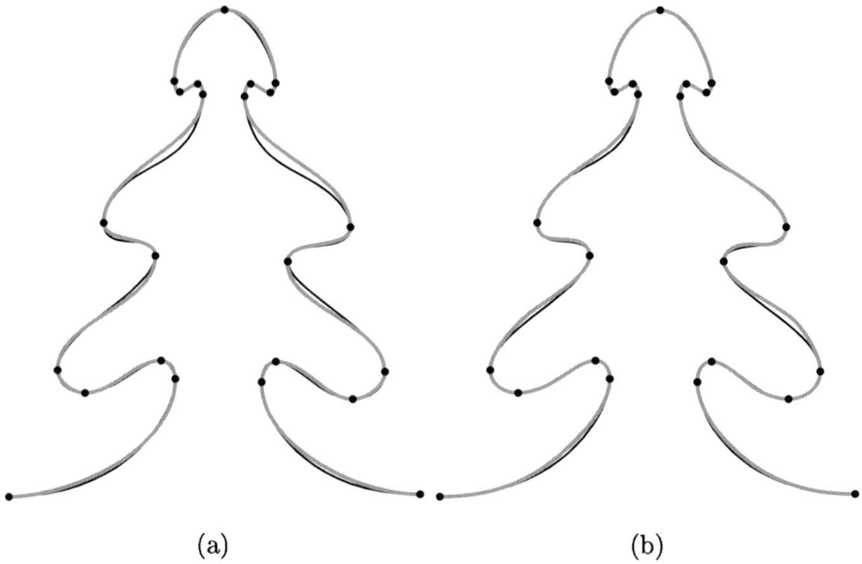

Fig. 11. Chordal cubic, C^1, $O(h^4)$, (a) and improved quintic, C^2, $O(h^6)$, (b).

8. Geometric Interpolation

An alternative approach to what we have discussed so far in this paper is so-called *geometric interpolation* as developed by [3,4,10,12–14]. These schemes, mainly Hermite, aim not only to retain full approximation order, but also to reduce the degree of the interpolant. The potential advantage of these schemes is that both the interpolant and parameterization are the simultaneous solutions to a set of equations. The disadvantage is that these equations are non-linear and only admit a solution under certain restrictions on the data points, and each scheme is dependent on the

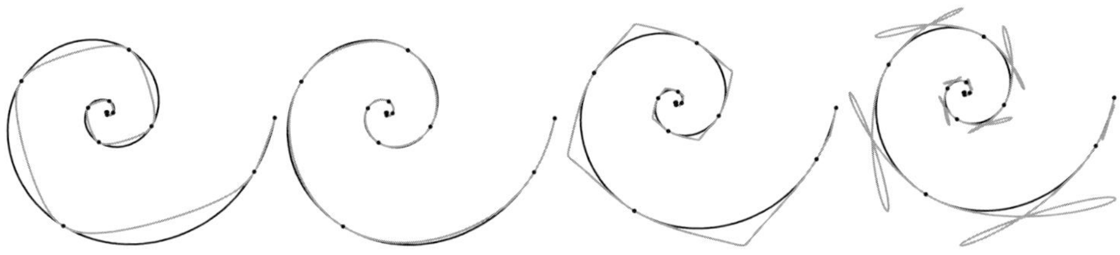

Fig. 12. Chord lengths multiplied by factors (a) 1/2, (b) 1, (c) 2, and (d) 5.

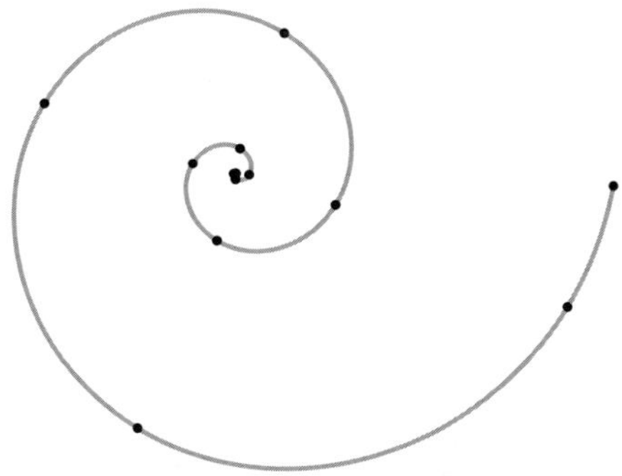

Fig. 13. BHS cubic, G^2.

dimension d.

Since for planar data ($d = 2$) the quintic Hermite $\mathbf{q}$ in (7) matches the tangents and curvatures of $\mathbf{f}$ at the two points, it is interesting to compare this quintic scheme with the scheme proposed for *planar* data by de Boor, Höllig, Sabin [3], which we will call the BHS scheme. When $\mathbf{f}$ is a curve in $\mathbb{R}^2$, the BHS scheme tries to fit a cubic to these tangents and curvatures. Thus the scheme attempts to satisfy

$$D_s^k \mathbf{p}(t_i) = \mathbf{f}^{(k)}(s_i), \qquad i = 0, 1, \quad k = 0, 1, 2,$$

where $\mathbf{p} : [t_0, t_1] \to \mathbb{R}^2$ is a cubic polynomial and D_s denotes differentiation with respect to arc length. The length $t_1 - t_0$ of the parameter interval is of no importance in this scheme and they set $t_0 = 0$, $t_1 = 1$. The approximation order is $O(h^6)$ around any point where the curvature is non-zero (a solution exists at such a point for small enough h). However there are examples of data sets for which there is no solution. Note that a method has been proposed in [15] for sampling points from a *given* curve $\mathbf{f}$ which supports the BHS scheme, in the sense that there is always a solution. For example, the BHS scheme has a solution for the data in Figure 13, generated by the method of [15], and as we can see in the figure, it is hard to distinguish the BHS (grey) curve from the original (black) curve, similar to the quintic in Figure 10(b).

Mørken and Scherer [12] view this scheme in a different way, in terms of reparameterization and degree reduction. First of all, instead of (7), we could find any quintic $\mathbf{q}$ which interpolates a reparameterization $\mathbf{g} = \mathbf{f} \circ \phi$ of the curve $\mathbf{f}$ where $\phi : [t_0, t_1] \to [s_0, s_1]$ is some increasing function with $\phi(t_i) = s_i$, $i = 0, 1$. Since

$$\mathbf{g}' = \phi'\mathbf{f}', \qquad \mathbf{g}'' = (\phi')^2\mathbf{f}'' + \phi''\mathbf{f}',$$

we can freely choose any values $\lambda_i, \mu_i \in \mathbb{R}$, $i = 0, 1$, and set

$$\mathbf{q}(t_i) = \mathbf{f}(s_i), \qquad i = 0, 1,$$
$$\mathbf{q}'(t_i) = \lambda_i \mathbf{f}'(s_i), \qquad i = 0, 1,$$
$$\mathbf{q}''(t_i) = \lambda_i^2 \mathbf{f}''(s_i) + \mu_i \mathbf{f}'(s_i), \qquad i = 0, 1.$$

The case $\lambda_i = 1$ and $\mu_i = 0$ reduces to (7), whereas the BHS scheme attempts to find λ_i and μ_i to reduce the degree of the quintic $\mathbf{q}$ to three. Mørken and Scherer applied this approach to interpolation of other degrees.

Grandine and Hogan [10] have recently proposed raising the degree of the BHS scheme to four in order to guarantee a solution in all (planar) cases.

Clearly it is an advantage to interpolate with a polynomial with as low a degree as possible. On the other hand the attraction of the quintic scheme (7) is its simplicity: it is constructed in just a few steps without having to solve any non-linear equations. Moreover, unlike the BHS scheme and the scheme of Grandine and Hogan, the quintic scheme (7) applies to curves in *any* space dimension, including the important case $\mathbb{R}^3$. Moreover, the resulting curve has C^2 continuity rather than just G^2 which can be an advantage for certain post-processes such as surface lofting.

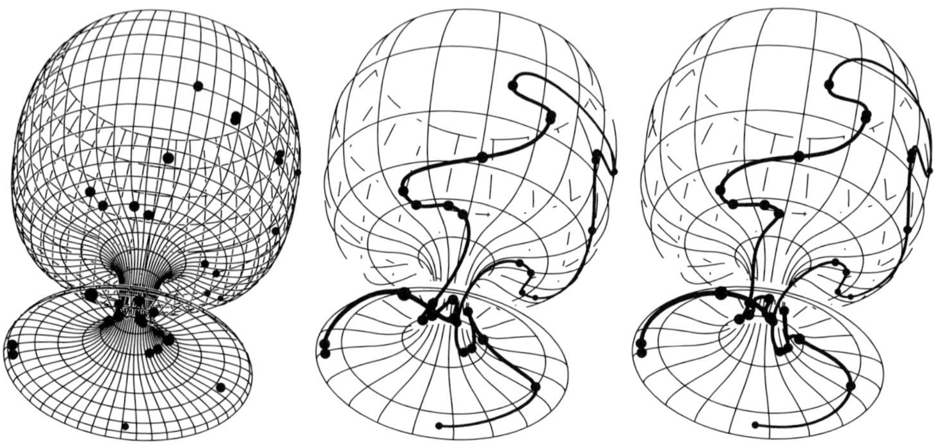

(a) (b) (c)

Fig. 14. Curve interpolation through points in $\mathbb{R}^3$.

9. Conclusion

We have obtained precise information about how the choice of parameter values affects the approximation order of curve interpolation in *any* space dimension $d \geq 2$, based on polynomials and piecewise polynomials. Chordal parameter values give full approximation order for cubic interpolation, but more accurate approximations to arc length are required (and can be found) for higher degrees. Figure 14 illustrates an application where points in $\mathbb{R}^3$ are sampled from a curve on a glass-shaped surface (a). Two interpolating curves are shown: (b) a chordal C^2 cubic spline curve, and (c) a C^2 quintic Hermite spline curve, using the improved parameterization (8).

Acknowledgement

This work was supported by the Bemata program of the Norwegian Research Council.

References

1. Ahlberg, J. H., Nilson, E. N. and Walsh, J. L., *The Theory of Splines and Their Applications*, Academic Press, New York, 1967.
2. de Boor, C., *A Practical Guide to Splines*, Springer, New York, 1978.
3. de Boor, C., Höllig, K. and Sabin, M., High accuracy geometric Hermite interpolation, *Computer Aided Geometric Design* **4** (1987), 269–278.
4. Degen, W., High accurate rational approximation of parametric curves, *Computer Aided Geometric Design* **10** (1993), 293–313.
5. Epstein, M. P., On the influence of parametrization in parametric interpolation, *SIAM J. Numer. Anal.* **13** (1976), 261–268.
6. Farin, G., *Curves and Surfaces for Computer Aided Geometric Design*, Academic Press, San Diego, 1988.
7. Floater, M. S., Arc length estimation and the convergence of parametric polynomial interpolation, preprint, CMA, Oslo (2005).
8. Floater, M. S., Chordal cubic spline interpolation is fourth order accurate, *IMA J. Numer. Anal.*, to appear.
9. Foley, T. A. and Nielson, G. M., Knot selection for parametric spline interpolation, in: *Mathematical Methods in Computer Aided Geometric Design*, Proc. Oslo 1988 (T. Lyche and L. L. Schumaker, Eds.), Academic Press, San Diego, 1989, pp. 261–272.
10. Grandine, T. A. and Hogan, T., A parametric quartic spline interpolant to position, tangent and curvature, *Computing* **72** (2004), 65–78.
11. Lee, E. T. Y., Choosing nodes in parametric curve interpolation, *Computer Aided Design* **21** (1989), 363–370.
12. Mørken, K. and Scherer, K., A general framework for high-accuracy parametric interpolation, *Math. Comp.* **66** (1997), 237–260.

13. Schaback, R., Interpolation with piecewise quadratic visually C^2 Bezier polynomials, *Computer Aided Geometric Design* **6** (1989), 219–233.
14. Schaback, R., Optimal geometric Hermite interpolation of curves, in: *Mathematical Methods for Curves and Surfaces II*, Proc. Lillehammer 1997 (M. Dæhlen, T. Lyche and L. L. Schumaker, Eds.), Vanderbilt University Press, Nashville, 1998, pp. 417–428.
15. Surazhsky, T. and Surazhsky, V., Sampling planar curves using curvature-based shape analysis, in: *Mathematical Methods for Curves and Surfaces*, Proc. Tromsø 2004 (M. Dæhlen, K. Mørken and L. L. Schumaker, Eds.), Nashboro Press, Brentwood, 2005, pp. 339–350.

Topics in Multivariate Approximation and Interpolation
K. Jetter et al., Editors

Refinable Multivariate Spline Functions

Tim Goodman [a], **Doug Hardin** [b,1]

[a]The University of Dundee, Dundee, Scotland
[b]Vanderbilt University, Nashville, Tennessee, USA

Abstract

We review recent developments concerning refinable, multivariate piecewise polynomials with compact support. We first consider uniform meshes, box splines, box spline wavelets and a generalisation of box splines called multi-box splines. Our next topic is spline functions on general triangulations, including continuous linear spline wavelets and hierarchical bases for C^1 splines based on macro-elements. Similar types of spline functions are then studied for meshes gained from triangulating a mesh of quadrilaterals.

Key words: splines, wavelets, refinable functions, triangulations

2000 MSC: 41A15, 41A63

1. Introduction

We shall consider spaces of spline functions on $\mathbb{R}^d$, $d \geq 2$, where by a spline function f we shall mean a piecewise polynomial. To make this more precise, the domain of f will be the union $D \subset \mathbb{R}^d$ of a collection T of regions, each of which is a union of a finite number of simplices. Distinct elements of T will intersect only in their common boundary, while any bounded subset of $\mathbb{R}^d$ will intersect only a finite

Email addresses: tgoodman@maths.dundee.ac.uk (Tim Goodman),
doug.hardin@vanderbilt.edu (Doug Hardin).
[1] Corresponding author

number of elements of T. Then f is a spline function over T if it coincides on each element of T with an algebraic polynomial with values in $\mathbb{R}$. We say that f has degree n if these polynomial pieces each have degree at most n.

For simplicity, we refer to a collection T, as above, as a *mesh.* We say that a mesh T is a *refinement* of a mesh S, denoted $S \prec T$, if every element of T is a subset of an element of S and if the union D of the elements of T equals the union of the elements of S. We shall consider a sequence of meshes T_j, $j = 0, 1, 2, \ldots$, with

$$T_j \prec T_{j+1}, \qquad j \geq 0,$$

and we consider, for each $j \geq 0$, a space V_j of spline functions over T_j, where

$$V_j \subset V_{j+1}, \quad j \geq 0. \tag{1}$$

Now suppose that for $j \geq 0$, V_j is a subspace of $L^2(D)$. Then we denote by W_j the orthogonal complement of V_j in V_{j+1}, i.e.

$$V_{j+1} = V_j + W_j, \qquad V_j \perp W_j. \tag{2}$$

Thus for any $j \geq 1$, V_j has an orthogonal decomposition

$$V_j = W_{j-1} + W_{j-2} + \cdots + W_0 + V_0. \tag{3}$$

We shall refer to the spaces W_j as wavelet spaces and the decomposition (3) as a wavelet decomposition. Such decompositions of functions with orthogonal components at different levels are useful in many applications such as signal processing, data compression, and multi-scale methods in numerical analysis.

More generally, we can replace $L^2(D)$ by some Banach space X of functions on D and suppose that for $j \geq 0$, V_j is a subspace of X. We then define

$$W_j = \{f \in V_{j+1} \,:\, P_j f = 0\}, \quad j \geq 0, \tag{4}$$

where P_j is a projection from X onto V_j satisfying

$$P_j P_{j+1} = P_j, \quad j \geq 0.$$

Then we still have the decomposition (3) but the components in the direct sum need not be orthogonal. It is usual to require that $\bigcup_{j=0}^{\infty} V_j$ is dense in X.

We have made the above definitions very general in order to cover all cases considered, but we shall study only special types of meshes and spaces. Section 2 considers the uniform case, i.e. when $D = \mathbb{R}^d$ and

$$f \in V_0 \Rightarrow f(\cdot - k) \in V_0, \quad k \in \mathbb{Z}^d, \tag{5}$$

$$V_j = \{f(M^j \cdot) \,:\, f \in V_0\}, \quad j \geq 1, \tag{6}$$

where M is a $d \times d$ dilation matrix, i.e. it has integer coefficients and $M^{-n}x \to 0$ as $n \to \infty$ for all x in $\mathbb{R}^d$. Although other choices of M may be of interest, we shall focus on the case $M = 2I$, where I denotes the identity matrix. We shall study spaces spanned by box splines and corresponding wavelet spaces, and then consider a generalisation of box splines to multi-box splines.

In Section 3 we shall consider the case when T_j, $j \geq 0$, is a triangulation of $D \subset \mathbb{R}^2$, i.e. T_j comprises triangles, and distinct elements of T_j can intersect only in a common edge or common vertex. We suppose that for $j \geq 0$, T_{j+1} is obtained from T_j by *mid-point subdivision*, i.e. each element of T_{j+1} is a sub-triangle of a triangle t in T_j and has as vertices either the mid-points of the edges of t, or a vertex v of t and the mid-points of the edges of t which meet in v. An extension to $\mathbb{R}^d$, $d \geq 3$, is also mentioned. We shall study continuous linear splines on these meshes and also C^1 quadratic splines gained through macro-elements. Also considered are C^1 cubic and quintic splines.

Finally, in Section 4, we consider the case where T_j, $j \geq 0$, is a triangulation of D gained from a mesh of quadrilaterals by inserting the diagonals of each quadrilateral. For $j \geq 0$, T_{j+1} is gained from T_j by mid-point subdivision of the quadrilateral, i.e. any quadrilateral for T_{j+1} lies in a quadrilateral for T_j and has as vertices the intersection of the diagonals of q, a vertex v of q, and the mid-points of the edges of q which meet in v. On these meshes we study continuous linear splines, C^1 quadratic splines and C^1 cubic splines.

2. Uniform Meshes

In this section we shall study some general constructions for spaces of spline functions over uniform meshes. Some further examples of spline functions over uniform meshes will be studied in the remaining two sections, where they appear as special cases of spaces over more general meshes. What we mean by our spaces being over uniform meshes is most easily described by saying that they are shift-invariant, as we proceed to describe.

A space V of real-valued functions on $\mathbb{R}^d$, $d \geq 1$, is *shift-invariant* if

$$f \in V \Rightarrow f(\cdot - j) \in V, \quad j \in \mathbb{Z}^d. \tag{7}$$

We shall say that V is *refinable* if

$$f \in V \Rightarrow f(M^{-1}\cdot) \in V, \tag{8}$$

where M is a dilation matrix, as described after (6). Defining

$$V_j := \{f(M^j\cdot) \, : \, f \in V\}, \quad j \geq 0, \tag{9}$$

we see that (8) is equivalent to

$$V_j \subset V_{j+1}, \quad j \geq 0,$$

as in (1). Now let W denote the orthogonal complement of V_0 in V_1. Then we see that for $j \geq 0$,

$$W_j := \{f(M^j\cdot) \, : \, f \in W\} \tag{10}$$

is the orthogonal complement of V_j in V_{j+1}, as in (2).

For integrable $\phi_1, \ldots, \phi_r \in V$ with compact support, $\phi = (\phi_1, \ldots, \phi_r)$ is called a *generator (of V)* if V comprises all linear combinations of shifts of ϕ, i.e. $f \in V$ if and only if

$$f = \sum_{j \in \mathbb{Z}^d} \phi(\cdot - j)c(j), \tag{11}$$

for a sequence c of $r \times 1$ matrices $c(j)$, $j \in \mathbb{Z}^d$. In this case we write $V = V(\phi)$ and call V a local finitely generated shift-invariant (local FSI) space. We shall be concerned with spline functions f of compact support and, for such functions, we wish the summation in (11) to comprise a finite number of terms. It will therefore be convenient to call ϕ a ***local generator*** of V if every f in V with compact support satisfies (11) for c with finite support.

Now suppose that ϕ is a local generator of a refinable local FSI space V. Since $\phi(M^{-1}\cdot) \in V$, it is a finite linear combination of shifts of ϕ, i.e.

$$\phi = \sum_{j \in \mathbb{Z}^d} \phi(M \cdot -j)a(j), \tag{12}$$

for a sequence a with finite support of $r \times r$ matrices $a(j)$, $j \in \mathbb{Z}^d$. Such an equation (12) is called a ***refinement equation*** (or two-scale equation) and a vector ϕ of functions satisfying such an equation is called ***refinable***. Conversely, if a generator ϕ of a local FSI space V is refinable, then for any $f \in V$, $f(M^{-1}\cdot)$ can be expressed as a linear combination of shifts of ϕ, i.e. V is refinable.

We say that a generator ϕ is ***linearly independent*** if its shifts are linearly independent, i.e.

$$\sum_{j \in \mathbb{Z}^d} \phi(\cdot - j)c(j) = 0 \Rightarrow c = 0.$$

This concept can be extended as follows. For a non-empty open subset U of $\mathbb{R}^d$, a generator ϕ is said to be ***linearly independent over*** U if

$$\sum_{j \in \mathbb{Z}^d} \phi(x - j)c(j) = 0, \qquad x \in U,$$

for $r \times 1$ matrices $c(j)$, $j \in \mathbb{Z}^d$, implies that $c(j)_i = 0$ whenever $\phi_i(\cdot - j) \not\equiv 0$ on U. Clearly if ϕ is linearly independent over $(0,1)^d$, then ϕ is linearly independent. The converse is not true; indeed it is shown in [42] that there is a local FSI space on $\mathbb{R}$ with a linearly independent generator $\phi = (\phi_1, \phi_2)$ of continuous functions, which has no generator which is linearly independent over $(0,1)$. We say that ϕ is ***locally linearly independent*** if it is linearly independent over any non-empty open subset in $\mathbb{R}^d$.

An elegant characterisation of linearly independent generators is given in [31], in terms of Fourier transforms. For a generator ϕ, its Fourier transform $\hat{\phi}$ is analytic in $\mathbb{R}^d$ and so can be extended to $\mathbb{C}^d$. It is shown in [31] that ϕ is linearly independent if and only if for each $z \in \mathbb{C}^d \setminus \{0\}$, there are $b_1, \ldots, b_r$ in $\mathbb{Z}^d$ for which the matrix

$$A := \left[\hat{\phi}_j(z + 2\pi b_k)\right]_{j,k=1}^{r} \tag{13}$$

is non-singular.

It is further shown in [31] that the shifts of ϕ form a Riesz basis in $L^2(\mathbb{R}^d)$ if the above condition holds for each $z \in \mathbb{R}^d \setminus \{0\}$. In this case we shall say that ϕ is ***stable***. Thus linear independence of ϕ implies its stability. The converse is not true: an example is given in [26] of a refinable local FSI space of univariate spline functions with a generator $\phi = (\phi_1, \phi_2)$ which is stable but not linearly independent.

Taking Fourier transforms of (11) shows that for $f \in S(\phi)$,

$$\hat{f}(u) = \hat{\phi}(u)P(e^{-iu}), \qquad u \in \mathbb{R}^d,$$

where P is the $p \times 1$ matrix of Laurent polynomials,

$$P(z) = \sum_{j \in \mathbb{Z}^d} c(j)z^j, \qquad z \in (\mathbb{C} \setminus \{0\})^d.$$

It follows that if $\phi = (\phi_1, \ldots, \phi_r)$ is a stable, local generator of S, then $\psi = (\psi_1, \ldots, \psi_r)$ is also a stable, local generator of S if and only if $s = r$ and

$$\hat{\psi}(u) = \hat{\phi}(u)A(e^{-iu}), \qquad u \in \mathbb{R}^d,$$

where A is an $r \times r$ matrix of Laurent polynomials which is *unimodular*, i.e. $\det A(z)$ is a non-trivial monomial. This allows us to define the following analogy of dimension of a vector space.

If V is a local FSI space with a stable, local generator $\phi = (\phi_1, \ldots, \phi_r)$, then we say V has *multiplicity* r.

Now suppose that the wavelet space W has a stable generator $\psi = (\psi_1, \ldots, \psi_r)$. Then there are constants $A, B > 0$ such that for any $f \in W$,

$$f = \sum_{j=1}^{r} \sum_{k \in \mathbb{Z}^d} a_{j,k} \psi_j(\cdot - k),$$

the following estimates hold true:

$$A \sum_{j=1}^{r} \sum_{k \in \mathbb{Z}^d} a_{j,k}^2 \leq \|f\|_2^2 \leq B \sum_{j=1}^{r} \sum_{k \in \mathbb{Z}^d} a_{j,k}^2. \tag{14}$$

Next suppose that $f \in L^2(\mathbb{R}^d)$ has a wavelet decomposition

$$f = \sum_{i=0}^{\infty} \sum_{j=1}^{r} \sum_{k \in \mathbb{Z}^d} a_{i,j,k} |\det M|^{i/2} \psi_j(M^i \cdot -k).$$

Then by the orthogonality between levels (2), (10), we have the same stability constants as in (14):

$$A \sum_{i=0}^{\infty} \sum_{j=1}^{r} \sum_{k \in \mathbb{Z}^d} a_{i,j,k}^2 \leq \|f\|_2^2 \leq B \sum_{i=0}^{\infty} \sum_{j=1}^{r} \sum_{k \in \mathbb{Z}^d} a_{i,j,k}^2. \tag{15}$$

We shall shortly state certain characterisations of local FSI spaces of spline functions with multiplicity 1, but in order to do this we must first introduce box splines. These were defined by de Boor and Höllig in [1] and have since been studied and applied by many authors. For a comprehensive study of box splines, see [2].

Take $n \geq 0$ and non-trivial vectors $v_1, \ldots, v_{n+d}$ in $\mathbb{Z}^d$ which span $\mathbb{R}^d$, $d \geq 1$, where for $j = 1, \ldots, n + d$, the components of v_j are coprime. One elegant way to define the *box spline* $B_n = B_n(\cdot|v_1, \ldots, v_{n+d})$ is by its Fourier transform

$$\hat{B}_n(u) = \prod_{j=1}^{n+d} \frac{1 - e^{-iuv_j}}{iuv_j}, \qquad u \in \mathbb{R}^d, \tag{16}$$

where for $u, v \in \mathbb{R}^d$, uv denotes their scalar product uv^t.

Putting $d = 1$, $v_1 = \ldots = v_{n+1}$, (16) reduces to

$$\hat{B}_n(u) = \left(\frac{1 - e^{-iu}}{iu}\right)^{n+1}, \qquad u \in \mathbb{R}^d,$$

which is a definition of the well-known B-spline of degree n with support on $[0, n+1]$ and simple knots at $0, 1, \ldots, n+1$, which was introduced by Schoenberg [51].

In order to describe the structure of the box spline B_n, $d \geq 2$, as a spline function, we define $H = H(v_1, \ldots, v_{n+d})$ as the set of all hyperplanes in $\mathbb{R}^d$ of the form

$$\left\{ k + \sum_{j=1}^{d-1} t_j u_j \, : \, t_j \in \mathbb{R}, \; j = 1, \ldots, d-1 \right\}, \tag{17}$$

for $k \in \mathbb{Z}^d$ and linearly independent elements $u_1, \ldots, u_{d-1}$ of $\{v_1, \ldots, v_{n+d}\}$. Then we define the mesh $T = T(v_1, \ldots, v_{n+d})$ as the collection of all regions which are bounded by but whose interiors are not intersected by elements of H. As an example we take $d = 2$ and $\{v_1, \ldots, v_{n+2}\} = \{(1,0), (0,1), (1,1)\}$. Then H comprises all lines through points in $\mathbb{Z}^2$ in the directions of $(1,0)$, $(0,1)$, and $(1,1)$, and T comprises all triangles with vertices $\{k, k+(1,0), k+(1,1)\}$ or $\{k, k+(0,1), k+(1,1)\}$, for $k \in \mathbb{Z}^2$. This choice of T is called a three-direction mesh or *type-1 triangulation*; and we shall consider this further in Section 3.

Theorem 1 ([1]). *The box spline $B(\cdot|v_1, \ldots, v_{n+d})$ is a spline function of degree n over $T(v_1, \ldots, v_{n+d})$ with support $\{\sum_{j=1}^{n+d} t_j v_j \, : \, 0 \leq t_j \leq 1, \, j = 1, \ldots, n+d\}$. Across any element of $H(v_1, \ldots, v_{n+d})$ of form (17) it has continuous derivatives of order $n+d-2-|J|$, where $J := \{1 \leq j \leq n+d \, : \, v_j$ lies in the span of $u_1, \ldots, u_{d-1}\}$.*

Now for $m \in \mathbb{Z}$, $m \geq 2$, we see from (16) that

$$\hat{B}_n(mu) = \prod_{j=1}^{n+d} \left(\frac{1 + e^{-iuv_j} + \cdots + e^{-i(m-1)uv_j}}{m} \right) \hat{B}_n(u), \qquad u \in \mathbb{R}^d,$$

and thus B_n is refinable with dilation matrix $M = mI$. The final property which we mention concerns the linear independence of the generator (B_n), i.e. the linear independence of the shifts of B_n.

Theorem 2 ([10,30]). *The following are equivalent.*

(a) (B_n) is linearly independent.
(b) (B_n) is locally linearly independent.
(c) (B_n) is stable.
(d) For any elements $u_1, \ldots, u_d$ in $\{v_1, \ldots, v_{n+d}\}$ which are linearly independent, the $d \times d$ matrix $[u_1^t, \ldots, u_d^t]$ has determinant 1 or -1.

We remark that for $d = 2$, condition (d) is equivalent to requiring that lines in $H(v_1, \ldots, v_{n+2})$ intersect only in points of $\mathbb{Z}^2$. In [54] it is shown that if a refinable local FSI space of spline functions is generated by a single function, then that function must be a homogeneous differential operator acting on a certain linear combination of translates of a box spline. This result did not consider stability, but stability was later studied in [27], which gives the following result. This result and the work in [27] are generalizations of results in [34] for the univariate case.

Theorem 3 ([27]). *If V is a local FSI space of spline functions on $\mathbb{R}^d$ which is refinable with respect to $M = mI$, $m \geq 2$, and has multiplicity one, then it has a generator of form*

$$\phi = B_n\left(\cdot - \frac{k}{m-1}\right),$$

for a box spline B_n and $k \in \mathbb{Z}^d$.

Henceforward we shall assume $M = 2I$. Suppose that V is as in Theorem 3, so that it is generated by a box spline B_n satisfying the conditions of Theorem 2. Let $V_0 = V \cap L^2(\mathbb{R}^d)$. As in (6) we define

$$V_j = \{f(2^j \cdot) : f \in V_0\}, \qquad j \geq 1, \tag{18}$$

and, as before, W denotes the orthogonal complement of V_0 in V_1. Let E denote the set of vertices of $[0,1]^d$. Then V_1 has as generator $(\phi(2\cdot - j) : j \in E)$ and so V_1 has multiplicity 2^d. Then W has multiplicity $2^d - 1$, and a generator of W is said to comprise *prewavelets* (the term wavelets being reserved for a generator whose shifts are orthogonal).

We now consider a construction of prewavelets due to Riemenschneider and Shen [46] and found independently in [5]. The construction depends on a function $\eta : E \to E$ satisfying $\eta(0) = 0$, $(\eta(\mu) + \eta(\nu))(\mu + \nu)$ is odd for $\mu \neq \nu$. For $d = 1$, such a map is given by $\eta(0) = 0$, $\eta(1) = 1$, while for $d = 2$ it can be given by $\eta(0) = 0$, $\eta(0,1) = (0,1)$, $\eta(1,0) = (1,1)$, $\eta(1,1) = (1,0)$. Such a mapping for $d = 3$ is given in [45] but, as remarked there, no such maps exist for $d > 3$.

We note that any element ψ of V_1 with compact support satisfies

$$\hat{\psi}(u) = Q(e^{-iu/2})\hat{B}_n\left(\frac{u}{2}\right), \quad u \in \mathbb{R}^d,$$

for some Laurent polynomial Q in $\mathbb{C}^d$. Now define Laurent polynomials P and H on $\mathbb{C}^d$ by

$$P(z) := \sum_{j \in \mathbb{Z}^d} z^j \int B_n B_n(\cdot - j), \tag{19}$$

$$H(z) := \prod_{k=1}^{n+d} (1 + z^{v_k}). \tag{20}$$

It is shown in [46] that a stable generator of W is given by $(\psi_j : j \in E \setminus \{0\})$,

$$\hat{\psi}_j(u) := H_j(e^{-iu/2})\hat{B}_n\left(\frac{u}{2}\right), \quad u \in \mathbb{R}^d, \tag{21}$$

where for $z = e^{-iu/2} = (e^{-iu_1/2}, \dots, e^{-iu_d/2})$,

$$H_j(z) := z^{\eta(j)} P((-1)^j z) H((-1)^j z). \tag{22}$$

The above construction, as we have said, does not work for $d \geq 4$. Constructions for prewavelets from box splines in general dimensions are given in [53]. Further results on box spline prewavelets appear in [35,36,55].

Now suppose $d = 1, 2$, or 3, as above, and $v_1, \dots, v_{n+d}$ lie in $\{-1, 0, 1\}^d$. For this case [3] gives a construction of prewavelets giving smaller support (and hence less computational cost in applications) than those above. Without loss of generality we may suppose

$$v_j \neq -v_k, \quad j, k = 1, \dots, n+d.$$

We suppose that the distinct elements of $\{v_1, \dots, v_{n+d}\}$ are $w_1, \dots, w_\ell$, occurring with multiplicities $n_1, \dots, n_\ell$, so that $n_1 + \dots + n_l = n + d$ and (16) becomes

$$\hat{B}_n(u) = \prod_{k=1}^{\ell} \left(\frac{1 - e^{-iuw_k}}{iuw_k} \right)^{n_k}, \qquad u \in \mathbb{R}^d.$$

We note that, for $d = 2$, we may suppose $w_1, \dots, w_\ell$ lie in $\{(1,0), (0,1), (1,1), (1,-1)\}$ and the condition (d) of Theorem 2 implies that $\ell = 2$ or 3.

It is shown in [3] that a stable generator of W is given by $(\psi_j : j \in E \setminus \{0\})$, where

$$\hat{\psi}_j(u) := G_j(e^{-iu/2}) \hat{B}_n\left(\frac{u}{2}\right), \quad u \in \mathbb{R}^d, \tag{23}$$

where for $z = e^{-iu}$,

$$\overline{G_j(z)} := z^{\eta(j)} P((-1)^j z) \prod_{\substack{k=1 \\ w_k j \text{ even}}}^{\ell} (1 - z^{w_k})^{n_k} \prod_{\substack{k=1 \\ w_k j \text{ odd}}}^{\ell} S_k(z^{w_k}), \tag{24}$$

and for $k = 1, \dots, \ell$, $w \in \mathbb{C}$, $r \in \mathbb{Z}$,

$$S_k(w) = \begin{cases} w^{-r}, & n_k = 2r, \\ w^{-r}(1 + w), & n_k = 2r - 1. \end{cases} \tag{25}$$

As an example of the above constructions, consider the case $d = 2$, $w_1 = (1,0)$, $w_2 = (0,1)$, $w_3 = (1,1)$ with multiplicities $n_1 = n_2 = n_3 = 2$. So $n = 4$ and B_4 is a C^2 quartic spline function. Here

$$H(z) = (1 + z_1)^2 (1 + z_2)^2 (1 + z_1 z_2)^2, \quad z \in \mathbb{C}^2.$$

Choosing q as above we have by (22),

$$H_j(z) = P((-1)^j z) \begin{cases} z_1 z_2 (1 - z_1)^2 (1 + z_2)^2 (1 - z_1 z_2)^2, \ j = (1,0), \\ z_2 (1 + z_1)^2 (1 - z_2)^2 (1 - z_1 z_2)^2, \quad j = (0,1), \\ z_1 (1 - z_1)^2 (1 - z_2)^2 (1 + z_1 z_2)^2, \quad j = (1,1), \end{cases}$$

while by (24), (25),

$$\overline{G_j(z)} = P((-1)^j z) \begin{cases} z_1^{-1}(1-z_2)^2 & j=(1,0), \\ z_1^{-1}z_2^{-1}(1-z_1)^2, & j=(0,1), \\ z_2^{-1}(1-z_1z_2)^2, & j=(1,1). \end{cases}$$

The coefficients of polynomials H_j or G_j are referred to as the *mask* of the corresponding prewavelet ψ_j. Here the masks for the first construction each have 91 non-zero coefficients, while those for the second construction each have 51 non-zero coefficients.

For the case of the continuous linear box spline B_1 on $\mathbb{R}^2$ with $v_1 = (1,0)$, $v_2 = (0,1)$, $v_3 = (1,1)$, the above two constructions give the same prewavelets, each with 19 non-zero coefficients in its mask. For this special case a construction of prewavelets with only 10 non-zero coefficients in each mask is given in [32]. We shall consider this further in Section 3, when we study extensions to non-uniform triangulations.

We do not know of any extension of Theorem 3 characterising refinable local FSI spline functions with multiplicity more than one, however we shall now consider a family of spaces of this form which give a partial generalisation of the spaces V in Theorem 3 generated by box splines B_n. Despite their many elegant properties, such spaces generated by box splines fail to generalise some of the basic properties of the space of all univariate C^{n-1} spline functions of degree n with knots in $\mathbb{Z}$, which is generated by a uniform B-spline. Except for the cases $n = 0$ and 1, the elements of V do not have maximal continuity C^{n-1}. Moreover V does not comprise all piecewise polynomials of degree n subject to some continuity conditions. Indeed the restriction of V to any region in $T(v_1, \ldots, v_{n+d})$ coincides with a space P which is a proper subspace of polynomials of degree n comprising the common null space of certain differential operators [1,10]. Moreover V does not comprise all piecewise polynomials with pieces coinciding with such spaces P subject to certain continuity conditions. In contrast, we now consider spaces of the following form.

Take integers $n \geq 0$, $r \geq 1$, and pairwise linearly independent vectors $v_0, \ldots, v_{n+r}$ in $\mathbb{Z}^2$. Let $S_n = S_n(v_0, \ldots, v_{n+r})$ comprise all functions f with continuous Fourier transforms of form

$$\hat{f}(u) = \frac{\sum_{|\alpha|=r-1} P_\alpha(e^{-iu})u^\alpha}{(uv_0)\cdots(uv_{n+r})}, \quad u \in \mathbb{R}^2,$$

where P_α is a Laurent polynomial for $\alpha \in \mathbb{Z}^2$, $|\alpha| = r-1$. It follows easily from the definition that the space S_n is refinable. Our next result characterises elements of S_n as spline functions. As in (17), $H = H(v_0, \ldots, v_{n+r})$ denotes the set of lines in $\mathbb{R}^2$ of the form $\{k + tv_j : t \in \mathbb{R}\}$, for $k \in \mathbb{Z}^2$ and $0 \leq j \leq n+r$. The mesh $T = T(v_0, \ldots, v_{n+r})$ comprises all regions which are bounded by, but whose interiors are not intersected by, lines in H.

Theorem 4 **([25]).** *The space S_n comprises all C^{n-1} spline functions of degree n over the mesh $T(v_0,\dots,v_{n+r})$ with compact support, such that the jump of any derivative of order n across any line in $H(v_0,\dots,v_{n+r})$ changes only at points in $\mathbb{Z}^2$.*

The condition on the discontinuities of the derivatives of a function f in S_n can be thought of as f having knots in $\mathbb{Z}^2$. Thus S_n can be thought of as a generalization of the space of all univariate C^{n-1} spline functions of degree n with knots in $\mathbb{Z}$ and compact support. Next we consider generators for S_n.

Theorem 5 **([25]).** *There is a generator $\phi = (\phi_1,\dots,\phi_r)$ such that any element of S_n is a finite linear combination of shifts of ϕ. Moreover ϕ is such a generator if and only if*

$$\hat{\phi}(u) = \frac{\tilde{u}M(e^{-iu})}{(uv_0)\cdots(uv_{n+r})}, \qquad u \in \mathbb{R}^2, \tag{26}$$

where $\tilde{u} := (u_1^{r-1}, u_1^{r-2}u_2, \dots, u_2^{r-1})$ and M is an $r \times r$ matrix of Laurent polynomials with

$$\det M(z) = cz^k \prod_{j=0}^{n+r}(1 - z^{v_j}), \quad z \in \mathbb{C}^2, \tag{27}$$

for some $k \in \mathbb{Z}^2$, $c \in \mathbb{R}$, $c \neq 0$.

We note that when $r = 1$, (26) and (27) show that any generating function ϕ satisfies

$$\hat{\phi}(u) = ce^{-iku}\frac{\prod_{j=0}^{n+r}(1 - e^{-iuv_j})}{(uv_0)\cdots(uv_{n+r})},$$

and so by (16), ϕ is a multiple of a shift of a box spline

$$\phi = cB_n(\cdot - k|v_0,\dots,v_{n+r}).$$

The case $r = 2$ was introduced in [23], where the generator was called a multi-box spline, and this case was studied further in [24]. Our next result analyses stability of the generators ϕ in Theorem 5.

Theorem 6 **([25]).** *For any generator ϕ as in Theorem 5, ϕ is stable if and only if at most r lines in $H(v_0,\dots,v_{n+r})$ intersect except at points in $\mathbb{Z}^2$.*

By the remark after Theorem 2, we see that the conditions of Theorem 6 generalise condition (d) of Theorem 2 for the case $d = 2$, $r = 1$.

Now consider the line $L = \{tu : t \in \mathbb{R}\}$ for $u = (u_1, u_2) \in \mathbb{Z}^2$. If u_1 and u_2 are odd, then putting $t = \frac{1}{2}$ show that L passes through a point equal to $(\frac{1}{2}, \frac{1}{2}) \bmod \mathbb{Z}^2$. Thus for stability of ϕ, the condition of Theorem 6 requires that at most r vectors in $\{v_0,\dots,v_{n+r}\}$ have both components odd. Similarly there are at most r vectors in $\{v_0,\dots,v_{n+r}\}$ with components of form (odd, even) and at most r vectors with components of form (even, odd). Thus stability of ϕ implies that $n \leq 2r-1$. It can be shown that for any $r \geq 1$, there is a choice of vectors $\{v_0, \ldots, v_{n+r}\}$ for which at most r lines in $H(v_0, \ldots, v_{n+r})$ intersect other than in points of $\mathbb{Z}^2$, and hence we have the following result.

Theorem 7 (**[25]**). *For any positive integer r, there is a space $S_{2r-1}(v_0, \ldots, v_{3r-1})$, as in Theorem 4, with a stable generator $\phi = (\phi_1, \ldots, \phi_r)$.*

Thus by choosing large enough r, we can construct stable multi-box splines of arbitrarily high degree n and smoothness C^{n-1}. Before giving some examples, we discuss the possible symmetry of multi-box splines ϕ for all $r \geq 1$. We say that $\phi = (\phi_1, \ldots, \phi_r)$ is *symmetric* if for $j = 1, \ldots, r$, there are $\sigma_j = \pm 1$, $\alpha_j \in \{0,1\}^2$, with

$$\phi_j(-\cdot) = \sigma_j \phi_j(\cdot + \alpha_j), \tag{28}$$

i.e. ϕ_j is even or odd about $\frac{1}{2}\alpha_j$ as $\sigma_j = 1$ or -1.

Theorem 8 (**[25]**). *If ϕ as in Theorem 5 is symmetric, then*

$$\sigma_1 \cdots \sigma_r = (-1)^{(n+1)(r+1)}, \quad \alpha_1 + \cdots + \alpha_r = v_0 + \cdots + v_{n+r} \bmod \mathbb{Z}^2.$$

For $r = 1$, this corresponds to the well-known fact that the box spline $B_n(\cdot|v_0, \ldots, v_{n+r})$ is even about $\frac{1}{2}(v_0 + \cdots + v_{n+1})$. Next we consider symmetry under the operator $T : \mathbb{R}^2 \to \mathbb{R}^2$, $Tx = (x_2, x_1)$. If we have

$$\{Tv_0, \ldots, Tv_{n+r}\} = \{t_0 v_0, \ldots, t_{n+r} v_{n+r}\}, \tag{29}$$

where $t_j = \pm 1$, $j = 0, \ldots, n+r$, then the space $S_n(v_0, \ldots, v_{n+r})$ is invariant under T and it is natural to require a corresponding multi-box spline to be symmetric under T, i.e. for $j = 1, \ldots, r$,

$$\phi_j(T\cdot) = \tau_j \phi_j, \quad \tau_j = \pm 1. \tag{30}$$

Theorem 9 (**[25]**). *Suppose that (29) holds and ϕ as in Theorem 5 satisfies (30). Then*

$$\tau_1 \cdots \tau_r = (-1)^s (t_0 \cdots t_{n+r})^{r-1}, \tag{31}$$

where $r = 2s$ or $2s+1$.

It may be that (30) is not satisfied for $j = k, l$, some $k \neq l$, but instead $\phi_k = \phi_l(T\cdot)$. In this case (31) holds with $\tau_k \tau_l$ replaced by -1.

Another possible symmetry is reflection in the x_2-axis, i.e. $R : \mathbb{R}^2 \to \mathbb{R}^2$, $Rx = (-x_1, x_2)$. If we have

$$\{Rv_0, \ldots, Rv_{n+r}\} = \{r_0 v_0, \ldots, r_{n+r} v_{n+r}\}, \tag{32}$$

where $r_j = \pm 1$, $j = 0, \ldots, n+r$, then $S_n(v_0, \ldots, v_{n+r})$ is invariant under R and it is natural to require that for $j = 1, \ldots, r$,

$$\phi_j(Rx) = p_j \phi_j(x_1 + (\alpha_j)_1, x_2), \quad x \in \mathbb{R}^2, \quad \rho_j = \pm 1. \tag{33}$$

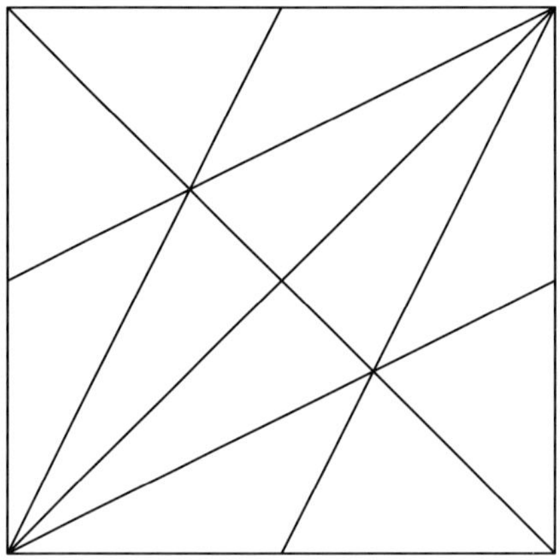

Fig. 1. $T(v_0, \ldots, v_5)$ on $[0,1]^2$.

Theorem 10 ([25]). *Suppose that (32) holds and ϕ as in Theorem 8 satisfies (33). Then for s as in Theorem 9*

$$\rho_1 \cdots \rho_r = (-1)^s (r_0 \cdots r_{n+r})^{r-1}. \tag{34}$$

As for Theorem 9, we replace $\rho_k \rho_l$ in (34) by -1 if $\phi_k = \phi_l(R\cdot)$. Clearly a corresponding result to Theorem 10 holds for reflection in the x_1-axis.

It is suggested in [25], from examples considered there, that there is always a symmetric multi-box spline which satisfies the above symmetry conditions where appropriate, but this is not proved in general.

We finish by considering some examples of multi-box splines. First we take $n = 1$, $r = 2$, $v_0 = (1,0)$, $v_1 = (0,1)$, $v_2 = (1,1)$, $v_3 = (1,-1)$. The mesh $T = T(v_0, \ldots, v_3)$ is called a four-direction mesh or *type-2 triangulation* and we shall consider this further in Section 4. Here the space S_1 comprises *all* continuous linear splines over T. There is a choice of symmetric generator $\phi = (\phi_1, \phi_2)$, where ϕ_1 has support $[0,1]^2$ and

$$\phi_2(x) = \phi_1\left(\frac{x_1 + x_2 + 1}{2}, \frac{x_1 - x_2 + 1}{2}\right), \quad x \in \mathbb{R}^2.$$

Since at most two lines in $H(v_0, \ldots, v_3)$ intersect other than in $\mathbb{Z}^2$, ϕ is stable. Equations (28), (30), and (33) hold with

$$\alpha_1 = (1,1), \quad \alpha_2 = (0,0), \quad \sigma_j = \tau_j = \rho_j = 1, \quad j = 1,2.$$

Next we take $n = 2$, $r = 3$, $v_0, \ldots, v_3$ as above, and $v_4 = (2,1)$, $v_5 = (1,2)$. The mesh $T(v_0, \ldots, v_5)$ restricted to the triangle with vertices $(0,0)$, $(1,0)$, $(1,1)$ gives the Powell-Sabin 6-split of this triangle, [44], and we have the corresponding split for the triangle with vertices $(0,0)$, $(1,0)$, $(1,1)$. This is illustrated in Figure 1. Here the space S_2 comprises *all* C^1 quadratic splines over T. A choice of symmetric generator for S_2 is given by the finite element basis $\phi = (\phi_1, \phi_2, \phi_3)$, which is defined as follows. For $k \in \mathbb{Z}^2$, define linear functionals on S_2 by

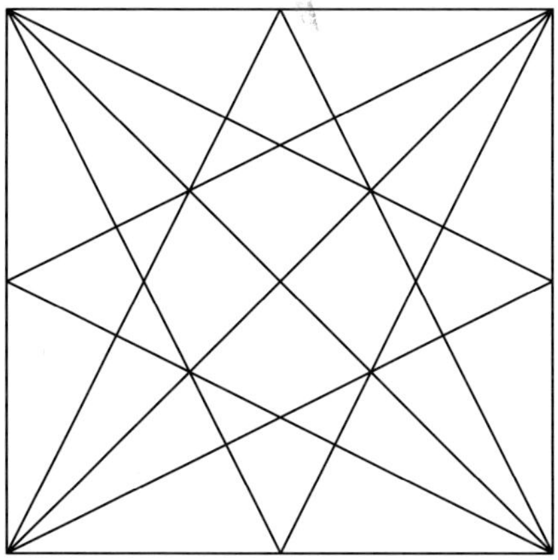

Fig. 2. $T(v_0, \ldots, v_7)$ on $[0,1]^2$.

$$\sigma_{1k} f = f(k), \quad \sigma_{2k} f = D_{v_2} f(k), \quad \sigma_{3k} f = D_{v_3} f(k), \tag{35}$$

where for $v \in \mathbb{R}^2$, D_v denotes the directional derivative in direction v. Then for $i = 1, 2, 3$, ϕ_i is defined as the unique function in S_2 satisfying

$$\sigma_{jk} \phi_i = \delta_{ij} \delta_{k0}, \quad j = 1, 2, 3, \quad k \in \mathbb{Z}^2. \tag{36}$$

These functions are considered in [9] (under a linear transformation of $\mathbb{R}^2$), where they are defined explicitly in terms of Bézier coefficients. Further properties of ϕ are given in [9] and will be considered in in Section 3. The support of ϕ_i, $i = 1, 2, 3$, is the hexagon with vertices $(-1,-1)$, $(-1,0)$, $(0,-1)$, $(1,0)$, $(0,1)$, $(1,1)$. Equations (28) and (30) hold with

$$\sigma_1 = 1, \quad \sigma_2 = \sigma_3 = -1, \quad \alpha_1 = \alpha_2 = \alpha_3 = (0,0),$$

$$\tau_1 = \tau_2 = 1, \quad \tau_3 = -1.$$

Since at most three lines in $H(v_0, \ldots, v_5)$ intersect other than in $\mathbb{Z}^2$, the above generator ϕ is stable. A more symmetric mesh $T(v_0, \ldots, v_7)$ is gained by taking $v_0, \ldots, v_5$ as before, and $v_6 = (2, -1)$, $v_7 = (1, -2)$, see Figure 2. There are still at most three lines in $H(v_0, \ldots, v_7)$ which intersect other than in $\mathbb{Z}^2$. Thus there is a stable generator for the case $n = 3$, $r = 4$, and for the case $n = 4$, $r = 3$.

3. General Triangulations

Let T denote a triangulation whose union D is a simply connected region in $\mathbb{R}^2$. Let $V(T)$ denote the set of all vertices of triangles in T and $E(T)$ the set of all edges of triangles in T. We suppose that any boundary vertex (i.e. vertex in the boundary of D) is the intersection of exactly two boundary edges. We shall first consider the space $L(T)$ of all continuous linear spline functions over T. This space has a natural basis of *nodal functions*, defined as follows. For each v in $V(T)$ we let ϕ_v be the unique element of $L(T)$ satisfying

$$\phi_v(v) = 1, \quad \phi_v(w) = 0, \quad w \in V(T), \quad w \neq v.$$

Then for any element f of $L(T)$,

$$f = \sum_{v \in V(T)} f(v)\phi_v.$$

The support of ϕ_v is the union of all triangles in T which have v as a vertex.

Now put $T_0 = T$ and let T_1 denote the triangulation gained from T_0 by mid-point subdivision as described in Section 1. Thus T_1 is gained by subdividing each element t of T into four congruent sub-triangles, each similar to t. For $j \geq 1$, we recursively obtain T_{j+1} from T_j by mid-point subdivision. Then we define

$$V_j = L(T_j) \cap L^2(D), \quad j \geq 0. \tag{37}$$

First we shall consider the uniform case where $D = \mathbb{R}^2$, T denotes the type-1 triangulation $T((1,0),(0,1),(1,1))$ and $L(T) = S_1((1,0),(0,1),(1,1))$. In this case $\phi_{(0,0)}$ is the box spline $B_1(\cdot|(1,0),(0,1),(1,1))$ which generates S_1. Now $L(T_1) = \{f(2\cdot) : f \in L(T)\}$ and so V_j in (37) satisfies (18). As before, we let W denote the orthogonal complement of V_0 in V_1.

In [32] there is given a construction of a generator $\psi = (\psi_1, \psi_2, \psi_3)$ for W where ψ_1 has support as in Figure 3 and

$$\psi_2(x,y) = \psi_1(y,x), \quad \psi_3(x,y) = \psi_1(y, y-x), \quad x, y \in \mathbb{R}. \tag{38}$$

Also ψ_1 is even about $(\frac{1}{2}, 0)$, i.e.

$$\psi_1(-x,-y) = \psi_1(x+1,y), \quad x,y \in \mathbb{R}. \tag{39}$$

The function ψ_1 is non-zero at all 13 points in $\frac{1}{2}\mathbb{Z}^2$ which lie inside its support. Equivalently, when ψ_1 is written in the form

$$\psi_1 = \sum_{j \in \mathbb{Z}^2} \psi_1\left(\frac{j}{2}\right) B_1(2 \cdot - j),$$

then the mask of ψ_1, i.e. the coefficient in this linear combination, has 13 non-zero terms. At the expense of the symmetry (39), [32] also constructs a generator $\tilde{\psi} = (\tilde{\psi}_1, \tilde{\psi}_2, \tilde{\psi}_3)$ of W which satisfies the analogue of (38) but such that the mask of $\tilde{\psi}_1$ has only 10 non-zero coefficients. In [29] a characterisation is given of prewavelets with the minimum of 10 non-zero coefficients in their masks for spaces comprising those elements of $L(T_1)$ with supports on given bounded regions.

In a series of papers [18–20], Floater and Quak generalise the construction of the prewavelet ψ above to the case of a general triangulation on a bounded domain D, as considered earlier. In order to gain an elegant construction, they define the space W_j, as in (2), as the orthogonal complement of V_j in V_{j+1} with respect to the inner product

$$\langle f, g \rangle = \sum_{t \in T} \frac{1}{2a(t)} \int_t fg, \qquad f, g \in C(D), \tag{40}$$

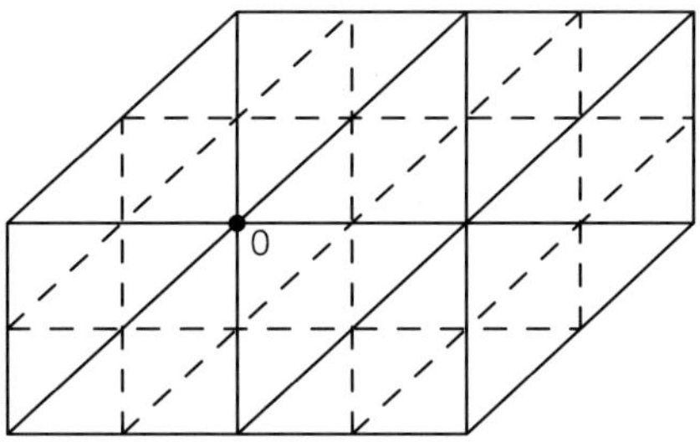

Fig. 3. Support of ψ_1.

where $a(t)$ denotes the area of triangle t. Thus W_j is defined by (4), where P_j denotes the orthogonal projection with respect to (40). For the uniform case above, (40) is the usual L^2 inner product.

Floater and Quak construct a basis $\{\psi_v : v \in V(T_1) \setminus V(T)\}$ for W_0. Take any $v \in V(T_1) \setminus V(T)$. Then $v = \frac{1}{2}(u_1 + u_2)$ for $u_1, u_2 \in V(T)$, and the support of ψ_v is the union of all triangles in T having u_1 or u_2 as a vertex. The prewavelets ψ_v satisfy the elegant symmetry property

$$\psi_v(u) = \psi_u(v), \quad u, v \in V(T_1) \setminus V(T).$$

For the case when T is the restriction of a type-1 triangulation to a domain D and the vertex $v \in V(T_1) \setminus V(T)$ is the centre of an edge whose end-points are interior vertices, the function ψ_v coincides with $\psi_j(\cdot - k)$, as above, for some $k \in \mathbb{Z}^2$, $1 \leq j \leq 3$.

By replacing T above by T_j, $j \geq 0$, there is a corresponding basis $\{\psi_v^j : v \in V_*^j\}$ for W_j, where $V_*^j := V(T_{j+1}) \setminus V(T_j)$, and we normalise by requiring $\|\psi_v^j\|_2 = 1$. For the uniform case [32] derives estimates for the stability constants as in (14). In [20] such estimates are derived for the constants A, B in the following expansion for the non-uniform case:

$$f = \sum_{v \in V(T)} a_v \phi_v + \sum_{j=0}^{\infty} \sum_{v \in V_*^j} b_{j,v} \psi_v^j,$$

$$A \left(\sum_{v \in V(T)} a_v^2 + \sum_{j=0}^{\infty} \sum_{v \in V_*^j} b_{j,v}^2 \right) \leq \|f\|_2^2 \leq B \left(\sum_{v \in V(T)} a_v^2 + \sum_{j=0}^{\infty} \sum_{v \in V_*^j} b_{j,v}^2 \right).$$

In [52] a construction of linear prewavelets is given which holds in $\mathbb{R}^d$ for any $d \geq 2$. In this case the meshes T_j, $j \geq 0$, comprise simplices in $\mathbb{R}^d$. Any such simplex has $d+1$ faces which are simplices of dimension $d-1$. These in turn have faces of dimension $d-2$, and proceeding recursively we have a collection $F(s)$ of simplices of dimension $0, \ldots, d-1$ in the boundary of s. We assume that distinct elements s, t of T_j can intersect only in an element of $F(s) \cap F(t)$. We suppose that for $j \geq 0$, T_{j+1} is gained from T_j by subdividing any simplex t in T_j into 2^d congruent sub-simplices, each similar to t, though for $d \geq 3$ there is no canonical way to do this.

As before, we denote by $L(T_j)$ the space of all continuous linear spline functions over T_j. We define V_j, $j \geq 0$, by (37) and define W_j, $j \geq 0$, by (2), this time with the usual inner product in $L^2(D)$. Then [52] gives a construction for a basis $\{\psi_v : v \in V(T_{j+1}) \setminus V(T_j)\}$ for W_j. For the case of a type-1 triangulation of $\mathbb{R}^2$ this leads to a generator $\bar{\psi} = (\bar{\psi}_1, \bar{\psi}_2, \bar{\psi}_3)$, which satisfies the analogues of (38) and (39). However, while ψ_1 has support of area 5 and mask with 13 non-zero coefficients, $\bar{\psi}_1$ has support of area 8 and mask with 23 non-zero coefficients.

Next we shall consider refinable spaces of C^1 quadratic splines constructed from Powell-Sabin macro-elements. First we study the uniform case where each triangle in the type-1 triangulation $T(v_0, v_1, v_2)$, for $v_0 = (1,0)$, $v_1 = (0,1)$, $v_2 = (1,1)$, is divided into six sub-triangles by inserting the medians, thus producing the triangulation $T = T(v_0, \ldots, v_5)$ for $v_3 = (1,-1)$, $v_1 = (2,1)$, $v_2 = (1,2)$, as discussed near the end of Section 2. As described there, the space $S_2 = S_2(v_0, \ldots, v_5)$ of all C^1 quadratic splines over T is generated by the fundamental functions for Hermite interpolation of values and first-order derivatives at $\mathbb{Z}^2$, i.e. the shifts of the multi-box splines $\phi = (\phi_1, \phi_2, \phi_3)$ defined by the conditions (35) and (36).

Now let $V_0 = S_2 \cap L^\infty(\mathbb{R}^2)$ and, as in (18),

$$V_j = \{f(2^j \cdot) : f \in V_j\}, \quad j \geq 1.$$

For $j \geq 0$, we let P_j denote the projection from $C^1(\mathbb{R}^2) \cap L^\infty(\mathbb{R}^2)$ onto V_j given by

$$P_j f(2^{-j}k) = f(2^{-j}k), \quad D_{v_i} P_j f(2^{-j}k) = D_{v_i} f(2^{-j}k), \quad i = 0,1, \quad k \in \mathbb{Z}^2.$$

Then $P_j P_{j+1} = P_j$, $j \geq 0$, and we define, as in (4),

$$W_j = \{f \in V_{j+1} : P_j f = 0\}, \quad j \geq 0.$$

Thus W_0 comprises all elements of V_1 whose values and first order derivatives vanish on $\mathbb{Z}^2$, and

$$W_j = \{f(2^j \cdot) : f \in W_0\}, \quad j \geq 0.$$

Then for $j \geq 0$, a basis for W_j is given by $\psi_{i,j,k} := \phi_i(2^{j+1} \cdot -k)$, $i = 1,2,3$, $k \in \mathbb{Z}^2 \setminus 2\mathbb{Z}^2$. Such functions comprise what is called a *hierarchical basis.* In [9] it is shown that this basis is 'weakly stable' in the sense that there are constants $A, B > 0$ such that for any $n \geq 0$ and $f \in V_{n+1}$ of the form

$$f = \sum_{j=0}^{n} \sum_{i=1}^{3} \sum_{k \in \mathbb{Z}^2} a_{i,j,k} \psi_{i,j,k},$$

$$A\|a\|_\infty \leq \|f\|_\infty \leq Bn\|a\|_\infty,$$

where $\|a\|_\infty = \sup\{|a_{i,j,k}| : j = 0, \ldots, n, \, i = 1,2,3, \, k \in \mathbb{Z}^2\}$. It is also shown in [9] that the operators P_j are uniformly bounded and that for any $f \in C^1(\mathbb{R}^2) \cap L^\infty(\mathbb{R}^2)$, $\lim_{j \to \infty} P_j f = f$ uniformly on bounded subsets of $\mathbb{R}^2$.

In [41] hierarchical bases are considered on general triangulations using Powell-Sabin elements based on splitting each triangle t into a mesh $T(t)$ of 12 sub-triangles formed by joining the mid-points of each edge with each other and with the opposite vertex, see Figure 4. Each C^1 quadratic spline over $T(t)$ is defined uniquely by the following 12 values: the values and first derivatives at the vertices A_1, A_2, A_3, and

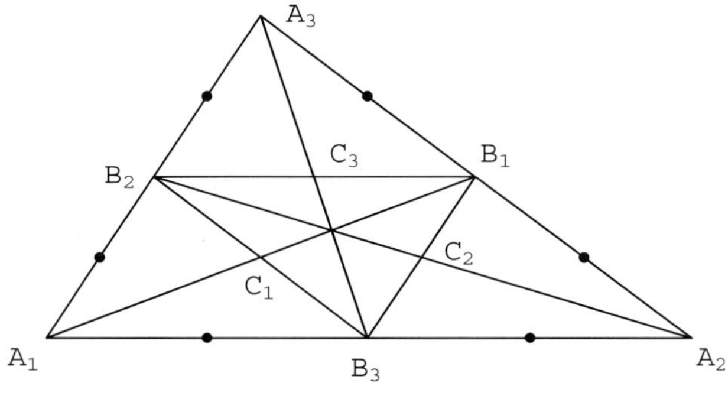

Fig. 4. The Powell-Sabin 12-split.

the normal derivatives at the mid-points of the edges B_1, B_2, B_3. Let T denote a triangulation with union D as described at the beginning of this section. For any such triangulation T we denote by T' the triangulation gained by splitting each element of T into 12 sub-triangles, as above. The space of all C^1 quadratic splines over T' will be denoted by $Q(T')$.

Now for any point $x \in \mathbb{R}^2$ we define the linear functionals on $C^1(D)$:

$$\sigma_{x1} f = f(x), \quad \sigma_{x2} f = D_{(1,0)} f(x), \quad \sigma_{x3} f = D_{(0,1)} f(x). \tag{41}$$

Also for any finite line segment e in $\mathbb{R}^2$, we denote by $\sigma_{et} f$ and $\sigma_{en} f$ derivatives of f at the mid-point of e in directions tangential and normal to e, respectively. We denote by ϕ_{vj}, $j = 1,2,3$, $v \in V(T)$, and ϕ_e, $e \in E(T)$, the unique fundamental functions in $Q(T')$ satisfying

$$\sigma_{ui}\phi_{vj} = \delta_{uv}\delta_{ij}, \quad \sigma_{ui}\phi_e = 0, \quad u \in V(T), \quad i = 1,2,3,$$

$$\sigma_{fn}\phi_{vj} = 0, \quad \sigma_{fn}\sigma_e = \delta_{ef}, \quad e \in E(T).$$

Then a nodal basis for $Q(T')$ is given by

$$\{\phi_{vj}, \phi_e \, : v \in V(T), j = 1,2,3, e \in E(T)\}.$$

The support of ϕ_{vj}, $j = 1,2,3$, is the union of all triangles in T having v as a vertex, and the support of ϕ_e is the union of the triangles in T having e as an edge.

The triangulation T' is refinable under mid-point subdivision of T, as we now describe. Put $T_0 = T$, $T'_0 = T'$, and let T_1 denote the triangulation gained from T_0 by mid-point subdivision. Then $T'_0 \prec T'_1$. More generally, for $j \geq 1$, we may recursively define T_{j+1} by mid-point subdivision of T_j, and we have

$$T'_j \prec T'_{j+1}, \quad j \geq 0,$$

and hence

$$Q(T'_j) \subset Q(T'_{j+1}), \quad j \geq 0.$$

For $j \geq 0$, we let P_j denote the projection from $C^1(D)$ onto $Q(T'_j)$ defined by

$$\sigma_{vi} P_j f = \sigma_{vi} f, \quad v \in V(T_j), \quad i = 1,2,3,$$

$$\sigma_{en} P_j f = \sigma_{en} f, \quad e \in E(T_j).$$

As before we have $P_j P_{j+1} = P_j$, $j \geq 0$, and define

$$W_j := \left\{ f \in Q(T'_{j+1}) \, : \, P_j f = 0 \right\}.$$

A basis for W_0 is given by the fundamental functions

$$\left\{ \psi_e, \tilde{\psi}_e, \psi_f \, : \, e \in E(T_0), \, f \in E(T_1) \right\},$$

where, with $\sigma_e \psi$ denoting the value of ψ at the mid-point of edge e,

$$\sigma_g \psi_e = \delta_{ge}, \quad \sigma_g \tilde{\psi}_e = 0, \quad \sigma_g \psi_f = 0, \quad g \in E(T_0),$$

$$\sigma_{gt} \psi_e = 0, \quad \sigma_{gt} \tilde{\psi}_e = \delta_{ge}, \quad \sigma_{gt} \psi_f = 0, \quad g \in E(T_0),$$

$$\sigma_{hn} \psi_e = 0, \quad \sigma_{hn} \tilde{\psi}_e = 0, \quad \sigma_{hn} \psi_f = \delta_{hf}, \quad h \in E(T_1).$$

Similarly, we may define a hierarchical basis for all W_j, $j \geq 0$.

In [41] there is also considered a hierarchical basis of C^1 cubic splines. The well-known Clough-Tocher element [8] is not refinable under mid-point subdivision and so they consider instead the space of all C^1 cubic splines on the Powell-Sabin 12-split of a triangle, as in Figure 4. Each such function is determined uniquely by the following 30 values: the values and first order derivatives at the points A_1, A_2, A_3, B_1, B_2, B_3, the values and tangential derivatives at C_1, C_2, C_3, and the normal derivatives at the six mid-points of the edges of sub-triangles, denoted by dots in Figure 4. For any triangulation T and refinement T', as for the previous example, [41] then considers the space of all C^1 cubic spline function on T' and studies the nodal basis and corresponding hierarchical basis derived from the above Hermite interpolation functionals.

The above hierarchical bases studied in [41] are introduced for preconditioning the finite element equations for fourth-order elliptic boundary value problems when using the conjugate gradient method. Motivated by the same problem, [11] constructs hierarchical bases of C^1 quintic splines, as we now briefly discuss.

As before we let T_0 denote a general triangulation with union D and define the triangulation T_j, $j \geq 1$, recursively by mid-point subdivision. For $j \geq 0$, let $S_5(T_j)$ denote the space of all C^1 quintic spline functions over T_j. In [37] a nodal basis for $S_5(T_j)$ is constructed comprising fundamental functions for Hermite interpolation functionals involving values, first-order and certain second-order directional derivatives at vertices in $V(T_j)$ and normal derivatives at mid-points of edges in $E(T_j)$. For the applications considered in [11], the authors study the subspace S^0 of $S_5(T_j)$ comprising all functions whose values and first-order derivatives vanish on the boundary of D. Since S^0 is not spanned by a subset of the above nodal basis for $S_5(T_j)$, [11] constructs a nodal basis for S^0 by keeping those of the above nodal functions whose supports lie in the interior of D but modifying the remaining elements of the above nodal basis.

4. Triangulated Quadrangulations

Let Q denote a mesh of quadrilaterals with union $D \subset \mathbb{R}^2$, each with interior angles less than π, such that distinct elements of Q intersect only in a common edge or a common vertex. We now divide each quadrilateral q in Q into four triangles, by inserting the diagonals of q, to produce a triangulation T. As in Section 3, we assume that D is a simply connected region in $\mathbb{R}^2$ and any boundary vertex of W is the intersection of exactly two boundary edges.

We now construct a mesh Q_1 of quadrilaterals from Q by mid-point subdivision, i.e. each element q of Q is divided into four elements of Q_1 by joining the mid-points of the edges of q to the intersection of the diagonals of q. We then define a triangulation T_1, which is a refinement of T, by inserting the diagonals of all the elements of Q_1. This is illustrated in Figure 5. We shall first consider the space $L(T)$ of all continuous linear spline functions over T.

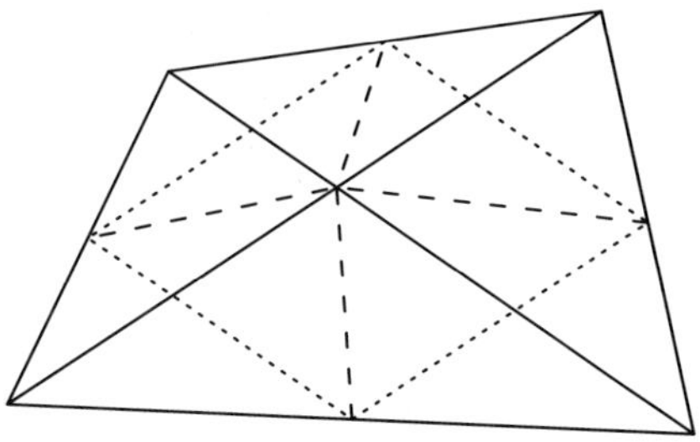

Fig. 5. Subdivision of T on a quadrilateral.

Now consider the uniform case when $D = \mathbb{R}^2$ and the vertices of Q comprise $\mathbb{Z}^2$. Then T is the type-2 triangulation $T(v_0, v_1, v_2, v_3)$, where $v_0 = (1,0)$, $v_1 = (0,1)$, $v_2 = (1,1)$, $v_3 = (1,-1)$, and T_1 is gained from T by shrinking by a factor of two, as in Section 2. In this case $L(T) = S_1(v_0, v_1, v_2, v_3)$, which is generated by the multi-box spline $\phi = (\phi_1, \phi_2)$, as in Section 2, where ϕ_1, ϕ_2 are the nodal functions (see Section 3),

$$\phi_1 = \phi_{(\frac{1}{2},\frac{1}{2})}, \quad \phi_2 = \phi_{(0,0)}.$$

Since the triangulation T_1 is not gained by mid-point subdivision of T, the general constructions of prewavelets of [20] and [52], as described in Section 3, do not hold in this case. In [21], Floater and Quak construct prewavelets for this case using the same approach as in [20], (i.e. each prewavelet is the sum of two 'semi-wavelets'). To be more precise, let T_0 denote the above type-2 triangulation restricted to a rectangle R with vertices in $\mathbb{Z}^2$ and sides of length at least three. Let T_1 denote the refinement of T_0, as above. As before, $V(T_0)$ and $V(T_1)$ denote the sets of vertices of all triangles in T_0 and T_1 respectively, while W denotes the orthogonal complement of $L(T_0)$ in $L(T_1)$. Then [21] constructs a basis $\{\psi_v : v \in V(T_1) \setminus V(T_0)\}$ for W. As for the previous construction in [20], the support of the prewavelet ψ_v is the union of all triangles in the T_0 having u_1 or u_2 as a vertex, where $v = \frac{1}{2}(u_1 + u_2)$, $u_1, u_2 \in V(T_0)$. There are two types of interior prewavelets, one with a mask of 17 non-zero coefficients and one with a mask of 13 coefficients. As in the prewavelet construction given in [29] on type-1 meshes, interior prewavelets with masks containing 11 and 17 terms, respectively, may also be constructed (see [28]).

Now consider the corresponding uniform case, where

$$V_0 = S_1(v_0, v_1, v_2, v_3) \cap L^2(\mathbb{R}^2), \tag{42}$$

$$V_1 = \{f(2\cdot) : f \in V_0\}, \tag{43}$$

and W denotes the orthogonal complement of V_0 in V_1. Since V_0 has multiplicity 2, W has multiplicity 6. A generator of W is given by

$$\psi = \left(\psi_v : v = (\frac{1}{2}, 0), (0, \frac{1}{2}), (\frac{1}{4}, \frac{1}{4}), (\frac{1}{4}, \frac{3}{4}), (\frac{3}{4}, \frac{1}{4}), (\frac{3}{4}, \frac{3}{4})\right).$$

Two examples of refinable, piecewise linear, orthogonal generators were constructed in [16] where by orthogonal we mean that the shifts of the components of these generators form orthogonal systems. The first generator is piecewise linear on a type-2 triangulation and the second on type-1 triangulation. Associated piecewise linear orthogonal wavelets were constructed in [14] (also see [17]), but here we restrict our discussion to the refinable generators. The construction relies on techniques introduced in [15] for constructing univariate orthogonal spline wavelets. The main observation is that if V is a refinable FSI space and $\tilde{V}$ is a local FSI space such that $V_0 \subset \tilde{V} \subset V_1$ for some $k \in \mathbb{Z}$ (where $(V_j)_{j\in\mathbb{Z}}$ is defined by (9) with dilation matrix $M = 2I$) then $\tilde{V}$ is also refinable since if $f \in \tilde{V}_0$ then $f \in V_1$ and so $f(\cdot/2) \in V_0 \subset \tilde{V}$. In this case the sequences of spaces $(V_j)_{j\in\mathbb{Z}}$ and $(\tilde{V}_j)_{j\in\mathbb{Z}}$ are said to form *intertwining multiresolution analyses* because of the relations

$$\cdots \subset \tilde{V}_{-1} \subset V_0 \subset \tilde{V}_0 := \tilde{V} \subset V_1 \subset \tilde{V}_1 \subset \cdots.$$

Note that $\tilde{V}$ inherits many of the properties of V. For example, if V is a spline space then $\tilde{V}$ is also a spline space with the same polynomial degree, smoothness, and approximation order.

We next outline the construction of the generator on the type-2 triangulation $T = T(v_0, v_1, v_2, v_3)$. Let $V_0 = L(T) \cap L^2(\mathbb{R}^2)$ and let ϕ_1 denote the continuous function that is piecewise linear on the triangulation of $[-1,1]^2$ consisting of the four triangles formed by the boundary of this square and its two diagonals and satisfying $\phi_1(0,0) = 1$ and $\phi(v) = 0$ for $v \in \{-1,1\}^2$. Furthermore, let $\phi_2 = \phi_1(2\cdot -(1,1))$. It is not difficult to verify that (ϕ_1, ϕ_2) is a generator for V_0. Next, for $i = 3, \ldots, 6$, let $\phi_i = \phi_2(2\cdot -k_i)$ where $\{k_3, k_4, k_5, k_6\} = \{0,1\}^2$ and let $\hat{V} = V(\phi_1, \ldots, \phi_6)$. Since $V_0 \subset \hat{V} \subset V_1$ it follows that $\hat{V}$ is refinable. Then it is shown that there exist functions $\phi_7, \phi_8 \in \hat{V}_1$ with support $[-1,1] \times [0,1]$ and $[0,1] \times [-1,1]$, respectively, and satisfying the conditions

(a) $\phi_7(-x,y) = \phi_7(x,y)$ and $\phi_8(x,y) = \phi_7(y,x)$ for $(x,y) \in \mathbb{R}^2$,
(b) $\phi_7 \perp V(\phi_2, \ldots \phi_6)$,
(c) $\phi_7 \perp V(\phi_8)$,
(d) $P_{V(\phi_2,\ldots,\phi_8)^\perp}\phi_1 \perp \phi_1(\cdot - k)$, for $k = (1,0)$ and $(0,1)$.

Condition (a) ensures symmetry of $V(\phi_1, \ldots, \phi_8)$ with respect to the symmetry group of T. The conditions in (b) are linear and, because the supports of the functions $\phi_2, \ldots \phi_6$ are contained in $[0, 1]$, are easy to characterise. The conditions (c) and (d) are the critical conditions and reduce to solving three homogeneous quadratic equations in 8 variables. Similarly functions ϕ_9 and ϕ_{10} in V_1 are found that satisfy the conditions

(e) $\phi_9(y, x) = \phi_9(x, y)$, $\phi_9(1 - y, 1 - x) = -\phi_9(x, y)$, and $\phi_{10}(x, y) = \phi_9(1 - x, y)$ for $(x, y) \in \mathbb{R}^2$,

(f) $\phi_9 \perp V(\phi_2, \ldots, \phi_8)$,

(g) $\phi_9 \perp V(\phi_{10})$,

(h) $P_{V(\phi_2,\ldots,\phi_{10})^\perp}\phi_1 \perp \phi_1(\cdot - k)$, for $k = (1, 1)$ and $(-1, -1)$.

Finally, an orthogonal generator $\tilde{\phi} = (\tilde{\phi}_1, \ldots, \tilde{\phi}_{10})$ may be chosen for the space $\tilde{V} = V(\phi_1, \ldots, \phi_{10})$ in the following way. Let $\tilde{\phi}_1 = P_{V(\phi_2,\ldots,\phi_{10})^\perp}\phi_1$, choose $(\tilde{\phi}_2, \ldots, \tilde{\phi}_6)$ so that it forms an orthogonal basis for the span of $(\phi_2, \ldots, \phi_6)$ and let $\tilde{\phi}_j = \phi_j$ for $j = 7, 8, 9, 10$. Then it follows from the above conditions (a–h) as well as properties of the supports of these functions that $\tilde{\phi}$ is an orthogonal generator. Since $\tilde{V}$ is refinable, then so is $\tilde{\phi}$.

The second piecewise linear orthogonal refinable generator $\tilde{\phi} = (\tilde{\phi}_1, \ldots, \tilde{\phi}_{14})$ given in [16] is on the three direction 'hexagonal' triangulation given by $v_1 = (1, 0)$, $v_2 = (\frac{1}{2}, \frac{\sqrt{3}}{2})$ and $v_2 = (-\frac{1}{2}, \frac{\sqrt{3}}{2})$. Then $V(\tilde{\phi})$ is invariant under the symmetry group $S(T)$ associated with the triangulation $T = T(v_1, v_2, v_3)$. In this case $\tilde{\phi}$ has 14 components. However, as shown in [28], one may generate $V(\tilde{\phi})$ from a 'macro-element' $\gamma = (\gamma_1, \ldots, \gamma_5)$ consisting of 5 functions supported on the triangle $(0,0), v_1, v_2$. Let $V_{S(T)}(\gamma)$ denote the algebraic span of the collection of functions $\{\gamma \circ \lambda : \lambda \in S(T)\}$. Then

$$V(\tilde{\phi}) = V_{S(T)}(\gamma) \cap C(\mathbb{R}^2) \cap L^2(\mathbb{R}^2)$$

and

$$\tilde{\phi}_i = \sum_{\lambda \in S(T)} a(\lambda)_{i,j} \gamma_j \circ \lambda$$

for some finitely supported collection of 14×5 matrices $(a(\lambda))_{\lambda \in S(T)}$. Because the essential orthogonality relations of $\tilde{\phi}$ occur triangle by triangle and are reflected in γ, the macro-element point of view provides a construction of continuous orthogonal wavelets on semi-regular sequence of triangulations generated by successively applying mid-point subdivision to an arbitrary initial triangulation (cf. [17]).

Next we consider the space $S_2 = S_2(v_0, v_1, v_2, v_3)$ of all C^1 quadratic splines on the type-2 triangulation $T = T(v_0, v_1, v_2, v_3)$. This is generated by the box spline $B_2 = B_2(\cdot|v_0, v_1, v_2, v_3)$, which is called the Zwart-Powell element, because it was first considered in [56,43]. This function has support in the octagon with vertices (0,-1), (1,-1), (2,0), (2,1), (1,2), (0,2), (-1,1), (-1,0). It shares all the symmetries of T, i.e. it satisfies (28), (30), (33) with

$$\sigma_1 = \tau_1 = \rho_1 = 1, \quad \alpha_1 = (1, 1).$$

Since two lines in $H(v_0, v_1, v_2, v_3)$ intersect in $(\frac{1}{2}, \frac{1}{2})$, the generator (B_2) is *not* stable. Nevertheless it has been much studied and generalized to a triangulation of a non-uniform rectangular mesh [6,7,47–49]. To be more precise we let $D = [a, b] \times [c, d]$ and let Q denote the mesh of rectangles $\{[x_{i-1}, x_i] \times [y_{i-1}, y_i] : 1 \leq i \leq m,\ 1 \leq j \leq n\}$, where $a = x_0 < x_1 < \ldots < x_m = b$, $c = y_0 < y_1 < \ldots < y_m = d$. Let T denote the triangulation formed from Q by inserting the diagonals of all rectangles in Q, and let S denote the space of all C^1 quadratic splines over T. Then S is generated by $\{B_{ij} : 0 \leq i \leq m+1,\ 0 \leq j \leq n+1\}$ for locally supported functions B_{ij}. These functions generalise B_2 in the sense that if $2 \leq i \leq m-1$, $2 \leq j \leq n-1$ and

$$x_{i-2+k} = x_{i-2} + kh,\ y_{j-2+k} = y_{j-2} + kh,\ k = 1, 2, 3,\ h > 0,$$

then

$$B_{ij}(x, y) = B_2\left(\frac{1}{h}(x - x_{i-1}), \frac{1}{h}(y - y_{j-1})\right), \quad (x, y) \in D.$$

Now consider again a triangulation T formed by inserting the diagonals into a general mesh Q of quadrilaterals, as in the beginning of this section. We let $C(T)$ denote the space of all C^1 cubic splines over T. To consider a basis for $C(T)$ we first consider the mesh T_1 comprising the triangulation of a single quadrilateral q divided by its diagonals. Then each element F of $C(T_2)$ is determined uniquely by the value and first order derivatives of f at each vertex of q, and the first order normal derivative of f at the mid-point of each edge of q. Thus $C(T)$ has a nodal basis, defined as follows. Let $V(Q)$ and $E(Q)$ denote the sets of all vertices and all edges of elements of Q respectively. For $v \in V(Q)$ we define the linear functionals $\sigma_{v_1}, \sigma_{v_2}, \sigma_{v_3}$ as in (41), while for $e \in E(Q)$, $\sigma_{en} f$ denotes, as before, the normal derivative of f at the mid-point of e. We denote by ϕ_{vj}, $v \in V(Q)$, $j = 1, 2, 3$, and ϕ_e, $e \in E(Q)$, the corresponding fundamental functions in $C(T)$ satisfying

$$\sigma_{ui}\phi_{vj} = \delta_{uv}\delta_{ij},\ \sigma_{ui}\phi_e = 0,\ u \in V(Q),\ i = 1, 2, 3,$$

$$\sigma_{fn}\phi_{vj} = 0,\ \sigma_{fn}\phi_e = \delta_{ef},\ e \in E(Q).$$

Then $\{\phi_{vj}, \phi_e : v \in V(Q),\ j = 1, 2, 3,\ e \in E(Q)\}$ forms a nodal basis for $C(T)$. This basis was first considered by Fraeijs de Veubeke [22] and Sander [50] and for this reason the type of triangulation T considered above is often called a Fraeijs de Veubeke–Sander (FVS) triangulation. The support of ϕ_{vj}, $j = 1, 2, 3$, is the union of all quadrilaterals in Q having v as a vertex, and the support of ϕ_e is the union of the quadrilaterals in Q having e as an edge.

We first consider the uniform case when $D = \mathbb{R}^2$ and $T = T(v_0, v_1, v_2, v_3)$, as before. Here the space $C(T)$ is shift-invariant with multiplicity 5 and generator $\phi = (\phi_1, \ldots, \phi_5)$, where

$$\phi_j = \phi_{(0,0),j},\ j = 1, 2, 3,\ \phi_4 = \phi_e,\ \phi_5 = \phi_f,$$

where e denotes the edge from $(0, 0)$ to $(0, 1)$ and f the edge from $(0, 0)$ to $(1, 0)$. The generator ϕ is linearly independent over $[0, 1]^2$ and satisfies (28), (30), and (33)

with

$$\sigma_1 = 1,\ \sigma_j = -1,\ j = 2, \ldots, 5,$$
$$\alpha_j = (0,0),\ j = 1,2,3,\ \alpha_4 = (0,1),\ \alpha_5 = (1,0),$$
$$\tau_1 = 1\ \rho_1 = \rho_3 = \rho_5 = 1,\ \rho_2 = \rho_4 = -1.$$

Moreover, with T as in (30),

$$\phi_3 = \phi_2(T\cdot),\ \phi_5 = \phi_4(T\cdot).$$

For $j = 1,2,3$, ϕ_j has support $[-1,1]^2$ and ϕ_4 has support $[-1,1] \times [0,1]$. The Bernstein-Bézier representation of ϕ is given in [33]. As noted in [4], we can form a generator of $C(T)$ with smaller support by replacing ϕ_2 and ϕ_3 by $\tilde{\phi}_2$ and $\tilde{\phi}_3$ where

$$\tilde{\phi}_2 = \phi_2 - \phi_3 - \frac{1}{4}(\phi_4 + \phi_4(\cdot + (0,1)) - \phi_5 - \phi_5(\cdot + (1,0)),$$
$$\tilde{\phi}_3(x,y) = \tilde{\phi}_2(-x,y), \quad x,y \in \mathbb{R}.$$

The support of $\tilde{\phi}_2$ is the hexagon with vertices $(-1,-1)$, $(-1,0)$, $(0,-1)$, $(1,0)$, $(0,1)$, $(1,1)$.

Now, in a similar manner to (42), (43), we define

$$V_0 = C(T) \cap L^2(\mathbb{R}^2), \quad V_1 = \{f(2\cdot) : f \in V_0\},$$

and let W denote the orthogonal complement of V_0 in V_1. In [4] a Riesz basis is constructed for W, as we now describe. First a Riesz basis $B = \{t_{i,k} : i = 1, \ldots, 5,\ k \in \mathbb{Z}^2\}$ for V_0 is defined by

$$t_{1,k} = \phi_1(\cdot - k),\ t_{j,k} = \tilde{\phi}_j(\cdot - k),\ j = 2,3,$$
$$t_{4,k} = \begin{cases} \phi_4(\cdot - k) + \phi_4(\cdot - k + (0,1)),\ k \in \mathbb{Z} \times 2\mathbb{Z}, \\ \phi_4(\cdot - k) - \phi_4(\cdot - k - (0,1)),\ k \in \mathbb{Z} \times (2\mathbb{Z} - 1), \end{cases}$$
$$t_{5,k} = t_{4,Tk}(T\cdot).$$

Next a set $\tilde{B} = \{\tilde{t}_{i,k} : i = 1, \ldots, 5,\ k \in \mathbb{Z}^2\}$ in V_1 is constructed which is defined uniquely by each function $\tilde{t}_{i,k}$ having the same support and symmetry as $t_{i,k}$, and $\tilde{B}$ being biorthogonal to B, i.e.

$$\langle t_{i,k}, \tilde{t}_{j,\ell} \rangle = \|t_{i,k}\|_2^2 \delta_{ij} \delta_{k\ell},$$

where $\langle\, , \rangle$ denotes the inner product on $L^2(\mathbb{R}^2)$. Finally a Riesz basis $\{\psi_{i,k} : i = 1, \ldots, 5,\ k \in \mathbb{Z}^2 \setminus 2\mathbb{Z}^2\}$ for W is defined by

$$\psi_{i,k} = t_{I,k}(2\cdot) - \sum_{j=1}^{5} \sum_{\ell \in \mathbb{Z}^2} \frac{\langle t_{i,k}(2\cdot), t_{j,\ell} \rangle}{\|t_{j,\ell}\|^2} \tilde{t}_{j,\ell}.$$

We now return to the non-uniform case and describe a generalisation of the FVS triangulation defined in [11]. Let Q_0 be a mesh comprising triangles and quadrilaterals. We define a triangulation T_0 by dividing each quadrilateral in Q_0 into four triangles by inserting the diagonals, as before, and dividing each triangle T in Q_0 into three sub-triangles by joining the vertices of t to its barycentre. We now construct a mesh Q_1 of quadrilaterals by dividing each quadrilateral in Q_0 into four sub-quadrilaterals, as before, and by dividing each triangle t in Q_0 into three quadrilaterals by joining the mid-points of the edges of t to its barycentre.

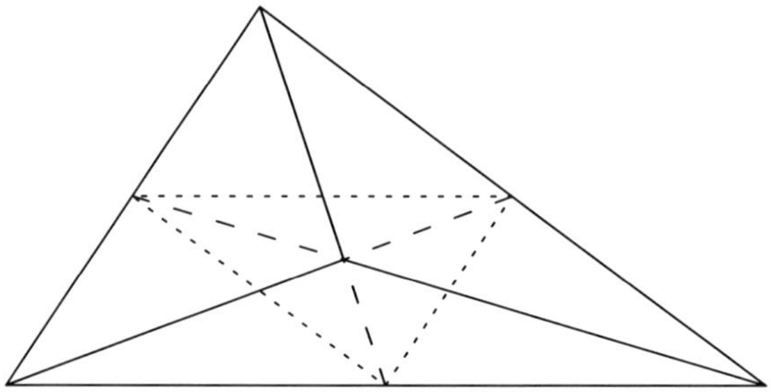

Fig. 6. Subdivision of a triangle in Q_0.

We then define a triangulation T_1, $T_0 \prec T_1$, by inserting the diagonals of all elements of Q_1. This process on a triangle in Q_0 is illustrated in Figure 6.

For $j \geq 1$, we may recursively define Q_{j+1} by mid-point subdivision of Q_j, and T_{j+1} with $T_j \prec T_{j+1}$, by inserting the diagonals of all elements of Q_{j+1}. It is noted in [11] that for $j \geq 2$, each quadrilateral in Q_j is either a parallelogram or similar to a quadrilateral in Q_1.

For T_j as above, $j \geq 0$, we denote by $C(T_j)$ the space of all C^1 cubic splines over T_j. Then

$$C(T_j) \subset C(T_{j+1}), \quad j \geq 0.$$

Now the restriction of any function f in $C(T_0)$ to a triangle t in Q_0 is determined uniquely, as in the Clough-Tocher element [8], by the value and first-order derivatives of f at each vertex of t and the normal derivative of f at the mid-points of each edge of t. Thus there is a nodal basis for $C(T_0)$ of fundamental functions for Hermite interpolation of value and first-order derivatives at each vertex $v \in V(Q_0)$ and normal derivatives at the mid-point of each edge $e \in E(Q_0)$. A corresponding hierarchical basis can be defined as before.

All C^1 nodal bases considered so far, in both this and the previous section, are based on Hermite interpolation. To finish, we mention some nodal bases of C^1 cubic splines, and some corresponding hierarchical bases, based on Lagrange interpolation. As before, let Q denote a mesh of quadrilaterals with union D, let T denote the triangulation formed by inserting the diagonals of the elements of Q, and let $C(T)$ denote the space of all C^1 cubic splines over T. In [38] it is assumed that Q is *checkerboard*, i.e. all interior vertices are of degree 4 and elements of Q can be coloured black and white in such a way that any two quadrilaterals sharing an edge have opposite colours. Then [38] constructs a basis of $C(T)$ with local support of fundamental functions for Lagrange interpolation at certain points. (Further study of bases of this type is given in [39,40]. A similar construction for a nodal basis of C^2 splines of degree 6 is given in [13].)

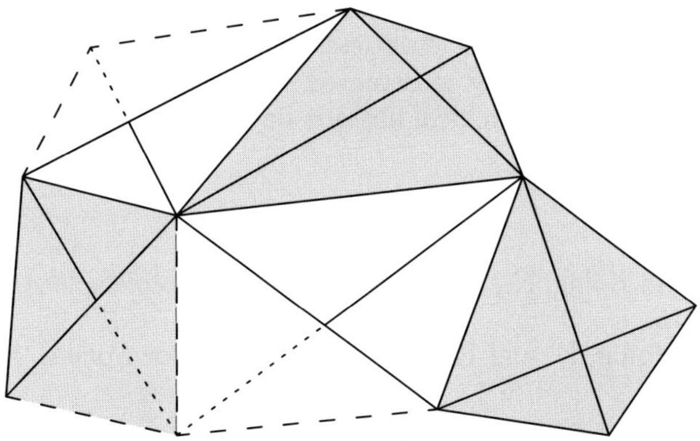

Fig. 7. Triangulation T_0.

Similar nodal bases, and corresponding hierarchical bases, are studied in the recent paper [12]. Here more general triangulations T are allowed, as we now describe. For certain boundary quadrilaterals q in Q, divide q into two triangles by inserting a diagonal and denote one of the triangles, which has at least one edge in the boundary of D, by $t(q)$. Then let Ω be the region gained by removing all triangles $t(q)$ from D. Let T_0 denote the triangulation of Ω comprising all triangles in T which lie in Ω, see Figure 7. It is shown in [12] that any domain with Lipschitz continuous piecewise linear boundary can be triangulated by such a triangulation T_0.

Unlike our previous constructions, we form a refinement Q_1 of Q by dividing each element of Q into *nine* sub-quadrilaterals by trisecting each edge and subdiagonal. This is illustrated in Figure 8, where we have shown the colouring of the sub-quadrilaterals for the case of a 'white quadrilateral.' For a 'black quadrilateral,' the colours are reversed. By inserting the diagonals of all elements of Q_1 we obtain a triangulation T_1 which is a refinement of T, each triangle in T being divided into nine sub-triangles, see Figure 8.

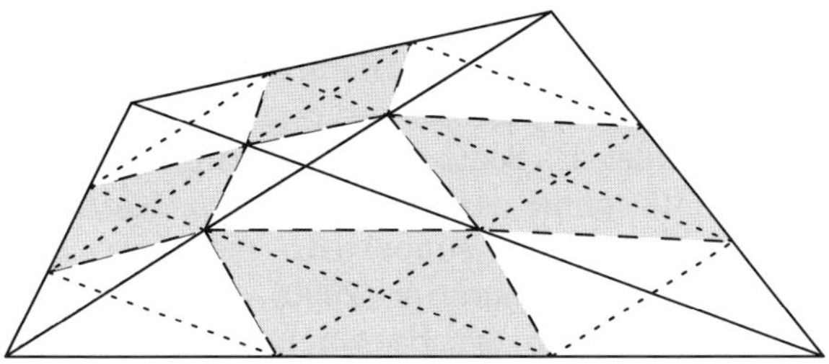

Fig. 8. Refinement of a 'white quadrilateral.'

A basis of $C(T_0)$ is given in [12] comprising fundamental functions for Lagrange interpolation at certain points $P(T_0)$, which depend on the colouring of the quadrilaterals. These points are chosen so that corresponding points $P(T_1)$ contain $P(T_0)$ and hence a corresponding hierarchical basis can be constructed. It is furthermore shown that the hierarchical basis is a Riesz basis for $H^s(\Omega)$, $1 < s < 5/2$, thus improving the stability of formerly constructed C^1 hierarchical bases, though further discussion of stability of hierarchical bases is beyond the scope of this paper.

References

1. de Boor, C. and Höllig, K., B-splines from parallelepipeds, *J. Analyse Math.* **42** (1982), 99–115.
2. de Boor, C., Höllig, K. and Riemenschneider, S., *Box Splines*, Springer Verlag, New York, 1993.
3. Buhmann, M. D., Davydov, O. and Goodman, T. N. T., Box spline prewavelets of small support, *J. Approx. Theory* **112** (2001), 16–27.
4. Buhmann, M. D., Davydov, O. and Goodman, T. N. T., Cubic spline prewavelets on the four-direction mesh, *Foundations of Comp. Math.* **3** (2003), 113–133.
5. Chui, C. K., Stöckler, J. and Ward, J. D., Compactly supported box spline wavelets, *Approx. Theory Appl.* **8** (1992), 77–100.
6. Chui, C. K., Schumaker, L. L. and Wang, R. H., On spaces of piecewise polynomials with boundary conditions III. Type 2 triangulations, in: *Proc. Canadian Math. Soc. Conf., American Math. Soc., Volume 3*, 1983, pp. 67–80.
7. Chui, C. K., Schumaker, L. L. and Wang, R. H., Concerning C^1 B-splines on triangulations of nonuniform rectangular partitions, *Approx. Theory Appl.* **1** (1984), 11–18.
8. Clough, R. and Tocher, J., Finite element stiffness matrices for analysis of plates in bending, in: *Proc. of Conference on Matrix Methods in Structural Mechanics*, Wright-Patterson Air Force Base, Ohio, 1965, pp. 515–545.
9. Daehlen, M., Lyche, T., Mørken, K., Schneider, R. and Seidel, H. P., Multiresolution analysis over triangles, based on quadratic Hermite interpolation, *J. Comput. Appl. Math.* **119** (2000), 97–114.
10. Dahmen, W. and Micchelli, C. A., Translates of multivariate splines, *Linear Algebra Appl.* **52** (1983), 217–234.
11. Dahmen, W., Oswald, P. and Shi, X.-Q., C^1-hierarchical bases, *J. Comput. Appl. Math.* **51** (1994), 37–56.
12. Davydov, O. and Stevenson, R., Hierarchical Riesz bases for $H^s(\Omega)$, *Constr. Approximation*, to appear.
13. Davydov, O. and Zeilfelder, F., Scattered data fitting by direct extension of local polynomials to bivariate splines, *Advances in Comp. Math.* **21** (2004), 223–271.
14. Donovan, G. C., Compactly supported, piecewise affine wavelets on triangulations, unpublished manuscript (2000).
15. Donovan, G. C., Geronimo, J. S. and Hardin, D. P., Intertwining multiresolution analyses and the construction of piecewise-polynomial wavelets, *SIAM J. Math. Anal.* **27** (1996), 1791–1815.

16. Donovan, G. C., Geronimo, J. S. and Hardin, D. P., Compactly supported, piecewise affine scaling functions on triangulations, *Constr. Approximation* **16** (2000), 201–219.
17. Donovan, G. C., Geronimo, J. S., Hardin, D. P. and Kessler, W. B., C^0 macro-element wavelets on triangulations, in preparation.
18. Floater, M. S. and Quak, E. G., A semi-prewavelet approach to piecewise linear prewavelets on triangulations, in: *Approximation Theory IX, Vol. 2: Computational Aspects* (C. K. Chui and L. L. Schumaker, Eds.), Vanderbilt University Press, Nashville, 1998, pp. 60–73.
19. Floater, M. S. and Quak, E. G., Piecewise linear prewavelets on arbitrary triangulations, *Numer. Math.* **82** (1999), 221–252.
20. Floater, M. S. and Quak, E. G., Linear independence and stability of piecewise linear prewavelets on arbitrary triangulations, *SIAM J. Numer. Anal.* **38** (2000), 58–79.
21. Floater, M. S. and Quak, E. G., Piecewise linear wavelets over type-2 triangulations, *Computing Supplement* **14** (2001), 89–103.
22. Fraeijs de Veubeke, G., Bending and stretching of plates, in: *Proc. Conf. on Matrix Methods in Structural Mechanics*, Wright Patterson Air Force Base, Ohio, 1965, pp. 863–886.
23. Goodman, T. N. T., Constructing pairs of refinable splines, in: *Surface Fitting and Multiresolution Methods* (A. Le Méhauté, C. Rabut and L. L. Schumaker, Eds.), Vanderbilt University Press, Nashville, 1997, pp. 151–158.
24. Goodman, T. N. T., Properties of bivariate refinable spline pairs, in: *Multivariate Approximation, Recent Trends and Results* (W. Haussmann, K. Jetter and M. Reimer, Eds.), Akademie-Verlag, Berlin, 1997, pp. 63–82.
25. Goodman, T. N. T., Multi-box splines, manuscript.
26. Goodman, T. N. T. and Lee, S. L., Some properties of refinable splines, in: *International Conference of Wavelet Analysis and its Applications*, AMS/IP *Studies in Advanced Mathematics* **25** (2002), pp. 93–107.
27. Guan, Y., Lu, S. and Tang, Y., Characterization of compactly supported refinable splines whose shifts form a Riesz basis, *J. Approx. Theory* **133** (2005), 245–250.
28. Hardin, D. P. and Hong, D., Construction of wavelets and prewavelets over triangulations, *J. Comput. Appl. Math.* **155** (2003), 91–109.
29. Hong, D. and Mu, A. Y., On construction of minimum supported piecewise linear prewavelets over triangulations, in: *Wavelet Analysis and Multiresolution Methods*, Lecture Notes for Pure and Applied Mathematics, Marcel Dekker, Inc., New York, NY, 2000, pp. 181–201.
30. Jia, R. Q., Local linear independence of the translates of a box spline, *Constr. Approximation* **1** (1985), 175–182.
31. Jia, R. Q. and Micchelli, C. A., On linear independence of integer translates of a finite number of functions, *Proc. Edinburgh Math. Soc.* **36** (1992), 69–85.
32. Kotyczka, U. and Oswald, P., Piecewise linear prewavelets of small support, in: *Approximation Theory VIII, Vol. 2* (C. K. Chui and L. L. Schumaker, Eds.), World Scientific, Singapore, 1995, pp. 235–242.

33. Lai, M. J., Approximation order from bivariate C^1 cubics on a four-direction mesh is full, *Computer Aided Geometric Design* **11** (1994), 215–223.
34. Lawton, W., Lee, S. L. and Shen, Z. W., Characterization of compactly supported refinable splines, *Advances in Comp. Math.* **3** (1995), 137–145.
35. Lorentz, R. A. and Madych, W. R., Wavelets and generalized box splines, *Appl. Analysis* **44** (1992), 51–76.
36. Lorentz, R. A. and Oswald, P., Nonexistence of compactly supported box spline prewavelets in Sobolev spaces, in: *Surface Fitting and Multiresolution Methods* (A. Le Méhauté, C. Rabut and L. L. Schumaker, Eds.), Vanderbilt University Press, Nashville, 1997, pp. 235–244.
37. Morgan, J. and Scott, R., A nodal basis for C^1 piecewise polynomials of degree $n \geq 5$, *Math. Comp.* **29** (1975), 736–740.
38. Nürnberger, G., Schumaker, L. L. and Zeilfelder, F., Local Lagrange interpolation by bivariate C^1 cubic splines, in: *Mathematical Methods in CAGD*, Proc. Oslo 2000 (T. Lyche and L. L. Schumaker, Eds.), Vanderbilt University Press, Nashville, 2001, pp. 393–404.
39. Nürnberger, G., Schumaker, L. L. and Zeilfelder, F., Lagrange interpolation by C^1 cubic splines on triangulations of separable quadrangulations, in: *Approximation Theory X: Splines, Wavelets, and Applications* (C. K. Chui, L. L. Schumaker and J. Stöckler, Eds.), Vanderbilt University Press, Nashville, 2002, pp. 405–424.
40. Nürnberger, G., Schumaker, L. L. and Zeilfelder, F., Lagrange interpolation by C^1 cubic splines on triangulated quadrangulations, *Advances in Comp. Math.* **21** (2004), 357–380.
41. Oswald, P., Hierarchical conforming finite element methods for the biharmonic equation, *SIAM J. Numer. Anal.* **29** (1992), 1610–1625.
42. Plonka, G. and Zhou, D. X., Properties of locally linearly independent refinable function vectors, *J. Approx. Theory* **122** (2003), 24–41.
43. Powell, M. J. D., Piecewise quadratic surface fitting in contour plotting, in: *Software for Numerical Mathematics* (D. J. Evans, Ed.), Academic Press, London, 1974, pp. 253–271.
44. Powell, M. J. D. and Sabin, M. A., Piecewise quadratic approximations on triangles, *ACM Trans. Math. Software* **3** (1977), 316–325.
45. Riemenschneider, S. and Shen, Z., Box splines, cardinal series, and wavelets, in: *Approximation Theory and Functional Analysis* (C. K. Chui, Ed.), Academic Press, New York, 1991, pp. 133–149.
46. Riemenschneider, S. and Shen, Z., Wavelets and prewavelets in low dimension, *J. Approx. Theory* **71** (1992), 18–38.
47. Sablonnière, P., Bernstein Bézier methods for the construction of bivariate spline approximants, *Computer Aided Geometric Design* **2** (1985), 29–36.
48. Sablonnière, P., On some multivariate quadratic spline quasi-interpolants on bounded domains, in: *Modern Developments in Multivariate Approximation*, Proc. Bommerholz 2002 (W. Haussmann, K. Jetter, M. Reimer and J. Stöckler, Eds.), ISNM, Vol. 145, Birkhäuser-Verlag, Basel, 2003, pp. 263–278.

49. Sablonnière, P., Quadratic spline quasi-interpolants on bounded domains of $\mathbb{R}^d$, $d = 1, 2, 3$, in: *Splines and Radial Functions, Rend. Sem. Univ. Pol. Torino* **61**, 2003, pp. 229–246.
50. Sander, G., Bornes supérieures et inférieures dans l'analyse matricielle des plaques en flexion-torsion, *Bull. Soc. Roy. Sci. Liège* **33** (1964), 456–494.
51. Schoenberg, I. J., Contributions to the problem of approximation of equidistant data by analytic functions, *Quart. Appl. Math.* **4** (1948), 45–99, 112–141.
52. Stevenson, R., Piecewise linear (pre-)wavelets on non-uniform meshes, in: *Multigrid Methods V* (W. Hackbusch and G. Wittum, Eds.), Springer-Verlag, Heidelberg, 1996, pp. 306–319.
53. Stöckler, J., Multivariate wavelets, in: *Wavelets: A Tutorial in Theory and Applications* (C. K. Chui, Ed.), Academic Press, San Diego, 1992, pp. 325–355.
54. Sun, Q., Refinable functions with compact support, *J. Approx. Theory* **86** (1996), 240–252.
55. Zhou, D. X., Construction of real-valued wavelets by symmetry, *J. Approx. Theory* **81** (1995), 323–331.
56. Zwart, P. B., Multivariate splines with non-degenerate partitions, *SIAM J. Numer. Anal.* **10** (1973), 665–673.

Topics in Multivariate Approximation and Interpolation
K. Jetter et al., Editors

Adaptive Wavelets for Sparse Representations of Scattered Data

Angela Kunoth

Institut für Angewandte Mathematik, Universität Bonn,
Wegelerstr. 6, D–53115 Bonn, Germany

Abstract

We consider large–scale scattered data problems where the information is given in form of nonuniformly distributed functional data. Seeking for a sparse representation of the data with a minimal amount of degrees of freedom, an adaptive scheme which operates in a coarse-to-fine fashion based on spline-(pre)wavelets is proposed. At each stage a least-squares approximation of the data is computed, taking into account different requests arising in scattered data fitting. We discuss the fast iterative solution of the least square systems, regularization and the treatment of outliers. A particular application concerns the approximate continuation of harmonic functions, an issue arising in geophysics.

Key words: Scattered data fitting, irregularly distributed data, adaptivity, wavelet, least squares, fast solution, regularization, outlier, harmonic function.

2000 MSC: 65D15, 65T60

1. Introduction

Suppose that for some large $N \in \mathbb{N}$ there is a set $X = \{x_i\}_{i=1,\ldots,N}$ of irregularly spaced and pairwise distinct points $x_i \in \overline{\Omega} := [0,1]^n$ where $n \in \{1,2,3\}$. For each

Email address: `kunoth@iam.uni-bonn.de` (Angela Kunoth).

i let $z_i \in \mathbb{R}$ be corresponding functional data which is allocated in the set Z. The problem of scattered data fitting is the following. Given the cloud of points $P := (X, Z) = \{(x_i, z_i),\ i = 1, \ldots, N\}$, find a function $f : \overline{\Omega} \to \mathbb{R}$ that approximates P in a least squares sense with regularization, that is, f minimizes the functional

$$\mathcal{J}(f) := \sum_{i=1}^{N} (z_i - f(x_i))^2 + \nu \mathcal{F}(f). \tag{1}$$

Here $\mathcal{F}(f)$ denotes some quadratic functional of f which will be specified below. The first term in (1) is denoted as *data fidelity term* which is responsible for approximating the data while the second term requires f to be in some smoothness space which is a subspace of $L_2(\Omega) \cap C^0(\overline{\Omega})$, therefore called *regularization term*. Finally, the parameter $\nu \geq 0$ balances these two requirements.

We will construct an expansion of the minimizer f of (1) of the form

$$f(x) = \sum_{\lambda \in \Lambda} d_\lambda \, \psi_\lambda(x), \quad x \in \overline{\Omega}. \tag{2}$$

Among the various possibilities which suggest themselves for the selection of the expansion functions ψ_λ, such as B- or box splines on uniform or nonuniform grids, trigonometric functions, radial basis functions, or the different classes of wavelets or multiwavelets, we want to choose those which best match the following 'wish list'. First, as the data are presumably large and irregularly spaced and as they may represent a nonsmooth function and may contain outliers, that is, false data, we insist on ψ_λ with a possibly small support in order to be able to localize the representation. Second, as we are envisaging iterative solution methods for the least squares problem arising from (1) in view of a large N, the ψ_λ should give rise to favorable spectral condition numbers of these systems. Third, we wish to devise a unified method for very different sets of data. Hence, approaches based, e.g., on triangulations of X as an auxiliary step or based on interpolating P are not considered, partly also due to the fact that one would have to adapt these ingredients repeatedly whenever an outlier in the data is identified. Fourth, the approximation scheme should allow for a data-driven multiscale approximation, allowing very different resolutions in the data and final error depending only on the data. In this context, the expansion coefficients d_λ should have some meaning in the sense that large values indicate important contributions and small coefficients are negligible. Fifth, the functions ψ_λ should in some sense characterize the smoothness space enforced by the functional $\mathcal{F}(f)$.

In view of this 'wish list', we propose to use as building blocks for ψ_λ tensor products of certain boundary-adapted spline-prewavelets from [57], shortly called 'wavelets' here, which are a special case of the construction in [13], see, e.g., [12] for a general introduction into wavelet theory. Their properties are listed below in Section 2. While these wavelets are defined on structured grids and the set of all these wavelets constitute a Riesz basis for $L_2(\Omega)$, our algorithm proposed below in Section 3 works with a subset constructed in an adaptive coarse-to-fine fashion in the course of the algorithm. Thus, the index set Λ employed in the representation (2) actually turns out to be a lacunary set of all possible indices so that the set $\{\psi_\lambda\}_{\lambda \in \Lambda}$

is a particular multiscale basis adapted to the distribution and vertical variety of the data. In this way, our scheme is of minimal complexity in the following sense. As we do *not* assume that the data lies on a particular uniform grid, we do not want to associate an (often artificially chosen) 'finest' grid Ω_J with the data. Therefore, any 'adaptive' method that requires in a first step to apply the Fast Wavelet Transform on such a uniform grid Ω_J followed by some thresholding (as it was done in [31] for data given on uniform grids) is in our sense not an appropriate strategy, since then the full complexity $\#\Omega_J$ of this grid would enter into the algorithm independent of the cardinality or distribution of the data points. A particular feature of our approach is that the selection of wavelet bases relevant for the approximation is independent of the particular P when compared to schemes based on, e.g., data dependent triangulations, see, e.g, [24]. Our scheme which was developed initially in [6] can be seen as one of the approaches on structured grids which are by many considered a universally applicable and well–established procedure in scattered data fitting, see, e.g., [30,33,34,36,38,39,46,48,53,54,58,66]. Multiscale data fitting with or without smoothing has been discussed widely in the literature. To name only a few, in [27] a coarse–to–fine strategy has been presented with hierarchical splines on structured grids which is suited for parameterized gridded data. In [46] scattered functional data is approximated by multilevel B–splines. For data fitting problems on unstructured grids, radial basis function methods provide good approximating properties [51], a rich theoretical characterization [3,50] and a mesh free approach, which appears quite natural when working with unorganized data [4]. However, corresponding data reduction strategies like thinning [25] or adaptive thinning [24] usually require the construction of some auxiliary triangulation in order to achieve a multiscale formulation of the problem, see also [26] or [42]. For data approximation using finite elements on unstructured grids, see [61]. In principle, our scheme allows also for approximating data in dimensions higher than $n = 3$ as long as storage permits. Interpolating multivariate scattered data in higher dimensions is a somewhat different task where radial basis functions or other meshless methods may be particularly useful, see, e.g., [3,50–52,56,62].

Our algorithm operates in a coarse–to–fine fashion solely governed by the distribution of the data and adaptively constructs least–squares approximations to the data. Followed by thresholding irrelevant wavelet coefficients, we achieve a representation of the data which is minimally sparse. We systematically exploit the concept of adaptivity which is by now a well–established and powerful methodology to finding sparse representations of large–scale scientific problems.

The remainder of the material is organized as follows. In the next section, we recall the main ingredients of wavelet bases which will be used here. Section 3 describes the adaptive data fitting algorithm followed by addressing issues of numerical performance, regularization and robust treatment of outliers in the data in Sections 4 through 6. We conclude with an example from geophysics where the goal is to reconstruct a harmonic function which is done by including the Laplacian into the functional $\mathcal{F}$ in (1).

Figures in this paper are black-and-white. Coloured graphics are available at http://www.iam.uni-bonn.de/~kunoth/papers/papers.html.

2. Wavelet Bases

The main ingredient of the derivation of adaptive wavelet methods for the problem class described above are appropriate wavelet bases for a Hilbert space $H \subseteq L_2(\Omega)$ living on a bounded domain $\Omega \subset \mathbb{R}^n$. A single multivariate wavelet function will be indexed by a parameter $\lambda = (j, \mathbf{k}, \mathbf{e})$ which encodes different information such as the *resolution level* or *scale* $|\lambda| := j$, the *spatial location* $\mathbf{k}$, and the type $\mathbf{e} \in \{0,1\}^n$ of wavelet for $n > 1$. This accounts for the fact that in the multivariate case we will be using tensor products, see, e.g., [18,21,57]. Since the domain Ω is bounded, there is a coarsest resolution level denoted by j_0. Functions with index $\mathbf{e} = \mathbf{0}$ only occur on the coarsest level; they are *scaling functions* and consist of tensor products of B-splines. Functions with index $\mathbf{e} \neq \mathbf{0}$ represent detail information of higher frequencies. The infinite set of all possible indices will be denoted by $\mathit{I\!I}$. We call the collection of functions

$$\Psi := \{\psi_\lambda : \lambda \in \mathit{I\!I}\} \subset H \tag{3}$$

a *wavelet basis* provided it has the following properties:

Riesz basis property (R): Ψ constitutes a *Riesz basis* for H: every $v \in H$ has a unique expansion in terms of Ψ,

$$v = \sum_{\lambda \in \mathit{I\!I}} v_\lambda \, \psi_\lambda =: \mathbf{v}^T \Psi, \qquad \mathbf{v} := (v_\lambda)_{\lambda \in \mathit{I\!I}}, \tag{4}$$

and its expansion coefficients satisfy a *norm equivalence* of the following form. There exist finite positive constants $c_H \leq C_H$ such that for any $\mathbf{v} = \{v_\lambda : \lambda \in \mathit{I\!I}\}$ one has

$$c_H \|\mathbf{v}\|_{\ell_2(\mathit{I\!I})} \leq \|\mathbf{v}^T \Psi\|_H \leq C_H \|\mathbf{v}\|_{\ell_2(\mathit{I\!I})}. \tag{5}$$

This means that wavelet expansions induce isomorphisms between the function space H and the sequence space $\ell_2(\mathit{I\!I})$. For the applications considered here, H will be a Sobolev space $H^\alpha(\Omega)$ of smoothness $\alpha \geq 0$. A particular feature of wavelet bases is that such norm equivalences hold, in fact, for a whole range of Sobolev spaces $\alpha \in (-\tilde{\gamma}, \gamma)$ depending on the smoothness of the bases and the cancellation properties (CP) mentioned below, with an appropriate scaling of the expansion coefficients by a diagonal matrix, see, e.g., [18].

Locality (L): The functions ψ_λ are compactly supported: for each $\lambda \in \mathit{I\!I}$ one has

$$\operatorname{diam}(\operatorname{supp} \psi_\lambda) \sim 2^{-|\lambda|}. \tag{6}$$

Cancellation Property (CP): There exists an integer $\tilde{m}$ such that

$$\langle v, \psi_\lambda \rangle \lesssim 2^{-|\lambda|(n/2+\tilde{m})} \, |v|_{W_\infty^{\tilde{m}}(\operatorname{supp} \psi_\lambda)}, \tag{7}$$

where $\langle \cdot, \cdot \rangle$ denotes the dual form between H and its topological dual H' and $|v|_{W_\infty^{\tilde{m}}(\operatorname{supp} \psi_\lambda)}$ is the seminorm containing all weak derivatives of v of order $\tilde{m}$ in $L_\infty(\operatorname{supp} \psi_\lambda)$.

Throughout this paper, the relation $a \sim b$ expresses $a \lesssim b$ and $a \gtrsim b$, where the latter relation means that b can be estimated from above by a constant multiple of a independent of all parameters on which a or b may depend.

It is known that (CP) holds for Ψ if Ψ has a *dual* or *biorthogonal* basis $\tilde{\Psi} \subset H'$, i.e.,

$$\langle \psi_\lambda, \tilde{\psi}_\mu \rangle = \delta_{\lambda,\mu}, \qquad \lambda, \mu \in \mathbb{I}, \tag{8}$$

which is a Riesz basis for H' and has members $\tilde{\psi}_\lambda$ such that the multiresolution spaces $\text{span}\{\tilde{\psi}_\lambda : |\lambda| < j\}$ contain all polynomials of order $\tilde{m}$. By duality arguments, one can show that (5) implies

$$C_H^{-1} \|\tilde{\mathbf{v}}\|_{\ell_2(\mathbb{I})} \leq \|\tilde{\mathbf{v}}^T \tilde{\Psi}\|_{H'} \leq c_H^{-1} \|\tilde{\mathbf{v}}\|_{\ell_2(\mathbb{I})} \tag{9}$$

for any $\tilde{v} = \tilde{\mathbf{v}}^T \tilde{\Psi} \in H'$, see, e.g., [14,18]. Thus, given such a pair $\Psi, \tilde{\Psi}$, the expansion coefficients in (9) are of the form $\tilde{v}_\lambda = \langle \tilde{v}, \psi_\lambda \rangle$. Recall that property (CP) means that integration against a wavelet acts like taking an $\tilde{m}$th order difference which annihilates the smooth part of $\tilde{v}$. Thus, the expansion coefficients are small where $\tilde{v}$ is smooth.

Together, properties (R), (L) and (CP) allow for the proof of numerous theoretical statements ranging from optimal condition number estimates of elliptic operators in wavelet bases [19] to convergence results of adaptive wavelets methods for linear and nonlinear variational problems [15,16] and PDE-constrained elliptic control problems [20,44]. At the same time, these properties still allow to work with piecewise polynomials which is an essential advantage for numerical computations. Among the various constructions of different classes of wavelets, it has turned out that for data fitting problems tensor products of the piecewise linear boundary–adapted B–spline pre–wavelets from [57] are a good compromise for meeting the 'wish list' from Section 1 since they have the additional property

Semi-Orthogonality (O): with respect to $L_2(\Omega)$, i.e., for $|\lambda| \neq |\mu|$ one has

$$\int_\Omega \psi_\lambda(x)\, \psi_\mu(x)\, dx = 0. \tag{10}$$

In this case, the dual basis $\tilde{\Psi}$ will have global support which will, however, not affect the computations since we never employ the dual basis. The semi–orthogonality between different resolution levels will be very useful for the fast iterative solution of the linear systems appearing in the scattered data context.

3. Basic Coarse-To-Fine Algorithm

Our algorithm described in this section which was originally developed in [6] is designed in such a way that we can handle very large data sets. Our goal is to work with as few wavelets as possible, taking into account only those basis functions which are necessary to resolve the data given by P. As only finitely many data are to be approximated, we expect that only finitely many indices from $\mathbb{I}$ are essential for the representation of the minimizer f of (1). For a given cloud of points P, the determination of an appropriate index set $\Lambda \subset I\!I$, $\#\Lambda < \infty$, is therefore the central issue. On the one hand, Λ should be rich enough to be able to represent all features of P while at the same time any *overfitting* as well as processing redundant data should be avoided. Our basic adaptive coarse-to-fine algorithm will be described for the case that P is not noisy and does not contain outliers, and we describe it for the case of the least squares functional (1) without regularization, i.e., for $\nu = 0$.

Before giving the technical description of our scheme, it may be helpful to list several properties. The choice of relevant indices in the index set Λ is determined in such a way that it progressively adapts to the data, without introducing at any point a 'finest' artificial full grid. The finest resolution is only determined by the data. In each step of the scheme, a least squares approximation is computed by solving the normal equations by conjugate gradient (CG) iterations (this issue is discussed in detail below in Section 4). After this step, an appropriate thresholding on the wavelet coefficients guarantees that only relevant wavelet coefficients are kept.

The scheme operates in a coarse-to-fine fashion as follows. On the coarsest level j_0, all indices contained in $I\!I$ are selected; these indices correspond to the generator functions on this level. Thanks to the scale localization property (L), we progressively construct 'from coarse to fine' a sequence of (nested) index sets $\Lambda_j \subset I\!I$, $j \geq j_0$, in such a way that we only consider those wavelets whose supports contain enough data points, parameterized by some number q. From a set of indices Λ_j we construct a new set Λ_{j+1} by considering only children of previously chosen wavelets, with the additional requirement that those children have to have at least q points in their support. In this way, we generate a tree as data structure for Λ. Denote by $\delta(\Lambda_j)$ the indices of newly allowable candidates and by $S_q(\delta(\Lambda_j)) \subset \delta(\Lambda_j)$ their subset of indices for wavelets having at least q points in their support. After an appropriate index set $\tilde{\Lambda}_{j+1}$ is found, the minimum of the least squares functional

$$\mathcal{J}(f_{j+1}) := \sum_{i=1}^{N} \left(z_i - \sum_{\lambda \in \tilde{\Lambda}_{j+1}} d_\lambda^{j+1} \, \psi_\lambda(x_i) \right)^2 \tag{11}$$

is computed by CG iterations on the normal equations. Since we aim at finding a sparse representation of P, we afterwards make in each step use of the norm equivalence (5) by *thresholding* to discard those wavelet coefficients which are smaller than a certain threshold ε.

ALGORITHM: FIX q AND A THRESHOLD ε AND DO FOR EACH $j = j_0, \ldots$

(I) DETERMINE $\delta(\Lambda_j)$ AND SELECT $S_q(\delta(\Lambda_j))$,
(II) IF $S_q(\delta(\Lambda_j)) = \emptyset$ STOP, ELSE SET $\tilde{\Lambda}_{j+1} := \Lambda_j \cup S_q(\delta(\Lambda_j))$,
(III) COMPUTE $\{d_\lambda^{j+1}\}_{\lambda \in \tilde{\Lambda}_{j+1}}$ BY MINIMIZING $\mathcal{J}(f_{j+1})$ DEFINED IN (11),
(IV) DETERMINE $\Lambda_{j+1}^{\varepsilon} := \{\lambda \in \tilde{\Lambda}_{j+1} \setminus \Lambda_j : |d_\lambda^{j+1}| \geq \varepsilon\}$,
(V) IF $\Lambda_{j+1}^{\varepsilon} = \emptyset$ STOP, ELSE SET $\Lambda_{j+1} = \Lambda_j \cup \Lambda_{j+1}^{\varepsilon}$ AND $j = j + 1$.

The algorithm terminates after finitely many levels at a *finest level* J with an index set Λ_J which only depends on the density and the distribution of the data. Note that J is neither user–specified nor previously chosen. In view of (L), for given finite data X the algorithm always terminates since either $S_q(\delta(\Lambda_j))$ or $\Lambda_{j+1}^{\varepsilon}$ are at some point empty. The algorithm grows exactly one level in each step so that the highest resolution level contained in the index set Λ_J is J. The final index set Λ_J has tree structure, and the representation is sparse in the sense that it provides a least squares approximation to the data with minimal degrees of freedom.

After the index set for possible wavelet candidates $\Lambda = \tilde{\Lambda}_{j+1}$ in (2) is fixed, the necessary conditions for minimizing $\mathcal{J}$ in (11) yield the *normal equations*

$$M\mathbf{d} = \mathbf{b}, \qquad M := A^T A, \quad \mathbf{b} := A^T\mathbf{z}, \tag{12}$$

where $\mathbf{d}$ and $\mathbf{z}$ are the column vectors for the coefficients and the given values in Z, respectively. The *observation matrix* $A \in \mathbb{R}^{N\times(\#\Lambda)}$ has entries $A_{i,\lambda} = \psi_\lambda(x_i)$ so that the *cross product matrix* $M = A^T A$ and right hand side $\mathbf{b} = (b_\lambda)_{\lambda\in\Lambda}$ are given as

$$M_{\lambda,\lambda'} = \sum_{i=1}^{N} \psi_\lambda(x_i)\psi_{\lambda'}(x_i), \qquad b_\lambda = \sum_{i=1}^{N} z_i\,\psi_\lambda(x_i). \tag{13}$$

Note that M can be interpreted as a discretized Gramian matrix since the $M_{\lambda,\lambda'}$ have the form of a discretization of the L_2 inner product. It is not immediately clear that M resulting from the above algorithm is indeed invertible. Sufficient conditions guaranteeing A to have full rank, which blend into the above scheme and which are based on the Theorem of Schoenberg and Whitney for B–Splines are provided in [5].

As an example, the performance of ALGORITHM is tested on the 3D data set from [5] displayed on the left of Fig. 1. The reconstruction with the full coarsest grid $j = 3$ provides already a good reconstruction in smooth areas but fails to do so in the neighborhood of the ring where high gradients occur. New wavelets on level $j = 4$ are only chosen in this area and, as can be seen on the right of Fig. 2, provide a more accurate reconstruction. Further examples are displayed in [5,43].

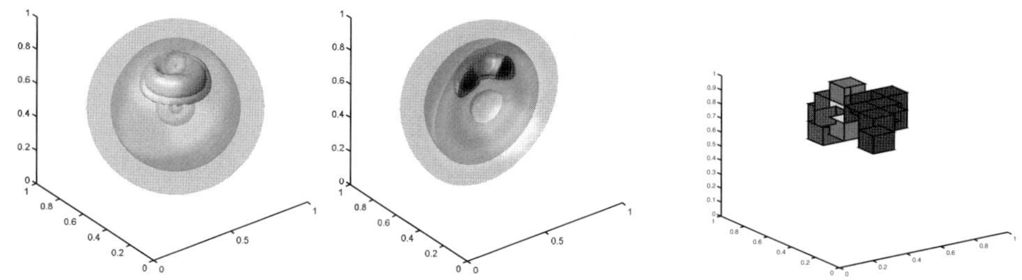

Fig. 1. Left: Original data set consisting of 41^3 points displaying the result of a simulation of the two-body distribution probability of nucleon-nucleon correlations for ^{16}0 (Courtesy of VolVis and SFB 382 of the DFG). Middle: Isosurface at $y = 0.5$. Right: Distribution and size of wavelet coefficients for level $j = 3$ and wavelet type $\mathbf{e} = (1, 1, 1)$; coefficients range from small (blue/dark) to large (red/light).

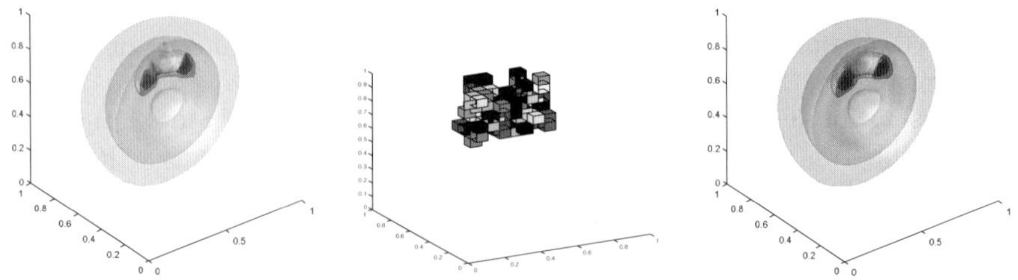

Fig. 2. Left: Isosurface of reconstruction at level $j = 3$. Middle: Distribution and size of wavelet coefficients for level $j = 4$ and wavelet type $\mathbf{e} = (1,1,1)$. Right: : Isosurface of reconstruction at level $j = 4$ with 2107 new degrees of freedom as opposed to the 35937 possible ones.

A further interesting application of the above least squares data fitting approach based on adaptive wavelets for the approximate evaluation of *nonlinearities* is studied in [5,10]. There the data points P are chosen depending on the type of nonlinearity to provide a reconstruction which is comparable to the results based on adaptive quadrature [1]. This scheme is then used to compute the solution of Burgers' equation by means of an adaptive wavelet viscosity method, see [5,10].

4. Numerical Solution

A crucial step in our adaptive ALGORITHM is the computation of the minimizer of the least squares functional (11). Such least squares problems are often solved by computing a QR decomposition of A. In the present context, we prefer to solve the resulting normal equations (12) for the following reasons.

First, for very large data sets as the one considered in Fig. 1, we usually get $\#\Lambda \ll N$ so that we need much less space for storing M which is of size $\mathbb{R}^{(\#\Lambda)\times(\#\Lambda)}$ than for A, where the amount of data N enters as the number of rows. In this sense, forming the normal equations entails in fact some *data compression* effect, as a large N only appears as the size of the sums in the entries of $M_{\lambda,\mu} = \sum_{i=1}^{N} \psi_\lambda(x_i)\psi_\mu(x_i)$ and not in the size of the matrix M.

Second, for the particular wavelet basis we use, the spectral condition number of M is moderate. This may be attributed to the fact that M is a *discretization* of the Gramian matrix with entries $(\psi_\lambda, \psi_\mu)_{L_2(\Omega)}$ which has in view of the L_2-stability of the wavelet basis (5) a uniformly bounded condition number independent of the finest level of resolution. For growing N, the entries of M 'approximate' the entries of the Gramian even better, meant as a trend. (Of course, a concrete approximation estimate between the matrix entries would depend on the distribution of points.) Consequently, iterative methods like a conjugate gradient (CG) scheme on M are expected to reduce the error in each step by a fixed amount and the number of iterations to reach a prescribed accuracy are envisaged to be very moderate. This has been confirmed in all numerical experiments we have conducted. We have also compared in [7] approximation errors of our results with the normal equations with approximation errors for a least squares solution computed using QR decomposition of A and found only negligible differences, while the iterations took much less time for the normal equations. Further details on the efficient numerical solution of least squares problems may be found in [2].

Third, in view of the tree structure generated by the adaptive algorithm, the matrix M on level j is a submatrix of M on level $j+1$; the latter only contains additional entries with respect to wavelet functions ψ_λ for $|\lambda| = j+1$. Thus, one expects a *nested iteration* algorithm which takes the solution vector $\mathbf{d}$ from level j, padded with zeroes, as initial guess for the CG iteration on level $j+1$ to perform particularly well. In addition, the 'almost' orthogonality between levels in M which is a result of the semi-orthogonality property (O) provides a strong decoupling between levels, so that a solution on level j is on its low frequencies only marginally improved by the iteration on level $j+1$.

We illustrate these particular features of our wavelet basis by comparing it with another standard multiscale basis, the *hierarchical basis* which has proven to be very popular when solving elliptic PDEs in variational form [65]. In one dimension, it consists of the scaled versions of the linear B-spline, also called 'hat function', where on higher levels the support decreases. This basis is not stable with respect to L_2 and, therefore, the spectral condition number of the Gramian grows like 2^j [5]. Comparisons between the hierarchical B–Spline basis and a wavelet basis were made in [32] in the context of geometric modeling. In computer graphics, a hierarchical B–spline approach was studied in [27]. As an example for the different performance of the wavelet basis used here and the hierarchical basis, Fig. 3 from [5,6] is displayed, where these issues have been discussed in detail. Very few CG iterations suffice for the wavelet basis to reach the saturation error 10^{-2}.

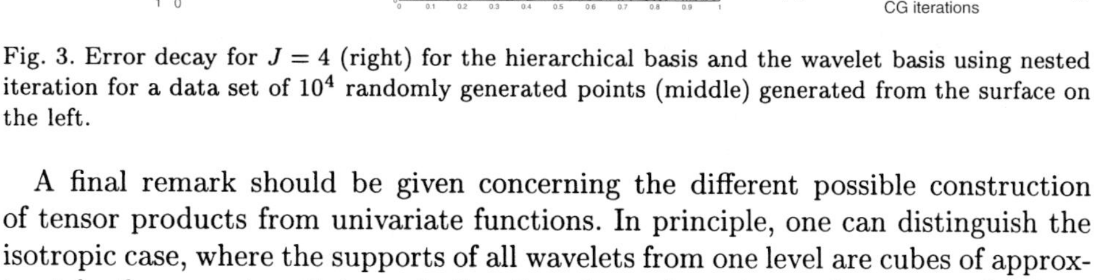

Fig. 3. Error decay for $J = 4$ (right) for the hierarchical basis and the wavelet basis using nested iteration for a data set of 10^4 randomly generated points (middle) generated from the surface on the left.

A final remark should be given concerning the different possible construction of tensor products from univariate functions. In principle, one can distinguish the isotropic case, where the supports of all wavelets from one level are cubes of approximately the same length in each direction, from the anisotropic case, see [5] for a precise description of the construction. In the numerical experiments also reported there, the spectral properties of M using one or the other construction are for the data fitting problem very similar and do not influence much the behavior of the CG scheme on the normal equation.

5. Multilevel Regularization

Next we extend the least squares fitting strategy to include into the objective functional a Sobolev seminorm with positive smoothness index α as a regularizing term in the least squares functional (1) to enforce a smooth approximation. This classical approach for $\alpha \in \{1,2\}$ was studied, e.g., in [23] for general splines, [55] for splines with free knots, [34] for hierarchical splines, [37] for splines on triangulations, or [32] in a wavelet reformulation of the spline problem. Here we choose more generally $\mathcal{F}(f) := \|f\|^2_{H^\alpha(\Omega)}$ so that we consider

$$\mathcal{J}(f) = \sum_{i=1}^{N}(z_i - f(x_i))^2 + \nu\|f\|^2_{H^\alpha(\Omega)}, \tag{14}$$

where the not necessary integral $\alpha > 0$ parameterizes the smoothing and $\nu > 0$ balances the fidelity of the data and the smoothness. In view of the statements made after (5), the wavelet formulation allows to represent functions in Sobolev norms in terms of sequence norms of weighted coefficients from their wavelet expansions. Accordingly, following [11], we replace $\|\cdot\|^2_{H^\alpha(\Omega)}$ by the corresponding sequence norm on the right hand side of (14),

$$\left\|\sum_{j,\mathbf{k},\mathbf{e}} d_{j,\mathbf{k},\mathbf{e}}\,\psi_{j,\mathbf{k},\mathbf{e}}\right\|^2_{H^\alpha(\Omega)} \sim \sum_{j,\mathbf{k},\mathbf{e}} 2^{2\alpha j}|d_{j,\mathbf{k},\mathbf{e}}|^2. \tag{15}$$

The minimization of (14) using this sequence norm is now equivalent to the solution of the normal equations

$$(M + \nu R)\mathbf{d} = \mathbf{b}. \tag{16}$$

Here R is a diagonal matrix with entries

$$R = \operatorname{diag}(2^{2\alpha j_0}I, \ldots, 2^{2\alpha J}I) \tag{17}$$

if J is the highest level of approximation and the sizes of the identity matrices depend on the cardinality of Λ_j. This representation of the smoothing term in the wavelet context has some interesting properties. First, the different dyadic scales are decoupled which means that the weak decoupling between levels in M is maintained, and the effect of the regularization term boils down to a penalization of the higher frequencies. Thus, the parameter ν controls the balance between fidelity and regularization while α controls the relative penalization across scales. Second, since (5) holds for the whole scale of fractional Sobolev spaces $\{H^\alpha(\Omega)\}_{\alpha\in[0,\gamma)}$, one has easy access to the entire scale which reduces to a simple diagonal scaling in the wavelet framework by means of the diagonal matrix R. Third, the formulation is independent of the wavelet family. Of course, the smoothness of the wavelets imposes a limit on the maximal value of α for which (15) holds.

We have now two parameters in (14), the smoothness index α and the weight parameter ν to balance the data approximation and the smoothness part which can be used to adjust as best as possible to the given data. An example of an application from photogrammetry with fixed $\alpha = 4$ and $\nu = 0.01$ is provided in Figs. 4 and 5. It should be mentioned that for the linear pre-wavelets we employ the norm equivalence (15) does not hold for this high value of α. Nevertheless, the reconstruction in Fig. 5 gives the best result for this case.

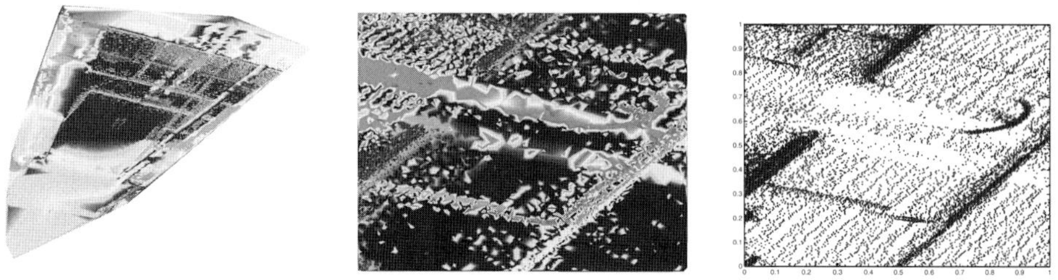

Fig. 4. Left: Original data, a 3D point set consisting of 330.000 points of an industrial site (taken by Leica Cyrax 2500, Prof. Staiger, GH Essen). Middle: Vertical view of a section of the original data. Right: Sampling geometry of this section.

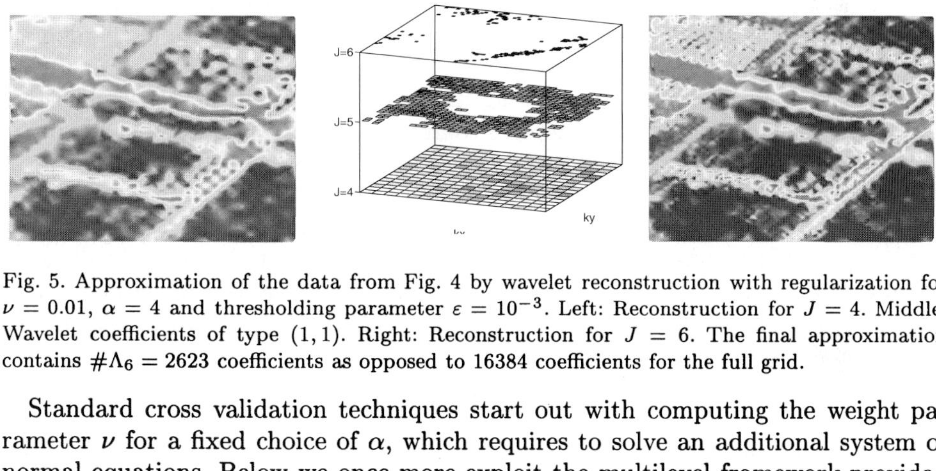

Fig. 5. Approximation of the data from Fig. 4 by wavelet reconstruction with regularization for $\nu = 0.01$, $\alpha = 4$ and thresholding parameter $\varepsilon = 10^{-3}$. Left: Reconstruction for $J = 4$. Middle: Wavelet coefficients of type $(1, 1)$. Right: Reconstruction for $J = 6$. The final approximation contains $\#\Lambda_6 = 2623$ coefficients as opposed to 16384 coefficients for the full grid.

Standard cross validation techniques start out with computing the weight parameter ν for a fixed choice of α, which requires to solve an additional system of normal equations. Below we once more exploit the multilevel framework provided by wavelets and introduce a multilevel version of the cross validation procedure. This scheme turns out to be both relatively inexpensive from a computational point of view as well as adjusting well to smoothness and to localization effects.

Let us first describe the classical situation where the smoothness parameter α is fixed. Different ways to select an appropriate regularization parameter ν in (14) were proposed, where one of the most popular choices are cross validation methods, see, e.g., [60]. Particularly, in the *Generalized Cross Validation* (GCV) method one computes ν as the minimizer of the *GCV potential* with respect to the Euclidean norm $\|\cdot\|$, that is,

$$GCV(\nu) := \frac{\|I - H(\nu)\mathbf{z}\|^2}{(\mathrm{tr}(I - H(\nu)))^2}, \tag{18}$$

where the *influence matrix* $H(\nu)$ is defined as $H(\nu) := A(M + \nu R)^{-1}A^T$. The influence matrix is never computed explicitly, since only the application of $H(\nu)$

to a vector needs to be carried out. This in turn requires the solution of a linear system with matrix $M + \nu R$ which is of a similar type as in (16). The trace in the denominator of (18) is usually estimated stochastically as $\text{tr}(B) = \mathbf{u}^T B \mathbf{u}$ where $\mathbf{u}$ is a random vector with entries 1 and -1 and B is a quadratic symmetric matrix, see [41]. Consequently, the evaluation of $GCV(\nu)$ for each ν requires the solution of two linear systems of the type (16). The total computation time then depends on the number of GCV evaluations which are needed to find the minimum of $GCV(\nu)$. In [63] an automatized computation of a GCV involving up to 15 evaluations of the functional is reported, together with the CPU time which is needed in the solution of a system of normal equations with penalization for a given ν. We have displayed in [7] a series of bivariate examples where we have applied these GCV techniques and where the solution of each linear system costed less than one second.

The GCV technique yields an optimal ν in the sense that it minimizes the variance under the assumption that the data has been corrupted by white noise, see, e.g., [59]. This theory developed for splines immediately carries over to the wavelet case [5]. Moreover, as shown in [7] the underlying heuristics can also be applied successfully to other data fitting approaches which require some regularization such as for data sets with a highly varying point distribution or for data with holes in the domain where typically overfitting artifacts occur.

The main issue in [7] is the following multilevel version of the GCV algorithm used within the context of the adaptive coarse-to-fine ALGORITHM. The idea is to simultaneously use the GCV with the hierarchical growth of the wavelet tree as the levels become higher. We have observed above that the wavelet representation penalizes the different dyadic levels separately with the same weight $2^{2\alpha j}$. In view of the structure of the diagonal matrix R given in (17), we prescribe instead a diagonal penalizing matrix with the same values $\nu_j \geq 0$ for all entries of each level j, where the ν_j are computed independently following a GCV criterion. Now at each level j the normal equations attain the form

$$(A_{\Lambda_j}^T A_{\Lambda_j} + R_j)d = A_{\Lambda_j}^T z, \tag{19}$$

where the diagonal matrix R_j is defined component–wise as $(R_j)_{\lambda,\lambda'} := \delta_{\lambda,\lambda'}\nu_j$ for some set of scalars $\{\nu_{j'}\}_{j_0 \leq j' \leq j}$. These scalars are computed inductively at each level following a GCV criterion. At level j_0, we define the influence matrix

$$H(\nu_{j_0}) := A_{\Lambda_{j_0}}(A_{\Lambda_{j_0}}^T A_{\Lambda_{j_0}} + R_{j_0})^{-1} A_{\Lambda_{j_0}}^T,$$

and ν_{j_0} is obtained by minimization of the corresponding GCV potential (18). At any subsequent level j the influence matrix is defined as

$$H(\nu_j; \nu_{j_0}, \ldots, \nu_{j-1}) := A_{\Lambda_j}(A_{\Lambda_j}^T A_{\Lambda_j} + R_j)^{-1} A_{\Lambda_j}^T \tag{20}$$

to have ν_j as only variable since the $\nu_{j'}$ from previous levels $j_0 \leq j' < j$ are already computed. ν_j is then determined by minimizing (18). Recall that the influence matrices are never computed explicitly as they only need to be applied to a vector. The new penalizing term can no more be interpreted as stemming from a smoothing term $\|f\|_{H^\alpha(\Omega)}^2$ since no relation between the penalizing parameters $\{\nu_j\}_{j \geq j_0}$ is prescribed. Nevertheless, this approach offers some interesting advantages. It is easily built into the coarse–to–fine growth of the tree. One can attain higher flexibility for the smoothing effect. Moreover, overfitting artifacts are typically localized in scale which enables us to disentangle the different scales. Last, the method is much cheaper as the GCV method with both α and ν as free parameters from a computational point of view.

As illustration, we apply the multilevel GCV algorithm to the data set in Fig. 6 obtained from [49] which corresponds to a bathymetrical study of part of the sea floor of the Dominican Republic. As can be seen from the left image, measurements of the sea floor depth are irregularly distributed, forming lines, clusters and holes. The central plot shows a visualization of the depth of the full set of data points using piecewise linear interpolation. For the classical GCV, there is no obvious way to predict which α may give a good reconstruction. In contrast, our multilevel GCV algorithm gives a sufficiently good reconstruction in the right plot.

Fig. 6. Left: Sampling geometry with $N = 10732$ points. Middle: Piecewise linear interpolation. Right: Reconstruction with multilevel GCV.

6. Treatment of Outliers

Next we discuss the issue how to treat *outliers* in the data where firstly the definition of an outlier has to be clarified. For the 'human eye', the presence of an outlier must create a 'cusp' in the approximating function; an outlier is recognized as an artifact which is extremely well localized in space *and* frequency. For this reason, the wavelet framework may be the method of choice for outlier detection. The method introduced in [5,8] exploits the least squares approach described above by building outlier detection into it in the following way. The result of the data fitting procedure is the construction of a function f in the form (2) which represents the data. Thus, it seems natural to transfer the problem of outlier detection from the raw data point of view to the approximation f. If f catches the data features, the presence of an outlier must consequently create a local *jump* (or oscillation or peak) in f. Otherwise, if f is not affected by the outlier, the scheme does not have to take the outlier into account. Thus, we face the issues how to define these local jumps in a rigorous mathematical way which is at the same time easy to implement, and how can we distinguish jumps created by outliers and jumps contained in the data? The left plot in Fig. 7 illustrates four areas of different prototypes of outliers and various types of data which we need to be able to handle. In the leftmost area we see one point (x_{100}, z_{100}) which is definitely classifiable as an outlier. The second area represents a cusp really represented in the data, and points in this area should not be marked as outliers as they represent a high frequency phenomenon in the data and accidental removal of points in this area would eliminate significant information about this local structure. The data in the third area present a spatially located feature of high frequency. Here the frequency is lower than in the second area and more data points are involved in its local representation of the local structure. A removal of some points from this area would not necessarily eliminate significant information, as the remaining points would certainly reproduce the local features of the data. The point (x_{800}, z_{800}) in the fourth area is an outlier embedded into a highly energetic zone. These kind of points pose the hardest difficulties to outlier detection algorithms. On one hand, the neighborhood can mask the effect of the outlier. On the other hand, neighboring points carry significant information about the local structure on the data, and false removal in this area should be avoided by any means.

The wavelet representation of the data allows for a very practical answer to the questions raised above. Concerning the identification of jumps, there is a vast amount of literature, see, e.g., [17,40,45]. There the basic idea is that the presence of a jump is reflected by large wavelet coefficients of f. However, jumps caused by outliers cannot be distinguished from data–inherent jumps solely by inspection of the wavelet representation of the data. In fact, one has to analyze how individual points influence the wavelet representation. In the approach described below, this task can be easily performed after having constructed the approximation since all the information needed to perform this analysis has already been processed in the data structures. Without treating the outliers, our ALGORITHM produces the approximation shown in the middle plot in Fig. 7. Clearly the outliers produce undesired artifacts in the first and the fourth area.

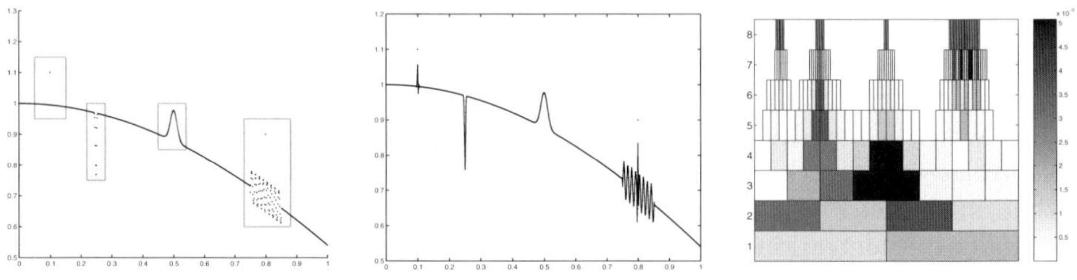

Fig. 7. Left: Synthetic data set with $N = 1000$ data points showing different types of outliers. Middle: Adaptive reconstruction of the data set with $\#\Lambda = 190$ wavelets. Right: Resulting wavelet coefficients of the approximation. The y–axis shows the resolution levels and the x–axis displays the spatial location of the wavelet coefficients. Larger wavelet coefficients are darker. Highest resolution level in Λ is $J = 8$.

After having carried out ALGORITHM, we have at our disposal the coefficients $\{d_\lambda\}_{\lambda\in\Lambda}$ of the constructed approximation f of the form (2) as the main product of the algorithm and an efficient encoding of the observation matrix $A = A_\Lambda$. The main computational effort lies in the construction of A, which implicitly contains a complete analysis of the data. We will exploit this in the outlier-finding approach by separating the outlier finding procedure into three steps. First, we *locate* in which area an outlier can be present by simple inspection of the coefficients $\{d_\lambda\}_{\lambda\in\Lambda}$, as the presence of outliers must cause large wavelet coefficients at high levels, see the right plot in Fig. 7. We then *extract* out of the original N points those contained in the support of all wavelets ψ_λ classified in the first step. Third, we test all the points (x_i, z_i) in the areas identified in the first two steps, accumulated in the subset $S \subset P$, based on a *merit criterion* $\omega(i)$ which serves as an 'outlierness indicator'. Points whose $\omega(i)$ is above a predefined threshold τ are discarded from the data set. Next we need to describe how to construct meaningful and computationally fast outlierness profiles $\omega(i)$.

The first criterion is a *global refitting criterion* defined as follows. Consider one of the points (x_i, z_i) for $i \in S$ and test if it is an outlier. The basis of this test is to measure to which extent its presence alters the shape of the approximation f which can be achieved in three steps. We first construct an approximation f to the whole data set. Then we construct an approximation to the whole set of data *except* the point (x_i, z_i) to be checked. In order to do so, we use the same configuration of wavelets which has been used to obtain the approximation to the whole data set. Thus, we compute

$$f^{[i]} := \arg \min_{g=\sum_{\lambda\in\Lambda} d_\lambda \psi_\lambda} \sum_{\ell\neq i} (z_\ell - g(x_\ell))^2 . \tag{21}$$

Then we compare the behavior of f and $f^{[i]}$ in the 'neighborhood' of x_i. The last step can be done in a natural way using the wavelet coefficients. We define the *influence set* of (x_i, z_i) starting from level j in the index set Λ as the subset of those λ which includes indices of all wavelets whose support contain x_i. We denote this neighborhood as

$$\Lambda_j^{[i]} := \{\lambda \in \Lambda,\ |\lambda| \geq j :\ x_i \in \text{supp}\ \psi_\lambda\}, \tag{22}$$

or shortly $\Lambda^{[i]}$. Then we compare the local behavior of f and $f^{[i]}$. The *local energy* of a function f by means of a weighted summation of a subset of its wavelet coefficients is defined as follows: for a function $g = \sum_{\lambda\in\Lambda} d_\lambda \psi_\lambda$ on Ω and a set $\Lambda' \subset \Lambda$, let

$$E_\alpha^{\Lambda'}(g) := \sum_j 2^{2j\alpha} \sum_{\mathbf{k},\mathbf{e}\in\Lambda'} |d_{j,\mathbf{k},\mathbf{e}}|^2. \tag{23}$$

Apparently this definition relies on the norm equivalence (15), compare [35] for the more general case of norm equivalences for Besov seminorms. If (x_i, z_i) is an outlier, in the neighborhood of x_i the local energy of $f^{[i]}$ should be much smaller than the local energy of f. We define the *merit profile* of (x_i, z_i) according to a *global criterion* as

$$\omega_{\text{global}}(i) := \log \left(\frac{E_\alpha^{\Lambda_j^{[i]}}(f)}{E_\alpha^{\Lambda_j^{[i]}}(f^{[i]})} \right). \tag{24}$$

In our experiments we found that a typical thresholding value τ should be in the order of magnitude of 1. Points (x_i, z_i) for which $\omega_{\text{global}}(i) \geq \tau$ are then classified as an outlier. Thus, in the proposed model we expect the presence of an outlier to cause a noticeable increase of the local energy.

Revisiting the synthetic data set from Fig. 7, we explore next how this method works for the different areas represented in the data. If we take the outlier (x_{100}, z_{100}) and compute the global approximations f and $f^{[100]}$, we obtain the wavelet coefficients displayed in Fig. 8. As expected, no difference is visible outside the indicated box in the upper left corner of the wavelet coefficients starting at level 4. The presence of the outlier really does act locally. This is shown in detail in Fig. 9. In

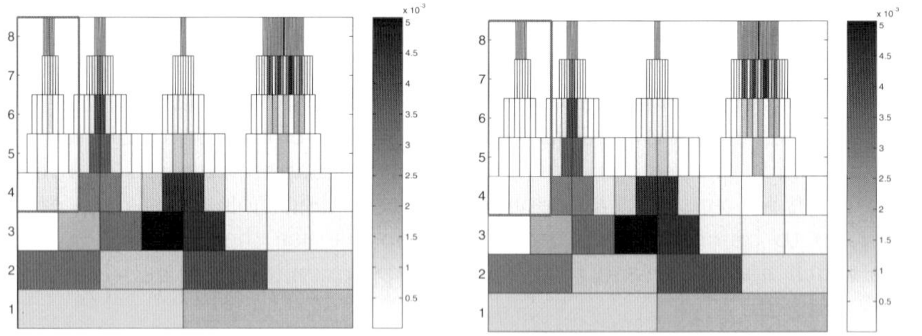

Fig. 8. Global removal criterion applied to (x_{100}, z_{100}). Wavelet coefficients of f (left) and of $f^{[100]}$ (right).

the left plot we see f (dashed line) and $f^{[100]}$ (solid line) in the neighborhood of (x_{100}, z_{100}), where f shows an oscillation created by the point (x_{100}, z_{100}). This is reflected by the coefficients of the wavelets in $\Lambda^{[100]}$. After eliminating (x_{100}, z_{100}) from the data to be fitted, the energy content of $\Lambda^{[100]}$ is practically empty. The criterion (24) here yields the value $\omega_{\text{global}}(100) = 7.02$. The same computation for the regular neighboring point (x_{102}, z_{102}) gives $\omega_{\text{global}}(102) = 0.0078$, as its removal does not critically vary the local energy.

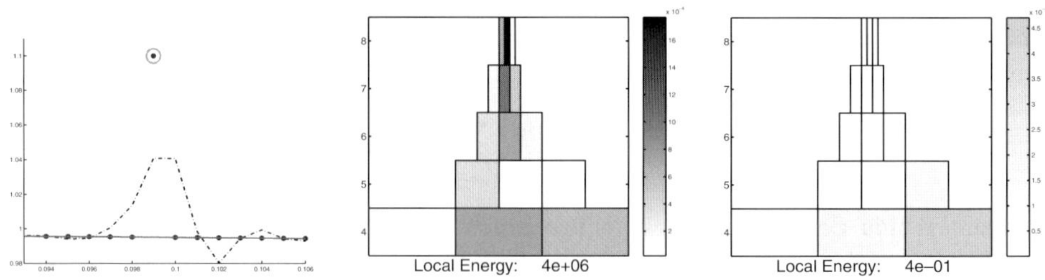

Fig. 9. Global removal criterion applied to (x_{100}, z_{100}). Left: Local view of f (dashed line) and $f^{[100]}$ (solid line). Middle: Coefficients of wavelets in $\Lambda^{[100]}$ for f. Right: Coefficients of wavelets in $\Lambda^{[100]}$ for $f^{[100]}$.

Although these results are convincing, the proposed strategy has an obvious drawback since it requires the computation of $f^{[i]}$ for every suspicious point (x_i, z_i): we have to construct and solve a different set of normal equations for every i. Fortunately, the structure of the system $M\mathbf{d} = \mathbf{b}$ allows to simplify the computations essentially by subtracting only a matrix of rank one as described in [8]. Moreover, the outlier detection strategy can be further sped up by considering *local energy criteria* which boils down to a *local reconstruction* of f and $f^{[i]}$, see the results displayed in Fig. 10. Point (x_{800}, z_{800}) is an outlier in the highly energetic environment in the fourth area from Fig. 7 while the neighboring point (x_{802}, z_{802}) is not. We dispense with a precise description of the local energy criteria here and refer to [8].

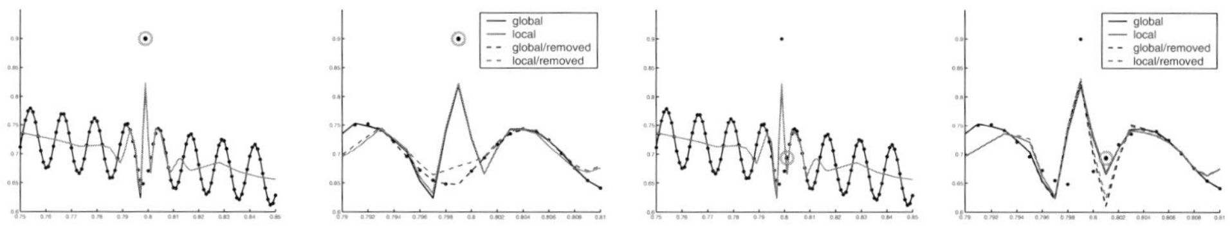

Fig. 10. Left two plots: Global and local reconstructions near (x_{800}, z_{800}). Local view of f and $f^{[800]}$ (left) and removal of (x_{800}, z_{800}) in global and local approximations. Right two plots: Global and local reconstructions near (x_{802}, z_{802}). Local view of f and $f^{[802]}$ (left) and removal of (x_{802}, z_{802}) in global and local approximations.

The above methodology can naturally be extended to deal in addition with noisy data and also works in higher spatial dimensions and for a large number of outliers, see [8], where also extensive tables displaying numerical results and percentages of outlier detection rates can be found.

As a final example, we analyze a geophysical data set taken again from [49]. The set includes 18634 points ordered in a square grid which is displayed in the left of Fig. 11. We add 1000 randomly distributed outliers to this data, yielding the data on the right. In the left plot of Fig. 12 we see the performance of our algorithm after a first run which eliminates 75% of the outliers, while the data eliminated by false detection does not appear to damage the reconstruction. A second run of the algorithm, that is, another iteration on the cleaned data results in the reconstruction on the right of the figure. As we start from a situation where the density of outliers has been reduced, further outliers that were previously conveiled by neighboring ones are now successfully detected.

7. Approximate Continuation of Harmonic Functions

Finally, first preliminary results from [9] on the following continuation problem arising in the context of the computation of the gravity potential are addressed. Assuming that the data set stems from a *harmonic function*, that is, a function $f : \overline{\Omega} \to \mathbb{R}$ satisfying $\Delta f = 0$ where Δ denotes the Laplacian, we study how well f can be reconstructed by computing the minimizer of

$$\mathcal{J}(f) := \sum_{i=1}^{N} (z_i - f(x_i))^2 + \nu \|\Delta f\|_{L_2(\Omega)}^2. \tag{25}$$

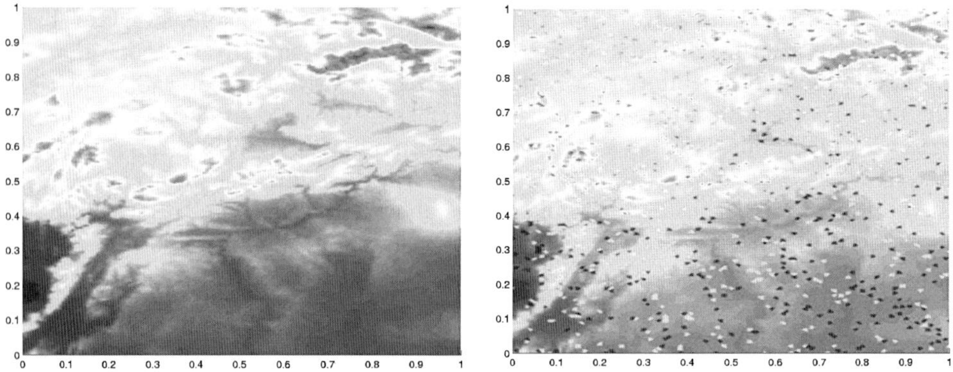

Fig. 11. Geophysical data set with $N = 18605$ data points. Left: Vertical view of original data. Right: Data set corrupted with 1000 outliers which amount to 5.6% of the data.

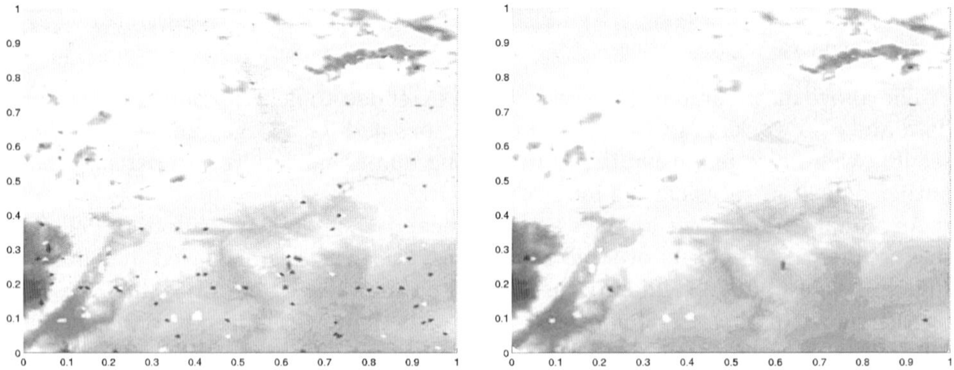

Fig. 12. Reconstruction of the geophysical data from Fig. 11 after robust cleaning. Left: Results after first run. Right: Approximation after second run.

Here $\mathcal{F}(f) := \|\Delta f\|^2_{L_2(\Omega)}$ in (1) is chosen. Usually such harmonic continuations are based on *interpolating* the data in terms of harmonic Ansatz functions like the spherical harmonics or wavelet-type approximations, see, e.g., [28,29]. These functions, however, are globally supported so that the corresponding linear equations stemming from the interpolation problem involve fully populated matrices. Consequently, their numerical solution is prohibitively expensive, in addition to the huge storage space required for the system matrix.

In our present approach, the idea is to employ tensor products of cubic B-splines based on a uniform grid of spacing 2^{-j} and enforce the harmonicity by the regularization term in (25). This was originally proposed in [47] where also first promising experiments based on higher order finite elements were conducted. The normal equations derived from (25) are $(M + \nu G)\mathbf{d} = \mathbf{b}$ where the matrix G represents the regularization term. The minimizer of (25) will be denoted by f_ν.

As an example, we consider

$$f(x,y) = \frac{1}{2}\left(e^{-10x}\sin(10y) + e^{-10(1-x)}\sin(10(1-y)) + 1\right) \tag{26}$$

on $\Omega = (0,1)^2$ from which a set of data P is extracted by sampling f on the boundary with respect to a sampling density $2^{-\ell}$ for $\ell \in \{6,7,8\}$ displayed in Fig. 13.

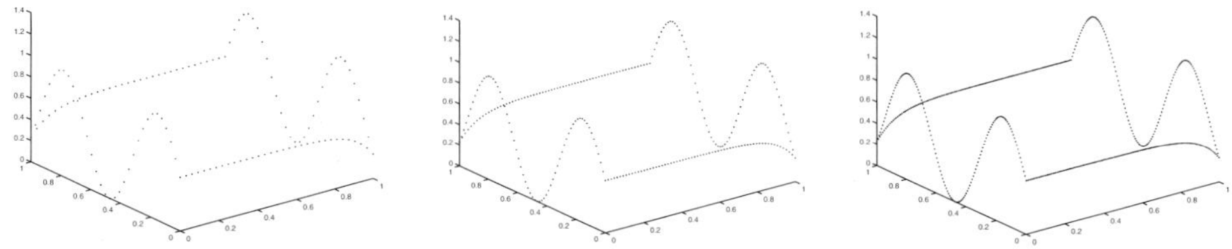

Fig. 13. Geometry of points for resolution $2^{-\ell}$, $\ell = 5, 6, 7$.

For tensor products of cubic B-splines on resolution 2^{-4}, the results for the reconstruction for data resolved at 2^{-5} are shown in Fig. 14. Naturally the reconstruction error is larger at the boundaries where the amplitude of f is also larger than in the interior.

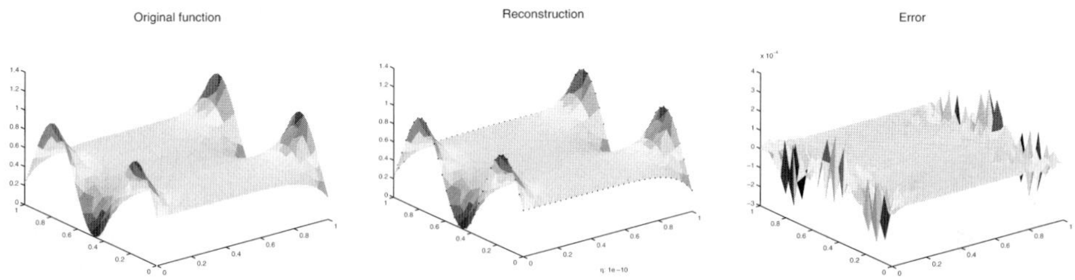

Fig. 14. Left: Original function f. Middle: Reconstruction f_ν for ν = 1.0e-10 and data points. Right: Reconstruction error $f - f_\nu$.

In Table 1 we list a number of numerical values for the reconstruction of data gridded at spacing 2^{-5} at the boundary. The first column shows the value of the regularization parameter ν in the (perhaps unnecessarily large) range $\nu \in [10^{-10}, 10^{10}]$. The second column contains the value for $||\Delta f_\nu||^2_{L_2(\Omega)}$ which can be computed exactly and which indicates how much f_ν deviates from being harmonic. In the third column we display the spectral condition number of $M_\nu := M + \nu G$. The next two columns show the errors E_P (partial) and E_T (total) between the exact and the reconstructed function on the given data on the boundary and on the whole domain, respectively, for resolution 2^{-5}, computed as $E_P = \sqrt{\frac{1}{N}\sum_{i=1}^N |f_\nu(x_i,y_i) - f(x_i,y_i)|^2}$, and correspondingly for E_T. The next column contains the values for $||f_\nu||^2_{L_2(\Omega)}$ while the last column shows the residual of the normal equations.

We see from the numerical results that a larger ν yields as expected a smaller value for $||\Delta f_\nu||^2_{L_2(\Omega)}$, at the expense of a larger spectral condition number of M_ν

Table 1. Numerical values for the reconstruction of the harmonic function f from (26).

ν	$\|\|\Delta f_\nu\|\|^2_{L_2}$	$\kappa(M_\nu)$	E_P	E_T	$\|\|f_\nu\|\|^2_{L_2}$	residual
1.00e–10	5.3930e-02	7.89e+10	4.7082e-06	1.1791e-06	2.8032e-01	4.87e-16
1.00e–08	5.3930e-02	7.89e+08	4.7082e-06	1.1791e-06	2.8032e-01	4.18e-16
1.00e–06	5.3921e-02	7.90e+06	4.7082e-06	1.1791e-06	2.8032e-01	4.85e-16
1.00e–04	5.3448e-02	9.20e+04	4.8271e-06	1.1872e-06	2.8032e-01	5.48e-16
1.00e–02	5.2648e-02	2.42e+04	1.2312e-05	1.8966e-06	2.8032e-01	1.19e-14
1.00e+00	4.7262e-02	9.65e+04	3.5240e-04	4.8718e-05	2.8006e-01	1.47e-12
1.00e+02	9.8129e-03	7.93e+06	7.5805e-03	1.2049e-03	2.7117e-01	1.48e-10
1.00e+04	5.1618e-05	7.71e+08	1.9241e-02	3.1836e-03	2.6478e-01	1.23e-08
1.00e+06	3.0610e-08	6.88e+10	2.1339e-02	3.6897e-03	2.8227e-01	1.34e-06
1.00e+08	4.0425e-12	6.84e+12	2.1427e-02	3.7169e-03	2.8344e-01	1.24e-04

and, correspondingly, a larger residual of the normal equations as well as a larger partial and total error of the data fit. Depending on the desired accuracy of the data fit or the harmonicity request, a value of ν between 1 and 100 may be a good compromise. Currently computations with real satellite data are under way.

Acknowledgements

This work was supported by the Deutsche Forschungsgemeinschaft, Grant KU 1028/7–1, and by the SFB 611, Universität Bonn. I would like to thank Daniel Castaño and Gabriela Constantinescu for their assistance during the preparation of this manuscript.

References

1. Barinka, A., Dahmen, W. and Schneider, R., Fast computation of adaptive wavelet expansions, preprint, 2004.
2. Björck, A., *Numerical Methods for Least Squares Problems*, SIAM, Philadelphia, 1996.
3. Buhmann, M., Radial basis functions: the state-of-the-art and new results, *Acta Numerica* **9** (2000), 1–37.
4. Carr, J. C., Beatson, R. K., Cherrie, J. B., Mitchell, T. J., Fright, W. R., McCallum, B. C. and Evans, T. R., Reconstruction and representation of 3D objects with radial basis functions, in: *Computer Graphics*, Proc. SIGGRAPH 2001 (L. Pocock et al., Eds.), ACM Press, New York, 2001, pp. 67–76.
5. Castaño, D., *Adaptive Scattered Data Fitting with Tensor Product Spline–Wavelets*, PhD Dissertation, Universität Bonn, 2005.

6. Castaño, D. and Kunoth, A., Adaptive fitting of scattered data by spline-wavelets, in: *Curves and Surfaces* (A. Cohen, J.-L. Merrien and L. L. Schumaker, Eds.), Nashboro Press, Brentwood, 2003, pp. 65–78.
7. Castaño, D. and Kunoth, A., Multilevel regularization of wavelet based fitting of scattered data – Some experiments, *Numer. Algorithms* **39** (1-3) (2005), 81–96.
8. Castaño, D. and Kunoth, A., Robust regression of scattered data with adaptive spline–wavelets, *IEEE Trans. Image Proc.*, to appear.
9. Castaño, D., Constantinescu, G., Kunoth, A. and Schuh, W. D., Approximate continuation of harmonic functions, manuscript, in preparation.
10. Castaño, D., Gunzburger, M. D. and Kunoth, A., An adaptive wavelet viscosity method for hyperbolic conservation laws, manuscript, in preparation.
11. Chambolle, A., DeVore, R. A., Lee, N.-Y. and Lucier, B. J., Nonlinear wavelet image processing: Variational problems, compression, and noise removal through wavelet shrinkage, *IEEE Trans. Image Proc.* **7**(3) (1998), 319–335.
12. Chui, C. K., *An Introduction to Wavelets*, Vol. 1, Academic Press, Boston, 1992.
13. Chui, C. K. and Quak, E. G., Wavelets on a bounded interval, in: *Numerical Methods of Approximation Theory*, Vol. 9, (D. Braess and L. L. Schumaker, Eds.), ISNM, vol. 105, Birkhäuser, Basel, 1992, pp. 53–75.
14. Cohen, A., *Numerical Analysis of Wavelet Methods*, Studies in Mathematics and its Applications, Vol. 32, North Holland / Elsevier, Amsterdam, 2003.
15. Cohen, A., Dahmen, W. and DeVore, R., Adaptive wavelet methods for elliptic operator equations – Convergence rates, *Math. Comp.* **70** (2001), 27–75.
16. Cohen, A., Dahmen, W. and DeVore, R., Adaptive wavelet schemes for nonlinear variational problems, *SIAM J. Numer. Anal.* **41** (2003), 1785–1823.
17. Coldwell, R. L., Robust fitting of spectra to splines with variable knots, In: *AIP Conf. Proc. 475(1)* (J. L. Duggan and I. L. Morgan, Eds.), Amer. Inst. Physics, New York, 1990, pp. 604–607.
18. Dahmen, W., Wavelet and multiscale methods for operator equations, *Acta Numerica* **6** (1997), 55–228.
19. Dahmen, W. and Kunoth, A., Multilevel preconditioning, *Numer. Math.* **63** (1992), 315–344.
20. Dahmen, W. and Kunoth, A., Adaptive wavelet methods for linear–quadratic elliptic control problems: Convergence rates, *SIAM J. Contr. Optim.* **43** (2005), 1640–1675.
21. Dahmen, W., Kunoth, A. and Urban, K., Biorthogonal spline-wavelets on the interval – Stability and moment conditions, *Applied Comp. Harmonic Analysis* **6** (1999), 132–196.
22. DeVore, R. A., Nonlinear approximation, *Acta Numerica* **7** (1998), 51–150.
23. Dierckx, P., *Curve and Surface Fitting with Splines*, Oxford University Press, Oxford, 1993.
24. Dyn, N., Floater, M. S. and Iske, A., Adaptive thinning for bivariate scattered data, *J. Comput. Appl. Math.* **145** (2002), 505–517.
25. Floater, M. S. and Iske, A., Thinning, inserting and swapping scattered data, in: *Surface Fitting and Multiresolution Methods* (A. Le Méhauté, C. Rabut and

L. L. Schumaker, Eds.), Vanderbilt University Press, Nashville, 1996, pp. 139–144.

26. Floater, M. S. and Iske, A., Multistep scattered data interpolation using compactly supported radial basis functions, *J. Comput. Appl. Math.* **73** (1996), 65–78.
27. Forsey, D. R. and Bartels, R. H., Hierarchical B–spline refinement, *Computer Graphics* **22** (1988), 205–212.
28. Freeden, W., Gervens, T. and Schreiner, M., *Constructive Approximation on the Sphere, With Applications to Geomathematics*, Oxford Science Publications, Clarendon Press, Oxford, 1998.
29. Freeden, W. and Michel, V., *Multiscale Potential Theory with Applications to Geoscience*, Birkhäuser, Basel, 2004.
30. Garcke, J. and Griebel, M., Data mining with sparse grids using simplicial basis functions, in: *Seventh ACM SIGKDD Int. Conf. on Knowledge Discovery and Data Mining*, Proc. San Francisco 2001, ACM Press, New York, pp. 87–96.
31. Gerstner, Th., Helfrich, H.-P. and Kunoth, A., Wavelet analysis of geoscientific data, in: *Dynamics of Multiscale Earth Systems* (H. J. Neugebauer and C. Simmer, Eds.), Lecture Notes in Earth Sciences, Springer, New York, 2003, pp. 69–88.
32. Gortler, S. J. and Cohen, M., Hierarchical and variational geometric modeling with wavelets, in: *Proc. Symposium on Interactive 3D Graphics* (M. Zyda et al., Eds.), ACM Press, New York, 1995, pp. 35–42.
33. Gregorski, B. F., Hamann, B. and Joy, K. I., Reconstruction of B-spline surfaces from scattered data points, in: *Proceedings of Computer Graphics International 2000* (N. Magnenat-Thalmann and D. Thalmann, Eds.), IEEE Computer Society, Washington DC, pp. 163–170.
34. Greiner, G. and Hormann, K., Interpolating and approximating scattered 3D-data with hierarchical tensor product splines, in: *Surface Fitting and Multiresolution Methods* (A. Le Méhauté, C. Rabut and L. L. Schumaker, Eds.), Vanderbilt University Press, Nashville, 1996, pp. 163–172.
35. Härdle, W., Kerkyacharian, G., Picard, D. and Tsybakov, A., *Wavelets, Approximation, and Statistical Applications*, Lecture Notes in Statistics, Springer, New York, 1998.
36. Hegland, M., Adaptive sparse grids, *Australian and New Zealand Industr. Appl. Math. J.* **44(E)** (2003), C335–C353.
37. Hegland, M., Roberts, S. and Altas, I., Finite element thin plate splines for surface fitting, in: *Computational Techniques and Applications*, Proc. CTAC97 (B. J. Noye, M. D. Teubner and A. W. Gill, Eds.), World Scientific, Singapore, 1997, pp. 289–296.
38. Hofierka, J., Parajka, J., Mitasova, H. and Mitas, L., Multivariate interpolation of precipitation using regularized spline with tension, *Transactions in Geographic Information Science* **6** (2) (2002), 135–150.
39. Höllig, K. and Reif, U., Nonuniform web-splines, *Computer Aided Geometric Design* **20** (2003), 277–294.
40. Huber, P. J., *Robust Statistics*, John Wiley & Sons, New York, 1981.

41. Hutchinson, M. F. and de Hoog, F. R., Smoothing noisy data with spline functions, *Numer. Math.* **47** (1985), 99–106.
42. Iske, A. and Levesley, J., Multilevel scattered data approximation by adaptive domain decomposition, *Numer. Algorithms* **39** (1-3) (2005), 187-198.
43. Kunoth, A., Two applications of adaptive wavelet methods, in: *Modern Developments in Multivariate Approximation* (W. Haussmann, K. Jetter, M. Reimer and J. Stöckler, Eds.), ISNM, Vol. 145, Birkhäuser, Basel, 2003, pp. 175–201.
44. Kunoth, A., Adaptive wavelet schemes for an elliptic control problem with Dirichlet boundary control, *Numer. Algorithms* **39** (1-3) (2005), 199-220.
45. Launer, R. L. and Wilkinson, G. N. (Eds.), *Robustness in Statistics*, Academic Press, New York, 1979.
46. Lee, S., Wolberg, G. and Shin, S. Y., Scattered data interpolation with multilevel B-splines, *IEEE Trans. Visualization and Computer Graphics* **3** (3) (1997), 228–244.
47. Meissl, P., The use of finite elements in physical geodesy, Report No. 313, Dept. of Geodetic Science, Ohio State University, 1981.
48. Pereyra, V. and Scherer, G., Large scale least squares scattered data fitting, *Appl. Numer. Math.* **44** (1–2) (2002), 73–86.
49. The Puerto Rico Tsunami Warning and Mitigation Program. Data obtainable at http://poseidon.uprm.edu
50. Powell, M. J. D., Radial basis functions for multivariate interpolation: a review, in: *Algorithms for Approximation* (J. C. Mason and M. G. Cox, Eds.), Clarendon Press, Oxford, 1987, pp. 143–167.
51. Schaback, R., Multivariate interpolation by polynomials and radial basis functions, *Constr. Approximation* **21** (2005), 293–317.
52. Schaback, R. and Wendland, H., Adaptive greedy techniques for approximate solution of large rbf systems, *Numer. Algorithms* **24** (2000), 239–254.
53. Scheib, V., Haber, J., Lin, M. C. and Seidel, H. P., Efficient fitting and rendering of large scattered data sets using subdivision surfaces, In: *Computer Graphics Forum*, Proc. Eurographics 2002, Eurographics Publ., 2002, pp. 353–362.
54. Schumaker, L. L., Fitting surfaces to scattered data, in: *Approximation Theory II* (G. G. Lorentz, C. K. Chui and L. L. Schumaker, Eds.), Academic Press, New York, 1976, pp. 203–268.
55. Schwetlick, H. and Schütze, T., Least squares approximation by splines with free knots, *BIT* **35** (3) (1995), 361–384.
56. Schweitzer, M. A., *A Parallel Multilevel Partition of Unity Method for Elliptic Partial Differential Equations*, Lecture Notes in Computational Science and Engineering, Vol. 29, Springer, New York, 2003.
57. Stollnitz, E. J., DeRose, T. D. and Salesin, D. H., *Wavelets for Computer Graphics*, Morgan Kaufmann Publishers, San Mateo, 2000.
58. Sulebak, J. R. and Hjelle, Ø., Multiresolution spline models and their applications in geomorphology, in: *Concepts and Modelling in Geomorphology: International Perspectives*, Proc. Tokyo 2003 (I. S. Evans, R. Dikau,

E. Tokunaga, H. Ohmori and M. Hirano, Eds.), Terra Scientific Publ. Company, Tokyo, 2003, pp. 221–237.

59. Vogel, C. R., *Computational Methods for Inverse Problems*, Frontiers in Applied Mathematics, SIAM, Philadelphia, 2002.
60. Wahba, G., *Spline Models for Observational Data*, CBMS-NSF Reg. Conf. Series in Applied Mathematics, Vol. 59, SIAM, Philadelphia, 1990.
61. Weimer H. and Warren, J., Fast approximating triangulation of large scattered data sets, *Advances in Engineering Software* **30** (6) (1999), 389–400.
62. Wendland, H., *Scattered Data Approximation*, Cambridge Monographs on Applied and Computational Mathematics, Vol. 17, Cambridge University Press, Cambridge, 2005.
63. Williams, A. and Burrage, K., Surface fitting using GCV smoothing splines on supercomputers, Article No. 11 in: *Supercomputing*, Proc. San Diego 1995, ACM / IEEE Comp. Soc., 1995.
64. Wu, Z. and Schaback, R., Local error estimates for radial basis function interpolation of scattered data, *IMA J. Numer. Anal.* **13** (1993), 13–27.
65. Yserentant, H., On the multilevel splitting of finite element spaces, *Numer. Math.* **49** (1986), 379–412.
66. Zeilfelder, F., Scattered data fitting with bivariate splines, in: *Tutorials on Multiresolution in Geometric Modelling, Mathematics and Visualization* (A. Iske, E. Quak and M. S. Floater, Eds.), Springer, New York, 2002, pp. 243–286.

Topics in Multivariate Approximation and Interpolation
K. Jetter et al., Editors

Ready-to-Blossom Bases in Chebyshev Spaces

Marie-Laurence Mazure

Laboratoire de Modélisation et Calcul, Université Joseph Fourier, Grenoble, France

Abstract

This paper gives a survey on blossoms and Chebyshev spaces, with a number of new results and proofs. In particular, Extended Chebyshev spaces are characterised by the existence of a certain type of bases which are especially suited to enable us to prove both existence and properties of blossoms under the weakest possible differentiability assumptions. We also examine the case of piecewise spaces built from different Extended Chebyshev spaces and connection matrices.

Key words: Hermite interpolation, Taylor interpolation, extended Chebyshev space, W-space, extended Chebyshev piecewise space, W-piecewise space, Bernstein basis, B-spline basis, blossom, geometric design

2000 MSC: 65D17, 65D07, 41A15, 41A50

1. Introduction

In order to stress the main purpose of the present paper, we shall start with an elementary problem. Consider the function $\Phi : \mathbb{R} \to \mathbb{R}^n$ defined by $\Phi(x) := (\Phi_1(x), \dots, \Phi_n(x))^T := (x, x^2, \dots, x^n)^T$. Given pairwise distinct $a_1, \dots, a_r \in \mathbb{R}$ and

Email address: `mazure@imag.fr` (Marie-Laurence Mazure).

positive integers $\mu_1, \ldots, \mu_r$, with $\mu_1 + \cdots + \mu_r = n$, how to prove that the r osculating flats $\mathrm{Osc}_{n-\mu_i}\Phi(a_i)$, $1 \le i \le r$, have a unique common point, given that, for any nonnegative integer k, the kth order osculating flat of Φ at $x \in \mathbb{R}$ is defined as the affine flat $\mathrm{Osc}_k\Phi(x)$ passing through $\Phi(x)$ and the direction of which is spanned by the first k derivatives of Φ at x, namely:

$$\mathrm{Osc}_k\Phi(x) := \left\{ \Phi(x) + \sum_{p=1}^{k} \lambda_p \Phi^{(p)}(x) \mid \lambda_1, \ldots, \lambda_k \in \mathbb{R} \right\} .$$

For the sake of simplicity let us first investigate the case $r = n$. We are then dealing with n osculating hyperplanes $\mathrm{Osc}_{n-1}\Phi(a_i)$, $1 \le i \le n$. One possible proof consists in using the directions orthogonal to the hyperplanes. Indeed, if $X \in \mathbb{R}^n$, we have the equivalence

$$X \in \bigcap_{i=1}^{n} \mathrm{Osc}_{n-1}\Phi(a_i) \Leftrightarrow \langle X, \Phi^\sharp(a_i)\rangle = \langle \Phi(a_i), \Phi^\sharp(a_i)\rangle, \ 1 \le i \le n, \qquad (1)$$

where $\langle .,. \rangle$ denotes the inner product in $\mathbb{R}^n$ and where, for all $x \in \mathbb{R}$, $\Phi^\sharp(x) := \Phi'(x) \wedge \ldots \wedge \Phi^{(n-1)}(x)$ is the cross product of the first $(n-1)$ derivatives at x. One can check that, up to multiplication by some nonzero real number, we have $\Phi^\sharp = (\Phi_1^\sharp, \ldots, \Phi_n^\sharp)^T$, where the kth component $\Phi_k^\sharp$ is given by $\Phi_k^\sharp(x) := (-1)^{n-k}\binom{n}{k}x^{n-k}$. The main fact is that the space $\mathcal{E}^\sharp$ spanned by the functions $\Phi_1^\sharp, \ldots, \Phi_n^\sharp$ is the polynomial space $\mathcal{P}_{n-1}$ of degree $(n-1)$, and, because any nonzero element of $\mathcal{E}^\sharp$ cannot have n distinct zeros, the n vectors $\Phi^\sharp(a_i)$, $1 \le i \le n$, are linearly independent. This is the reason why the linear system appearing in (1) has a unique solution. In the general case, that is, when allowing $r < n$, the point X belongs to all osculating flats $\mathrm{Osc}_{n-\mu_i}\Phi(a_i)$, $1 \le i \le r$, iff it satisfies

$$\langle X, \Phi^{\sharp(k)}(a_i)\rangle = \langle \Phi(a_i), \Phi^{\sharp(k)}(a_i)\rangle, \ 0 \le k \le \mu_i - 1, \ 1 \le i \le r. \qquad (2)$$

Again, existence and uniqueness of the solution to the system (2) is justified by the fact that any nonzero element of $\mathcal{E}^\sharp$ has at most $(n-1)$ zeros, but this time the zeros being counted with their multiplicities.

The unique solution to our system (1) is nothing but the point $\varphi(a_1, \ldots, a_n)$, where φ denotes the blossom of the degree n polynomial function Φ [20,16]. More generally, the unique solution to (2) is the point $\varphi(a_1{}^{[\mu_1]}, \ldots, a_r{}^{[\mu_r]})$, where the symbolic power notation $x^{[k]}$ stands for x repeated k times. Clearly the arguments developed above are not limited to polynomial spaces. They obviously remain valid whenever we deal with an $(n+1)$-dimensional space $\mathcal{E}$ of sufficiently differentiable functions which contains constants and for which the directions orthogonal to the osculating hyperplanes of some "mother" function Φ generate an n-dimensional space $\mathcal{E}^\sharp$ which possesses the same properties concerning (simple as well as multiple) zeros as $\mathcal{P}_{n-1}$. This means whenever $\mathcal{E}^\sharp$ is an Extended Chebyshev space. After H. Pottmann [19], it is now classical to define Chebyshevian blossoms this way, that is; by means of intersections of osculating flats.

The fact that $\mathcal{E}^\sharp$ is an Extended Chebyshev space implies that the space $\mathcal{E}$ itself is an Extended Chebyshev space. However, when it is of dimension $(n+1)$, such a space $\mathcal{E}$ is generally supposed to be composed of functions which are only C^n. Unfortunately, this does not guarantee a priori the elements of $\mathcal{E}^\sharp$ to be sufficiently differentiable to make it possible for it to be an Extended Chebyshev space. Our previous arguments thus implicitly assume too much differentiability and they may therefore be impossible to use. Nevertheless, blossoms do exist even in case we limit ourselves to the strict C^n differentiability in $\mathcal{E}$: their existence can be shown for instance using Theorem 3.1 of [5], but it is not a trivial proof.

Blossoms and EC-spaces are so intimately connected that it is a bit paradoxical that existence and uniqueness of a common point to relevant osculating flats are not somehow visible at first sight. This paradox is the actual motivation of the present work. Furthermore, if the blossom φ of our initial polynomial function Φ is known to be affine in each variable, in the Chebyshevian case, this property is replaced by a more general one, which we are used to refer to as pseudoaffinity in each variable. Again this property was initially proved using more differentiability than necessary. We established it under the strict C^n differentiability assumption in [14], but the latter proof was far from being direct. This difficulty too seemed somewhat paradoxical.

The paper brings an answer to the points mentioned above: it is indeed possible to make existence of blossoms visible at first sight, provided that we use relevant bases. To put the reader on the right track, we shall come back to the example of polynomial spaces in the first case we examined. Let us introduce the function $\Psi : \mathbb{R} \to \mathbb{R}^n$ as $\Psi := (\Psi_1, \ldots, \Psi_n)^T$, with $\Psi_i(x) := (x - a_i)^n$, $1 \leq i \leq n$. It is obvious that $x_i = 0$ is an equation of the osculating hyperplane $\mathrm{Osc}_{n-1}\Psi(a_i)$. Accordingly, the origin of $\mathbb{R}^n$ belongs to $\cap_{i=1}^n \mathrm{Osc}_{n-1}\Psi(a_i)$ and it is the only point in this intersection. Both $(\mathbb{1}, \Phi_1, \ldots, \Phi_n)$ and $(\mathbb{1}, \Psi_1, \ldots, \Psi_n)$ being bases of the polynomial space $\mathcal{P}_n$, there exists a unique affine map $h := \mathbb{R}^n \to \mathbb{R}^n$ such that $\Phi = h \circ \Psi$, and h is injective. Therefore

$$\bigcap_{i=1}^{n} \mathrm{Osc}_{n-1}\Phi(a_i) = h\Big(\bigcap_{i=1}^{n} \mathrm{Osc}_{n-1}\Psi(a_i)\Big) = \{h(0_{\mathbb{R}^n})\}.$$

Not only is this new proof very short, but it also has the advantage to make no use of the space $\mathcal{E}^\sharp$. What exactly made it possible? The answer is: the fact that the functions $\Psi_i'(x) = n(x - a_i)^{n-1}$, $1 \leq i \leq n$, form a basis of the space $\mathcal{P}_{n-1}$, each of them vanishing the appropriate number of times at the concerned point to make the corresponding osculating hyperplane as easy to express as possible. Similar arguments can be developed in the general polynomial case we considered later on. The simple example of polynomial spaces brings out the importance of choosing the appropriate basis to solve a given problem concerning blossoms: for other bases the result will follow by taking images under affine maps. This is exactly what we shall extend to the Chebyshevian framework.

This work was an excellent opportunity to revisit both Extended Chebyshev spaces and their links with blossoms. It is intended as a survey on the topic, even though it also presents new results and proofs. Section 2 gathers various characterisations of Extended Chebyshev spaces, either classical ones recalled in a way to prepare the rest of the article, or recent ones which can be considered our first step towards Section 3. In the latter section we characterise these spaces by the existence of bases generalising our previous polynomial basis $(\Psi'_1, \ldots, \Psi'_n)$ (see Theorem 23). To achieve this, we first establish some interesting technical results emerging from Sylvester's identity for determinants. The fourth section shows the advantage of such bases for blossoms: not only do they make their existence obvious, but they are also the relevant bases to achieve their crucial pseudoaffinity property. They actually permit a direct proof of it, that is, unlike previous papers on the same subject, with no need to involve a "dual space", which required either more differentiability than necessary or complicated demonstrations. This section also makes a general survey of all important consequences of pseudoaffinity, along with the fundamental links between blossoms and Bernstein or B-spline bases, so far stated under more differentiability assumptions (see [13]). Finally the last two sections are devoted to the piecewise case, obtained by connecting different Extended Chebyshev spaces via lower triangular matrices with positive diagonal elements. We first investigate how to adapt the various properties reviewed in the second section to Extended Chebyshev Piecewise spaces. Then we show that, as in the nonpiecewise case, this provides us with bases tailor-made for blossoms.

2. Extended Chebyshev Spaces Revisited

In this section we survey both classical and recently obtained results about Chebyshev spaces. In particular we lay emphasis on various ways to characterise the possibility of doing Taylor and Hermite interpolation in a given functional space: number of zeros, nonvanishing determinants, or existence of special bases. The latter ones will be the starting point for the fundamental result presented in the next section. For further classical results on Chebyshev spaces, we refer to [6,21].

2.1. Taylor interpolation and W-spaces

Throughout the first three sections, I denotes a real interval with a nonempty interior. The possibility of solving interpolation problems in a unique way is classically characterised by bounding the number of zeros. This requires us to start with a few preliminary comments on how to count multiple zeros and on the notations we shall use. Given $k \leq n+1$, a function $U \in C^n(I)$ is said to ***vanish k times at** $a \in I$*, or a is said to be ***a zero of multiplicity** k*, if $U(a) = U'(a) = \cdots = U^{(k-1)}(a) = 0$. Given $k \leq n$, U is said to ***vanish exactly k times at** a*, or a is said to be ***a zero of exact multiplicity** k*, if $U(a) = U'(a) = \cdots = U^{(k-1)}(a) = 0$ and $U^{(k)}(a) \neq 0$. We denote by $Z_{n+1}(U)$ the total number of zeros of U in I, counting multiplicities up to $n+1$. Similarly, if S is a subset of I, we denote by $Z^S_{n+1}(U)$ the total number of zeros of U on the points of S.

In the rest of the section, $\mathcal{E}$ denotes an $(n+1)$-dimensional subspace of $C^n(I)$. Selecting a basis $(F_0, \ldots, F_n)$ of $\mathcal{E}$, we set $\mathbb{F} := (F_0, \ldots, F_n)^T$. On the other hand, for $0 \leq k \leq n+1$, and $a \in I$, we introduce the subspace

$$\mathcal{E}_k(a) := \{F \in \mathcal{E} \mid F \text{ vanishes } k \text{ times at } a\}. \tag{3}$$

We can first address *Taylor interpolation problems in* $n+1$ *data in the space* $\mathcal{E}$, that is, problems of the following form:

$$\text{Find } F \in \mathcal{E} \text{ such that } F^{(i)}(a) = \alpha_i\,, \quad 0 \le i \le n, \tag{4}$$

where a is given in I, and $\alpha_0, \ldots, \alpha_n$ are any real numbers.

Definition 1. The $(n+1)$-dimensional space $\mathcal{E} \subset C^n(I)$ is said to be a *W-space on* I if it satisfies the following property:
(i) *Any Taylor interpolation problem* (4) *has a unique solution in* $\mathcal{E}$.

As recalled below, W-spaces can be characterised in terms of Wronskians, which justifies the name we adopted.

Proposition 2. *The* $(n+1)$*-dimensional subspace* $\mathcal{E}$ *of* $C^n(I)$ *is a W-space on* I *if and only if it meets any of the following equivalent requirements:*
(ii) *Any nonzero element* F *of* $\mathcal{E}$ *vanishes at most* n *times at any given point of* I*, i.e.,* $Z_{n+1}^{\{a\}}(F) \le n$ *for all* $a \in I$.
(iii) *The Wronskian*

$$W(F_0, \ldots, F_n)(x) := \det\bigl(\mathbb{F}(x), \mathbb{F}'(x), \ldots, \mathbb{F}^{(n)}(x)\bigr),$$

never vanishes on I.
(iv) *For any* $a \in I$*, there exists a basis* $(\Psi_0^a, \Psi_1^a, \ldots, \Psi_n^a)$ *in* $\mathcal{E}$ *such that, for* $0 \le i \le n$*,* Ψ_i^a *vanishes exactly* i *times at* a.
(v) *For any* $a \in I$*,* $\mathcal{E}_{n+1}(a) = \{0\}$.

Although classical, we insist on the following remarks because of their fundamental importance, in particular in this article.

Remark 3. (1) Given a function $\omega \in C^n(I)$, the Leibniz' rule to differentiate products leads to

$$W(\omega F_0, \ldots, \omega F_n) = \omega^{n+1} W(F_0, \ldots, F_n). \tag{5}$$

Hence, if ω never vanishes on I, the space $\mathcal{E}$ is a W-space on I if and only if the space $\omega\mathcal{E} := \{\omega F \mid F \in \mathcal{E}\}$ is a W-space on I.

(2) We denote by D the ordinary differentiation. On account of the obvious equality

$$W(\mathbb{1}, F_1, \ldots, F_n) = W({F_1}', \ldots, {F_n}'), \tag{6}$$

when the space $\mathcal{E}$ contains constants, it is a W-space on I if and only if the space $D\mathcal{E} := \{F' \mid F \in \mathcal{E}\}$ is a W-space on I (of dimension n).

(3) The latter two remarks provide an easy way to obtain W-spaces by successive integration of a given W-space, and with, at each step, possible multiplication by a sufficiently differentiable nonvanishing function. For instance, setting either

$U(x) := \cos x$ and $V(x) := \sin x$, or $U(x) := \cosh x$ and $V(x) := \sinh x$, the space these two functions span is a two-dimensional W-space on $I = \mathbb{R}$, because in either case they satisfy $W(U,V)(x) = 1$ for all $x \in \mathbb{R}$. It follows that, for all $n \geq 2$, the space $\mathcal{E}_n(U,V)$ spanned by the $(n+1)$ functions $1, x, \ldots, x^{n-2}, U(x), V(x)$ is an $(n+1)$-dimensional W-space on $\mathbb{R}$.

Remark 4. One can also say that the space $\mathcal{E}$ is a W-space on I if and only if, for all $a \in I$, the sequence

$$\{0\} = \mathcal{E}_{n+1}(a) \subset \mathcal{E}_n(a) \subset \cdots \subset \mathcal{E}_1(a) \subset \mathcal{E}_0(a) = \mathcal{E}$$

is a strictly increasing one, *i.e.*, for $0 \leq k \leq n+1$, $\mathcal{E}_k(a)$ is $(n-k+1)$-dimensional. In other words, it means that, given any $a \in I$, and any nonzero $U \in \mathcal{E}$, we are able to give the exact multiplicity k, $0 \leq k \leq n$, of a as a zero of U.

2.2. Hermite interpolation and EC-spaces

More generally one can consider *Hermite interpolation problems in* $n+1$ *data in the space* $\mathcal{E}$, that is, any problem of the following form:

$$\text{Find } F \in \mathcal{E} \text{ such that } F^{(j)}(\tau_i) = \alpha_{i,j}\ , \quad 1 \leq i \leq r,\ 0 \leq j \leq \mu_i - 1\ , \tag{7}$$

in which $\tau_1, \ldots, \tau_r$ are pairwise distinct points in I, $\mu_1, \ldots, \mu_r$ are positive numbers such that $\sum_{i=1}^{r} \mu_i = n+1$, and $\alpha_{i,j}$, $1 \leq i \leq r$, $0 \leq j \leq \mu_i - 1$, are any real numbers. We say that the problem (7) is *based on the r points* $\tau_1, \ldots, \tau_r$. Hermite interpolation problems based on one point are thus Taylor interpolation problems.

Definition 5. The $(n+1)$-dimensional space $\mathcal{E} \subset C^n(I)$ is said to be an *Extended Chebyshev space (in short, EC-space) on* I if it satisfies the following property:
(i)′ *Any Hermite interpolation problem* (7) *has a unique solution in* $\mathcal{E}$.

Below we recall some classical characterisations of EC-spaces.

Proposition 6. *The $(n+1)$-dimensional subspace $\mathcal{E}$ of $C^n(I)$ is an EC-space on I if and only if it meets any of the following equivalent requirements:*
(ii)′ *Any nonzero element F of $\mathcal{E}$ vanishes at most n times on I, counting multiplicities, i.e., $Z_{n+1}(F) \leq n$.*
(iii)′ *For any $r \geq 1$, any positive integers $\mu_1, \ldots, \mu_r$ such that $\sum_{i=1}^{r} \mu_i = n+1$, and any pairwise distinct $a_1, \ldots, a_r \in I$, the determinant*

$$\det\big(\mathbb{F}(a_1), \mathbb{F}'(a_1), \ldots, \mathbb{F}^{(\mu_1-1)}(a_1), \mathbb{F}(a_2), \ldots, \mathbb{F}(a_r), \ldots, \mathbb{F}^{(\mu_r-1)}(a_r)\big), \tag{8}$$

is not equal to zero.

Remark 7. (1) Due to Leibniz' formula, when multiplying each function F_i by $\omega \in C^n(I)$, the determinant (8) is multiplied by $\prod_{i=1}^{r} \omega(a_i)^{\mu_i}$. Therefore, as in the case of W-spaces, if ω is sufficiently differentiable and if it never vanishes on I, then the space $\mathcal{E}$ is an EC-space on I iff and only if the space $\omega\mathcal{E}$ is an EC-space on I.

(2) If the space $\mathcal{E}$ contains constants and if the space $D\mathcal{E}$ is an n-dimensional EC-space on I, then $\mathcal{E}$ is an $(n+1)$-dimensional EC-space on I. This results from the following inequality

$$Z_n(U') \geq Z_{n+1}(U) - 1, \tag{9}$$

which is valid for any $U \in C^n(I)$ as a direct consequence of Rolle's theorem. However, unlike the case of W-spaces, the converse is not true. Let us illustrate the latter observations by considering the same spaces $\mathcal{E}_n(U,V)$ as in Remark 3. In the hyperbolic case, the space $\mathcal{E}_1(U,V)$ spanned by U, V is a 2-dimensional EC-space on $\mathbb{R}$ because, in addition to their nonvanishing Wronskian, the two functions satisfy, for any distinct a_1, a_2, $\begin{vmatrix} U(a_1) & U(a_2) \\ V(a_1) & V(a_2) \end{vmatrix} = \sinh(a_2 - a_1) \neq 0$. Hence, for all $n \geq 2$, the space $\mathcal{E}_n(U,V)$ too is an EC-space on $\mathbb{R}$. In the trigonometric case, the latter determinant is equal to $\sin(a_2 - a_1)$. We can thus assert that it is nonzero for any distinct a_1, a_2 in I only if we require the interval I to be strictly contained in some $[\alpha, \alpha+\pi]$. Hence the space $\mathcal{E}_1(U,V) = D\mathcal{E}_2(U,V)$ is a 2-dimensional EC-space only on such an interval. Still, one can check that, for $n = 2$ (hence, for $n > 2$ too), the trigonometric space $\mathcal{E}_n(U,V)$ is an EC-space on any interval strictly contained in any $[\alpha, \alpha + 2\pi]$.

2.3. Complete W-spaces

While being an EC-space on I clearly implies being a W-space on I, the converse is not true as shown by the example of the trigonometric spaces mentioned in Remarks 3 and 7. However there exist further important links between W-spaces and EC-spaces, as recalled subsequently.

Definition 8. The $(n+1)$-dimensional space $\mathcal{E} \subset C^n(I)$ is said to be a *Complete W-space (in short, CW-space) on* I if there exists a nested sequence

$$\mathcal{E}_0 \subset \mathcal{E}_1 \subset \cdots \subset \mathcal{E}_{n-1} \subset \mathcal{E}_n := \mathcal{E}, \tag{10}$$

where, for $0 \leq i \leq n$, $\mathcal{E}_i$ is an $(i+1)$-dimensional W-space on I. A sequence $(U_0, \ldots, U_n)$ in $C^n(I)$ is said to be a CW-system on I if it satisfies

$$W(U_0, \ldots, U_k)(x) \neq 0 \quad x \in I\,, \quad 0 \leq k \leq n. \tag{11}$$

A sequence (10) automatically provides us with CW-systems by picking a function U_i in $\mathcal{E}_i \setminus \mathcal{E}_{i-1}$ for $0 \leq i \leq n$, with $\mathcal{E}_{-1} := \{0\}$. Accordingly, $\mathcal{E}$ is a CW-space on I iff it possesses a basis which is a CW-system on I. Another basis of $\mathcal{E}$ is not necessarily a CW-system on I.

The following lemma is a key-point in the proof of the well-known important result stated in Theorem 10 below.

Lemma 9. *Given a sequence $(U_0,\dots,U_n)$ in $C^n(I)$, the following two properties are equivalent:*

1- *$(U_0,\dots,U_n)$ is a CW-system on I.*

2- *U_0 does not vanish on I and $(DL_0U_1,\dots,DL_0U_n)$ is a CW-system on I, where $L_0F := F/U_0$.*

Proof. The equivalence readily follows from (5) and (6), □

Theorem 10. *If $\mathcal{E}$ is a CW-space on I, then it is an EC-space on I.*

Proof. Although classical, we need to give a brief proof of this theorem. It is done by induction on n. Being an EC-space or being a W-space is clearly the same for $n = 0$. Suppose that $n \geq 1$ and that the result has been proved for $n-1$. Let $\mathcal{E}$ be a CW-space and let $(U_0,\dots,U_n)$ be a CW-system in it. According to Lemma 9, $(DL_0U_1,\dots,DL_0U_n)$ is a CW-system. The recursive assumption proves that the n-dimensional CW-space $DL_0\mathcal{E}$ is an EC-space on I. The fact that $\mathcal{E}$ is an $(n+1)$-dimensional EC-space on I follows by applying Remark 7, (1) and (2). □

Given a nested sequence (10) in $C^n(I)$, in which each $\mathcal{E}_i$ is $(i+1)$-dimensional, requiring all spaces $\mathcal{E}_i$ to be W-spaces on I is thus the same as requiring them all to be EC-spaces on I.

A given CW-system $(U_0,\dots,U_n)$ naturally generates differential operators $L_0,\dots,$ L_n on $C^n(I)$ by iteration of Lemma 9. This is a well-known fact [6] but it is desirable to recall it. We have already introduced L_0 in Lemma 9. Applying Lemma 9 to the CW-system $(DL_0U_1,\dots,DL_0U_n)$, we know that the function DL_0U_1 does not vanish on I and that $(DL_1U_2,\dots,DL_1U_n)$ is a CW-system on I, where $L_1F := DL_0F/DL_0U_1$, and so forth. The differential operators are thus defined on $C^n(I)$ in a recursive way as:

$$L_0F := \frac{F}{U_0}\,, \quad L_iF := \frac{DL_{i-1}F}{DL_{i-1}U_i}\,, \quad 1 \leq i \leq n\,. \tag{12}$$

The functions $w_0 := U_0$, $w_i := DL_{i-1}U_i$, $1 \leq i \leq n$, are classically called the *weight functions* associated with the CW-system $(U_0,\dots,U_n)$. Each w_i belongs to $C^{n-i}(I)$ and it does not vanish on I. The space $\mathcal{E}$ spanned by $U_0,\dots,U_n$ can then be described as

$$\mathcal{E} = \{F \in C^n(I) \mid L_nF \text{ is constant on } I\}.$$

Conversely, given non vanishing functions $w_0,\dots,w_n$, with $w_i \in C^{n-i}(I)$ for $0 \leq i \leq n$, one can consider the differential operators on $C^n(I)$ $L_0F := F/w_0$, $L_iF := DL_{i-1}F/w_i$, $1 \leq i \leq n$. It is well-known that the functions $F \in C^n(I)$ for which L_nF is constant on I form a CW-space $\mathcal{E}$: indeed, setting $\mathcal{E}_i := \mathrm{Ker} DL_i$ for $0 \leq i \leq n-1$ and $\mathcal{E}_n := \mathcal{E}$, one can show, via Remark 3, that these spaces form a strictly nested sequence of W-spaces on I. In such a case, we will write $\mathcal{E} = CW(w_0,\dots,w_n)$. This is a classical way to built EC-spaces from weight functions.

2.4. EC spaces and Hermite interpolation based on at most two points

As stated in (iv) of Proposition 2, W-spaces are characterised by the existence of certain type of bases with prescribed numbers of zeros at any given point in I. Among other things, in this subsection, we are interested in bases with prescribed numbers of zeros at any two distinct points in I, according to the definition below, obviously inspired by the polynomial Bernstein basis of degree n, $B_i^n(x) := \binom{n}{i}(1-x)^{n-i}x^i$, $0 \leq i \leq n$.

Definition 11. Given two distinct points $a, b \in I$, and $(n+1)$ elements $B_0, \ldots, B_n \in \mathcal{E}$, we say that $(B_0, \ldots, B_n)$ is a Bernstein-like basis relative to (a, b) if, for $0 \leq i \leq n$, the function B_i vanishes exactly i times at a and exactly $(n-i)$ times at b.

We recently pointed out [14] how intimately Bernstein-like bases are connected with EC-spaces, as stated in the following theorem. The proof we give here differs from [14] : it is intended to prepare for the next section along with the piecewise case examined in Sections 5 and 6.

Theorem 12. *The $(n+1)$-dimensional space $\mathcal{E} \subset C^n(I)$ is an EC-space on I if and only if it satisfies any of the following equivalent properties:*

(i)'' *Any Hermite interpolation problem based on any one or two points of I has a unique solution in $\mathcal{E}$.*

(ii)'' *For any distinct $a, b \in I$, any nonzero element $F \in \mathcal{E}$ vanishes at most n times on $\{a, b\}$, counting multiplicities, i.e., $Z_{n+1}^{\{a,b\}}(F) \leq n$.*

(iii)'' *Given any distinct $a, b \in I$ and any integers $i, j \geq 0$ with $i + j = n + 1$, the determinant*

$$\det\left(\mathbb{F}(a), \mathbb{F}'(a), ..., \mathbb{F}^{(i-1)}(a), \mathbb{F}(b), \mathbb{F}'(b)..., \mathbb{F}^{(j-1)}(b)\right) \tag{13}$$

is not equal to zero.

(iv)'' *$\mathcal{E}$ possesses a Bernstein-like basis relative to any pair of distinct points of I.*

(v)'' *Given any $a \in I$, for $0 \leq k \leq n$, the space $\mathcal{E}_k(a)$ is an $(n-k+1)$-dimensional W-space on any interval $J \subset I \setminus \{a\}$.*

Proof. The equivalence between the first four properties is easy to prove (see [14]). Suppose that (i)'' holds and let us prove that (v)'' is then satisfied. Given any $a \in I$, and any integer k, $0 \leq k \leq n$, the space $\mathcal{E}_k(a)$ is of dimension $(n-k+1)$ on any nontrivial subinterval of I (see Remark 4). Choose $b \in I \setminus \{a\}$, and any real numbers $\alpha_0 \ldots, \alpha_{n-k}$. By (i)'', we know that there exists a unique $F \in \mathcal{E}$ such that

$$F(a) = \cdots = F^{(k-1)}(a) = 0, \quad F^{(j)}(b) = \alpha_j, \quad 0 \leq j \leq n-k,$$

which means the existence of a unique element $F \in \mathcal{E}_k(a)$ satisfying $F^{(j)}(b) = \alpha_j$ for $0 \leq j \leq n-k$. In other words, any Taylor interpolation problem in $(n-k+1)$ data at any $b \in I \setminus \{a\}$ has a unique solution. It follows that $\mathcal{E}_k(a)$ is a W-space on any J contained in $I \setminus \{a\}$.

Suppose now that (v)″ is satisfied and let us prove that $\mathcal{E}$ is an EC-space on I. Given any $a \in I$, consider the strictly nested sequence

$$\mathcal{E}_n(a) \subset \mathcal{E}_{n-1}(a) \subset \cdots \subset \mathcal{E}_1(a) \subset \mathcal{E}_0(a) := \mathcal{E}.$$

Taking the restrictions to a given interval $J \subset I \setminus \{a\}$ of all subspaces involved in the latter sequence shows that each $\mathcal{E}_k(a)$ is a CW-space on J, hence an $(n-k+1)$-dimensional EC-space on J (see Theorem 10). As a special case, it follows that the space $\mathcal{E}$ itself is an EC-space on any interval strictly contained in I. If the interval I is not closed and bounded, this means that $\mathcal{E}$ is an EC-space on the whole of I too. Suppose now that I is a closed bounded interval $[a,b]$. The space $\mathcal{E}$ being an EC-space on any interval strictly contained in I, it is a W-space on I. Let F be any nonzero element of $\mathcal{E}$. According to Remark 4, we can consider the exact multiplicity $i \in \{0,\dots,n\}$ of a as a zero of U. Then, F belongs to $\mathcal{E}_i(a)$, which is an $(n-i+1)$-dimensional EC-space on $]a,b]$. Therefore, $Z_{n-i+1}^{]a,b]}(F) \le n-i$. The count of multiplicities implies that $Z_{n+1}^{]a,b]}(F) \le n-i$ as well. This proves that F satisfies $Z_{n+1}(F) = Z_{n+1}^{]a,b]}(F) + Z_{n+1}^{\{a\}}(F) \le n$. □

Although it is far from being the main point, Theorem 12 contains the fact that an EC-space on I is a CW-space on any strict subinterval of I. This gives us an opportunity to mention an important consequence of the latter fact ([19]).

Proposition 13. *Suppose that the interval I is closed and bounded. Then, as soon as $\mathcal{E}$ is an EC-space on I, it is a CW-space on I.*

Proof. We can assume that the space $\mathcal{E}$ is defined on the whole of $\mathbb{R}$, and that it is known to be an EC-space only on the closed bounded interval I. If so, let us set, for any $r \ge 0$, any positive integers $\mu_1,\dots,\mu_r$ with $\sum_{i=1}^r \mu_i = n+1$, and any pairwise distinct $a_1,\dots,a_r \in I$

$$\Lambda({a_1}^{[\mu_1]},\dots,{a_r}^{[\mu_r]}) := \frac{\Gamma_r(\mu_1,\dots,\mu_r)}{\prod_{1\le k<\ell\le r}(a_\ell - a_k)^{\mu_k\mu_\ell}}$$
$$\times \det\left(\mathbb{F}(a_1),\mathbb{F}'(a_1),\dots,\mathbb{F}^{(\mu_1-1)}(a_1),\mathbb{F}(a_2),\dots,\mathbb{F}(a_r),\dots,\mathbb{F}^{(\mu_r-1)}(a_r)\right), \tag{14}$$

with $\Gamma_r(\mu_1,\dots,\mu_r) := 1/\prod_{\ell=1}^r 1!2!\dots(\mu_\ell-1)!$. The function $\Lambda : \mathbb{R}^{n+1} \to \mathbb{R}$ so defined is symmetric and continuous on $\mathbb{R}^{n+1}$. As a matter of fact, this function Λ emerges as the continuous extension to the whole of $\mathbb{R}^n$ of the symmetric expression

$$\det\left(\mathbb{F}(a_1),\mathbb{F}(a_2),\dots,\mathbb{F}(a_n)\right) / \prod_{1\le k<\ell\le n}(a_\ell - a_k),$$

well-defined only for pairwise distinct $a_1,\dots,a_n$ (see [10]). Our assumption that $\mathcal{E}$ is an EC-space on I implies that Λ does not vanish on I^{n+1} (see (iii)′ of Proposition 6). Therefore, out of continuity, Λ does not vanish on J^{n+1}, where J is some interval strictly containing I. Hence, according to Proposition 6, (iii)′, $\mathcal{E}$ is an EC-space on the interval J. It follows from (v)″ that $\mathcal{E}$ is a CW-space on any interval strictly contained in J, so in particular on I. □

Therefore, whenever $\mathcal{E}$ is an $(n+1)$-dimensional EC-space on a closed bounded interval I, we can find nonvanishing weight functions $w_0, \dots, w_n$ such that $\mathcal{E} = CW(w_0, \dots, w_n)$. Since they all are continuous functions, if needed, we can assume them to be positive on I.

To conclude this section, we want to draw the reader's attention to the fact that each property (i)″, (ii)″, and (iii)″ of Theorem 12 is the "two point version" respectively of each property (i)′, (ii)′, and (iii)′ of Definition 5 and Proposition 6. It is thus natural to try and imagine what should be the "multiple point version" of either (iv)″ or (v)″. Below we state a natural (iv)′, for the proof of which we refer to [14].

Theorem 14. *The space $\mathcal{E}$ is an EC-space on I, if and only if it satisfies:*
(iv)′ *Given any sequence $x_0 \leq x_1 \leq x_2 \leq \cdots \leq x_n < x_{n+1} \leq x_{2n} \leq x_{2n+1}$ in I, there exists a basis $(V_0, \dots, V_n)$ of $\mathcal{E}$ such that, for $0 \leq i \leq n$, V_i vanishes on $(x_{i+1}, \dots, x_{i+n})$, but it vanishes neither on $(x_i, \dots, x_{i+n})$ nor on $(x_{i+1}, \dots, x_{i+n+1})$.*

In (iv)′, supposing that $(x_{i+1}, \dots, x_{i+n}) = (\xi_1{}^{[\mu_1]}, \dots, \xi_r{}^{[\mu_r]})$ with $\xi_1 < \xi_2 < \cdots < \xi_r$ and positive $\mu_1, \dots, \mu_r$, the expression "V_i vanishes on $(x_{i+1}, \dots, x_{i+n})$" means that the function V_i vanishes μ_k times at ξ_k for $1 \leq k \leq r$. If they generalise in a natural way the notion of Bernstein-like basis, the bases involved in (iv)′ are of no real use for blossoms. It is the "multiple point version" of (v)″, achieved in the next section, that will bring us closer to blossoming.

3. On the Way to Blossoms via Sylvester's Identity

The aim of this section is to characterise EC-spaces by the existence of a certain type of bases giving a generalisation of property (v)″ of Theorem 12 (see Theorem 23 below). In the next section the latter bases will prove to be especially suited to establish both existence and properties of blossoms. We shall achieve them via Sylvester's identity and a number of preliminary results proved in the first subsection.

3.1. Some consequences of Sylvester's identity

Let us start by recalling Sylvester's identity for determinants (see, for instance, [1]).

Lemma 15. (Sylvester's identity). *Let M be a square matrix of order $n+1$, and let N be the square submatrix of order $n-1$ obtained by deleting the first and last rows and columns of M. Let A, B, C, D be the four square submatrices of order n obtained as follows:*

- *A: by deleting the last row and the last column of M.*
- *B: by deleting the last row and the first column of M.*
- *C: by deleting the first row and the last column of M.*
- *D: by deleting the first row and the first column of M.*

The following equality holds

$$\det M \ \det N = \det A \ \det D - \det B \ \det C \ . \tag{15}$$

Formula (15) is the exact generalisation of the rule to calculate determinants of order 2, namely $\begin{vmatrix} a & b \\ c & d \end{vmatrix} = ad - bc$, with, by convention, $\det\emptyset = 1$. Sylvester's identity has plenty of interesting implications. The following lemma is an example for an application of (15).

Lemma 16. *Given $k \geq 0$, let $V_1, \ldots, V_k, U, V$ be $(k+1)$ times differentiable functions on I. Then*

$$\begin{aligned} W(V_1, ..., V_k)(x) W(V_1, ..., V_k, U, V)(x) = \\ W(V_1, ..., V_k, V)'(x) W(V_1, ..., V_k, U)(x) \\ - W(V_1, ..., V_k, V)(x) W(V_1, ..., V_k, U)'(x) \end{aligned} \tag{16}$$

Proof. Let us set $\Gamma := (U, V_1, \ldots, V_k, V)^T$. Then, for any $x \in I$, formula (16) is obtained by applying (15) to the determinant $D(x)$ of order $(k+2)$ defined by:

$$D(x) := \det\big(\Gamma^{(k+1)}(x), \Gamma(x), \Gamma'(x), \ldots, \Gamma^{(k-1)}(x), \Gamma^{(k)}(x)\big) \ .$$

□

As a direct consequence of (16), we obtain the following result.

Lemma 17. *Given $k+2$ functions $V_1, \ldots, V_k, U, V$ supposed to be $k+1$ times differentiable on the interval I, consider the function u defined by*

$$u := \frac{W(V_1, \ldots, V_k, V)}{W(V_1, \ldots, V_k, U)}$$

on the subset Ω of I where $W(V_1, \ldots, V_k, U)$ does not vanish. Its derivative is then given by

$$u'(x) = \frac{W(V_1, \ldots, V_k)(x) W(V_1, \ldots, V_k, U, V)(x)}{[W(V_1, \ldots, V_k, U)(x)]^2}, \quad x \in \Omega. \tag{17}$$

One important consequence of formula (17) is that it provides us with nice expressions in terms of Wronskians for the differential operators (12) associated with a given CW-system.

Lemma 18. *Let $(U_0, \ldots, U_n)$ be a CW-system on I, and let $L_0, \ldots, L_n$ denote the associated differential operators, defined according to* (12). *Then we have, for all $F \in C^n(I)$,*

$$L_i F = \frac{W(U_0, ..., U_{i-1}, F)}{W(U_0, ..., U_{i-1}, U_i)}, \quad 0 \le i \le n. \tag{18}$$

Proof. Given that $L_0 F := F/U_0$, equality (18) is clearly satisfied for $i = 0$. Consider an integer i such that $0 \le i \le n-1$, and suppose that (18) holds for i. By application of (17), we can derive

$$DL_i F = \frac{W(U_0, ..., U_{i-1})\, W(U_0, ..., U_{i-1}, U_i, F)}{W(U_0, ..., U_i)^2}. \tag{19}$$

The equality $L_{i+1}F := (DL_iF)/(DL_iU_{i+1})$ then proves (18) for $i+1$. □

With the convention $W(\emptyset) = \mathbb{1}$, equality (19) shows that the weight functions classically associated to a given CW-system are given by (see [6])

$$w_i := \frac{W(U_0, ..., U_{i-2})\, W(U_0, ..., U_i)}{W(U_0, ..., U_{i-1})^2}, \quad 0 \le i \le n.$$

Using (18) and (1) of Remark 3, iteration of Lemma 9 yields the following result.

Lemma 19. *Given $(n+1)$ functions $U_0, \ldots, U_n \in C^n(I)$, we fix an integer k, $0 \le k \le n$ and we consider the functions*

$$F_i^k := W(U_0, ..., U_k, U_i), \quad k+1 \le i \le n.$$

Then, $(U_0, \ldots, U_n)$ is a CW-system on I if and only if both $(U_0, \ldots, U_k)$ and $(F_{k+1}^k, \ldots, F_n^k)$ are CW-systems on I.

The right-hand side of equality (16) is nothing but the Wronskian of the two functions $W(U_0, ..., U_k, U)$ and $W(U_0, ..., U_k, V)$. Formula (16) is thus a special case of the following result which will prove useful in Subsection 4.3 and with which we conclude the present subsection. Note that it also gives another proof of Lemma 19.

Lemma 20. *Given $(n+1)$ functions $U_0, \ldots, U_n \in C^n(I)$, with the notations of the previous lemma, we have, for $0 \le k \le n$ and for all $x \in I$,*

$$W(F_{k+1}^k, ..., F_n^k)(x) = \left[W(U_0, ..., U_k)(x)\right]^{n-k-1} W(U_0, ..., U_n)(x). \tag{20}$$

Proof. Note that formula (20) is independent of the order between $U_0, \ldots, U_k$ on the one hand, and between $U_{k+1}, \ldots, U_n$ on the other. Furthermore, for any k, $1 \le k \le n-1$, for any x such that $F_k^{k-1}(x) \neq 0$, by application of (5) and (6), we obtain

$$W(F_k^{k-1}, \ldots, F_n^{k-1})(x) = \left[F_k^{k-1}(x)\right]^{n-k+1} W\left(\left(\frac{F_{k+1}^{k-1}}{F_k^{k-1}}\right)', \ldots, \left(\frac{F_n^{k-1}}{F_k^{k-1}}\right)'\right)(x).$$

Using (17) to calculate the functions $\left(F_i^{k-1}/F_k^{k-1}\right)'$, we obtain

$$W(F_k^{k-1},\dots,F_n^{k-1})(x) = \frac{\left[W(U_0,\dots,U_{k-1})(x)\right]^{n-k}}{\left[W(U_0,\dots,U_k)(x)\right]^{n-k-1}} W(F_{k+1}^k,\dots,F_n^k)(x), \quad (21)$$

subject that $F_k^{k-1}(x) \neq 0$. We shall prove (20) by induction on $k \geq -1$. For $k = -1$, it is trivially satisfied independently of n. Suppose that $k \geq 0$ and that the result holds for any integer less than or equal to $k-1$.

1) Let us first consider a point $x_0 \in I$ such that $W(U_0,...,U_k)(x_0) = 0$.

We want to prove that $W(F_{k+1}^k,...,F_n^k)(x_0) = 0$. This is clearly satisfied as soon as $F_i^k(x_0) = 0$ for any integer i such that $k+1 \leq i \leq n$. Suppose that there exists an integer i, $k+1 \leq i \leq n$, such that $F_i^k(x_0) \neq 0$. We may as well assume that $i = k+1$. By continuity, F_{k+1}^k does not vanish on some neighbourhood I_0 of x_0 contained in I. Therefore, equality (21) in which we replace k by $k+1$ is valid for $x \in I_0$. This proves that $W(F_{k+1}^k,...,F_n^k)(x_0) = 0$.

2) Let us now consider a point $x_0 \in I$ such that $W(U_0,...,U_k)(x_0) \neq 0$.

In this case, we can find an integer i, $0 \leq i \leq k$, such that $W(U_0,...,U_{i-1},U_{i+1},...,U_k)(x_0) \neq 0$. Without loss of generality we may assume that $i = k$. Our recursive assumption enables us to write, for all $x \in I$,

$$W(F_k^{k-1},...,F_n^{k-1})(x) = \left[W(U_0,...,U_{k-1})(x)\right]^{n-k} W(U_0,...,U_n)(x). \quad (22)$$

Since $W(U_0,...,U_{k-1})(x_0) \neq 0$, the two equalities (21) and (22) prove that (20) is satisfied for $x = x_0$. □

3.2. A new characterisation of EC-spaces

The various consequences of Sylvester's identity developed in the previous subsection will provide us with interesting bases in the space $\mathcal{E}$, which later on will prove to be tailor-made for blossoms. Such bases will generalise Lemma 21 below which is nothing but a different way of stating property (v)″ of Theorem 12.

Lemma 21. *The $(n+1)$-dimensional space $\mathcal{E} \subset C^n(I)$ is an EC-space on I if and only if the following two properties hold simultaneously:*

(v)$_1''$ *$\mathcal{E}$ is a W-space on I, i.e., (according to Proposition 2) for any $a \in I$, there exists a basis $(\Psi_n^a,\dots,\Psi_0^a)$ in $\mathcal{E}$ such that, for $0 \leq i \leq n$, Ψ_i^a vanishes exactly i times at a.*

(v)$_2''$ *For any $a \in I$, $(\Psi_n^a,\dots,\Psi_0^a)$ is a CW-system on any interval contained in $I \setminus \{a\}$, i.e., it satisfies*

$$W(\Psi_n^a,\dots,\Psi_k^a)(x) \neq 0, \quad x \in I \setminus \{a\},\ 1 \leq k \leq n.$$

Note that part (v)$_1''$ of Lemma 21 guarantees that the Wronskian $W(\Psi_n^a,\dots,\Psi_0^a)$ never vanishes on the whole of I.

Example 22. In the polynomial case, take $\Psi_i^a(x) := (x-a)^i$, for $0 \le i \le n$. Then, choosing an integer k between 0 and n, by application of (5), it is easy to see that

$$\begin{aligned} W(\Psi_k^a, \ldots, \Psi_n^a)(x) &= [\Psi_k^a(x)]^{n-k+1} W(\Psi_0^a, \ldots, \Psi_{n-k}^a)(x) \\ &= \Big(\prod_{j=1}^{n-k} j!\Big)(x-a)^{k(n-k+1)}, \quad x \in \mathbb{R}. \end{aligned} \tag{23}$$

Below we state the "multiple point version" of Lemma 21, that is, of (v)″ of Theorem 12.

Theorem 23. *The space $\mathcal{E}$ is an EC-space on I if and only if it satisfies the following property:*

(v)′ *given any integer $r \ge 1$, any positive integers $\mu_1, \ldots, \mu_r$ such that $\sum_{i=1}^r \mu_i = n+1$, and any pairwise distinct $a_1, \ldots, a_r \in I$, $\mathcal{E}$ possesses a basis $\Psi_n^{a_1}, \ldots, \Psi_{n-\mu_1+1}^{a_1}, \Psi_n^{a_2}, \ldots, \Psi_n^{a_r}, \ldots, \Psi_{n-\mu_r+1}^{a_r}$ such that, for $1 \le i \le r$ and $n-\mu_i+1 \le k \le n$, the function $\Psi_k^{a_i}$ vanishes exactly k times at a_i.*

Moreover, when $\mathcal{E}$ is an EC-space on I, each such basis is a CW-system on any interval J contained in $I \setminus \{a_1, \ldots, a_r\}$, i.e., with the same data as in (v)′ *except that now we assume $\sum_{i=1}^r \mu_i \le n$, we have, for all $x \in I \setminus \{a_1, \ldots, a_r\}$,*

$$W(\Psi_n^{a_1}, \ldots, \Psi_{n-\mu_1+1}^{a_1}, \Psi_n^{a_2}, \ldots, \Psi_n^{a_r}, \ldots, \Psi_{n-\mu_r+1}^{a_r})(x) \ne 0. \tag{24}$$

Proof. Let us first suppose that (v)′ holds. The case $r=1$ proves that the space $\mathcal{E}$ satisfies (iv) of Proposition 2, hence it is a W-space on I. Let us now consider the case $r=2$, that is, consider distinct $a_1, a_2 \in I$ and positive integers μ_1, μ_2 such that $\mu_1 + \mu_2 = n+1$. Since $\mathcal{E}$ is a W-space on I, the Wronskian of the corresponding basis $(\Psi_n^{a_1}, \ldots, \Psi_{n-\mu_1+1}^{a_1}, \Psi_n^{a_2}, \ldots, \Psi_{n-\mu_2+1}^{a_2})$ provided by (v)′ never vanishes on I. Now, from the zeros of these functions it is easy to derive

$$\begin{aligned} W(\Psi_n^{a_1}, \ldots, \Psi_{n-\mu_1+1}^{a_1}, \Psi_n^{a_2}, \ldots, \Psi_{n-\mu_2+1}^{a_2})(a_2) &= W(\Psi_n^{a_1}, \ldots, \Psi_{n-\mu_1+1}^{a_1})(a_2) \\ &\quad \times W(\Psi_n^{a_2(\mu_1)}, \ldots, \Psi_{n-\mu_2+1}^{a_2}{}^{(\mu_1)})(a_2). \end{aligned}$$

The latter relation enables us to conclude that, for any $a_1 \in I$ and any positive integer $\mu_1 \le n$,

$$W(\Psi_n^{a_1}, \ldots, \Psi_{n-\mu_1+1}^{a_1})(a_2) \ne 0 \quad \text{for all } a_2 \in I \setminus \{a_1\}.$$

Lemma 21 says that $\mathcal{E}$ is an EC-space on I.

Conversely, let us now assume that $\mathcal{E}$ is an EC-space on I. Let us prove (v)′ along with (24) by induction on $r \ge 1$. For $r=1$, it is nothing but Lemma 21. Suppose that $r \ge 2$, and that both (v)′ and (24) hold for $r-1$ in any EC-space on any interval.

Consider positive integers $\mu_1, \ldots, \mu_r$ such that $\mu := \sum_{i=1}^r \mu_i \le n+1$ and pairwise distinct $a_1, \ldots, a_r \in I$. We may as well assume that $a_2, \ldots, a_r$ all belong to some interval I_1 contained in $I \setminus \{a_1\}$. Let $\widetilde{\mathcal{E}}_1$ denote the restriction to I_1 of the space $\widetilde{\mathcal{E}}$ defined by

$$\widetilde{\mathcal{E}} := \{W(\Psi_n^{a_1}, \ldots, \Psi_{n-\mu_1+1}^{a_1}, F) \mid F \in \mathcal{E}\}.$$

From Lemma 21, we know that $(\Psi_n^{a_1}, \ldots, \Psi_0^{a_1})$ is a CW-system on I_1. Therefore, as a consequence of Lemma 19 and Theorem 10, we know that the space $\widetilde{\mathcal{E}}_1$ is an $(\widetilde{n}+1)$-dimensional EC-space on I_1, where $\widetilde{n} := n - \mu_1$. It is to this space $\widetilde{\mathcal{E}}_1$, and to $a_2, \ldots, a_r$ and $\mu_2, \ldots, \mu_r$, that we shall apply the recursive assumption. Given $b \in I \setminus \{a_1\}$ and $k \le \widetilde{n}$, we denote by $\widetilde{\Psi}_k^b$ any element of $\widetilde{\mathcal{E}}$ which vanishes exactly k times at b. Due to $(\mathrm{v})_2''$ of Lemma 21 and to Lemma 24 below, for any $b \in I_1$, we can choose

$$\widetilde{\Psi}_k^b := W(\Psi_n^{a_1}, \ldots, \Psi_{n-\mu_1+1}^{a_1}, \Psi_{k+\mu_1}^b), \quad 0 \le k \le \widetilde{n}. \tag{25}$$

According to (20), such a choice yields:

$$\begin{aligned} W(\widetilde{\Psi}_{\widetilde{n}}^{a_2}, \ldots, \widetilde{\Psi}_{\widetilde{n}-\mu_2+1}^{a_2}, \ldots, \widetilde{\Psi}_{\widetilde{n}}^{a_r}, \ldots, \widetilde{\Psi}_{\widetilde{n}-\mu_r+1}^{a_r})(x) = \\ \left[W(\Psi_n^{a_1}, \ldots, \Psi_{n-\mu_1+1}^{a_1}(x)\right]^{\mu-\mu_1-1} \\ \times W(\Psi_n^{a_1}, \ldots, \Psi_{n-\mu_1+1}^{a_1}, \Psi_n^{a_2}, \ldots, \Psi_n^{a_r}, \ldots, \Psi_{n-\mu_r+1}^{a_r})(x), \quad x \in I. \end{aligned}$$

On the other hand, the recursive assumption tells us that the left hand side of the latter equality does not vanish on $I_1 \setminus \{a_2, \ldots, a_r\}$ if $\mu - \mu_1 \le \widetilde{n}$, and on the whole of I_1 when $\mu - \mu_1 = \widetilde{n}$. Accordingly, in any case, the Wronskian $W(\Psi_n^{a_1}, \ldots, \Psi_{n-\mu_1+1}^{a_1}, \Psi_n^{a_2}, \ldots, \Psi_n^{a_r}, \ldots, \Psi_{n-\mu_r+1}^{a_r})$ never vanishes on $I_1 \setminus \{a_2, \ldots, a_r\}$. For $\mu = n+1$, this implies the linear independence of the functions $\Psi_n^{a_1}, \ldots, \Psi_{n-\mu_1+1}^{a_1}, \Psi_n^{a_2}, \ldots, \Psi_n^{a_r}, \ldots, \Psi_{n-\mu_r+1}^{a_r}$ on I, and for $\mu \le n$, it proves (24). □

Theorem 23 says in particular that, when $\mu = n+1$, (24) is valid not only on $I \setminus \{a_1, \ldots, a_r\}$, but even on the whole of I.

Lemma 24. *Given $k+1$ functions $V_1, \ldots, V_k, U$ supposed to be $N-1 \ge k$ times differentiable on I, we assume that U vanishes N times at a point $a \in I$. Then, the Wronskian $H := W(V_1, \ldots, V_k, U)$ vanishes $N-k$ times at a. Moreover, if the functions are N times differentiable on I, H vanishes exactly $N-k$ times at a iff U vanishes exactly N times at a and $W(V_1, \ldots, V_k)(a) \neq 0$.*

Proof. We denote by $\langle .,. \rangle$ the inner product in $\mathbb{R}^{k+1}$. Let us introduce the following notation $\Delta_k U := (U, U', \ldots, U^{(k)})^T$, and let us set $\Theta(x) := \Delta_k V_1(x) \wedge \ldots \wedge \Delta_k V_k(x)$ (cross product). For $x \in I$, we then have $H(x) = \langle \Theta(x), \Delta_k U(x) \rangle$, which gives, for $0 \le i \le N-k-1$,

$$H^{(i)}(x) = \sum_{\ell=0}^{i} \binom{i}{\ell} \langle \Theta^{(i-\ell)}(x), (\Delta_k U)^{(\ell)}(x) \rangle, \quad x \in I.$$

Now, $(\Delta_k U)^{(\ell)}(x) = (U^{(\ell)}(x), U^{(\ell+1)}, \ldots, U^{(\ell+k)}(x))^T$. Therefore, on account of the assumption $U(a) = \cdots = U^{(N-1)}(a) = 0$, we can derive from the previous equality that $H(a) = \cdots = H^{(N-1-k)}(a) = 0$. Moreover, if the functions are N times differentiable, the same equality is valid for $i = N - k$, which yields

$$H^{(N-k)}(a) = U^{(N)}(a)\ W(V_1, \ldots, V_k)(a).$$

Whence the second part of the announced result. □

Example 25. Consider again the polynomial case, with, for all $a \in \mathbb{R}$, $\Psi_i^a(x) := (x-a)^i$. One can generalise formula (23), as follows: for any positive $\mu_1, \ldots, \mu_r$ with $\mu := \sum_{i=1}^r \mu_i \leq n+1$, and any $a_1, \ldots, a_r, x \in \mathbb{R}$,

$$W(\Psi^{a_1}_{n-\mu_1+1}, \ldots, \Psi^{a_1}_{n}, \ldots \Psi^{a_r}_{n-\mu_r+1}, \ldots, \Psi^{a_r}_{n})(x) = \\ B\prod_{i=1}^{r}(x-a_i)^{\mu_i(n-\mu+1)} \prod_{1\leq i<j\leq r}(a_i - a_j)^{\mu_i\mu_j}, \tag{26}$$

where $B := A/\prod_{i=1}^r \binom{n}{1}\binom{n}{2}\ldots\binom{n}{\mu_i-1}$, and $A := 1!2!\ldots(\mu-1)!\binom{n}{1}\binom{n}{2}\ldots\binom{n}{\mu-1}$. Formula (26) can be obtained by applying the results of [10] (see also proof of Proposition 13) to the following easy to obtain equality:

$$W(\Psi^{a_1}_{n}, \Psi^{a_2}_{n}, \ldots, \Psi^{a_\mu}_{n})(x) = A\prod_{i=1}^{\mu}(x-a_i)^{n-\mu+1} \prod_{1\leq i<j\leq \mu}(a_j - a_i), \quad x \in \mathbb{R},$$

valid for any $\mu \leq n+1$, and any pairwise distinct $a_1, \ldots, a_\mu \in \mathbb{R}$ (see (14)). In particular, when $\mu = n+1$, we can verify that, as expected, (26) never vanishes on $\mathbb{R}$: indeed, it is a nonzero constant. Further interesting properties of such polynomial bases can be found in [18].

4. Blossoms in W-spaces

This section is devoted to blossoms: existence, properties, and consequences. As a matter of fact, the results stated in subsections 4.2 and 4.3 below, which concern existence and properties of blossoms, are not new, but so far they were achieved either under more differentiability assumptions [19,9], or via rather difficult (although interesting) proofs [5].

4.1. Blossoms and Bézier points

From now on, $\mathcal{E}$ denotes an $(n+1)$-dimensional subspace of $C^n(I)$, assumed to contain constants. Choosing n elements $\Phi_1, \ldots, \Phi_n$ in $\mathcal{E}$ so that $(\mathbb{1}, \Phi_1, \ldots, \Phi_n)$ is a basis of $\mathcal{E}$, we set

$$\Phi := (\Phi_1, \ldots, \Phi_n)^T.$$

Such a function Φ play the rôle of a *mother-function*, in the sense that any $F \in \mathcal{E}^d$ $(d \geq 1)$ can be obtained as its image $F = h \circ \Phi$ under an affine map $h : \mathbb{R}^n \to \mathbb{R}^d$

and this affine map is uniquely determined. It is essential to be aware that, in the space $\mathcal{E}$, all results and all mathematical objects which we obtain subsequently from a given mother-function Φ will be independent of our choice of Φ.

Since $\mathcal{E} \subset C^n(I)$, for any $x \in \mathbb{R}$ and for $k \leq n$, we can consider osculating flats of Φ up to order n (see Section 1). Up to a permutation, a given n-tuple $(x_1, \ldots, x_n) \in I^n$ can always be written as $(a_1{}^{[\mu_1]}, \ldots, a_r{}^{[\mu_r]})$, where $\mu_1, \ldots, \mu_r$ are positive integers such that $\sum_{i=1}^r \mu_i = n$, and where $a_1, \ldots, a_r$ are pairwise distinct points of I. We then set

$$\{\varphi(x_1, \ldots, x_n)\} := \bigcap_{i=1}^r \mathrm{Osc}_{n-\mu_i}\, \Phi(a_i) \tag{27}$$

whenever the osculating flats $\mathrm{Osc}_{n-\mu_i}\, \Phi(a_i)$, $1 \leq i \leq r$, intersect at a single point. Let $\mathcal{D}(\varphi)$ denote the domain of definition of the function φ. It is a symmetric subset of I^n independent of our chosen mother-function Φ, on which φ is clearly symmetric. The set $\mathcal{D}(\varphi)$ obviously contains the diagonal of I^n and, by restriction to the latter diagonal, the function φ gives Φ in the sense that $\varphi(x^{[n]}) = \Phi(x)$ for all $x \in I$. With any $F \in \mathcal{E}^d$ we associate the function f defined on $\mathcal{D}(\varphi)$ by $f := h \circ \varphi$, where $h : \mathbb{R}^n \to \mathbb{R}^d$ is the unique affine map such that $F = h \circ \Phi$. Let us stress again that f does not depend on the mother-function Φ we start with.

Definition 26. We say that *Bézier points exist in the space* $\mathcal{E}$ when, for all distinct $a, b \in I$, the points

$$\Pi_i(a,b) := \varphi(a^{[n-i]}, b^{[i]}), \quad 0 \leq i \leq n, \tag{28}$$

called *the Bézier points of* Φ *relative to* (a,b), are well-defined, that is, when the set $\mathcal{D}(\varphi)$ contains $\{a,b\}^n$ for all distinct $a, b \in I$. We say that *blossoms exist in the space* $\mathcal{E}$ when $\mathcal{D}(\varphi) = I^n$, in which case the function φ, defined on the whole of I^n, is called *the blossom of* Φ.

According to (28) and to (27), when they exist the Bézier points of Φ relative to (a,b) are thus defined by

$$\Pi_0(a,b) := \Phi(a), \quad \Pi_n(a,b) := \Phi(b), \tag{29}$$

$$\{\Pi_i(a,b)\} := \mathrm{Osc}_i\Phi(a) \cap \mathrm{Osc}_{n-i}\Phi(b), \quad 1 \leq i \leq n-1. \tag{30}$$

More generally, the Bézier points of $F \in \mathcal{E}^d$ relative to (a,b) are then defined as the points $f(a^{[n-i]}, b^{[i]})$, $0 \leq i \leq n$. If blossoms exist in $\mathcal{E}$, *i.e.*, if $\mathcal{D}(\varphi) = I^n$, the function f is called the blossom of F.

4.2. Existence of blossoms in EC-spaces

In this subsection we shall see that the choice of relevant bases makes the proof of existence of blossoms a child's play!

Theorem 27. *Let us suppose that the space $\mathcal{U} := D\mathcal{E}$ is an EC-space on I. Then, blossoms exist in the space $\mathcal{E}$. Moreover, for any positive integers $\mu_1, \ldots, \mu_r$ such that $\sum_{i=1}^r \mu_i = n$ and any pairwise distinct $a_1, \ldots, a_r \in I$, the value of the blossom f of any $F \in \mathcal{E}$ at $(a_1{}^{[\mu_1]}, \ldots, a_r{}^{[\mu_r]})$ is given by*

$$f(a_1{}^{[\mu_1]}, \ldots, a_r{}^{[\mu_r]}) = \frac{W(\Psi_n^{a_2}, \ldots, \Psi_{n-\mu_2+1}^{a_2}, \ldots, \Psi_n^{a_r}, \ldots, \Psi_{n-\mu_r+1}^{a_r}, F)(a_1)}{W(\Psi_n^{a_2}, \ldots, \Psi_{n-\mu_2+1}^{a_2}, \ldots, \Psi_n^{a_r}, \ldots, \Psi_{n-\mu_r+1}^{a_r}, \mathbb{1})(a_1)}, \tag{31}$$

and all other expressions obtained by permutation of $a_1, \ldots, a_r$.

Proof. Suppose that the space $\mathcal{U} := D\mathcal{E}$ is an EC-space on I. Then the space $\mathcal{E}$ itself is an EC space on I. As previously, given $a \in I$, we can thus denote by Ψ_k^a an element of $\mathcal{E}$ which vanishes exactly k times at a.

Consider positive $\mu_1, \ldots, \mu_r$ such that $\sum_{i=1}^r \mu_i = n$, and pairwise distinct $a_1, \ldots, a_r \in I$. In order first to prove that all osculating flats $\mathrm{Osc}_{n-\mu_i}\Phi(a_i)$, $1 \leq i \leq n$, have a unique common point, then to obtain the value (31) of the blossom of any $F \in \mathcal{E}$, we can use any suitable mother-function. Now, for $0 \leq k \leq n-1$, the function $U_k^a := D\Psi_{k+1}^a$ vanishes exactly k times at a. Applying Theorem 23 to the space $\mathcal{U}$, we thus know that the functions $U_{n-1}^{a_1}, \ldots, U_{n-\mu_1}^{a_1}, \ldots, U_{n-1}^{a_r}, \ldots, U_{n-\mu_r}^{a_r}$ form a basis of $\mathcal{U}$. This is the exact reason why, as the mother-function, we are allowed to choose

$$\Phi := (\Psi_n^{a_1}, \ldots, \Psi_{n-\mu_1+1}^{a_1}, \Psi_n^{a_2}, \ldots, \Psi_{n-\mu_2+1}^{a_2}, \ldots, \Psi_n^{a_r}, \ldots, \Psi_{n-\mu_r+1}^{a_r})^T.$$

1) Existence of blossoms.

The latter choice makes it obvious that, for $1 \leq i \leq r$, the $(n-\mu_i)$-dimensional osculating flat $\mathrm{Osc}_{n-\mu_i}\Phi(a_i)$ is composed of all points $X = (X_1, \ldots, X_n)^T$ such that $X_{\mu_1+\cdots+\mu_{i-1}+1} = \cdots = X_{\mu_1+\cdots+\mu_i} = 0$. Therefore the origin of $\mathbb{R}^n$ is the unique point in $\bigcap_{i=1}^r \mathrm{Osc}_{n-\mu_i}\Phi(a_i)$.

2) Expressions of blossoms.

Formula (31) being trivially satisfied for $r = 1$, we assume that $r \geq 2$ and we set $\mu := \sum_{i=2}^r \mu_i = n - \mu_1 \leq n-1$. Now that we have proved existence of blossoms, we know that

$$\varphi(a_1{}^{[\mu_1]}, \ldots, a_r{}^{[\mu_r]}) = \Phi(a_1) + \sum_{\ell=1}^{\mu} \lambda_\ell \Phi^{(\ell)}(a_1), \tag{32}$$

where the real numbers $\lambda_1, \ldots, \lambda_\mu$ are uniquely determined by the following condition

$$\Phi(a_1) + \sum_{\ell=1}^{\mu} \lambda_\ell \Phi^{(\ell)}(a_1) \in \bigcap_{i=2}^{r} \mathrm{Osc}_{n-\mu_i}\Phi(a_i). \tag{33}$$

The value of the blossom f of any $F \in \mathcal{E}$ at the n-tuple $(a_1{}^{[\mu_1]}, \ldots, a_r{}^{[\mu_r]})$ is then obtained from (32) via affine maps. This yields:

$$f(a_1{}^{[\mu_1]}, \ldots, a_r{}^{[\mu_r]}) = F(a_1) + \sum_{\ell=1}^{\mu} \lambda_\ell F^{(\ell)}(a_1), \tag{34}$$

the λ_ℓ's being those which satisfy (33). Due to our choice for Φ, solving (33) just consists in solving the following linear system of μ equations in μ unknowns:

$$\sum_{\ell=1}^{\mu} \lambda_\ell \Psi_k^{a_i(\ell)}(a_1) = -\Psi_k^{a_i}(a_1), \quad n \geq k \geq n - \mu_i + 1, \ i = 2, \ldots, r, \tag{35}$$

the determinant of which is

$$W(U_{n-1}^{a_2}, \ldots, U_{n-\mu_2}^{a_2}, \ldots, U_{n-1}^{a_r}, \ldots, U_{n-\mu_r}^{a_r})(a_1) = W(\mathbf{1}, \Psi_n^{a_2}, \ldots, \Psi_{n-\mu_r+1}^{a_r})(a_1).$$

Considering (35) and (34) as a linear system of $(\mu+1)$ equations in the μ unknowns $\lambda_1, \ldots, \lambda_\mu$ which possesses a unique solution yields the announced formula (31) $\square$

Remark 28. We would like to draw the reader's attention to the fact that the previous arguments provide us with another interesting way to prove that, as soon as (v)' of Theorem 23 holds, then (24) automatically holds too. Indeed, if a given n-dimensional space $\mathcal{U} \subset C^{n-1}(I)$ satisfies (v)', then blossoms exist in the $(n+1)$-dimensional space $\mathcal{E} := \{F \in C^n(I) \mid F' \in \mathcal{U}\}$. This in turn guarantees that, for any pairwise distinct $a_2, \ldots, a_r$ and any positive $\mu_2 \ldots \mu_r$ such that $\mu_2 + \cdots + \mu_r < n$, all quantities $W(U_{n-1}^{a_2}, \ldots, U_{n-\mu_2}^{a_2}, \ldots, U_{n-1}^{a_r}, \ldots, U_{n-\mu_r}^{a_r})(a_1)$, $a_1 \in I \setminus \{a_2, \ldots, a_r\}$, are not zero, as determinants of regular systems (35).

4.3. Pseudoaffinity

Their geometrical definition makes it obvious that, as soon as blossoms exist in the space $\mathcal{E}$, they are symmetric on I^n and they give the associated function by restriction to the diagonal. They possess another crucial property, which generalises the affinity in each variable satisfied by polynomial blossoms, and which we refer to as *pseudoaffinity* in each variable. Again Theorem 23 will prove to be the key-point to achieve the latter property.

Theorem 29. *Suppose that $\mathcal{U} := D\mathcal{E}$ is an EC-space on I. Then, blossoms are pseudoaffine with respect to each variable, in the sense that, for any $x_1, \ldots, x_{n-1}$, $\gamma, \delta \in I$ with $\gamma < \delta$, for all $x \in I$, and any $F \in \mathcal{E}^d$, we have*

$$\begin{aligned} f(x_1, \ldots, x_{n-1}, x) = \big[1 - \beta(x_1, \ldots, x_{n-1}; \gamma, \delta; x)\big] f(x_1, \ldots, x_{n-1}, \gamma) \\ + \beta(x_1, \ldots, x_{n-1}; \gamma, \delta; x) f(x_1, \ldots, x_{n-1}, \delta), \end{aligned} \tag{36}$$

where $\beta(x_1, \ldots, x_{n-1}; \gamma, \delta; .) : I \to \mathbb{R}$ is a C^1 strictly increasing function (depending on $x_1, \ldots, x_{n-1}, \gamma, \delta$ but independent of F) with $\beta(x_1, \ldots, x_{n-1}; \gamma, \delta; \gamma) = 0$ and $\beta(x_1, \ldots, x_{n-1}; \gamma, \delta; \delta) = 1$.

Proof. Given $(x_1, \ldots, x_{n-1}) \in I^{n-1}$, in order to prove (36) for any $F \in \mathcal{E}^d$, it is sufficient to prove it when F is the mother-function Φ, with the announced properties for the function $\beta(x_1, \ldots, x_{n-1}; \gamma, \delta; .)$. This actually amounts to showing that the function

$$\widehat{\Phi}(x) := \varphi(x_1, \ldots, x_{n-1}, x), \quad x \in I,$$

is one-to-one and C^1 on I with values in an affine line.

1) $\widehat{\Phi}$ is C^1 on I, with values in an affine line. This can be proved using any mother-function Φ. In order to facilitate the proof, we shall adapt our choice to the $(n-1)$-tuple $(x_1, \ldots, x_{n-1})$. Suppose that, up to a permutation, $(x_1, \ldots, x_{n-1}) = (a_1{}^{[\mu_1]}, \ldots, a_r{}^{[\mu_r]})$, with $a_1 < a_2 < \cdots < a_r$ and with positive $\mu_1, \ldots, \mu_r$. Then, we choose our mother-function as

$$\Phi := (\Psi_n^{a_1}, \ldots, \Psi_{n-\mu_1+1}^{a_1}, \ldots, \Psi_n^{a_r}, \ldots, \Psi_{n-\mu_r+1}^{a_r}, \Phi_n)^T, \tag{37}$$

where Φ_n is any element of $\mathcal{E}$ which does not belong to $\text{span}(\Psi_n^{a_1}, \ldots, \Psi_{n-\mu_1+1}^{a_1}, \ldots, \Psi_{n-\mu_r+1}^{a_r}, \mathbb{1})$. According to Theorem 23, this choice is indeed allowed because $\mathcal{U}$ is assumed to be an EC-space on I.

Due to (27), all points $\widehat{\Phi}(x)$, $x \in I$, belong to $\cap_{i=1}^r \text{Osc}_{n-\mu_i} \Phi(a_i)$. This is why any such choice leads to

$$\widehat{\Phi}(x) = (\underbrace{0, \ldots, 0}_{(n-1)\text{ times}}, \widehat{\varphi}_n(x))^T, \quad x \in I,$$

where $\widehat{\varphi}_n := \varphi_n(a_1{}^{[\mu_1]}, \ldots, a_r{}^{[\mu_r]}, .)$. Accordingly, we just have to prove that the function $\widehat{\varphi}_n$ is C^1 on I.

Now, formula (31) yields, for any $x \in I \setminus \{a_1, \ldots, a_r\}$:

$$\widehat{\varphi}_n(x) = \frac{W(\Psi_n^{a_1}, \ldots, \Psi_n^{a_r}, \ldots, \Psi_{n-\mu_r+1}^{a_r}, \Phi_n)(x)}{W(\Psi_n^{a_1}, \ldots, \Psi_n^{a_r}, \ldots, \Psi_{n-\mu_r+1}^{a_r}, \mathbb{1})(x)}. \tag{38}$$

Both the numerator and the denominator in (38) are C^1 on I. Moreover, due to (24), the denominator of (38) never vanishes on $I \setminus \{a_1, \ldots, a_r\}$ since, up to the sign, it is equal to $W(U_{n-1}^{a_1}, \ldots, U_{n-1}^{a_r}, \ldots, U_{n-\mu_r}^{a_r})(x)$. Accordingly, $\widehat{\varphi}_n$ is C^1 on $I \setminus \{a_1, \ldots, a_r\}$. Let us now examine what happens at any of the a_i's. Again, we shall adapt the choice of Φ_n to the point a_i. For instance, consider the case $i = r$. Then, formula (31) gives

$$\widehat{\varphi}_n(a_r) = \frac{W(\Psi_n^{a_1}, \ldots, \Psi_n^{a_{r-1}}, \ldots, \Psi_{n-\mu_{r-1}+1}^{a_{r-1}}, \Phi_n)(a_r)}{W(\Psi_n^{a_1}, \ldots, \Psi_n^{a_{r-1}}, \ldots, \Psi_{n-\mu_{r-1}+1}^{a_{r-1}}, \mathbb{1})(a_r)}. \tag{39}$$

Let I_r denote the largest interval which contains a_r and which is contained in $I \setminus \{a_1, \ldots, a_{r-1}\}$, and let us choose $\Phi_n := \Psi_{n-\mu_r}^{a_r}$. This is allowed by the fact that $\mathcal{U}$ is an EC-space on I. With this choice, by application of (20) to (38) and (39), the values of $\widehat{\varphi}_n$ are given by

$$\widehat{\varphi}_n(x) = \frac{N(x)}{D(x)} \quad \text{for } x \in I_r \setminus \{a_r\}, \quad \widehat{\varphi}_n(a_r) = \frac{F_1^{a_r}(a_r)}{F_0^{a_r}(a_r)}, \tag{40}$$

the functions N and D being defined on the whole of I_r by

$$N(x) := W(F_1^{a_r}, F_2^{a_r}, \ldots, F_{\mu_r+1}^{a_r})(x), \qquad D(x) := W(F_0^{a_r}, F_2^{a_r}, \ldots, F_{\mu_r+1}^{a_r})(x),$$

where

$$F_0^{a_r} := W(\Psi_n^{a_1}, ..., \Psi_n^{a_{r-1}}, ..., \Psi_{n-\mu_{r-1}+1}^{a_{r-1}}, \mathbb{1}) ,$$

$$F_i^{a_r} := W(\Psi_n^{a_1}, ..., \Psi_n^{a_{r-1}}, ..., \Psi_{n-\mu_{r-1}+1}^{a_{r-1}}, \Psi_{i+n-\mu_r-1}^{a_r}) , \quad 1 \le i \le \mu_r + 1.$$

On account of Theorem 23 applied to both spaces $\mathcal{E}$ and $\mathcal{U}$ (that is, in particular, on account of (24)), and of the equality $\sum_{i=1}^r \mu_i = n-1$, Lemma 24 ensures that, for $0 \le i \le \mu_r + 1$, the function $F_i^{a_r}$ vanishes exactly i times at a_r. Lemma 31 below thus guarantees the existence of two nonzero real numbers A, B, such that

$$N(x) \sim A(x-a_r)^{\mu_r+1} , \quad D(x) \sim B(x-a_r)^{\mu_r} \quad \text{for } x \in I_r \setminus \{a_r\} \text{ close to } a_r.$$

Therefore, the left equality in (40) shows that

$$\widehat{\varphi}_n(x) \sim \frac{A}{B}(x-a_r) \quad \text{for } x \in I_r \setminus \{a_r\} \text{ close to } a_r. \tag{41}$$

On the other hand, the right equality in (40) proves that $\widehat{\varphi}_n(a_r) = 0$. Accordingly, the latter relation shows that $\widehat{\varphi}_n$ is differentiable (hence continuous) at a_r. Differentiating $\widehat{\varphi}_n$ on $I_r \setminus \{a_r\}$ through (17) gives

$$\widehat{\varphi}_n'(x) = \frac{W(F_2^{a_r}, F_3^{a_r}, ..., F_{\mu_r+1}^{a_r})(x) W(F_0^{a_r}, F_1^{a_r}, ..., F_{\mu_r+1}^{a_r})(x)}{[W(F_1^{a_r}, F_2^{a_r}, ..., F_{\mu_r+1}^{a_r})(x)]^2}, \quad x \in I_r \setminus \{a_r\}.$$

Using Lemma 31 as previously shows that $\lim_{x\to a_r} \widehat{\varphi}_n'(x)$ exists, which proves that $\widehat{\varphi}_n$ is C^1 at a_r.

2) injectivity of $\widehat{\Phi}$.

For any $F \in \mathcal{E}$ and any $x \in I$, $f(x_1, \ldots, x_{n-1}, x)$ is obtained as the image of $\widehat{\Phi}(x)$ under an affine map. Therefore, in order to prove the injectivity of $\widehat{\Phi}$, it is sufficient to check that, for any $r \ge 2$, any nonnegative μ_1, μ_r, and any positive $\mu_2, \ldots, \mu_{r-1}$ such that $\sum_{i=1}^r \mu_i = n-1$, it is possible to find a function $F \in \mathcal{E}$ such that

$$f({a_1}^{[\mu_1+1]}, {a_2}^{[\mu_2]}, ..., {a_{r-1}}^{[\mu_{r-1}]}, {a_r}^{[\mu_r]}) \neq \\ f({a_1}^{[\mu_1]}, {a_2}^{[\mu_2]}, ..., {a_{r-1}}^{[\mu_{r-1}]}, {a_r}^{[\mu_r+1]}). \tag{42}$$

By application of (31), we know that, whether $\mu_1, \mu_r = 0$ or not, for any $F \in \mathcal{E}$, $f({a_1}^{[\mu_1+1]}, {a_2}^{[\mu_2]}, ..., {a_r}^{[\mu_r]})$ is equal to

$$\frac{W(\Psi_n^{a_2}, ..., \Psi_{n-\mu_2+1}^{a_2}, ..., \Psi_n^{a_r}, ..., \Psi_{n-\mu_r+1}^{a_r}, F)(a_1)}{W(\Psi_n^{a_2}, ..., \Psi_{n-\mu_2}^{a_2}, ..., \Psi_n^{a_r}, ..., \Psi_{n-\mu_r}^{a_r}, \mathbb{1})(a_1)}, \tag{43}$$

while $f({a_1}^{[\mu_1]}, ..., {a_{r-1}}^{[\mu_{r-1}]} {a_r}^{[\mu_r+1]})$ is equal to

$$\frac{W(\Psi_n^{a_1}, ..., \Psi_{n-\mu_1+1}^{a_1}, ..., \Psi_n^{a_{r-1}}, ..., \Psi_{n-\mu_{r-1}+1}^{a_{r-1}}, F)(a_r)}{W(\Psi_n^{a_1}, ..., \Psi_{n-\mu_1}^{a_1}, ..., \Psi_n^{a_{r-1}}, ..., \Psi_{n-\mu_{r-1}}^{a_{r-1}}, \mathbb{1})(a_r)}. \tag{44}$$

Choose $F := \Psi_{n-\mu_r}^{a_r}$. In (44) the numerator is then equal to 0 (see part 1 of the proof). To the contrary, Theorem 23 guarantees that, in (43), it is not equal to 0. Therefore, with this choice of F, (42) is proven. □

Example 30. Consider the polynomial case $\mathcal{E} = \mathcal{P}_n$, with, as usual, $\Psi_k^a(x) := (x-a)^k$. We choose $\Phi_n := \Psi_{n-\mu_r}^{a_r}$. Then, according to (26), there exist two constants Γ_1, Γ_2 such that, for any $x \in \mathbb{R}$,

$$W(\Psi_n^{a_1}, ..., \Psi_{n-\mu_1+1}^{a_1}, ..., \Psi_n^{a_r}, ..., \Psi_{n-\mu_r+1}^{a_r}, \Phi_n)(x) = \Gamma_1 (x-a_r)^{\mu_r+1} \prod_{i=1}^{r-1} (x-a_i)^{\mu_i}$$

and

$$W(U_{n-1}^{a_1}, ..., U_{n-\mu_1}^{a_1}, ..., U_{n-1}^{a_r}, ..., U_{n-\mu_r}^{a_r}, \mathbb{1})(x) = \Gamma_2 \prod_{i=1}^{r} (x-a_i)^{\mu_i}.$$

Take the mother-function Φ as in (37) with our previous choice for Φ_n. Then the previous equalities show that, instead of (41), formulae (38) and (39) lead to

$$\widehat{\phi}_n(x) = \Gamma(x-a_r), \quad x \in I_r,$$

for some non-zero constant Γ. Blossoms for polynomials, defined geometrically as we did, are thus proved to be affine in each variable, which fits in with the algebraic definition by L. Ramshaw [20].

The previous proof made use of the following result an illustration of which is given by formula (26) in the polynomial case.

Lemma 31. *Consider* $(n+1)$ *integers* $k_0, \ldots, k_n$ *such that* $k_0 \geq 0$, $k_i \geq k_{i-1}+1$ *for* $1 \leq i \leq n$, *and* $(n+1)$ *functions* $U_0, \ldots, U_n \in C^{k_n}(I)$. *Suppose that, for* $0 \leq i \leq n$, U_i *vanishes exactly* k_i *times at* $a \in I$. *Then we have, for* $x \in I$ *close to* a *and with* $N := \sum_{i=0}^{n}(k_i - i)$,

$$W(U_0, \ldots, U_n)(x) \sim (x-a)^N \prod_{i=0}^{n} \frac{U_i^{(k_i)}(a)}{k_i!} \prod_{0 \leq i < j \leq n} (k_j - k_i). \tag{45}$$

Proof. The result trivially holds for $n = 0$. Suppose that it is proven for $n-1 \geq 0$ and let us prove it for n. Since U_0 vanishes exactly $k_0 \geq 0$ times at a, we can choose a nontrivial interval I_0 contained in I and containing a such that $U_0(x) \neq 0$ for any $x \in I_0 \setminus \{a\}$. For $1 \leq i \leq n$, we define the function $\widetilde{U}_i$ on $I_0 \setminus \{a\}$ by

$$\widetilde{U}_i(x) := \left(\frac{U_i}{U_0}\right)'(x), \quad x \in I_0 \setminus \{a\}.$$

Each function U_i/U_0 is clearly C^{k_n} on $I_0 \setminus \{a\}$, and according to (5) and (6), we can write

$$H(x) := W(U_0, \ldots, U_n)(x) = [U_0(x)]^{n+1}\, W(\widetilde{U}_1, \ldots, \widetilde{U}_n)(x), \quad x \in I_0 \setminus \{a\}. \tag{46}$$

Now, one can actually prove that, for $0 \leq i \leq n$, the function $\overline{U}_i(x)$ defined by

$$\overline{U}_i(x) := \frac{U_i(x)}{(x-a)^{k_0}} \quad \text{if } x \in I_0 \setminus \{a\}, \quad \overline{U}_i(a) := \frac{U_i^{(k_0)}(a)}{k_0!},$$

is $C^{k_n-k_0}$ on I_0. The fact that $\overline{U}_i$ is $(k_n - k_0)$-times continuously differentiable at a requires a long technical proof that can found in [8]. Clearly, $\overline{U}_i$ vanishes exactly $k_i - k_0$ times at a. For $1 \le i \le n$, we can thus extend $\widetilde{U}_i$ by continuity on the whole of I_0 by setting

$$\widetilde{U}_i(x) := \left(\frac{\overline{U}_i}{\overline{U}_0}\right)'(x), \quad x \in I_0.$$

The function $\widetilde{U}_i$ thus belongs to $C^{k_n-k_0-1}(I_0)$. We can therefore introduce the function

$$\widetilde{H}(x) := W(\widetilde{U}_1, \ldots, \widetilde{U}_n)(x) \ , \quad x \in I_0.$$

It is continuous on I_0 and equality (46) can now be extended by continuity at a, so that

$$H(x) = [U_0(x)]^{n+1} \ \widetilde{H}(x), \quad x \in I_0. \tag{47}$$

Moreover, for $1 \le i \le n$, the function $\widetilde{U}_i$ vanishes exactly $(k_i - k_0 - 1)$ times at a. We can apply the recursive assumption to $\widetilde{H}$. It guarantees that

$$\widetilde{H}(x) \sim (x-a)^{\widetilde{N}} \widetilde{A} \quad \text{for } x \in I_0 \text{ close to } a, \tag{48}$$

with $\widetilde{N} := \sum_{i=1}^{n} (k_i - k_0 - i)$ and with

$$\widetilde{A} := \prod_{i=1}^{n} \frac{\widetilde{U}_i^{(k_i-k_0-1)}(a)}{(k_i - k_0 - 1)!} \prod_{1 \le i < j \le n} (k_j - k_i)$$

The two relations (47) and (48) yield (45) after observing that

$$\frac{\widetilde{U}_i^{(k_i-k_0-1)}(a)}{(k_i - k_0 - 1)!} = (k_i - k_0) \frac{U_i{}^{(k_i)}(a)}{k_i!} \frac{k_0!}{U_0{}^{(k_0)}(a)} \ , \quad 1 \le i \le n,$$

which easily follows from the definition of the $\widetilde{U}_i$'s. □

Remark 32. The arguments used in the proof of Lemma 31 actually state that any sufficiently differentiable functions $U_0, \ldots, U_n$ on I which vanish exactly $k_0, \ldots, k_n$ times at $a \in I$, respectively, with $k_0 \le k_1 \le \cdots \le k_n$, form a CW-system on $]a, a+\eta[$ and/or $[a-\eta, a[$ for some sufficiently small positive η. Note that relation (45) is consistent with the particular case $U_i(x) = (x-a)^{k_i}$, $0 \le i \le n$, for which we have $W(U_0, \ldots, U_n)(x) = (x-a)^N \prod_{0 \le i < j \le n} (k_j - k_i)$ for all $x \in \mathbb{R}$.

4.4. Blossoms, Bézier points, and Bernstein bases

Clearly, if blossoms exist in $\mathcal{E}$, then Bézier points exist in $\mathcal{E}$. As a consequence of Theorem 12, Theorem 34 below states the converse, along with the equivalence between existence of blossoms and existence of Bernstein bases [14], according to the following definition.

Definition 33. In $\mathcal{E}$, a Bernstein-like basis $(B_0, \ldots, B_n)$ relative to $(a, b) \in I^2$, $a \neq b$, is said to be the *Bernstein basis relative to* (a, b) if it is normalized (*i.e.*, if it satisfies $\sum_{i=0}^n B_i = \mathbb{1}$), and if each B_i is positive strictly between a and b.

As is classical, when writing any $F \in \mathcal{E}^d$ as $F(x) = \sum_{i=0}^n F_i(x)P_i$, $x \in I$, where $(F_0, \ldots, F_n)$ is a normalised basis of $\mathcal{E}$, we call the points $P_0, \ldots, P_n \in \mathbb{R}^d$ *the control points* of F relative to $(F_0, \ldots, F_n)$.

Theorem 34. *Let $\mathcal{E} \subset C^n(I)$ be an $(n+1)$-dimensional W-space on I containing constants, and let $\mathcal{U}$ denote the space $D\mathcal{E}$. The following four properties are then equivalent:*

(a) *The space $\mathcal{U}$ is an EC-space on I.*
(b) *Blossoms exist in the space $\mathcal{E}$.*
(c) *Bézier points exist in the space $\mathcal{E}$.*
(d) *The space $\mathcal{E}$ possesses a normalised Bernstein-like basis relative to any pair of distinct points of I.*

Moreover, if the latter properties are satisfied, then:

(e) *Relative to any given $(a, b) \in I^2$, $a \neq b$, the normalised Bernstein-like basis $(B_0, \ldots, B_n)$ is automatically positive strictly between a and b, i.e., it is the Bernstein basis. The control points of any $F \in \mathcal{E}^d$ relative to $(B_0, \ldots, B_n)$ are its Bézier points relative to (a, b).*

Proof. Suppose that $\mathcal{E}$ is a W-space on I, or, equivalently, that $\mathcal{U}$ is a W-space on I (see Remark 3,(2)). Then, for any given $(a, b) \in I^2$, $a \neq b$, and for $1 \leq i \leq n-1$, the Bézier point $\varphi(a^{[n-i]}, b^{[i]})$ exists iff there exit unique real numbers $\lambda_1, \ldots, \lambda_i$, $\varrho_1, \ldots, \varrho_{n-i}$ such that

$$\Phi(b) - \Phi(a) = \sum_{k=1}^{i} \lambda_k \Phi^{(k)}(a) + \sum_{\ell=1}^{n-i} \varrho_\ell \Phi^{(\ell)}(b).$$

that is, iff the n vectors $\Phi'(a), \ldots, \Phi^{(i)}(a), \Phi'(b), \ldots, \Phi^{(n-i)}(b)$ are linearly independent. Accordingly, the equivalence between (a) and (c) results from applying Theorem 12 to the space $\mathcal{U}$. Taking account of Theorem 27, and of the implication (b) $\Rightarrow$ (c) being obvious, we thus have proved that (a) $\Leftrightarrow$ (b) $\Leftrightarrow$ (c).

Supposing that (a) holds, let us now briefly recall why both (d) and (e) are then satisfied. We know not only that blossoms exist, but also that they are symmetric and pseudoaffine in each variable (Theorem 29). Given $a, b, x \in I$, with $a < b$, applying (36) with $(x_1, \ldots, x_{n-1}) := (a^{[n-i-k]}, b^{[i]}, x^{[k-1]})$ for some integers k, i such that $1 \leq k \leq n$ and $0 \leq i \leq n-k$, provides us with the existence of a one-to-one function $\alpha_{i,k} : I \to \mathbb{R}$ (depending on a, b, x), satisfying

$$\alpha_{i,k}(a) = 0, \ \alpha_{i,k}(b) = 1, \ \alpha_{i,k}(y) > 0 \quad \text{for } y \in]a, b[, \tag{49}$$

and such that, for all $F \in \mathcal{E}^d$, we have in particular:

$$f(a^{[n-i-k]},b^{[i]},x^{[k]}) = (1-\alpha_{i,k}(x))\ f(a^{[n-i-k+1]},b^{[i]},x^{[k-1]})$$
$$+\alpha_{i,k}(x)\ f(a^{[n-i-k]},b^{[i+1]},x^{[k-1]})\ , \quad x \in I. \quad (50)$$

Equality (50) describes the well-known *de Casteljau algorithm* relative to $[a,b]$. For $k=n$, it gives, for any $x \in I$, the point $f(x^{[n]}) = F(x)$ as an affine combination of the $(n+1)$ points Bézier points of F relative to (a,b), *i.e.*,

$$F(x) = \sum_{i=0}^{n} B_i(x)\ f(a^{[n-i]},b^{[i]})\ , \quad \sum_{i=0}^{n} B_i(x) = 1\ , \quad x \in I\ , \quad (51)$$

the coefficients being independent of F. Relations (51) prove that the functions $B_0,\dots,B_n$ provided by the de Casteljau algorithm form a normalised basis of the space $\mathcal{E}$. On the other hand, from (49) we can deduce that each affine combination (50) is a strictly convex one when $x \in]a,b[$, which proves the positivity of $B_0,\dots,B_n$ on $]a,b[$. Applying (51) to a mother-function Φ, the geometrical meaning (29) and (30) of its Bézier points guarantees that $(B_0,\dots,B_n)$ is a Bernstein-like basis relative to (a,b). It is thus the Bernstein basis relative to any (a,b). Hence, both (d) and (e) are satisfied.

Given distinct $a,b \in I$, suppose there exists of a normalised Bernstein-like basis $(B_0,\dots,B_n)$ in the space $\mathcal{E}$ relative to (a,b). Due to the normalisation, the functions $\widetilde{B}_i := B_0' + \cdots B_i'$, $0 \le i \le n-1$, form a Bernstein-like basis relative to (a,b) in the space $\mathcal{U}$. Accordingly, when (d) holds, the space $\mathcal{U}$ possesses Bernstein-like bases relative to any pair of distinct points of I. Theorem 12 ensures that (a) holds, and the proof of Theorem 34 is now complete. □

4.5. Blossoms, B-spline bases, and Total Positivity

Throughout this subsection we assume that $\mathcal{E}$ is a W-space as in Theorem 34. In the previous subsection we saw the equivalence between existence of blossoms and existence of Bernstein bases in the space $\mathcal{E}$. For the sake of completeness, we shall now briefly see the links between blossoms and B-spline bases of which Bernstein bases can actually be considered a special case. We shall also recall why these bases have nice shape preserving properties.

Given a sequence of *knots* $\eta_0 < \eta_1 < \cdots < \eta_q < \eta_{q+1}$ in the interval I, and an associated sequence of *multiplicities* $m_0,\dots,m_{q+1}$, with $1 \le m_i \le n$ for $1 \le i \le q$ and $m_0 = m_{q+1} = n+1$, we define the corresponding *knot vector* as the $(m+2n+2)$-tuple (where $m := \sum_{i=1}^{q} m_i$ is the sum of the multiplicities of all interior knots)

$$\mathcal{K} := (\xi_{-n},\dots,\xi_{m+n+1}) := ({\eta_0}^{[m_0]}, {\eta_1}^{[m_1]},\dots,{\eta_q}^{[m_q]},{\eta_{q+1}}^{[m_{q+1}]})\ . \quad (52)$$

Associated with the knot vector $\mathcal{K}$, we define the *spline space based on* $\mathcal{E}$, as the $(n+m+1)$-dimensional space $\mathcal{S}(\mathcal{E},\mathcal{K})$ composed of all functions $S : [\eta_0,\eta_{q+1}] \to \mathbb{R}$ such that

(S)$_1$ For $1 \le i \le q$, S is C^{n-m_i} at η_i;

(S)$_2$ For $0 \le i \le q$, there exists a function $F_i \in \mathcal{E}$ such that

$$S(x) = F_i(x)\ , \quad x \in [\eta_i, \eta_{i+1}]\ .$$

With the knot vector $\mathcal{K}$ we also associate the set $\mathcal{A}(\mathcal{K})$ composed of all n-tuples $(x_1, \ldots, x_n) \in [\eta_0, \eta_{q+1}]^n$ which are *admissible relative to* $\mathcal{K}$, that is, such that, whenever $\min(x_1, \ldots, x_n) < \eta_\ell < \max(x_1, \ldots, x_n)$ for some integer ℓ, $1 \le \ell \le q$, then at least m_ℓ among the points $x_1, \ldots, x_n$ are equal to η_ℓ. The set $\mathcal{A}(\mathcal{K})$ is symmetric and it contains the diagonal of $[\eta_0, \eta_{q+1}]^n$. In the particular case $q = 0$, $\mathcal{A}(\mathcal{K}) = [\eta_0, \eta_1]^n$. Exactly as we did with the space $\mathcal{E}$, we select a mother-spline function $\Sigma := (\Sigma_1, \ldots, \Sigma_{n+m})^T$, meaning that $(\mathbb{1}, \Sigma_1, \ldots, \Sigma_{n+m})$ is a basis of $\mathcal{S}(\mathcal{E}, \mathcal{K})$. Then, any spline $S \in \mathcal{S}(\mathcal{E}, \mathcal{K})^d$ is the image of Σ under a unique affine map on $\mathbb{R}^{n+m}$. For $0 \le \ell \le q$, denote by Φ_ℓ the unique element of $\mathcal{E}^{n+m}$ such that

$$\Sigma(x) = \Phi_\ell(x), \quad x \in [\eta_\ell, \eta_{\ell+1}],\ 0 \le \ell \le q. \tag{53}$$

Given an admissible n-tuple $(x_1, \ldots, x_n)$, equal, up to a permutation, to $(\tau_1{}^{[\mu_1]}, \ldots, \tau_r{}^{[\mu_r]})$, with $\tau_1 < \cdots < \tau_r$ and positive μ_i's, we then have (see [9])

$$\bigcap_{i=1}^{r} \mathrm{Osc}_{n-\mu_i}\Sigma(\tau_i) = \bigcap_{i=1}^{r} \mathrm{Osc}_{n-\mu_i}\Phi_j(\tau_i)\ , \quad j \in \mathcal{J}(x_1, \ldots, x_n)\ , \tag{54}$$

where $\mathcal{J}(x_1, \ldots, x_n)$ is some nonempty set composed of consecutive integers. More precisely, in case there exists at least one integer $\ell \in \{1, \ldots, q\}$ such that η_ℓ appears at least m_ℓ times in the sequence $(x_1, \ldots, x_n)$, denoting by ℓ_1 and ℓ_2 the smallest and greatest such integer, respectively, we have $\mathcal{J}(x_1, \ldots, x_n) := \{\ell_1 - 1, \ldots, \ell_2\}$. Otherwise, $x_1, \ldots, x_n$ all belong to the same interval $[t_{\ell_0}, t_{\ell_0+1}]$ and we then have $\mathcal{J}(x_1, \ldots, x_n) := \{\ell_0\}$. Due to the admissibility, all osculating flats $\mathrm{Osc}_{n-\mu_i}\Sigma(\tau_i)$ involved in (54) are well defined, except possibly for the first and last ones, which, if necessary, must be interpreted as $\mathrm{Osc}_{n-\mu_1}\Sigma(\tau_1^+)$ and $\mathrm{Osc}_{n-\mu_r}\Sigma(\tau_r^-)$, respectively.

Suppose that blossoms exist in the space $\mathcal{E}$. Then, equality (54) shows that its left-hand side consists of a single point which is labeled as $\sigma(x_1, \ldots, x_n)$. The function σ, called the blossom of the mother spline function Σ, is then defined on the set $\mathcal{A}(\mathcal{K})$. Blossoms for other spline functions are defined from σ via affine maps. Due to (54), except that instead of being defined on I^n they are defined only on $\mathcal{A}(\mathcal{K})$, blossoms for splines possess the same properties as blossoms in the space $\mathcal{E}$: they are symmetric, pseudoaffine within $\mathcal{A}(\mathcal{K})$, and possess the diagonal property: $\Sigma(x) = \sigma(x^{[n]})$ for all $x \in [\eta_0, \eta_{q+1}]$. Classically, given x in some $[\eta_\ell, \eta_{\ell+1}]$, $0 \le \ell \le q$, for any spline $S \in \mathcal{S}(\mathcal{E}, \mathcal{K})^d$, the latter properties make it possible to develop an n-step *de Boor algorithm* to evaluate $S(x)$ as a strictly convex combination of at most $(n+1)$ consecutive *poles of* S, given that the $(n+m+1)$ poles of S are defined as the points

$$s(\xi_{i+1}, \ldots, x_{i+n}), \quad -n \le i \le m. \tag{55}$$

where s is the blossoms of S. This generalises the de Casteljau algorithm (50) and it eventually leads to

$$S(x) = \sum_{i=-n}^{m} N_i(x) s(\xi_{i+1}, \ldots, x_{i+n}), \quad \sum_{i=-n}^{m} N_i(x) = 1, \quad x \in [\eta_0, \eta_{q+1}]. \tag{56}$$

Due to (56), the functions N_i, $-n \le j \le m$, so obtained form a normalised basis of the spline space $\mathcal{S}(\mathcal{E}, \mathcal{K})$. Taking account of the geometrical definition of the blossom σ of the mother-spline function Σ through (54) and (27), they form *the B-spline basis* of $\mathcal{S}(\mathcal{E}, \mathcal{K})$, according to the definition below.

Definition 35. Given $(n + m + 1)$ functions $N_{-n}, \ldots, N_m$ in the spline space $\mathcal{S}(\mathcal{E}, \mathcal{K})$, we say that they form ***the B-spline basis*** if they satisfy the following four properties:

(i) $(\mathrm{BSB})_1$ *support property:* For all $j \in \{-n, \ldots, m\}$, the support of N_j is $[\xi_j, \xi_{j+n+1}]$.

(ii) $(\mathrm{BSB})_2$ *end point property:* For all $j \in \{-n, \ldots, m\}$, the function N_j vanishes exactly $(n - m_k + p + 1)$ times at the left end point $\xi_j = \eta_k$ of its support, and exactly $(n - m_{k'} + p' + 1)$ times at its right end point $\xi_{j+n+1} = \eta_{k'}$, where

$$p := \sharp\{\ell < j \mid \xi_\ell = \xi_j\}, \quad p' := \sharp\{\ell > j + n + 1 \mid \xi_\ell = \xi_{j+n+1}\}.$$

(iii) $(\mathrm{BSB})_3$ *normalisation property:* $\sum_{j=-n}^{m} N_j(x) = 1$ for all $x \in I$;

(iv) $(\mathrm{BSB})_4$ *positivity property:* for all $j \in \{-n, \ldots, m\}$, N_j is positive on the interior of its support.

If $N_{-n}, \ldots, N_m$ satisfy only the first two properties $(\mathrm{BSB})_1$ and $(\mathrm{BSB})_2$, we say that they form a ***B-spline-like basis.***

When the knot vector is $\mathcal{K} = (\eta_0{}^{[m_0]}, \eta_1{}^{[m_1]})$, *i.e.*, when $q = 0$, the spline space $\mathcal{S}(\mathcal{E}, \mathcal{K})$ is just the space $\mathcal{E}$ restricted to $[\eta_0, \eta_1]$, and in the latter space being a B-spline-like basis (resp. the B-spline basis) is the same as being a Bernstein-like basis (resp. the Bernstein basis) relative to (η_0, η_1).

On account of the previous observation and of Theorem 34, we can thus state the equivalence between existence of blossoms and existence of B-spline bases as follows:

Theorem 36. *The space $\mathcal{U}$ is an EC space on I if and only if B-spline bases exist in the sense that*

(f) *for any knot vector $\mathcal{K}$, the space $\mathcal{S}(\mathcal{E}, \mathcal{K})$ possesses a normalised B-spline-like basis.*

Moreover, when (f) *holds, all normalised B-spline-like bases are B-spline bases, that is, they satisfy the positivity property* $(\mathrm{BSB})_4$.

For the sake of completeness, we shall briefly mention further essential properties of the bases obtained in Theorems 34 and 36. Let us recall that a basis $(U_0, \ldots, U_n)$ of a given functional space on some interval J is said to be *Totally Positive* when, for any sequence $x_0 < x_1 < \cdots < x_{n-1} < x_n$ in J, the matrix $\big(U_i(x_j)\big)_{0 \le i,j \le n}$ is totally positive, *i.e.*, all its minors are nonnegative [7]. The utmost importance of Normalised Totally Positive (NTP) bases in geometric design, due to the fact that they guarantee good shape preserving properties, is clearly pointed out in [4,3] for instance. Let us just recall that when a functional space on $[a,b]$, $a<b$, possesses an NTP basis, it automatically possesses an optimal one, that is, a unique NTP basis $(U_0,\dots,U_n)$ from which any other NTP basis $(V_0,\dots,V_n)$ can be obtained via multiplication by a totally positive matrix, *i.e.*, such that the matrix M of the equality $(V_0,\dots,V_n)=(U_0,\dots,U_n)M$ is totally positive. The optimal NTP basis has optimal shape preserving properties (see [3]).

Let us get back to the de Casteljau algorithm (50) relative to $(a,b)\in I^2$, $a<b$. We have already mentioned that each affine combination (50) is a strictly convex one whenever x belongs to $]a,b[$. This implies that the de Casteljau algorithm is a so-called *corner cutting algorithm* on $[a,b]$. The importance of this remark is not limited to the fact that it proves the positivity on $]a,b[$ of the corresponding Bernstein basis: indeed, it also implies its total positivity on $[a,b]$. Hence, it is an NTP basis on $[a,b]$. As a matter of fact, the number of zeros of its elements at the end points even guarantees its optimality. The same arguments are more generally valid for de Boor algorithms and B-spline bases, as stated in the following theorem as a conclusion of this section (see details in [11]).

Theorem 37. *Suppose that any of the properties* (a)-(d) *of Theorem* 34 *is satisfied. Then, for any knot vector* $\mathcal{K}$, *the B-spline basis is the optimal NTP basis of the space* $\mathcal{S}(\mathcal{E},\mathcal{K})$. *In particular, the Bernstein basis relative to any* $(a,b)\in I^2$, $a<b$, *is the optimal NTP basis of the space* $\mathcal{E}$ *restricted to* $[a,b]$.

5. Extended Chebyshev Piecewise Spaces

From a design point of view, in order to take full advantage of the richness of shape effects EC-spaces provide, it is desirable to be able to consider curves with sections in different such spaces while maintaining existence of blossoms. This is why the last two sections are devoted to the extension of the results described so far to the piecewise framework.

5.1. Counting zeros for piecewise functions

Let $\mathcal{Z}$ denote any (finite, infinite, or bi-infinite) subset of $\mathbb{Z}$ composed of consecutive integers (at least two). From now on, we consider a fixed sequence of intervals, $\mathcal{T}=([t_\ell,t_{\ell+1}])_{\ell\in\mathcal{Z}}$, with $t_\ell<t_{\ell+1}$, and the interval $I:=\cup_{\ell\in\mathcal{Z}}[t_\ell,t_{\ell+1}]$. Let $\mathcal{Z}_0$ denote the set of all integers $\ell\in\mathcal{Z}$ such that t_ℓ is in the interior of I. The word "piecewise" will always refer to the sequence $\mathcal{T}$. We shall say that F is a *piecewise function on I* if F is defined separately on each interval $[t_\ell,t_{\ell+1}]$, implying in particular that, for any $\ell\in\mathcal{Z}_0$, both $F(t_\ell^-)$ and $F(t_\ell^+)$ are defined while a priori $F(t_\ell)$ is not. In such a case, unless explicitly mentioned, F is not a function on I. We shall deliberately

use the abusive notation $F : \cup_{\ell\in\mathcal{Z}}[t_\ell, t_{\ell+1}] \to \mathbb{R}$ to stress this fact. We shall say that F is a *piecewise C^n function on I* if F is C^n on each interval $[t_\ell, t_{\ell+1}]$. Given two piecewise functions F and G, the equality $F = G$ thus means that $F(x) = G(x)$ for all $x \in I \setminus \{t_\ell\,,\, \ell \in \mathcal{Z}_0\}$ and both $F(t_\ell^-) = G(t_\ell^-)$ and $F(t_\ell^+) = G(t_\ell^+)$ for all $\ell \in \mathcal{Z}_0$, which we shall sometimes sum up as $F(x^\varepsilon) = G(x^\varepsilon)$ for all $x \in I$ and for $\varepsilon \in \{-,+\}$. Of course in case x is an end point of I, then ε must be the appropriate sign.

Using notations similar to those used in Lemma 24, let us set, for any piecewise C^n function F, any $x \in I$ and for $\varepsilon \in \{-,+\}$,

$$\Delta_n F(x^\varepsilon) := \left(F(x^\varepsilon), F'(x^\varepsilon), \ldots, F^{(n)}(x^\varepsilon)\right)^T.$$

Given a sequence $\mathcal{M} := \left(\mathcal{M}_\ell\right)_{\ell\in\mathcal{Z}_0}$, where each $\mathcal{M}_\ell := \left(m_{i,j}^\ell\right)_{0\le i,j\le n}$ is a lower triangular matrix with positive diagonal elements, we introduce the space $PC^n(\mathcal{T},\mathcal{M})$ of all piecewise C^n functions $F : \cup_{\ell\in\mathcal{Z}}[t_\ell, t_{\ell+1}] \to \mathbb{R}$ satisfying the connection conditions

$$\Delta_n F(t_\ell^+) = \mathcal{M}_\ell \,.\, \Delta_n F(t_\ell^-)\ , \quad \ell \in \mathcal{Z}_0. \tag{57}$$

In this section the space $PC^n(\mathcal{T},\mathcal{M})$ will play the same rôle as the space $C^n(I)$ in the previous ones. The interest of considering such lower triangular matrices with positive diagonal elements will be justified geometrically in Subsection 6.2. At this stage, let us just point out that the regular lower triangular structure enables us to count zeros in the space $PC^n(\mathcal{T},\mathcal{M})$, for, if $\ell \in \mathcal{Z}_0$, t_ℓ^+ is a zero of (exact) multiplicity k iff so is t_ℓ^-. Hence, for any $F \in PC^n(\mathcal{T},\mathcal{M})$, we can introduce the number $Z_{n+1}(F)$ with the same meaning as in subsection 2.1. Moreover, for any $a \in I$, all subspaces $\mathcal{E}_k(a)$ introduced in (3) are well-defined. The interest of the additional positivity assumption on the diagonal elements lies in the fact that it makes possible piecewise versions of Rolle's theorem and (9) (see Lemma 38 below).

From now on, D denotes the piecewise differentiation. Given a piecewise function $F \in PC^n(\mathcal{T},\mathcal{M})$, its piecewise derivative DF is obviously a piecewise C^{n-1} function. However, we generally cannot count its zeros as we can for F unless we are in the special case examined in the lemma below.

Lemma 38. *Given a sequence $\mathcal{M}$ of lower triangular matrices with positive diagonal elements, we suppose that each $\mathcal{M}_\ell$, $\ell \in \mathcal{Z}_0$, has $(m_{0,0}^\ell, 0\ldots,0)^T$ as its first column and we denote by $\widetilde{\mathcal{M}_\ell}$ the square matrix of order $n-1$ obtained by deleting the first row and column of $\mathcal{M}_\ell$. Then, the piecewise derivative DF of any $F \in PC^n(\mathcal{T},\mathcal{M})$ belongs to $PC^{n-1}(\mathcal{T},\widetilde{\mathcal{M}})$, where $\widetilde{\mathcal{M}} := \left(\widetilde{\mathcal{M}_\ell}\right)_{\ell\in\mathcal{Z}_0}$, and we have*

$$Z_n(DF) \ge Z_{n+1}(F) - 1\ .$$

Proof. The first statement is obvious. Due to our observations above, if F vanishes k times at a given $a \in \mathbb{R}$, $1 \le k \le n+1$, DF vanishes $k-1$ times at a. On the other hand, for each $\ell \in \mathcal{Z}_0$, we have $DF(t_\ell^+) = m_{1,1}^\ell DF(t_\ell^-)$. All m_{11}^ℓ's being positive, Rolle's theorem is valid for F: indeed, whenever $F(a) = F(b) = 0$ with $a < b$, this implies the existence of some $\xi \in]a,b[$ such that $DF(\xi) = 0$, the latter equality meaning $DF(t_\ell^-) = DF(t_\ell^+) = 0$ in case $\xi = t_\ell$ for some $\ell \in \mathcal{Z}_0$. □

Lemma 39. *Let $w : \cup_{\ell \in \mathcal{Z}}[t_\ell, t_{\ell+1}] \to \mathbb{R}$ be a piecewise C^n function satisfying $w(t_\ell^-)w(t_\ell^+) > 0$ for all $\ell \in \mathcal{Z}_0$, and let $\mathcal{M} := (\mathcal{M}_\ell)_{\ell \in \mathcal{Z}_0}$ be a sequence of lower triangular matrices with positive diagonal elements. Then, for all $F \in PC^n(\mathcal{T}, \mathcal{M})$, the piecewise function wF belongs to $PC^n(\mathcal{T}, \mathcal{R})$, where $\mathcal{R}$ stands for the sequence composed of all matrices $\mathcal{R}_\ell$ (lower triangular with positive diagonal elements) defined by*

$$\mathcal{R}_\ell := \mathcal{C}_n(w, t_\ell^+) \,.\, \mathcal{M}_\ell \,.\, \mathcal{C}_n(w, t_\ell^-)^{-1}, \quad \ell \in \mathcal{Z}_0, \tag{58}$$

where, for $x \in I$ and $\varepsilon \in \{-,+\}$, $\mathcal{C}_n(w, x^\varepsilon) = \big(\mathcal{C}_n(w, x^\varepsilon)_{p,q}\big)_{0 \le p,q \le n}$ stands for the lower triangular square matrix defined by $\mathcal{C}_n(w, x^\varepsilon)_{p,q} := \binom{p}{q}\, w^{(p-q)}(x^\varepsilon)$ for $0 \le q \le p \le n$.

Proof. Equality (58) is a straightforward consequence of both (57) and of the Leibniz' formula to differentiate wF up to order n at t_ℓ^- and t_ℓ^+, $\ell \in \mathcal{Z}_0$. The positivity of the diagonal elements of each lower triangular matrix $\mathcal{R}_\ell$ is due to our assumption $w(t_\ell^-)w(t_\ell^+) > 0$. □

Lemma 40. *With the same notations as in Lemma 39, suppose that a given piecewise function $U_0 \in PC^n(\mathcal{T}, \mathcal{M})$ never vanishes on I. Then for any $F \in PC^n(\mathcal{T}, \mathcal{M})$, the piecewise function $D(F/U_0)$ belongs to $PC^{n-1}(\mathcal{T}, \mathcal{N})$, where each matrix $\mathcal{N}_\ell$, $\ell \in \mathcal{Z}_0$, (lower triangular with positive diagonal elements) is obtained from the matrix*

$$\mathcal{R}_\ell := \mathcal{C}_n(U_0, t_\ell^+)^{-1} .\, \mathcal{M}_\ell \,.\, \mathcal{C}_n(U_0, t_\ell^-)$$

by deleting its first row and column. Moreover, any $F \in PC^n(\mathcal{T}, \mathcal{M})$ satisfies

$$Z_n\big(D(F/U_0)\big) \ge Z_{n+1}(F) - 1 \,. \tag{59}$$

Proof. First note that, due to the positivity of all diagonal elements $m_{0,0}^\ell$, the piecewise function U_0 keeps the same strict sign on $\cup_{\ell \in \mathcal{Z}}[t_\ell, t_{\ell+1}]$. Let F belong to $PC^n(\mathcal{T}, \mathcal{M})$. From Lemma 39 we know that F/U_0 belongs to $PC^n(\mathcal{T}, \mathcal{R})$, where each matrix $\mathcal{R}_\ell$ is given by (58), with $w := 1/U_0$. The announced equality follows, taking account of the equality

$$\mathcal{C}_n(w_1, x^\varepsilon) \,.\, \mathcal{C}_n(w_2, x^\varepsilon) = \mathcal{C}_n(w_1 w_2, x^\varepsilon),$$

valid for any piecewise C^n functions w_1, w_2. The special case $f := U_0$ yields $(1, 0, \ldots, 0)^T = \mathcal{R}_\ell \,.\, (1, 0, \ldots, 0)^T$ for all $\ell \in \mathcal{Z}_0$, which means that the first column of each $\mathcal{R}_\ell$ is equal to $(1, 0, \ldots, 0)^T$. The fact that $D(F/U_0)$ belongs to $PC^{n-1}(\mathcal{T}, \mathcal{N})$ results from Lemma 38, and so does relation (59). □

5.2. W-Piecewise spaces, EC-Piecewise spaces

We shall use the expression *piecewise space* for a linear space composed of piecewise functions. In this subsection, we suppose that $\mathcal{E}$ is an $(n+1)$-dimensional subspace of $PC^n(\mathcal{T},\mathcal{M})$. We now consider *piecewise Hermite interpolation problems in* $n+1$ *data in the piecewise space* $\mathcal{E}$, that is, problems of the following form:

$$\text{Find } F\in\mathcal{E} \text{ such that } F^{(j)}(\tau_i^{\varepsilon_i})=\alpha_{i,j}\ , \quad 1\le i\le r,\ 0\le j\le \mu_i-1\ , \qquad (60)$$

for given pairwise distinct $\tau_1,\dots,\tau_r\in\mathbb{R}$, given $\varepsilon_1,\dots\varepsilon_r\in\{-,+\}$, given positive integers $\mu_1,\dots,\mu_r$ such that $\sum_{i=1}^r\mu_i=n+1$, and given real numbers $\alpha_{i,j}$, $1\le i\le r$, $0\le j\le\mu_i-1$. Of course, whenever τ_i is not one of the interior t_ℓ's we can omit ε_i in (60). Piecewise Taylor interpolation problems correspond to the case $r=1$. We can state piecewise versions of Definitions 1 and 5 and Propositions 2 and 6.

Proposition and Definition 41. The $(n+1)$-dimensional piecewise space $\mathcal{E}\subset PC^n(\mathcal{T},\mathcal{M})$ is said to be a *W-Piecewise space* (W-Pspace) on $I=\cup_{\ell\in\mathcal{Z}}[t_\ell,t_{\ell+1}]$ if it meets any of the following equivalent requirements:

(Pi) Any piecewise Taylor interpolation problem has a unique solution in $\mathcal{E}$.

(Pii) Any nonzero element $F\in\mathcal{E}$ vanishes at most n times at any given point of I, i.e., $Z_{n+1}^{\{a\}}(F)\le n$ for all $a\in I$.

(Piii) The (possibly left or right) Wronskian $W(F_0,\dots,F_n)$ never vanishes on I, i.e., for all $x\in I$ and for $\varepsilon\in\{-,+\}$,

$$W(F_0,\dots,F_n)(x^\varepsilon):=\det\big(\mathbb{F}(x^\varepsilon),\mathbb{F}'(x^\varepsilon),\dots,\mathbb{F}^{(n)}(x^\varepsilon)\big)\ne 0\ .$$

(Piv) For any $a\in I$, there exists a basis $(\Psi_0^a,\Psi_1^a,\dots,\Psi_n^a)$ in $\mathcal{E}$ such that, for $0\le i\le n$, the piecewise function Ψ_i^a vanishes exactly i times at a.

(Pv) For any $a\in I$, $\mathcal{E}_{n+1}(a)=\{0\}$.

Proposition and Definition 42. The $(n+1)$-dimensional piecewise space $\mathcal{E}\subset PC^n(\mathcal{T},\mathcal{M})$ is said to be an *EC-Piecewise space* (EC-Pspace) on $I=\cup_{\ell\in\mathcal{Z}}[t_\ell,t_{\ell+1}]$ if it meets any of the following equivalent requirements:

(Pi)′ Any piecewise Hermite interpolation problem (60) has a unique solution in $\mathcal{E}$.

(Pii)′ Any nonzero element $F\in\mathcal{E}$ vanishes at most n times on I, counting multiplicities, i.e., $Z_{n+1}(F)\le n$.

(Piii)′ For any $r\ge 1$, any positive integers $\mu_1,\dots,\mu_r$ such that $\sum_{i=1}^r\mu_i=n+1$, and any pairwise distinct $a_1,\dots,a_r\in I$, and any $\varepsilon_1,\dots,\varepsilon_r\in\{-,+\}$, the determinant

$$\det\big(\mathbb{F}(a_1^{\varepsilon_1}),\mathbb{F}'(a_1^{\varepsilon_1}),...,\mathbb{F}^{(\mu_1-1)}(a_1^{\varepsilon_1}),...,\mathbb{F}(a_r^{\varepsilon_r}),...,\mathbb{F}^{(\mu_r-1)}(a_r^{\varepsilon_r})\big), \qquad (61)$$

is not equal to zero.

Remark 43. (1) The $(n+1)$-dimensional subspace $\mathcal{E}$ of $PC^n(\mathcal{T},\mathcal{M})$ is a W-Pspace on I if and only, for each $\ell\in\mathcal{Z}$, its restriction $\mathcal{E}_\ell$ to $[t_\ell,t_{\ell+1}]$ is a W-space on $[t_\ell,t_{\ell+1}]$.

(2) If $\mathcal{E}$ is an EC-Pspace on I, then for each $\ell \in \mathcal{Z}$, $\mathcal{E}_\ell$ is an EC-space on $[t_\ell, t_{\ell+1}]$. The converse is not true (see example below). Supposing that, for all $\ell \in \mathcal{Z}$, $\mathcal{E}_\ell$ is an EC-space on $[t_\ell, t_{\ell+1}]$, according to Proposition 13, $\mathcal{E}_\ell$ is an CW-space on $[t_\ell, t_{\ell+1}]$, hence there exists positive weight functions $w_0^\ell, \dots, w_n^\ell$ (with $w_i^\ell \in C^{n-i}([t_\ell, t_{\ell+1}])$ such that $\mathcal{E}_\ell = CW(w_0^\ell, \dots, w_n^\ell)$. Denoting by $L_0^\ell, \dots, L_n^\ell$ the associated differential operators on $C^n([t_\ell, t_{\ell+1}])$, at an interior t_ℓ the connection condition (57) can also be expressed in terms of the differential operators as

$$\left(L_0^\ell F(t_\ell^+), \dots, L_n^\ell F(t_\ell^+)\right)^T = \mathcal{Q}_\ell \,.\, \left(L_0^{\ell-1} F(t_\ell^-), \dots, L_n^{\ell-1} F(t_\ell^-)\right)^T, \tag{62}$$

where the matrix $\mathcal{Q}_\ell$ is lower triangular and has positive diagonal elements $q_{i,i}^\ell = m_{i,i}^\ell \prod_{0 \le k \le i} w_k^{\ell-1}(t_\ell^-)/w_k^\ell(t_\ell^+)$, $0 \le i \le n$. After P.J. Barry's work ([2], see also [9]) it is known that, if all matrices $\mathcal{Q}_\ell$, $\ell \in \mathcal{Z}_0$, are totally positive, then the space $\mathcal{E}$ is an EC-Pspace on I. However this is only a sufficient condition as pointed out by the following elementary example. Consider the space $\mathcal{E}$ composed of all piecewise affine functions F on $I := [-1, 0] \cup [0, 1]$ with values in $\mathbb{R}$ which satisfy the connection condition

$$\left(F(0^+), F'(0^+)\right)^T = \mathcal{M}_0 \,.\, \left(F(0^-), F'(0^-)\right)^T, \quad \mathcal{M}_0 := \begin{pmatrix} 1 & 0 \\ \alpha & 1 \end{pmatrix}.$$

In this example $\mathcal{Q}_0 = \mathcal{M}_0$, so the total positivity sufficient condition is satisfied if and only if $\alpha \ge 0$. Still, using (Piii)$'$ of Proposition 42, one can check that for $\mathcal{E}$ to be an EC-Pspace on I it is necessary and sufficient to take $\alpha > -2$.

As consequences of the previous propositions and of the lemmas established in Subsection 5.1, we obtain the piecewise versions of Remarks 3 and 7.

Proposition 44. *Given a piecewise function $U_0 \in PC^n(\mathcal{T}, \mathcal{M})$ $(n > 0)$ supposed to never vanish on I, we set $L_0 F := F/U_0$ for any $F \in PC^n(\mathcal{T}, \mathcal{M})$. Let $\mathcal{E}$ be an $(n+1)$-dimensional subspace of $PC^n(\mathcal{T}, \mathcal{M})$ containing U_0. Then*

1) *$\mathcal{E}$ is a W-Pspace on I if and only if the piecewise space*

$$DL_0\mathcal{E} := \{DL_0 F \;,\; F \in \mathcal{E}\}$$

is a W-Pspace on I.

2) *If $DL_0\mathcal{E}$ is an EC-Pspace on I, then $\mathcal{E}$ is an EC-Pspace on I.*

Proof. Let us first observe that, in our present piecewise context, each of the two equalities (5) and (6) holds as a piecewise equality, given that the function $\mathbf{1}$ belongs to some piecewise space $PC^n(\mathcal{T}, \mathcal{R})$ iff the first columns of each matrix $\mathcal{R}_\ell$ is equal to $(1, 0, \dots, 0)^T$. Our first statement thus follows from (Piii). The second one results from (Pii)$'$ and Lemma 40. □

5.3. Complete W-Pspaces

Definition 45. A sequence $(U_0,\dots,U_n)$ in $PC^n(\mathcal{T},\mathcal{M})$ is said to be a *Complete W-Psystem* (CW-Psystem) on I if it satisfies, for $\varepsilon\in\{-,+\}$,

$$W(U_0,\dots,U_k)(x^\varepsilon)\neq 0 \quad x\in I\,,\quad 0\leq k\leq n. \tag{63}$$

If (63) is satisfied, the $(n+1)$-dimensional piecewise space spanned by $U_0,\dots,U_n$ is said to be *a Complete W-Piecewise space* (in short, CW-Pspace) on I.

With a given CW-Psystem $(U_0,\dots,U_n)$ in $PC^n(\mathcal{T},\mathcal{M})$, it is possible to associate differential operators $L_0,\dots,L_n$, but of course piecewise ones. This fact presents a theoretical interest for specialists. It is why we give some details about it.

As previously, for any piecewise C^n function F, we set $L_0F(x^\varepsilon)=F(x^\varepsilon)/U(x^\varepsilon)$ for all $x\in\mathbb{R}$ and for $\varepsilon\in\{-,+\}$. Using the piecewise version of (5) and (6) along with Lemmas 39 and 40, we can state that:
- on the one hand, $(\mathbb{1},L_0U_1,\dots,L_0U_n)$ is a continuous CW-Psystem in $PC^n(\mathcal{T},\mathcal{R})$, where the matrices $\mathcal{R}_\ell$ are defined as in Lemma 40;
- on the other hand, $(DL_0U_1,\dots,DL_0U_n)$ is a CW-Psystem in $PC^{n-1}(\mathcal{T},\mathcal{M}^{\{1\}})$, each matrix $\mathcal{M}_\ell^{\{1\}}$ being obtained by deleting the first row and column in $\mathcal{R}_\ell$. Iteration of the process leads to (18), but the latter formulae are now meant piecewisely, defining piecewise differential operators $L_0,\dots,L_n$ on the set of all piecewise C^n functions. This provides us with two sequences of sequences of matrices: $\mathcal{R}^{\{i\}}=(\mathcal{R}_\ell^{\{i\}})_{\ell\in\mathcal{Z}_0}$ and $\mathcal{M}^{\{i\}}=(\mathcal{M}_\ell^{\{i\}})_{\ell\in\mathcal{Z}_0}$, $i\geq 0$, where the upper index $^{\{i\}}$ means that we are dealing with square matrices of order $(n+1-i)$. The two sequences are defined recursively by $\mathcal{M}^{\{0\}}:=\mathcal{M}$, $\mathcal{R}^{\{0\}}:=\mathcal{R}$ and, for $1\leq i\leq n$,

$$\mathcal{R}_\ell^{\{i\}}:=\mathcal{C}_{n-i}(DL_{i-1}U_i,t_\ell^+)^{-1}\,.\,M_\ell^{\{i\}}\,.\,\mathcal{C}_{n-i}(DL_{i-1}U_i,,t_\ell^-),\quad \ell\in\mathcal{Z}_0, \tag{64}$$

while $\mathcal{M}_\ell^{\{i\}}:=\left(m_{i,j}^{\ell\{i\}}\right)_{0\leq i,j\leq n-i}$ is obtained by deleting the first row and column in $\mathcal{R}_\ell^{\{i-1\}}$. All matrices $\mathcal{R}_\ell^{\{i\}}$ are lower triangular with positive diagonal elements and their first columns are equal to $(1,0,\dots,0)^T$. From (64), for instance, we can easily derive the diagonal elements of each matrix $\mathcal{M}_\ell^{\{i\}}$:

$$m_{k,k}^{\ell\{i\}}:=m_{k+1,k+1}^{\ell\{i-1\}}/m_{0,0}^{\ell\{i-1\}}=m_{k+i,k+i}^{\ell}/m_{i-1,i-1}^{\ell}\,,\quad 0\leq k\leq n-i,\ 1\leq i\leq n.$$

Each space $PC^{n-i}(\mathcal{T},\mathcal{R}^{\{i\}})$ is thus composed of functions which are continuous and piecewise C^{n-i} on I and, for $0\leq i\leq n$, we have

$$L_iPC^n(\mathcal{T},\mathcal{M})\subset PC^{n-i}(\mathcal{T},\mathcal{R}^{\{i\}}),\ DL_iPC^n(\mathcal{T},\mathcal{M})\subset PC^{n-i-1}(\mathcal{T},\mathcal{M}^{\{i+1\}}).$$

Moreover, $(\mathbb{1},L_iU_{i+1},\dots,L_iU_n)$ is a continuous CW-Psystem in $PC^{n-i}(\mathcal{T},\mathcal{R}^{\{i\}})$ and $(DL_iU_{i+1},\dots,DL_iU_n)$ is a CW-Psystem in $PC^{n-i-1}(\mathcal{T},\mathcal{M}^{\{i+1\}})$. The space $\mathcal{E}$ is the set of all piecewise functions $F\in PC^n(\mathcal{T},\mathcal{M})$ such that L_nF is a constant function on the whole of I.

As is classical in CW-spaces, we can thus also associate weights with the CW-Psystem $(U_0,\dots,U_n)$. The weights are not functions, but piecewise functions, defined by the piecewise equalities $w_0 := U_0$, $w_i := DL_{i-1}U_i$ for $1 \leq i \leq n$. The weight piecewise functions do not vanish on I and they satisfy:

$$w_i \in PC^{n-i}(\mathcal{T}, \mathcal{M}^{\{i\}}), \quad 0 \leq i \leq n.$$

Conversely, it would be interesting to find how to choose the matrices of the initial sequence $\mathcal{M}$ so as to provide us with a non vanishing function w_i in $PC^{n-i}(\mathcal{T}, \mathcal{M}^{\{i\}})$ at each stage i. The same arguments as in the nonpiecewise case would then lead to a CW-Pspace on I, hence to an EC-Pspace according to Theorem 46 below.

Theorem 46. *If $\mathcal{E}$ is a CW-Pspace on I, then it is an EC-Pspace on I.*

Proof. As in the nonpiecewise case, for a 1-dimensional piecewise space, being an EC-Pspace is equivalent to being a W-Pspace on I since both mean that any nonzero element never vanishes on I. Let $\mathcal{E}$ be the CW-Psubspace of $PC^n(\mathcal{T}, \mathcal{M})$ spanned by some CW-Psystem $(U_0, \ldots, U_n)$. Then, part 1 of Proposition 44 shows that $DL_0\mathcal{E}$ is a CW-Pspace on I. Supposing that the result holds for $n - 1 \geq 0$, we can thus assert that the piecewise space $DL_0\mathcal{E}$ is an EC-Pspace on I. Part 2 of Proposition 44 implies that $\mathcal{E}$ itself is an EC-Pspace on I. □

Consider a nested sequence (10) of subspaces in $PC^n(\mathcal{T}, \mathcal{M})$, in which each $\mathcal{E}_i$ is $(i + 1)$-dimensional. Similarly to the nonpiecewise case, Theorem 46 says that requiring all $\mathcal{E}_i$'s to be W-Pspaces on I is thus the same as requiring them all to be EC-Pspaces on I.

6. EC-Pspaces and Blossoms

Blossoms in piecewise spaces with connection matrices were first considered by H.-P. Seidel in the case of geometrically continuous polynomial splines [22]. In the more general framework of W-Pspaces, existence of blossoms was proved to be equivalent to existence of B-spline bases in [12], and to existence of Bézier points in [15]. However, in either case the spaces were supposed to be piecewise C^∞. Prior to that, in [2], P. J. Barry was the first to point out existence of B-spline bases with sections in different EC-spaces and with connections by means of totally positive regular lower triangular matrices. As we recalled in Remark 43, (2), the total positivity assumption concerned the differential operators associated with each section. He, too, supposed more differentiability than really needed since his proof relied on the fact that the dual spaces were connected so as to form an EC-Pspace (with the meaning of our present paper).

In this subsection as in the nonpiecewise context, existence and properties of blossoms will be derived from using relevant bases in EC-Pspaces, to which the first subsection is devoted. The result of it is that the equivalences stated in [12] and [15] will be valid with no extra differentiability assumption. However, this does not provide new practical conditions ensuring existence of blossoms.

6.1. Various ways to characterise EC-Pspaces

All relations obtained in Subsection 3.1 as consequences of Sylvester's identity are still valid in the piecewise context, apart from the fact that, of course, they are now valid piecewisely, that is, they give rise to two different equalities at each interior t_ℓ, one at t_ℓ^-, the other at t_ℓ^+. This, in particular, will enable us to extend Theorem 23 to EC-Pspaces. To achieve this, we need to first show that the various two point characterisations of EC-spaces given in Theorem 12 are valid in the piecewise context too. All these extensions are gathered in Theorem 47 below. We do not give details of the proofs, we just point out the exact facts which make the extensions possible.

Given piecewise functions $B_0, \dots, B_n$ in some $PC^n(\mathcal{T}, \mathcal{M})$, we say that they form a *piecewise Bernstein-like basis* of the space they span, relative to $(a, b) \in I^2$, $a \neq b$, if they satisfy the same conditions of zeros at a and b as Bernstein-like bases.

Theorem 47. *Let $\mathcal{E}$ be an $(n+1)$-dimensional subspace of $PC^n(\mathcal{T}, \mathcal{M})$. Then, $\mathcal{E}$ is an EC-Pspace on I if and only if it meets any of the following equivalent requirements:*

(Pi)″ *Any piecewise Hermite interpolation problem based on any one or two points of I has a unique solution in $\mathcal{E}$.*

(Pii)″ *For any distinct $a, b \in I$, any nonzero element $F \in \mathcal{E}$ vanishes at most n times on $\{a, b\}$, counting multiplicities, i.e., $Z_{n+1}^{\{a,b\}}(F) \leq n$.*

(Piii)″ *Given any distinct $a, b \in I$, any integers $i, j \geq 0$ with $i + j = n + 1$, and any $\varepsilon, \varepsilon' \in \{-, +\}$, the determinant*

$$\det\big(\mathbb{F}(a^\varepsilon), \mathbb{F}'(a^\varepsilon), \dots, \mathbb{F}^{(i-1)}(a^\varepsilon), \mathbb{F}(b^{\varepsilon'}), \mathbb{F}'(b^{\varepsilon'}) \dots, \mathbb{F}^{(j-1)}(b^{\varepsilon'})\big), \qquad (65)$$

is not equal to zero.

(Piv)″ *$\mathcal{E}$ possesses a piecewise Bernstein-like basis relative to any pair of distinct points of I.*

(Pv)″ *Given any $a \in I$, for $0 \leq k \leq n$, the space $\mathcal{E}_k(a)$ is an $(n-k+1)$-dimensional W-Pspace on any interval $J \subset I \setminus \{a\}$.*

(Pv)′ *Given any integer $r \geq 1$, any positive integers $\mu_1, \dots, \mu_r$ such that $\sum_{i=1}^r \mu_i = n + 1$, and any pairwise distinct $a_1, \dots, a_r \in I$, $\mathcal{E}$ possesses a basis*

$$\Psi_n^{a_1}, \dots, \Psi_{n-\mu_1+1}^{a_1}, \Psi_n^{a_2}, \dots, \Psi_n^{a_r}, \dots, \Psi_{n-\mu_r+1}^{a_r}$$

such that, for $1 \leq i \leq r$, and $n - \mu_i + 1 \leq k \leq n$, the piecewise function $\Psi_k^{a_i}$ vanishes exactly k times at a_i.

Proof. On account of Theorem 46, of Propositions 41 and 42, and of Remark 43, the fact that any among the properties (Pi)″ – (Pv)″ is equivalent to $\mathcal{E}$ being an EC-Pspace on I can be proved by following the same arguments we used for Theorem 12, including the case where I is a closed bounded interval, that is when the set $\mathcal{Z}$ is finite.

A piecewise version of Lemma 19 does exist, as a consequence of the (now piecewise) equality (20). This makes it possible to also extend to the present piecewise context the proof by induction of Theorem 23, starting the induction from (Pv)″. This enables us to show that if $\mathcal{E}$ is an EC-Pspace on I, it satisfies (Pv)′. The proof of the converse is similar to that of Theorem 23. □

6.2. Blossoms in EC-Pspaces

In this subsection we start with a sequence $\mathcal{M} = (\mathcal{M}_\ell)_{\ell\in\mathcal{Z}}$ of lower triangular matrices of order n with positive diagonals, and with a sequence of $(n+1)$-dimensional spaces $\mathcal{E}_\ell$, $\ell \in \mathcal{Z}$, each $\mathcal{E}_\ell$ being assumed both to be an EC-space on $[t_\ell, t_{\ell+1}]$ and to contain constants. We define the $(n+1)$-dimensional space $\mathcal{E}$ as the space of all continuous functions F on I the restrictions of which to each $[t_\ell, t_{\ell+1}]$ belong to $\mathcal{E}_\ell$ and which satisfy the connection conditions

$$\left(F'(t_\ell^+), \ldots, F^{(n)}(t_\ell^+)\right)^T = \mathcal{M}_\ell \cdot \left(F'(t_\ell^-), \ldots, F^{(n)}(t_\ell^-)\right)^T, \quad \ell \in \mathcal{Z}_0.$$

In other words, the space $\mathcal{E}$ itself contains constants and the space $\mathcal{U} := D\mathcal{E}$ is contained in $PC^{n-1}(\mathcal{T}, \mathcal{M})$.

We can choose a mother-function $\Phi \in \mathcal{E}^n$ the same way as in Section 4. Our requirements on the connection matrices $\mathcal{M}_\ell$ are exactly what is needed to ensure the minimum geometric continuity to be wished for, namely the fact that the parametric curve produced by Φ has the same left and right Frénet frames at each point t_ℓ, $\ell \in \mathcal{Z}_0$. Note that this also guarantees existence of osculating flats of any order $k \le n$ at each point t_ℓ, $\ell \in \mathcal{Z}_0$, defined as $\mathrm{Osc}\Phi_k(t_\ell) := \mathrm{Osc}\Phi_k(t_\ell^-) = \mathrm{Osc}\Phi_k(t_\ell^+)$. As a consequence, we can still associate with Φ the function φ defined by (27). In particular, the definitions of blossoms and Bézier points do not differ from Subsection 4.1.

We need to say how to define the spline spaces based on $\mathcal{E}$. First of all, if $\mathcal{T}'$ is a new sequence of intervals obtained from $\mathcal{T}$ by splitting the interval $[t_{\ell_0}, t_{\ell_0+1}]$ into two nontrivial subintervals $[t_{\ell_0}, \tau]$ and $[\tau, t_{\ell_0+1}]$, we observe that $PC^{n-1}(\mathcal{T}', \mathcal{M}') = PC^{n-1}(\mathcal{T}, \mathcal{M})$ provided that $\mathcal{M}'$ is obtained by inserting at the right place the identity matrix of order n in the sequence of matrices $\mathcal{M} = (\mathcal{M}_\ell)_{\ell\in\mathcal{Z}_0}$. As in Subsection 4.5, we consider a finite sequence of knots $\eta_0 < \eta_1 < \cdots < \eta_q < \eta_{q+1}$ in the interval I, and an associated sequence of multiplicities $m_0, \ldots, m_{q+1}$. We assume that $m_0 = m_{q+1} = n+1$ as previously, but now, for $1 \le i \le q$, we suppose that $0 \le m_i \le n$. Due to our observation above, in order to describe the spline space $\mathcal{S}(\mathcal{E}, \mathcal{K})$ associated with the corresponding knot vector (52), we may assume the existence of an integer ℓ such that $\eta_i = t_{\ell+i}$ for $0 \le i \le q+1$. If so, $\mathcal{S}(\mathcal{E}, \mathcal{K})$ is composed of all continuous functions $S : [\eta_0, \eta_{q+1}] \to \mathbb{R}$ satisfying property $(\mathrm{S})_2$ of Subsection 4.5 along with the connection conditions:

$$\left(S'(\eta_i^+), \ldots, S^{(n-m_i)}(\eta_i^+)\right)^T = \mathcal{C}_i \cdot \left(S'(\eta_i^-), \ldots, S^{(n-m_i)}(\eta_i^-)\right)^T, \quad 1 \le i \le q,$$

where $\mathcal{C}_i$ is the square matrix of order $(n - m_i)$ obtained by deleting the m_i last rows and columns in $\mathcal{M}_{\ell+i}$. Subject to existence, the B-spline basis of the space $\mathcal{S}(\mathcal{E}, \mathcal{K})$

is the unique basis satisfying the support, end point, positivity, and normalisation properties of Definition 35. As a special case, Bernstein bases have the meaning of Definition 33.

Below is a simplified version of the extensions of Theorem 34 and 36 to our present piecewise context.

Theorem 48. *Let $\mathcal{E}$ be the $(n+1)$-dimensional space described at the beginning of the present subsection and let $\mathcal{U}$ denote the piecewise space $D\mathcal{E}$. The following five properties are equivalent:*

(Pa) *The space $\mathcal{U}$ is an EC-Pspace on I.*
(Pb) *Blossoms exist in the space $\mathcal{E}$.*
(Pc) *Bézier points exist in the space $\mathcal{E}$.*
(Pd) *The space $\mathcal{E}$ possesses a normalised Bernstein basis relative to any pair of distinct points of I.*
(Pf) *For any knot vector $\mathcal{K}$, the space $\mathcal{S}(\mathcal{E},\mathcal{K})$ possesses a B-spline basis.*

Proof. Suppose that (Pa) holds. Existence of blossoms can easily be obtained using the relevant bases presented in (Pv)$'$, exactly as in the proof of Theorem 27. Moreover formula (31) is valid, except that we have to replace a_1 by ${a_1}^{\varepsilon_1}$ in both the numerator and the denominator. Existence of Bernstein bases, or more generally, of B-spline bases, will follow from the pseudoaffinity property of blossoms. In our piecewise context, equality (36) is still valid, except that $\beta(x_1,\ldots,x_{n-1};\gamma,\delta;.)$ is now a continuous strictly increasing function on I which is piecewise C^1 on I. To achieve this, we need a piecewise version of Lemma 31, which we give below.

Taking account of Theorem 47, the rest of the proof results from the same arguments as in the nonpiecewise case. □

Lemma 49. *Let a be an interior point of a given interval I, let $(n+1)$ integers $k_0,\ldots,k_n$ satisfy $k_0 \geq 0$, $k_i \geq k_{i-1}+1$ for $1 \leq i \leq n$, and let Q denote a lower triangular matrix of order (k_n+1), with positive diagonal elements. We consider $(n+1)$ piecewise C^{k_n} functions $U_0,\ldots,U_n$ on I (here, in the sense that they are C^{k_n} functions separately on $I^- := \{x \in I, x \leq a\}$ and $I^+ := \{x \in I, x \geq a\}$) assumed to satisfy*

$$\Delta_{k_n} U_i(a^+) = Q \,.\, \Delta_{k_n} U_i(a^-), \quad 0 \leq i \leq n. \tag{66}$$

Suppose that, for $0 \leq i \leq n$, U_i vanishes exactly k_i times at a. Then, we have, for $\varepsilon \in \{-,+\}$,

$$W(U_0,\ldots,U_n)(x) \sim (x-a)^N A_\varepsilon \quad \textit{for } x \in I^\varepsilon \textit{ close to } a \tag{67}$$

where $N := \sum_{i=0}^{n}(k_i - i)$ and where the two real numbers A_-, A_+ have the same strict sign.

Proof. We can apply Lemma 31 separately on I^- and I^+. This yields (67) with:

$$A_\varepsilon := \prod_{i=0}^{n} \frac{U_i{}^{(k_i)}(a^\varepsilon)}{k_i!} \prod_{0\le i<j\le n} (k_j - k_i). \tag{68}$$

Denote by $q_{0,0}, \ldots, q_{k_n,k_n}$ the (positive) diagonal elements of Q. Taking account of relation (66) along with the fact that each U_i vanishes exactly k_i times at a, we have

$$U_i{}^{(k_i)}(a^+) = q_{k_i,k_i} U_i{}^{(k_i)}(a^-)\ , \quad 0 \le i \le n.$$

As a consequence $A_+ = A_- \prod_{i=0}^n q_{k_i,k_i}$, and the proof is complete. □

Acknowledgements

This work was partially supported by the EC-IHP network HPRN-CT-2002-00286, "Breaking Complexity".

References

1. Aitken, A. C., *Determinants and Matrices*, Oliver and Boyd, Edinburgh and London, 1949.
2. Barry, P. J., de Boor-Fix dual functionals and algorithms for Tchebycheffian B-splines curves, *Constr. Approximation* **12** (1996), 385–408.
3. Carnicer, J.-M. and Peña, J.-M., Total positivity and optimal bases, in: *Total Positivity and its Applications* (M. Gasca and C. A. Micchelli, Eds.), Kluwer Academic Pub., Dordrecht, 1996, pp. 133–155.
4. Goodman, T. N. T., Total positivity and the shape of curves, in: *Total Positivity and its Applications* (M. Gasca and C. A. Micchelli, Eds.), Kluwer Academic Pub., Dordrecht, 1996, pp. 157–186.
5. Goodman, T. N. T. and Mazure, M.-L., Blossoming beyond extended Chebyshev spaces, *J. Approx. Theory* **109** (2001), 48–81.
6. Karlin, S. J. and Studden, W. J., *Tchebycheff Systems: with Applications in Analysis and Statistics*, Wiley Interscience, New York, 1966.
7. Karlin, S. J., *Total Positivity*, Stanford Univ. Press, Stanford, 1968.
8. Mazure, M.-L., Chebyshev blossoming, Report RR 953M IMAG, Université Joseph Fourier, Grenoble, January 1996.
9. Mazure, M.-L., Blossoming: a geometrical approach, *Constr. Approximation* **15** (1999), 33–68.
10. Mazure, M.-L., Bernstein bases in Müntz spaces, *Numer. Algorithms* **22** (1999), 285–304.
11. Mazure, M.-L., Blossoms and optimal bases, *Adv. Comput. Math.* **20** (2004), 177–203.
12. Mazure, M.-L., On the equivalence between existence of B-spline bases and existence of blossoms, *Constr. Approximation* **20** (2004), 603–624.

13. Mazure, M.-L., Various characterisations of Extended Chebyshev spaces via blossoms, *Compt. Rend. Acad. Sci. Paris, Ser. I* **339** (2004), 815–820.
14. Mazure, M.-L., Chebyshev spaces and Bernstein bases, *Constr. Approximation*, to appear.
15. Mazure, M.-L., Towards existence of piecewise Chebyshevian B-spline bases, *Numer. Algorithms* **39** (2005), 399–414.
16. Mazure, M.-L. and Laurent, P.-J., Affine and non affine blossoms, in: *Computational Geometry* (A. Conte, V. Demichelis, F. Fontanella and I. Galligani, Eds.), World Scientific Pub., Singapore, 1993, pp. 201–230.
17. Mazure, M.-L. and Pottmann, H., Tchebycheff curves, in: *Total Positivity and its Applications* (M. Gasca and C. A. Micchelli, Eds.), Kluwer Academic Pub., Dordrecht, 1996, pp. 187–218.
18. Mørken, K., Total positivity and splines, in: *Total Positivity and its Applications* (M. Gasca and C. A. Micchelli, Eds.), Kluwer Academic Pub., Dordrecht, 1996, pp. 47–84.
19. Pottmann, H., The geometry of Tchebycheffian splines, *Computer Aided Geometric Design* **10** (1993), 181–210.
20. Ramshaw, L., Blossoms are polar forms, *Computer Aided Geometric Design* **6** (1989), 323–358.
21. Schumaker, L. L., *Spline Functions: Basic Theory*, Wiley Interscience, New York, 1981.
22. Seidel, H.-P., New algorithms and techniques for computing with geometrically continuous spline curves of arbitrary degree, *Math. Modelling and Numer. Analysis* **26** (1992), 149–176.

Topics in Multivariate Approximation and Interpolation
K. Jetter et al., Editors

Structural Analysis of Subdivision Surfaces – A Summary

Ulrich Reif[a,1], **Jörg Peters**[b]

[a]Technische Universität Darmstadt, Fachbereich Mathematik,
D–64289 Darmstadt, Germany
[b]Department of CISE, University of Florida,
Gainesville, FL 32611–6120, USA

Abstract

This paper summarizes the structure and analysis of subdivision surfaces and characterizes the inherent similarities and differences to parametric spline surfaces. Besides presenting well known results in a unified way, we introduce new ideas for analyzing schemes with a linearly dependent generating system, and a significantly simplified test for the injectivity of the characteristic map.

Key words: subdivision, extraordinary point, characteristic map, central surface

2000 MSC: 65D17, 65D07, 68U07

1. Introduction

For a graphics designer, subdivision is a recipe for generating a finer and finer sequence of polyhedra that converges to a visually smooth limit surface after a few iteration steps (Figure 1). While this intuitive view accounts to a large extent for the

Email addresses: `reif@mathematik.tu-darmstadt.de` (Ulrich Reif),
`jorg@cise.ufl.edu` (Jörg Peters).
[1] Corresponding author

success of subdivision in applications, it fails to provide a framework explaining the unique analytical structure of subdivision surfaces vis-a-vis other representations. In particular, this view, which was predominant in the early subdivision literature, fails to characterize the inherent similarities and differences to standard parametric spline surfaces.

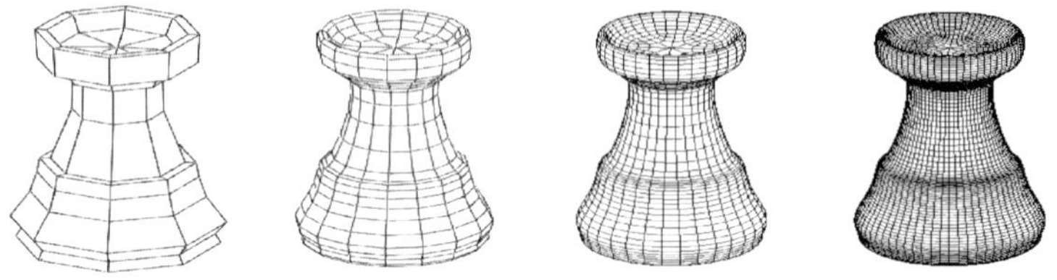

Fig. 1. Four steps of Catmull-Clark subdivision (from [88]).

To highlight the similarities, we use the term 'spline' in a much generalized sense. In the following, a spline is any function consisting of a finite or even infinite number of pieces, each of which is defined on an indexed copy of a standard domain. This definition covers in particular linear combinations of B-splines or box-splines. We focus on piecewise continuous functions defined on a union of *unit squares*; the analysis of spline surfaces over other, say triangular standard domains (see Figure 2), is analogous and need not be developed separately. To characterize continuity of a spline, its domain is endowed with the topological structure of a two-dimensional manifold. This avoids a more involved characterization by means of matching smoothness conditions for abutting patches.

To highlight the differences between spline surfaces and subdivision surfaces, we focus on the neighborhood of extraordinary points, e.g. points where $n \neq 4$ quadrilateral domains join. Here, the surface has the structure of a union of *spline*

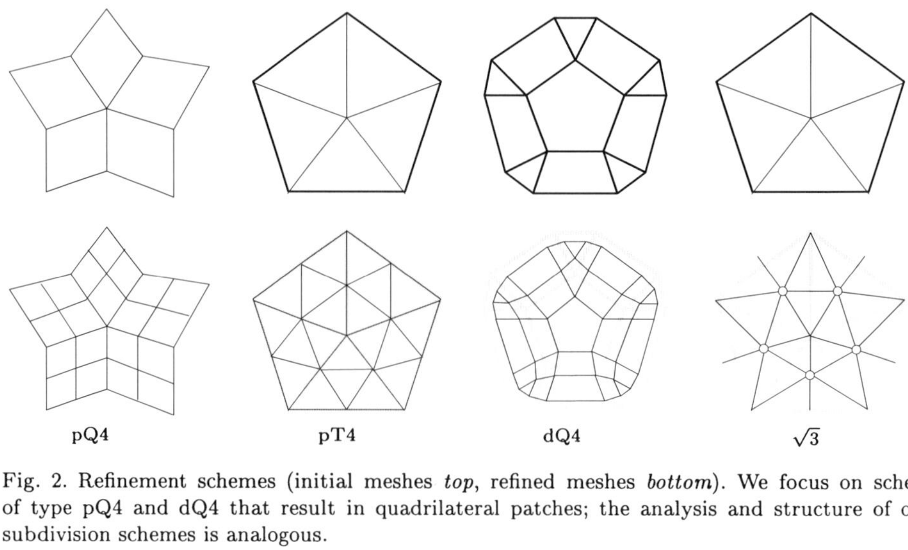

Fig. 2. Refinement schemes (initial meshes *top*, refined meshes *bottom*). We focus on schemes of type pQ4 and dQ4 that result in quadrilateral patches; the analysis and structure of other subdivision schemes is analogous.

rings, i.e. circular annuli formed by matching up the boundaries of the spline patches (see Figure 5,*right*). (The word 'ring' will not lead to confusion since no rings in the algebraic sense will be considered in this paper.) The infinite sequence of nested surface rings no longer shares all properties of the underlying splines. For example, since these rings contract *ad infinitum*, it is necessary to use, in the limit, a differential geometric characterization of smoothness: smoothness is measured in a natural local coordinate system. Injectivity with respect to this coordinate system is crucial but not always present in subdivision schemes; and the lack of second-order differentiability with respect to the coordinate system presents a challenge for characterizing shape.

1.1. Contents

After defining splines, subdivision surfaces and algorithms, the paper characterizes increasing levels of continuity and correspondingly increasing restrictions on admissible schemes. The table of contents is as follows.

1.2. Sources

This article is the summary of a book on the structural analysis of subdivision surfaces [78], which is currently in preparation. At the end of each section, we give some bibliographical notes pointing to the most relevant and first-time references, without claiming completeness. In addition to the citations given in the article, we also include some further suggested reading on the topic in the list of references.

1.3. Notation

We use greek letters for objects in $\mathbb{R}^2$ and maps into $\mathbb{R}^2$ such as planar curves and reparametrizations. Bold face is used, in particular, for points and functions in the embedding space $\mathbb{R}^d$, $d > 2$. These points are understood as *row vectors* so that, following established practice in the literature, we apply $n \times n$ subdivision matrices

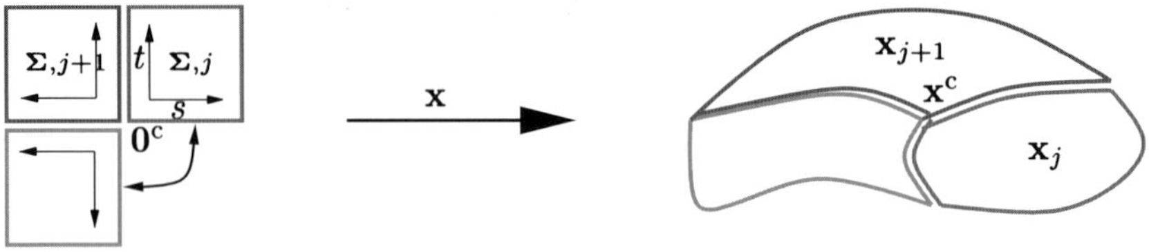

Fig. 3. (*left*) Domain manifold $\mathbf{\Sigma} \times \mathbb{Z}_3$ and (*right*) spline manifold $\mathbf{x}$.

from the left to a vector of n control points in $\mathbb{R}^d$. As in Matlab, elements in a row of a matrix or vector are separated by a comma, while rows are separated by a semicolon. For example,

$$[1, 2, 3; 4, 5, 6] = \begin{bmatrix} 1 & 2 & 3 \\ 4 & 5 & 6 \end{bmatrix}.$$

2. Spline Surfaces Near Extraordinary Points

To investigate surfaces with n quadrilateral pieces joining at a point, we define the unit interval in $\mathbb{R}$ and the unit square in $\mathbb{R}^2$, respectively:

$$U := [0, 1] \quad \text{and} \quad \mathbf{\Sigma} := U \times U.$$

Then the domain $\mathbf{S}$,

$$\mathbf{S} := \mathbf{\Sigma} \times \mathbb{Z}_n, \quad \mathbb{Z}_n := \mathbb{Z} \mod n,$$

of a *spline* $\mathbf{x} : \mathbf{S} \to \mathbb{R}^d$ consists of n indexed copies of $\mathbf{\Sigma}$. Points in $\mathbf{\Sigma}$ and $\mathbf{S}$ are typically denoted

$$\sigma = (s, t) \in \mathbf{\Sigma}, \quad \mathrm{s} = (\sigma, j) = (s, t, j) \in \mathbf{S}.$$

The restriction of $\mathbf{x}$ to a single unit square with index j is called a *patch* and denoted $\mathbf{x}_j$:

$$\mathbf{x} : \mathbf{S} \ni (\sigma, j) \mapsto \mathbf{x}_j(\sigma) \in \mathbb{R}^d.$$

Now, pairs of edges of the unit squares are set equal according to

$$(0, u, j) = (u, 0, j+1), \quad u \in U, \ j \in \mathbb{Z}_n.$$

The common origin of all patch domains is

$$\mathbf{0}^{\mathrm{c}} := (\mathbf{0}, 0) = \cdots = (\mathbf{0}, n-1),$$

see Figure 3. The superscript "c" for "center" is used to tell the origin $\mathbf{0}^{\mathrm{c}}$ of $\mathbf{S}$ apart from the origin $\mathbf{0}$ of $\mathbf{\Sigma}$. By identifying edges of adjacent unit squares, the domain $\mathbf{S}$ becomes a simply connected topological space. Thus, there is a well-defined notion of continuity for splines. Due to the identification $(0, u, j) = (u, 0, j+1)$, the patches have to satisfy the *consistency conditions*

$$\mathbf{x}_j(0, u) = \mathbf{x}_{j+1}(u, 0), \quad j \in \mathbb{Z}_n,$$

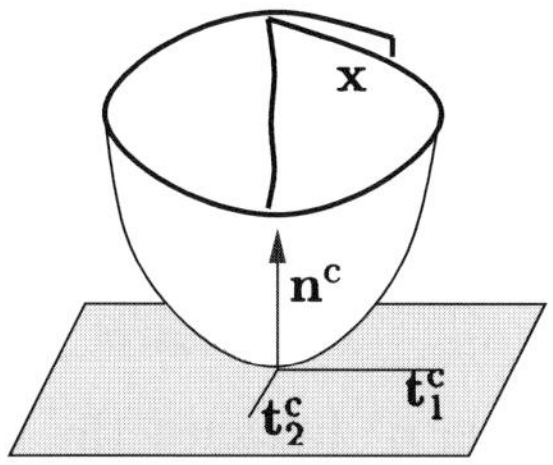

Fig. 4. A normal continuous surface that is not single-sheeted.

and, in particular,

$$\mathbf{x}^c := \mathbf{x}(\mathbf{0}^c) = \mathbf{x}_0(\mathbf{0}) = \cdots = \mathbf{x}_{n-1}(\mathbf{0})$$

is called the *center* of $\mathbf{x}$. Subsequently, patches are always assumed to be continuous. Together with the consistency conditions, this implies that the spline $\mathbf{x} : \mathbf{S} \to \mathbb{R}^d$ is a continuous map. We will be interested in several choices of the dimension d of the image space. For $d = 3$, $\mathbf{x}$ is a *spline surface*, and this is the case we are eventually interested in. For $d = 2$, we obtain planar splines which are used for analytical purposes. Especially, the characteristic map to be defined later is of that type. Finally, for $d = 1$, we obtain real valued splines. We will use such splines to define generating systems for finite dimensional spline spaces in $\mathbb{R}^3$, just as B-splines are used to model standard tensor product spline surfaces. Next, we consider smoothness.

Definition 1. A continuous spline $\mathbf{x} : \mathbf{S} \to \mathbb{R}^d$ is called C_0^k, if all patches $\mathbf{x}_j$ are C^k-functions on $\mathbf{\Sigma}\backslash\{\mathbf{0}\}$, and if the corresponding cross boundary derivatives at common edges are equal up to sign, i.e.,

$$\partial_s^\nu \mathbf{x}_j(0,u) = (-1)^\nu \partial_t^\nu \mathbf{x}_{j+1}(u,0), \quad \nu = 0,\ldots,k, \quad u \in (0,1]. \tag{1}$$

The space of all C^k-splines in $\mathbb{R}^d$ is denoted $C_0^k(\mathbf{S},\mathbb{R}^d)$.

Crucially, the central point is excluded from all smoothness conditions since, for $u = 0$, consistency implies, for $n \neq 4$ that either the partial derivatives of all patches vanish at the origin, or that the projection of $\mathbf{x}$ to the tangent plane at the origin is not injective. In the first case of a singular parametrization, the geometrical smoothness of the spline does not follow from the analytical smoothness of the parametrization; in the second case $\mathbf{x}$ cannot be a smooth surface in the sense of manifolds. As a consequence, we do not impose smoothness conditions at the center a priori. In essence, the rest of the paper deals with the challenge of reconciling the lack of smoothness in the parametrization of $\mathbf{x}$ with the geometric smoothness of its image. As a first step, we define normal continuity in the following way:

Definition 2. A spline surface $\mathbf{x} \in C_0^1(\mathbf{S}, \mathbb{R}^3)$ is called *normal continuous* at the center, if the limit

$$\mathbf{n}^c := \lim_{\mathbf{s} \to \mathbf{0}^c} \mathbf{n}(\mathbf{s})$$

of the *Gauss map*

$$\mathbf{n}(\mathbf{s}) := \tilde{\mathbf{n}}(\mathbf{s}) / \|\tilde{\mathbf{n}}(\mathbf{s})\|, \quad \tilde{\mathbf{n}}(\mathbf{s}) = \tilde{\mathbf{n}}(\sigma, j) := \partial_s \mathbf{x}_j \times \partial_t \mathbf{x}_j$$

exists and is unique. In this case, $\mathbf{n}^c$ is called the *central normal*, and the plane through $\mathbf{x}^c$ perpendicular to $\mathbf{n}^c$ is called the *central tangent plane.*

It is easily shown that for a normal continuous spline surface $\mathbf{x} \in C_0^k(\mathbf{S}, \mathbb{R}^3)$, the Gauss map extended by $\mathbf{n}(\mathbf{0}^c) := \mathbf{n}^c$ is a spline in $C_0^{k-1}(\mathbf{S}, S^2)$, where S^2 is the unit sphere in $\mathbb{R}^3$. Normal continuity is not sufficient for a spline surface to be smooth in the sense of manifolds since the projection of $\mathbf{x}$ into the central tangent plane may not be injective. To address this problem, we choose two orthonormal vectors $\mathbf{t}_1^c, \mathbf{t}_2^c$ in the central tangent plane and collect them in a (2×3)-matrix $\mathbf{T}^c$. Then we define the *projection* $\boldsymbol{\xi}$ of $\mathbf{x}$ to that plane by

$$\boldsymbol{\xi} : \mathbf{S} \ni \mathbf{s} \mapsto (\mathbf{x}(\mathbf{s}) - \mathbf{x}^c) \cdot \mathbf{T}^c \in \mathbb{R}^2. \tag{2}$$

Here and subsequently, a dot denotes multiplication by the transpose,

$$\mathbf{A} \cdot \mathbf{B} := \mathbf{A}\mathbf{B}^{\mathrm{T}}.$$

$\mathbf{x}$ is called *single-sheeted* if $\boldsymbol{\xi}$ is injective when restricted to a sufficiently small neighborhood $\mathbf{S}'$ of the origin. In this case, we can use the inverse function $\mathbf{s} = \mathbf{s}(\boldsymbol{\xi})$ to define the *central height function* h on the set $\Xi' := \boldsymbol{\xi}(\mathbf{S}') \subset \mathbb{R}^2$ by

$$h : \Xi' \ni \boldsymbol{\xi} \mapsto (\mathbf{x}(\mathbf{s}(\boldsymbol{\xi})) - \mathbf{x}^c) \cdot \mathbf{n}^c \in \mathbb{R}. \tag{3}$$

With these settings, points on the spline surface near the center can be written as

$$\mathbf{x}(\mathbf{s}) = \mathbf{x}^c + \boldsymbol{\xi}\mathbf{T}^c + h(\boldsymbol{\xi})\mathbf{n}^c, \quad \mathbf{s} \in \mathbf{S}', \ \boldsymbol{\xi} \in \Xi',$$

and the geometrical smoothness of $\mathbf{x}$ at the center is just the analytical smoothness of h at the origin. If $\mathbf{x} \in C_0^1(\mathbf{S}, \mathbb{R}^3)$, then h is continuous on Ξ' and continuously differentiable on $\Xi' \backslash \{\mathbf{0}\}$. If, moreover, $\mathbf{x}$ is normal continuous, then one can show using the mean value theorem that h is also differentiable at the origin. Both value and gradient vanish there,

$$h(\mathbf{0}) = 0, \quad Dh(\mathbf{0}) = \mathbf{0}. \tag{4}$$

Definition 3. A spline surface $\mathbf{x} \in C_0^k(\mathbf{S}, \mathbb{R}^3)$ is called C_r^k if it is single-sheeted, and if the central height function is r-times differentiable at the origin. The space of all C_r^k-splines is denoted $C_r^k(\mathbf{S}, \mathbb{R}^3)$.

One should keep in mind that the superscript k refers to the smoothness of the parametrization, while the subscript r refers to the smoothness of the central height function at the origin. Strictly speaking, $\mathbf{x}$ can be a smooth manifold even if it is neither normal continuous in the sense of Definition 2 nor single-sheeted in the sense of Definition 3. As an example, consider the 'flat' spline $\mathbf{x}$ with patches

$$\mathbf{x}_0(\boldsymbol{\xi}) = \cdots = \mathbf{x}_{n-1}(\boldsymbol{\xi}) = r\sin(1/r)\,[\cos 4\varphi,\, \sin 4\varphi,\, 0],$$

where $\boldsymbol{\xi} = r(\cos\varphi, \sin\varphi)$. The normal vector, computed as the normalized cross product of partial derivatives, alternates between $[0,0,1]$ and $[0,0,-1]$; and the projection of $\mathbf{x}$ to the xy-plane is not injective. Nevertheless, the image of $\mathbf{x}$ is simply a part of the xy-plane, hence a smooth manifold. We accept that such highly degenerate cases are not contained in the spaces $C^k_r(\mathbf{S}, \mathbb{R}^3)$.

Bibliographic notes

- The focus on spline surfaces consisting of quadrilateral patches is motivated by the Catmull-Clark-algorithm [9], the Doo-Sabin-algorithm [24], or the tensor-product four-point scheme [46,41]. A completely analogous theory can be developed for triangular patches, as obtained for instance by Loop's scheme [52], the butterfly scheme [30], or $\sqrt{3}$-subdivision [44].
- Viewing the domain $\mathbf{S}$ as a topological space appears natural. Nevertheless, it was not explicitly introduced prior to [77,98].
- An example for a surface which is normal continuous, but not single-sheeted can be found in [73].
- The conditions (1) are not necessary for a smooth join of the patches. Rather, it suffices to require coincidence of geometric quantities such as normal vectors, principal curvatures and directions, etc., at common edges. For a survey on the concept of so-called *geometric continuity* see, for instance, [58].
- The example of a degenerate parametrization of smooth manifolds elaborates on a remark in [99].

3. Subdivision Surfaces Defined

So far we have derived a general framework for splines near extraordinary points. Now we specialize it to the subdivision setting. In practice, subdivision surfaces are obtained by iterated refinement of control meshes. This refinement process enlarges the regular parts of the mesh, and scales down the central n-sided region near the extraordinary vertex. Since the limit surface corresponding to the regular parts of a mesh can at least in principle be determined explicitly, iterative refinement corresponds to the generation of a sequence of larger and larger parts of the final limit surface. Equally, one can represent the limit surface $\mathbf{x}$ as the union of the initially known regular part $\mathbf{x}^0$ and a sequence of ring-shaped parts $\mathbf{x}^m, m \in \mathbb{N}$, which are added by subsequent refinement steps. Skipping the details, this process corresponds to a partition of the domains $\boldsymbol{\Sigma}$ and $\mathbf{S}$ in the following way. Let

$$\boldsymbol{\Sigma}^0 := [0,2]^2 \backslash [0,1)^2,\ \boldsymbol{\Sigma}^m := 2^{-m}\boldsymbol{\Sigma}^0,\ \mathbf{S}^m := \boldsymbol{\Sigma}^m \times \mathbb{Z}_n,\ m \in \mathbb{N}_0,$$

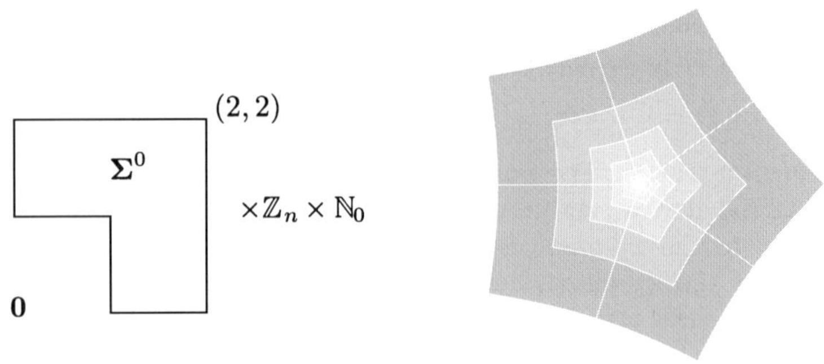

Fig. 5. (*left*) Domain $\mathbf{S}\backslash\mathbf{0}^c$ for (*right*) the union $\mathbf{x}\backslash\mathbf{x}^c$ of spline rings.

then

$$\boldsymbol{\Sigma} = \bigcup_{m\in\mathbb{N}} \boldsymbol{\Sigma}^m \cup \mathbf{0}, \quad \mathbf{S} = \bigcup_{m\in\mathbb{N}} \mathbf{S}^m \cup \mathbf{0}^c,$$

see Figure 5. Splines $\mathbf{x} \in C_0^k(\mathbf{S}, \mathbb{R}^d)$ are partitioned accordingly. For $m \in \mathbb{N}$ and $j \in \mathbb{Z}_n$, the *segment* $\mathbf{x}_j^m$ is defined by

$$\mathbf{x}_j^m : \boldsymbol{\Sigma}^0 \ni \boldsymbol{\sigma} \mapsto \mathbf{x}_j(2^{-m}\boldsymbol{\sigma}),$$

and the *spline ring* $\mathbf{x}^m$ is defined by

$$\mathbf{x}^m : \mathbf{S}^0 \ni \mathbf{s} \mapsto \mathbf{x}(2^{-m}\mathbf{s}),$$

where we used the convention that a scalar factor applies only to the continuous variables of $\mathbf{s}$,

$$a\mathbf{s} = a(s,t,j) := (as, at, j), \quad a \in \mathbb{R}.$$

The space of all C^k-spline rings is denoted by

$$C^k(\mathbf{S}^0, \mathbb{R}^d).$$

The segment $\mathbf{x}_j^m$ corresponds to the restriction of the patch $\mathbf{x}_j$ to the set $\boldsymbol{\Sigma}^m$, and the spline ring $\mathbf{x}^m$ corresponds to the restriction of the spline $\mathbf{x}$ to the set $\mathbf{S}^m$, i.e.,

$$\mathbf{x}_j^m(\boldsymbol{\Sigma}^0) = \mathbf{x}_j(\boldsymbol{\Sigma}^m), \quad \mathbf{x}^m(\mathbf{S}^0) = \mathbf{x}(\mathbf{S}^m),$$

where re-scaling facilitates the use of a common domain for all m. This implies

$$\mathbf{x}_j(\boldsymbol{\Sigma}) = \bigcup_{m\in\mathbb{N}} \mathbf{x}_j^m(\boldsymbol{\Sigma}^0) \cup \mathbf{x}^c, \quad \mathbf{x}(\mathbf{S}) = \bigcup_{m\in\mathbb{N}} \mathbf{x}^m(\mathbf{S}^0) \cup \mathbf{x}^c.$$

The partition of a spline into spline rings and segments leads to the notion of subdivision. It refers to a special way of representing splines rather than to a new class of objects.

Definition 4. $\mathbf{x} \in C_0^k(\mathbf{S}, \mathbb{R}^d)$ represented as

$$\mathbf{x} : \mathbf{S} \ni \mathbf{s} \mapsto \begin{cases} \mathbf{x}^m(2^m\mathbf{s}) & \text{if} \quad \mathbf{s} \in \mathbf{S}^m \\ \mathbf{x}^c & \text{if} \quad \mathbf{s} = \mathbf{0}^c \end{cases}$$

is called a *spline in subdivision form*. For $d = 3$, $\mathbf{x}$ is also called a *subdivision surface*.

Expressing a spline in subdivision form is a straightforward segmentation process. This point of view becomes relevant if we proceed in the opposite direction and assume that the sequence of spline rings $\mathbf{x}^m$ is generated iteratively by some algorithm. Then the task is to analyze properties of the spline obtained by gluing all these pieces together. The following lemma characterizes the relation between spline rings and splines.

Lemma 5. *A sequence $(\mathbf{x}^m)_m$ of spline rings constitutes a spline $\mathbf{x} \in C_0^k(\mathbf{S}, \mathbb{R}^d)$ if and only if*

- *all segments are k-times continuously differentiable,*

$$\mathbf{x}_j^m \in C^k(\mathbf{\Sigma}^0, \mathbb{R}^d), \tag{5}$$

- *all pairs of neighboring segments $\mathbf{x}_j^m, \mathbf{x}_{j+1}^m$ satisfy*

$$\partial_s^\nu \mathbf{x}_j^m(0, 1+u) = (-1)^\nu \partial_t^\nu \mathbf{x}_{j+1}^m(1+u, 0), \quad u \in U, \tag{6}$$

- *all pairs of consecutive segments $\mathbf{x}_j^m$, $\mathbf{x}_j^{m+1}$ satisfy*

$$\begin{aligned} \partial_s^\nu \mathbf{x}_j^m(1, u) &= 2^\nu \partial_s^\nu \mathbf{x}_j^{m+1}(2, 2u) \\ \partial_t^\nu \mathbf{x}_j^m(u, 1) &= 2^\nu \partial_t^\nu \mathbf{x}_j^{m+1}(2u, 2), \quad u \in U, \end{aligned} \tag{7}$$

- *there exists $\mathbf{x}^c \in \mathbb{R}^d$ such that for any sequence $\mathbf{s}^m \in \mathbf{S}^0$*

$$\mathbf{x}^c = \lim_{m \to \infty} \mathbf{x}^m(\mathbf{s}^m). \tag{8}$$

We omit the details of the proof, which essentially reduces to an application of the chain rule. The following theorem summarizes conditions for normal continuity and single-sheetedness.

Theorem 6. *Let $\mathbf{x}^m$ be a sequence of spline rings satisfying all conditions of Lemma 5 and $\mathbf{x}$ the corresponding spline. Then*

- *$\mathbf{x}$ is normal continuous if and only if there exists $\mathbf{n}^c \in S^2$ such that for any sequence $\mathbf{s}^m \in \mathbf{S}^0$*

  $$\mathbf{n}^c = \lim_{m \to \infty} \mathbf{n}^m(\mathbf{s}^m),$$

  where the spline rings of the Gauss map $\mathbf{n}$ are denoted by $\mathbf{n}^m$.

- *$\mathbf{x}$ is C_1^k, if and only if it is normal continuous and if there exists $m_0 \in \mathbb{N}$ such that*

  $\mathbf{n}^c \cdot \mathbf{n}^m > 0$ for all $m \geq m_0$, and

  the planar spline ring $\boldsymbol{\xi}_{m_0} := (\mathbf{x}_{m_0} - \mathbf{x}^c) \cdot \mathbf{T}^c$ corresponding to the projection $\boldsymbol{\xi}$ as defined in (2) is injective on the outer boundary of its domain

  $$\partial_+ \mathbf{S}^0 := \{(s, t, j) \in \mathbf{S}^0 : \max(s, t) = 2\}.$$

While the proof of the first part is straightforward, the second part is nontrivial and requires techniques of differential topology. For details, we refer to [78].

All subdivision algorithms currently in use and a large class of generalizations are characterized by the fact that all spline rings generated by them lie in a common space which is the d-fold Cartesian product of a finite-dimensional space of real-valued functions. For instance, for the Doo-Sabin algorithm as described in Section 7, the spline rings are C^1 and consist of n segments of three biquadratic pieces each. The dimension of this space is therefore $9dn$. In general,

$$G := [g_0, \ldots, g_{\bar{\ell}}], \quad g_\ell \in C^k(\mathbf{S}^0, \mathbb{R}), \ \ell = 0, \ldots, \bar{\ell},$$

is a row-vector of scalar-valued spline rings and we assume that they form a partition of unity,

$$\sum_{\ell=0}^{\bar{\ell}} g_\ell(\mathbf{s}) = 1, \quad \mathbf{s} \in \mathbf{S}^0. \tag{9}$$

The spline space spanned by these functions is denoted

$$C^k(\mathbf{S}^0, \mathbb{R}^d, G) := \left\{ \sum_{\ell=0}^{\bar{\ell}} g_\ell \mathbf{q}_\ell : \mathbf{q}_\ell \in \mathbb{R}^d \right\} \subset C^k(\mathbf{S}^0, \mathbb{R}),$$

and G is called the *generating system* of $C^k(\mathbf{S}^0, \mathbb{R}^d, G)$. In many applications, G is linearly independent. This is explicitly not assumed here so that the analysis covers cases like generalized box spline subdivision or matrix subdivision schemes.

$C^k(\mathbf{S}^0, \mathbb{R}^d, G)$ is a linear function space of dimension $\leq d(\bar{\ell} + 1)$. We endow it with the max-norm

$$\|\mathbf{x}^m\|_\infty := \max_{\mathbf{s} \in \mathbf{S}^0} |\mathbf{x}^m(\mathbf{s})|,$$

where $|\cdot|$ denotes the Euclidean norm in $\mathbb{R}^d$. Limits of sequences of spline rings are always understood with respect to this norm. The coefficients $\mathbf{q}_\ell^m \in \mathbb{R}^d$ of a spline ring

$$\mathbf{x}^m = \sum_{\ell=0}^{\bar{\ell}} g_\ell \mathbf{q}_\ell^m$$

are its *control points*. Collecting them in an $((\bar{\ell}+1) \times d)$-matrix $\mathbf{Q}^m := [\mathbf{q}_0^m; \ldots; \mathbf{q}_{\bar{\ell}}^m]$, we obtain

$$\mathbf{x}^m(\mathbf{s}) = \mathbf{x}_j^m(\sigma) = G(\mathbf{s})\mathbf{Q}^m, \quad \mathbf{s} = (\sigma, j) \in \mathbf{S}^0,$$

or, omitting arguments, simply $\mathbf{x}^m = G\mathbf{Q}^m$.

Bibliographic notes

- The idea of representing a subdivision surface as a union of spline rings dates back to [71,73]. Early attempts to analyze smoothness [3,4] were based on investigating sequences of finer and finer meshes converging to the subdivision surface. This approach, however, ultimately fails to capture important aspects of smoothness.
- Examples of subdivision surfaces which are not generated by a finite set of functions are, for instance, variational subdivision [40,42] or schemes based on geometrical procedures as in [31].

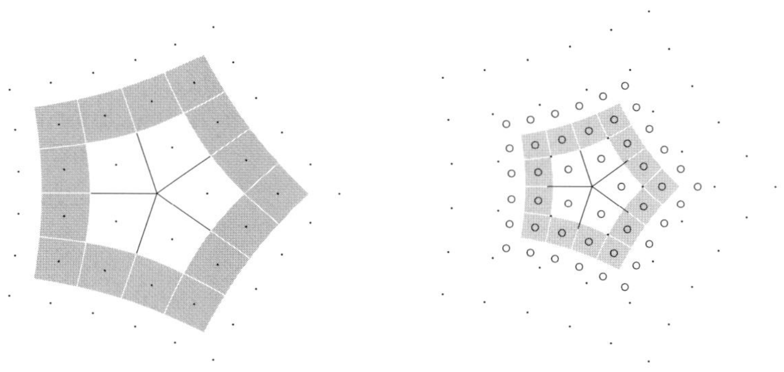

Fig. 6. (*left*) Control points $\mathbf{Q}$ (dots) and (*right*) refined control points $A\mathbf{Q}$ ($\circ$) for Doo-Sabin subdivision [24]. The corresponding spline rings are shaded grey.

4. Subdivision Algorithms

From an abstract point of view, a subdivision algorithm is a rule to compute sequences of spline rings from an initial set of control points. Here, we focus on the following special case: a (linear stationary) ***subdivision algorithm*** (A, G) is characterized by a square matrix A with all rows summing to 1 and a generating system G of according dimension. For a given set $\mathbf{Q}$ of control points, also referred to as *initial data*, the sequence of spline rings is computed by iterated application of the matrix A,

$$\mathbf{Q}^m := A^m\mathbf{Q}, \quad \mathbf{x}^m = GA^m\mathbf{Q}.$$

Since the rows of A, as well as the functions in G, sum to 1, the representation of the spline rings x^m is affine invariant. That is, applying an affine transformation to the initial control points $\mathbf{Q}$ is equivalent to applying this transformation to the spline rings $\mathbf{x}^m$, and hence to the complete spline $\mathbf{x}$.

The analysis of a subdivision algorithm can be split into three parts. First, the smoothness of the generating system has to be determined by verifying (5) and (6). Second, the contact conditions (7) between neighboring and consecutive segments have to be verified. Typically, this task is readily accomplished. Third, and this is the focus of this work, continuity and smoothness properties at the center have to be determined.

We start with continuity as defined by (8). Since the rows of A sum to one, $\lambda_0 := 1$ is the eigenvalue corresponding to the eigenvector $e := [1, \ldots, 1]^{\mathrm{T}}$, i.e., $Ae = e$. To ensure (8), i.e.

$$\lim_{m\to\infty} \mathbf{x}^m = \mathbf{x}^c,$$

it seems natural to demand that this eigenvalue be dominant. However, in general, this is by no means necessary. The reason for this is that the generating system need not be linearly independent. For example, assume that A has eigenvalues $\lambda_0 = 1$, $\lambda = 2$ and $\mu = 1/2$ corresponding to the eigenvectors e, v and w, respectively. Let $\mathbf{Q} = v\mathbf{p} + e\mathbf{q} + w\mathbf{r}$ for some points $\mathbf{p}, \mathbf{q}, \mathbf{r} \in \mathbb{R}^d \backslash \{\mathbf{0}\}$. Then $\mathbf{Q}^m = A^m\mathbf{Q} = 2^m v\mathbf{p} + e\mathbf{q} + 2^{-m} w\mathbf{r}$ is certainly divergent. But if v happens to be annihilated by

G, i.e. $Gv = 0$, then $\mathbf{x}^m = G\mathbf{Q}^m = \mathbf{q} + 2^{-m}Gw\mathbf{r} \to \mathbf{q} =: \mathbf{x}^c$ is convergent. We say that v is an *ineffective eigenvector.*

This shows that, in general, it is not possible to relate spectral properties of A to smoothness properties of the generated surfaces in a straightforward way. To address this issue in a consistent fashion, we proceed as follows. We define the linear spaces

$$N := \{v \in \mathbb{R}^{\bar{\ell}+1} : Gv = 0\}, \quad M_A := \bigcap_{m \in \mathbb{N}} A^m N.$$

That is, M_A consists of all vectors that do not escape the nullspace of G when A is applied. Further, two matrices $A, \tilde{A}$ are called *G-equivalent* if they generate identical sequences of control points, i.e., if $GA^m = G\tilde{A}^m$ for all $m \in \mathbb{N}_0$. Then the following holds:

Lemma 7 (Removal of ineffective eigenvectors). *a) For given $(\tilde{A}, G)$, let $\Pi_0 : \mathbb{R}^{\bar{\ell}+1} \to M_{\tilde{A}}$ be a projection on $M_{\tilde{A}}$ with $\Pi_0 e = 0$, and $\Pi_1 := \mathrm{Id} - \Pi_0$ its complement. Then the matrices $A := \Pi_1 \tilde{A}$ and $\tilde{A}$ are G-equivalent, and $M_A = \{0\}$.*

b) If $M_A = \{0\}$, then A does not have ineffective eigenvectors. That is, if $Av = \lambda v \neq 0$ then $Gv \neq 0$.

Proof. a) Since $Ge = 1$ by (9), we have $e \notin M_{\tilde{A}}$, and a projection of the desired form exists. By definition, $AM_A = M_A$ and $\tilde{A}M_{\tilde{A}} = M_{\tilde{A}}$. Hence,

$$\tilde{A}(M_{\tilde{A}} + M_A) = M_{\tilde{A}} + \Pi_0 \tilde{A} M_A + \Pi_1 \tilde{A} M_A = M_{\tilde{A}} + M_A.$$

Since $M_{\tilde{A}}$ is the largest $\tilde{A}$-invariant subspace of N, $M_A \subset M_{\tilde{A}}$. Further $M_{\tilde{A}} \subset \ker \tilde{A} \subset \ker A$ and $AM_A = M_A$ so that $M_A = \{0\}$. Finally, $Ae = \Pi_1 \tilde{A} e = e$. It remains to show that A and $\tilde{A}$ are equivalent. To this end, we consider $G(A^m - \tilde{A}^m) = G\big((\tilde{A} - \Pi_0\tilde{A})^m - \tilde{A}^m\big) =: G\Delta$. The matrix Δ is the sum of products of matrices $\tilde{A}$ and Π_0 all of which include the factor Π_0. With $\tilde{A}\Pi_0 = \Pi_0\tilde{A}\Pi_0$, it is clear that Δ can be represented in the form $\Delta = \Pi_0 \Delta'$. Therefore $G(A^m - \tilde{A}^m) = G\Pi_0\Delta' = \mathbf{0}$.

b) If $Av = \lambda v \neq 0$, then for all $m \in \mathbb{N}$ we have $Gv = \lambda^{-m} GA^m v = 0$. Hence, $A^m v \in N$ and $v \in M_A$ contradicting $M_A = \{0\}$. □

The property $M_A = \{0\}$, which is trivial for linearly independent G, can now be assumed for the general setting without loss of generality. The following theorem crucially depends on this property.

Theorem 8. *Let (A, G) be a subdivision algorithm with $M_A = \{0\}$. Then the continuity condition*

$$\mathbf{x}^c = \lim_{m \to \infty} \mathbf{x}^m$$

is satisfied for any set $\mathbf{Q}$ of initial data if and only if the eigenvalue $\lambda_0 = 1$ of A is strictly dominant, i.e., if $|\lambda| < 1$ for all other eigenvalues λ of A.

Proof. Let $\lambda_0 = 1$ be strictly dominant. With $\tilde{e}^{\mathrm{T}} = \tilde{e}^{\mathrm{T}} A$ the left eigenvector to λ_0 normalized by $\tilde{e}^{\mathrm{T}} e = 1$, we can decompose $\mathbf{Q}$ in the form $\mathbf{Q} = e\tilde{e}^{\mathrm{T}}\mathbf{Q} + \mathbf{R}$, where $\mathbf{R}$ is a linear combination of generalized eigenvectors of A corresponding to eigenvalues λ_i with $|\lambda_i| < r$ for some constant $r < 1$. Hence, using $Ge = 1$, we obtain convergence according to

$$\mathbf{x}^m = GA^m\mathbf{Q} = \tilde{e}^{\mathrm{T}}\mathbf{Q} + o(r^m) \to \tilde{e}^{\mathrm{T}}\mathbf{Q} =: \mathbf{x}^{\mathrm{c}}.$$

If $\lambda_0 = 1$ is not strictly dominant, we have to distinguish two cases. If $\lambda_0 = 1$ has geometric multiplicity 1, but algebraic multiplicity > 1, then there exists a generalized eigenvector v with $Av = v + e$. We set $\mathbf{q} := [1, \ldots, 1]$ and $\mathbf{Q} := v\mathbf{q}$ to obtain the divergent sequence

$$\mathbf{x}^m = GA^m\mathbf{Q} = GA^m v\mathbf{q} = G(v + me)\mathbf{q} = G\mathbf{Q} + m\mathbf{q}.$$

Otherwise, there exists an eigenvector v, which is linearly independent of e, to an eigenvalue λ with $|\lambda| \geq 1$. We set $\mathbf{q} := [1, \ldots, 1]$, $\mathbf{Q} := \operatorname{Re} v\mathbf{q}$, and $\lambda =: |\lambda| e^{it}$ to obtain

$$\mathbf{x}^m = GA^m\mathbf{Q} = |\lambda|^m \operatorname{Re}(e^{imt} Gv\mathbf{q}).$$

Since $Gv \neq 0$, this expression can converge to a constant $\mathbf{x}^{\mathrm{c}}$ only if $|\lambda| = 1$ and $t = 0$, i.e., if the eigenvalue $\lambda = \lambda_0 = 1$ is at least double, and if $Gv =: \alpha \in \mathbb{R}$ is constant. In this case, $v' := \alpha e - v$ is an ineffective eigenvector because $Av' = v' \neq 0$ and $Gv' = G(\alpha e - v) = 0$ contradicting Lemma 7. □

The results obtained so far suggest confining our considerations to subdivision schemes with $M_A = \{0\}$ and a strictly dominant eigenvalue $\lambda_0 = 1$. The next definition accounts for that.

Definition 9. Let (A, G) be a subdivision scheme with the following properties:
- The generating system is C^k, i.e., $G \in C^k(\mathbf{S}^0, \mathbb{R}^{\bar{\ell}+1})$.
- The conditions (7) are satisfied.
- A has no ineffective eigenvectors, i.e., $M_A = \{0\}$.
- The eigenvalue $\lambda_0 = 1$ to the eigenvector $e = Ae$ is strictly dominant.

Then (A, G) is called a C^k_0*-scheme*, and A is called the ***subdivision matrix***.

We recall our convention that the superscript k refers to the smoothness of the parametrization, while the subscript 0 indicates continuity at the center. We summarize our previous findings in the following theorem:

Theorem 10. *Let (A, G) be a C^k_0-scheme and $\tilde{e} = \tilde{e}A$ the dominant left eigenvector normalized by $\tilde{e}e = 1$. Then, for any $\mathbf{Q}$, the spline rings $\mathbf{x}^m := GA^m\mathbf{Q}$ together with the center $\mathbf{x}^{\mathrm{c}} := \tilde{e}\mathbf{Q}$ constitute a C^k_0-spline.*

We continue by defining B-spline-like functions for subdivision algorithms. Denote the unit vectors in $\mathbb{R}^{\bar{\ell}+1}$ by $e_0, \dots, e_{\bar{\ell}}$, and define the row-vector $B = [b_0, \dots, b_{\bar{\ell}}]$ of real-valued splines $b_\ell \in C_0^k(\mathbf{S}, \mathbb{R})$ by

$$b_\ell(\mathbf{s}) := \begin{cases} G(2^m\mathbf{s})A^m e_\ell & \text{if} \quad \mathbf{s} \in \mathbf{S}^m \\ \tilde{e} e_\ell & \text{if} \quad \mathbf{s} = \mathbf{0}^c. \end{cases} \tag{10}$$

Then, by linearity of subdivision, the relation between arbitrary initial data $\mathbf{Q}$ and the corresponding spline $\mathbf{x}$ can simply be written as

$$\mathbf{x} = B\mathbf{Q}. \tag{11}$$

Let us briefly discuss some properties of the functions in B. They

- *span the the space of splines* generated by the subdivision algorithm (A, G).
- form a *partition of unity* since

$$\sum_{\ell=0}^{\bar{\ell}} b_\ell(\mathbf{s}) = \begin{cases} G(2^m\mathbf{s})A^m e = G(2^m\mathbf{s})e = 1 & \text{if} \quad \mathbf{s} \in \mathbf{S}^m \\ \tilde{e}e = 1 & \text{if} \quad \mathbf{s} = \mathbf{0}^c. \end{cases}$$

- are *linearly independent*, if the generating system G is linearly independent. To show this, it suffices to consider the initial spline ring $\mathbf{x}_0 = G\mathbf{Q} = G\mathbf{Q}_0$ which, for linearly independent G, vanishes if and only if $\mathbf{Q} = \mathbf{0}$.
- satisfy the *scaling relation*

$$B(2^{-m}\mathbf{s}) = B(\mathbf{s})A^m, \quad \mathbf{s} \in \mathbf{S}, \ m \in \mathbb{N}_0.$$

  This can be proven as follows. For $\mathbf{s} \in \mathbf{S}^{m'}$, (10) yields $B(\mathbf{s}) = G(2^{m'}\mathbf{s})A^{m'}$. For $m \in \mathbb{N}_0$, it is $2^{-m}\mathbf{s} \in \mathbf{S}^{m+m'}$, and accordingly $B(2^{-m}\mathbf{s}) = G(2^{m'}\mathbf{s})A^{m+m'}$. Comparison of the two equations, which hold for any $m' \in \mathbb{N}$, verifies the claim.

The functions in B are important for many applications like solving interpolation or approximation problems for subdivision surfaces. In view of (11), the similarity with B-splines in the standard setting is evident. The only conceptional differences concern possible linear dependencies and a lack of parametric smoothness at the center. We will focus on geometric smoothness properties in the next sections.

Bibliographic notes

- Dominance of the eigenvector $\lambda_0 = 1$ was always considered a necessary condition for subdivision algorithms. The intriguing phenomenon of ineffective eigenvectors was first discussed in [77]. An example which shows that requiring linear independence of the generating system implies a loss of generality can be found in [78].
- A constructive procedure to efficiently compute a subdivision matrix A from a given matrix $\tilde{A}$ can be found in [78].
- The representation (11) of a subdivision surface as a finite linear combination of control points $\mathbf{q}_\ell$ and functions b_ℓ is most useful for computational purposes. For instance, it was used in [19] to compute subdivision surfaces which minimize a certain fairness functional while interpolating a given set of points.

- There exists a well-developed theory for the analysis of subdivision curves and surfaces when the domain manifold **S** is homeomorphic to the plane, see for instance [27,10,53,54,43].
- The linear independence and (lack of) local linear independence of subdivision functions has been analyzed in detail in [65].

5. C_1^k-Schemes and the Characteristic Map

In this section, we derive necessary and sufficient conditions for normal continuity and single-sheetedness of subdivision surfaces. As already mentioned above, the spectrum of A is crucial for the properties of a subdivision scheme. We sort the eigenvalues λ_i of A in descending order,

$$1 = \lambda_0 > |\lambda_1| \geq |\lambda_2| \geq \cdots |\lambda_{\bar{\ell}}|.$$

To simplify the exposition, we focus on a subclass of subdivision algorithms that covers all cases of practical relevance.

Definition 11. A C_0^k-scheme (A, G) according to Definition 9 is called a ***standard scheme***, if $k \geq 1$, and
- A has a double ***subdominant eigenvalue*** λ, i.e.,

$$1 > \lambda := \lambda_1 = \lambda_2 > |\lambda_3|,$$

- there exist two linearly independent eigenvectors v_1, v_2 to λ, i.e.,

$$A\mathbf{v} = \lambda\mathbf{v}, \quad \mathbf{v} = [v_1, v_2].$$

Such an eigenstructure is not really special, but typical for schemes with certain natural symmetry properties, as discussed in Section 6. Let $A =: VJV^{-1}$ denote the Jordan decomposition of the subdivision matrix of a standard scheme. Then $V = [e, v_1, v_2, \tilde{V}]$ and $J = \mathrm{diag}(1, \lambda, \lambda, \tilde{J})$. We define the *eigencoefficients* $\mathbf{P} = [\mathbf{p}_0; \ldots; \mathbf{p}_{\bar{\ell}}]$ and the *eigenfunctions* $F = [f_0, \ldots, f_{\bar{\ell}}]$ by

$$\mathbf{P} := V^{-1}\mathbf{Q}, \quad F := GV.$$

With $\mathbf{p}_0 = \tilde{e}^{\mathrm{T}}\mathbf{Q} = \mathbf{x}^{\mathrm{c}}$ the center and $f_0 = Ge = 1$ the 1-function we obtain the representation

$$\mathbf{x}^m = GA^m\mathbf{Q} = FJ^m\mathbf{P} = \mathbf{x}^{\mathrm{c}} + \lambda^m(f_1\mathbf{p}_1 + f_2\mathbf{p}_2) + o(\lambda^m).$$

To efficiently deal with such ***asymptotic expansions***, we introduce an equivalence relation for sequences of functions with coinciding leading terms. We write

$$a^m \stackrel{c^m}{=} b^m \quad \text{iff} \quad a^m - b^m = o(c^m),$$

where $o(c^m)/c^m$ converges uniformly to zero as $m \to \infty$. For example, $a^m \stackrel{1}{=} a$ means that a^m converges to a. For vector-valued expressions, the equivalence relation is understood component-wise. For simplicity, $\stackrel{c^m}{=}$ is mostly replaced by the symbol $\doteq$ with the understanding that the dot refers to the lowest order term specified explicitly on the right hand side of a relation. Hence, the expansion of the sequence of spline rings above now simply reads $\mathbf{x}^m \doteq \mathbf{x}^c + \lambda^m(f_1\mathbf{p}_1 + f_2\mathbf{p}_2)$, meaning that the omitted remainder term decays faster than λ^m. In the following, the two-dimensional spline ring built from the subdominant eigenfunctions f_1, f_2 plays a central role.

Definition 12. For a standard scheme (A, G) with subdominant eigenvectors $\mathbf{v} = [v_1, v_2]$ and eigenfunctions $F = [1, f_1, f_2, \ldots, f_{\bar{\ell}}]$ the *characteristic map* is defined by

$$\psi := G\mathbf{v} = [f_1, f_2] \in C^k(\mathbf{S}^0, \mathbb{R}^2, G).$$

With this definition, the sequence of spline rings becomes

$$\mathbf{x}^m \doteq \mathbf{x}^c + \lambda^m\, \psi[\mathbf{p}_1; \mathbf{p}_2].$$

Convergence towards the center $\mathbf{x}^c$ is evident. In order to compute normal vectors, we define the cross product of vectors in $\mathbb{R}^3$ as usual, and for vectors in $\mathbb{R}^2$ as the real number $\boldsymbol{\alpha} \times \boldsymbol{\beta} := \det(\boldsymbol{\alpha}, \boldsymbol{\beta})$. Accordingly, for spline rings in $\mathbb{R}^2$ or $\mathbb{R}^3$, we define the differential operator ${}^\times\!D := \partial_s \times \partial_t$ and obtain

$${}^\times\!D\mathbf{x}^m = \partial_s\mathbf{x}^m \times \partial_t\mathbf{x}^m \doteq \lambda^{2m}\, {}^\times\!D\psi\, (\mathbf{p}_1 \times \mathbf{p}_2),$$

where, by definition,

$${}^\times\!D\psi = \partial_s f_1\, \partial_t f_2 - \partial_t f_1\, \partial_s f_2$$

is the Jacobian determinant of the characteristic map. It is easily shown that ${}^\times\!D\psi \in C^{k-1}(\mathbf{S}^0, \mathbb{R}^2)$ if $\psi \in C^k(\mathbf{S}^0, \mathbb{R}^2)$. In order to distinguish degenerate cases, we say that the initial data $\mathbf{Q}$ are *generic*, if any three of the eigencoefficients $\mathbf{p}_1, \ldots, \mathbf{p}_{\bar{\ell}}$ are linearly independent. In this section, it would be sufficient to demand only $\mathbf{p}_1 \times \mathbf{p}_2 \neq \mathbf{0}$; the generality of the definition anticipates the requirements in the next sections. We say that a subdivision scheme is *normal continuous* or *single-sheeted,* if so are all surfaces generated from generic initial data.

Theorem 13. *A standard scheme is*
- *normal continuous with central normal*

$$\mathbf{n}^c = \operatorname{sign}({}^\times\!D\psi)\, \frac{\mathbf{p}_1 \times \mathbf{p}_2}{|\mathbf{p}_1 \times \mathbf{p}_2|},$$

if the characteristic map is regular, i.e., if ${}^\times\!D\psi \neq 0$.
- *not normal continuous, if* ${}^\times\!D\psi$ *changes sign.*

Proof. The first part of the statement follows immediately from $\mathbf{n}^m := {}^\times\!D\mathbf{x}^m / |{}^\times\!D\mathbf{x}^m|$ and the observation that $1/{}^\times\!D\psi$ is continuous, hence uniformly bounded, on the compact domain $\mathbf{S}^0$. To prove the second part, let us assume that ${}^\times\!D\psi(\mathbf{s}_1)\, {}^\times\!D\psi(\mathbf{s}_2) < 0$ for some arguments $\mathbf{s}_1, \mathbf{s}_2 \in \mathbf{S}^0$. Here, we obtain

$$\mathbf{n}^m(\mathbf{s}_i) \doteq \operatorname{sign}({}^{\times}\!D\psi(\mathbf{s}_i))\,\frac{\mathbf{p}_1\times\mathbf{p}_2}{|\mathbf{p}_1\times\mathbf{p}_2|},\quad i\in\{1,2\},$$

and see that $\mathbf{n}^m$ cannot converge to a constant limit since $|\mathbf{n}^m(\mathbf{s}_1)-\mathbf{n}^m(\mathbf{s}_2)|\doteq 2$. □

The conditions of this theorem are almost comprehensive. Only the exceptional case, where ${}^{\times}\!D\psi$ has zeros without changing sign remains open. Here, the behavior of ${}^{\times}\!D\mathbf{x}^m$ depends on higher order eigencoefficients and cannot be determined a priori. Now, the issue of single-sheetedness has to be addressed, and again, the characteristic map provides necessary and sufficient conditions.

Theorem 14. *A standard C_0^k-scheme with a regular characteristic map ψ is*
– *single-sheeted and moreover C_1^k, if ψ is injective.*
– *not single sheeted, if ψ restricted to the interior of $\mathbf{S}^0$ is not injective.*

Proof. Let ψ be regular and injective. Then we know that $\mathbf{x}$ is normal continuous with $\mathbf{n}^m \doteq \mathbf{n}^c = \operatorname{sign}({}^{\times}\!D\psi(\mathbf{s}_i))(\mathbf{p}_1\times\mathbf{p}_2)/|\mathbf{p}_1\times\mathbf{p}_2|$. Hence, $\mathbf{n}^c\cdot\mathbf{n}^m \doteq 1$, and in particular $\mathbf{n}^c\cdot\mathbf{n}^m > 0$ for m sufficiently large. Next, we consider a rescaled sequence of projections of spline rings,

$$\tilde{\boldsymbol{\xi}}^m := \lambda^{-m}\boldsymbol{\xi}^m \doteq \psi L,\quad L := [\mathbf{p}_1;\mathbf{p}_2]\cdot\mathbf{T}^c.$$

For generic initial data, ψL is injective because the (2×2)-matrix L has full rank. Since the set of regular injective mappings is open with respect to the C^1-norm, we conclude that $\tilde{\boldsymbol{\xi}}^m$ and eventually $\boldsymbol{\xi}^m$ is injective for m sufficiently large. Thus, all conditions of Theorem 6 are fulfilled, and $\mathbf{x}$ is C_1^k. Similar arguments show that also non-injectivity of ψ at interior points is inherited by $\tilde{\boldsymbol{\xi}}^m$ and $\boldsymbol{\xi}^m$. □

Again, the theorem is almost comprehensive. Only the exceptional case, when ψ restricted to the boundary of its domain is not injective, remains open. Theorems 13 and 14 suggest to focus on standard schemes with a regular and injective characteristic map. The following definition accounts for that observation.

Definition 15. A standard C_0^k-scheme (A,G) with a characteristic map ψ that is regular and injective is called a *standard C_1^k-scheme.* ψ is called *normalized* if

$$\psi(2,2,0) = (1,0)\quad\text{and}\quad {}^{\times}\!D\psi > 0.$$

As we have shown, C_1^k-schemes generate C_1^k-splines from generic initial data. The notion of normalization is introduced to select from the variety of possible characteristic maps a special class of representatives which is convenient for the forthcoming considerations. We prepare our discussion of that issue by the following observation. For initial data $\mathbf{Q} := [v_1, v_2]$ the corresponding two-dimensional spline is $\mathbf{x} = B\mathbf{v} \in C_0^k(\mathbf{S},\mathbb{R}^2)$, where the spline rings are just scaled copies of the characteristic map, $\mathbf{x}^m = \lambda^m\psi$. By (7), this implies for the segments ψ_j

$$\begin{aligned}\psi_j(1,u) &= \lambda\psi_j(2,2u)\\ \psi_j(u,1) &= \lambda\psi_j(2u,2),\end{aligned}\tag{12}$$

and in general

$$\begin{aligned}\partial_s^\nu \psi_j(1,u) &= \lambda\, 2^\nu \partial_s^\nu \psi(2,2u)\\ \partial_t^\nu \psi_j(u,1) &= \lambda\, 2^\nu \partial_t^\nu \psi(2u,2), \quad 0 \le \nu \le k.\end{aligned} \tag{13}$$

Now, we can prove that normalization is always possible if the characteristic map is regular and injective.

Lemma 16. *Let $\tilde{\psi} = F\tilde{\mathbf{v}}$ be the characteristic map of a standard C_1^k-scheme with $s := \operatorname{sign} \tilde{\psi}(2,2,0)$ and $[a,b] := \tilde{\psi}(2,2,0)$. Then $[a,b] \neq [0,0]$, and*

$$\psi := F\mathbf{v}, \quad \mathbf{v} := \tilde{\mathbf{v}}R, \quad R := \frac{1}{a^2+b^2}\begin{bmatrix} a & -sb \\ b & sa \end{bmatrix}$$

defines a normalized characteristic map of the scheme.

Proof. By (13) and injectivity, $\tilde{\psi}(1,1,0) = \lambda\tilde{\psi}(2,2,0) \neq \tilde{\psi}(2,2,0)$. Hence, $[a,b] \neq [0,0]$, and R is well defined. Since R has full rank, the columns of $\mathbf{v}$ are linearly independent eigenvectors of A to λ implying that $\psi = F\mathbf{v} = \tilde{\psi}R$ is a characteristic map as well as $\tilde{\psi}$. In particular, ψ is regular and injective. We find $\psi(2,2,0) = \tilde{\psi}(2,2,0)R = [1,0]$, and

$$^{\times}D\psi(2,2,0) = {}^{\times}D\tilde{\psi}(2,2,0)\det R = \frac{|^{\times}D\tilde{\psi}(2,2,0)|}{a^2+b^2} > 0.$$

Since $^{\times}D\psi$ is continuous and has no zeros, $^{\times}D\psi > 0$ follows showing that ψ is normalized. □

Bibliographic notes

- As pointed out in [63], shift and flip invariance (see Section 6) of a C_1^k-scheme imply a double subdominant Jordan block. For instance, for $n = 3$, simplest subdivision [62] yields a 6-fold subdominant eigenvalue $\lambda_1 = 1/4$, with two Jordan blocks of size 2 and two Jordan blocks of size 1. Still, C_1^k-smoothness is guaranteed. The analysis of this more general setting is only slightly more difficult, but requires considerably complex notation. The standard case discussed here covers most algorithms currently in use.
- Complete lists of possible leading eigenvalues campatible with C_1^k-schemes are provided in [77,98].
- The concept of the characteristic map was introduced in [73]. In some sense, it is related to the ***natural configuration*** defined in [92].
- In [73], it is shown that regularity and injectivity of the characteristic map are sufficient for smoothness. Necessity was proven in [63].
- An elegant computational way to verify regularity of the characteristic map even for non-polynomial schemes is described in [99].

6. Symmetry and Fourier Analysis

We continue the analysis of schemes with standard symmetry properties. According to the partition of splines into segments, vectors $\mathbf{Q}$ of control points can typically be partitioned into blocks $\mathbf{Q} = [\mathbf{Q}_0; \ldots; \mathbf{Q}_{n-1}]$, where all blocks $\mathbf{Q}_j$ have equal structure and size $\tilde{\ell} := (\bar{\ell}+1)/n$. If, as for the Catmull-Clark scheme, a central control point is common to all blocks, one can use n identical copies of it to achieve the desired structure. Shift invariance of a subdivision scheme refers to the fact that the shape of a subdivision surface does not depend on the special choice of the starting point when indexing the blocks of a given set of initial data $\mathbf{Q}$. More precisely, with E the identity matrix of dimension $\tilde{\ell}$, let

$$S := \begin{bmatrix} 0 & 0 & \cdots & 0 & E \\ E & 0 & \cdots & 0 & 0 \\ & & \ddots & & \\ 0 & 0 & \cdots & E & 0 \end{bmatrix}$$

denote the n-block shift matrix. A subdivision scheme (A, G) is called *shift invariant*, if

$$AS = SA \quad \text{and} \quad G(\cdot, j) = G(\cdot, j+1)S, \quad j \in \mathbb{Z}_n.$$

In this case, for any $\mathbf{Q}$ and $\tilde{\mathbf{Q}} := S^k\mathbf{Q}$, the segments of the splines $\mathbf{x} := B\mathbf{Q}$ and $\tilde{\mathbf{x}} := B\tilde{\mathbf{Q}}$ differ, just as the blocks $\mathbf{Q}_j = \tilde{\mathbf{Q}}_{j+k}$ of the initial data, only by an index shift:

$$\mathbf{x}_j = G(\cdot, j)A^m\mathbf{Q} = G(\cdot, j+k)S^k A^m \mathbf{Q} = G(\cdot, j+k)A^m S^k \mathbf{Q} = \tilde{\mathbf{x}}_{j+k}.$$

Flip invariance of a subdivision scheme refers to the fact that the shape of a subdivision surface does not depend on the orientation when indexing a given set of initial control points $\mathbf{Q}$. More precisely, a subdivision scheme (A, G) is called *flip invariant*, if there exists a matrix R with $R = R^{-1}$ such that

$$AR = RA \quad \text{and} \quad G(s,t,j) = G(t,s,-j)R, \quad (s,t,j) \in \mathbf{S}^0. \tag{14}$$

In this case, for any $\mathbf{Q}$ and $\tilde{\mathbf{Q}} := R\mathbf{Q}$, the splines $\mathbf{x} := B\mathbf{Q}$ and $\tilde{\mathbf{x}} := B\tilde{\mathbf{Q}}$ differ only by a flip $(s,t,j) \to (t,s,-j)$ of arguments,

$$\mathbf{x}_j(s,t) = G(s,t,j)A^m\mathbf{Q} = G(t,s,-j)RA^m\mathbf{Q} = G(t,s,-j)A^m R\mathbf{Q} = \tilde{\mathbf{x}}_{-j}(t,s).$$

From now on, we focus on schemes which respect both invariance principles.

Definition 17. A subdivision scheme is called *symmetric*, if it is both shift and flip invariant.

Let us continue by discussing the implications of symmetry on the eigenstructure of A. $SA = AS$ implies a block-circulant structure for the subdivision matrix,

$$A = \begin{bmatrix} A_0 & A_{n-1} & \cdots & A_1 \\ A_1 & A_0 & \cdots & A_2 \\ \vdots & \vdots & \ddots & \vdots \\ A_{n-1} & A_{n-2} & \cdots & A_0 \end{bmatrix}.$$

The key tool for handling such matrices is the *Discrete Fourier Transform (DFT)*. With

$$w_n := \exp(2\pi i/n),$$

the primitive n-th root of unity, we define the *Fourier block matrix* W as the Kronecker product of E and the Fourier matrix, i.e.

$$W := (w_n^{-jk}E)_{j,k\in\mathbb{Z}_n} = \begin{bmatrix} E & E & E & \cdots & E \\ E & w_n^{-1}E & w_n^{-2}E & \cdots & w_n^{1}E \\ E & w_n^{-2}E & w_n^{-4}E & \cdots & w_n^{2}E \\ \vdots & \vdots & \vdots & \ddots & \vdots \\ E & w_n^{1}E & w_n^{2}E & \cdots & w_n^{-1}E \end{bmatrix}.$$

It is easily verified by inspection that the inverse of W is given by

$$W^{-1} = \frac{1}{n}\,(w_n^{+jk}E)_{j,k\in\mathbb{Z}_n} = \frac{1}{n}\overline{W}.$$

The DFT of the matrix A is defined by $\hat{A} := WAW^{-1}$, and a standard computation shows that

$$\hat{A} = \operatorname{diag}(\hat{A}_0, \ldots, \hat{A}_{n-1})$$

is block diagonal with entries obtained by applying the Fourier matrix to the first block column of A,

$$\begin{bmatrix} \hat{A}_0 \\ \vdots \\ \hat{A}_{n-1} \end{bmatrix} := W \begin{bmatrix} A_0 \\ \vdots \\ A_{n-1} \end{bmatrix}, \quad \text{that is} \quad \hat{A}_k := \sum_{j\in\mathbb{Z}_n} w_n^{-jk} A_j.$$

By definition, A and $\hat{A}$ are similar, and in particular, they have equal eigenvalues. More precisely, if λ' is an eigenvalue of A, then there exists an index $k \in \mathbb{Z}_n$ such that λ' is an eigenvalue of $\hat{A}_k$. The set of all such indices is called the *Fourier index* of λ' and denoted

$$\mathcal{F}(\lambda') := \{k \in \mathbb{Z}_n : \det(\hat{A}_k - \lambda' E) = 0\}.$$

For the dominant eigenvalue $\lambda_0 = 1$ of A, we obtain $\mathcal{F}(1) = \{0\}$. Now we consider the double subdominant eigenvalue λ of a standard scheme. If $k \in \mathcal{F}(\lambda)$, then

$$\det(\hat{A}_k - \lambda E) = \det(\hat{A}_{n-k} - \lambda E) = 0$$

since λ is real and the diagonal blocks $\hat{A}_k$ and $\hat{A}_{n-k}$ are complex conjugate (unless $k = 0$ or $k = n - k$). Hence, the Fourier index of λ has the form $\mathcal{F} = \{k, n-k\}$ for some $k \in \mathbb{Z}_n$. If $\hat{A}_k \hat{v} = \lambda \hat{v}$, then the corresponding complex eigenvector of A is given by

$$v = W^{-1} \begin{bmatrix} \delta_{0,k}\hat{v} \\ \delta_{1,k}\hat{v} \\ \vdots \\ \delta_{n-1,k}\hat{v} \end{bmatrix} = \frac{1}{n} \begin{bmatrix} w_n^0 \hat{v} \\ w_n^k \hat{v} \\ \vdots \\ w_n^{(n-1)k}\hat{v} \end{bmatrix}. \tag{15}$$

The eigenfunction corresponding to v is just a complex version of the characteristic map ψ.

Definition 18. Let v be the complex eigenvector of a symmetric standard C_1^k-scheme (A, G) to the subdominant eigenvalue λ as defined above. Then the *complex characteristic map* of the scheme is defined as the spline ring

$$f := Gv \in C^k(\mathbf{S}^0, \mathbb{C}, G).$$

This definition is justified as follows: The real and imaginary part of v are the real eigenvectors v_1, v_2 as introduced in Definition 11. Hence,

$$f = G(v_1 + iv_2) = f_1 + if_2 \quad \text{and} \quad \psi = G[v_1, v_2] = [\operatorname{Re} f, \operatorname{Im} f].$$

Further, with ${}^{\times}\!\mathcal{D}f := \operatorname{Im}(\overline{\partial_s f}\, \partial_t f)$,

$${}^{\times}\!\mathcal{D}f = {}^{\times}\!\mathcal{D}\psi.$$

Due to the close relation between ψ and f, also f will shortly be referred to as the characteristic map of the scheme. Using (15), we obtain for the segments of the characteristic map

$$\begin{aligned} f_j &= \frac{1}{n} \sum_{\ell \in \mathbb{Z}_n} G_\ell(\cdot, j) w_n^{\ell k} \hat{v} = \frac{1}{n} \sum_{\ell \in \mathbb{Z}_n} G_{\ell - j}(\cdot, 0) w_n^{\ell k} \hat{v} \\ &= \frac{w_n^{jk}}{n} \sum_{\ell \in \mathbb{Z}_n} G_\ell(\cdot, 0) w_n^{\ell k} \hat{v} = w_n^{jk} f_0. \end{aligned}$$

This means that, due to shift invariance, all segments can be obtained from the first one by rotation. This observation leads immediately to a result concerning the appropriate Fourier index of the subdominant eigenvalue.

Theorem 19. *The characteristic map of a symmetric standard scheme can be injective only if* $\mathcal{F}(\lambda) = \{1, n-1\}$.

The proof is based on computing the winding number of curves in the image of f depending on k. Instead of going through the technical details, we refer to Figure 7, which gives a good impression of the consequences of a wrong Fourier index. In the

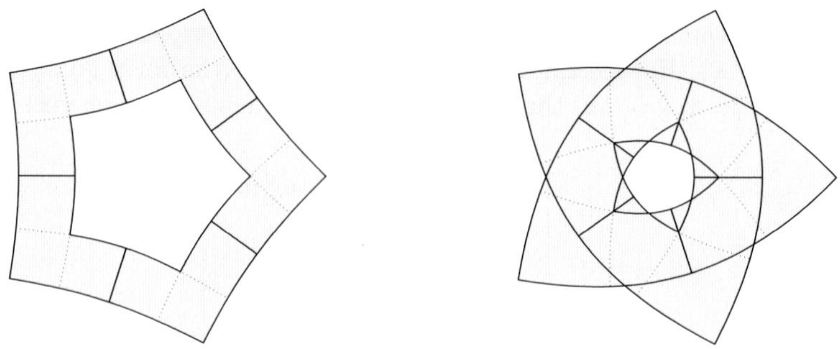

Fig. 7. Characteristic map with Fourier index (*left*) $\mathcal{F}(\lambda) = \{1, n-1\}$ and (*right*) $\mathcal{F}(\lambda) = \{2, n-2\}$.

following, we will always assume that the complex eigenvector v according to (15) corresponds to the index $k = 1$. The index $n-1$ leads to the complex conjugate eigenvector $\overline{v}$.

For a normalized characteristic map, flip invariance is exploited as follows: $AR = RA$ implies that also Rv is an eigenvector of A to λ, i.e., $Rv = av + b\overline{v}$ for some complex constants a, b. On one hand, by (14),

$$1 = G(2,2,0)v = G(2,2,0)Rv = a + b.$$

On the other hand, v and $\overline{v}$ are also eigenvectors of the shift matrix. With $Sv = w_n^{-1}v$ and $S\overline{v} = w_n\overline{v}$, we obtain

$$1 = G(2,2,0)v = G(2,2,0)SRSv = w_n^{-2}a + b.$$

The two conditions $a + b = w_n^{-2}a + b = 1$ have the unique solution $a = 0, b = 1$, i.e., $Rv = \overline{v}$. Hence,

$$f_0(s,t) = G(s,t,0)v = G(t,s,0)Rv = G(t,s,0)\overline{v} = \overline{f_0(t,s)}.$$

We summarize our findings concerning symmetry properties of the characteristic map as follows:

Theorem 20. *Let $f = Gv$ be the normalized characteristic map of a symmetric standard scheme (A, G) derived via (15) from the eigenvector $\hat{v}$ of the block $\hat{A}_1$ to the subdominant eigenvalue λ. Then, for $j \in \mathbb{Z}_n$,*

$$f_j(s,t) = w_n^j f_0(s,t) = w_n^j \overline{f_0(t,s)} = \overline{f_{-j}(t,s)}. \tag{16}$$

The theorem tells us that the complete information on the characteristic map is essentially contained in one half of the first segment. More precisely, we define the *half domain*

$$\mathbf{\Sigma}_\mathrm{h} := \{(s,t) \in \mathbf{\Sigma} : s \le t\},$$

and the *half segment* f_h as the restriction of the segment f_0 to this set. Obviously,

$$^{\times}\!Df(s,t,j) = \begin{cases} ^{\times}\!Df_\mathrm{h}(s,t) & \text{if } s \le t \\ ^{\times}\!Df_\mathrm{h}(t,s) & \text{if } s > t. \end{cases}$$

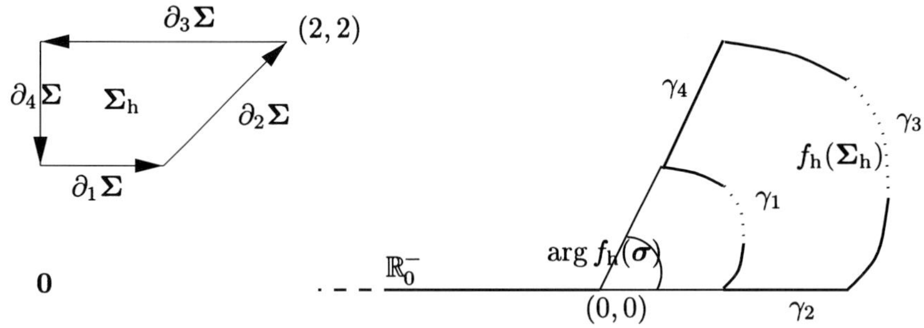

Fig. 8. (*left*) Domain halfsegment Σ_h and (*right*) spline image $f_h(\Sigma_h)$.

Thus, it suffices to verify ${}^{\times}\!\mathcal{D} f_h > 0$ to ensure regularity of the complete characteristic map. Since in a concrete setting ${}^{\times}\!\mathcal{D} f_h$ can be evaluated either numerically or even analytically, the required check of sign is typically easy to accomplish. By contrast, verifying injectivity seems to be a much harder task, and in most known proofs for specific algorithms, much effort is spent on that issue. The following theorem provides a significant simplification of the injectivity test.

Theorem 21. *Let (A, G) be a symmetric standard scheme with Fourier index $\mathcal{F} = \{1, n-1\}$, and assume that the half segment f_h of the complex characteristic map f is normalized and regular. Then f is regular and injective if and only if all real points on the curve $\gamma_1(u) := f_h(u, 1), u \in U$, are positive:*

$$\gamma_1(U) \cap \mathbb{R}_0^- = \emptyset.$$

Proof. As already shown above, f is regular if and only if f_h is regular. If f is injective and $f_h(u,1) \in \mathbb{R}_0^-$, then, by Theorem 20, $f_h(u,1) = f(1,u,0) = \overline{f(u,1,0)} = f(u,1,0)$. This implies $u = 1$, and with (12), the contradiction $0 > f(1,1,0) = \lambda f(2,2,0) = \lambda > 0$ follows. If all real points in $\gamma_1(U)$ are positive, we argue as follows:

First, we show that in this case also all real points in the image $f_h(\Sigma_h)$ are positive. To this end, we denote the four boundary segments of Σ_h by $\partial_1\Sigma, \ldots, \partial_4\Sigma$, see Figure 8, *left*. The restrictions of f_h to these sets yield four boundary curves which we parametrize over $U = [0, 1]$ by

$$\gamma_1(u) := f_h(u, 1), \qquad \gamma_2(u) := f_h(1+u, 1+u)$$
$$\gamma_3(u) := f_h(2-2u, 2), \qquad \gamma_4(u) := f_h(0, 2-u),$$

see Figure 8,*right*. By (12), the curves γ_1 and γ_3 are related according to $\gamma_1(u) = \overline{\lambda\gamma_3(1-u)}$. By (16), $\gamma_2(u) = f_0(1+u, 1+u) = \overline{f_0(1+u, 1+u)}$ is real. Since f is regular, $\gamma_2' = \mathcal{D} f_h[1;1] \neq 0$. The endpoints are $\gamma_2(0) = \lambda$, $\gamma_2(1) = 1$. Hence, we conclude that $\gamma_2(u)$ is strictly monotone increasing and positive for all $u \in U$. Also by (16),

$$\gamma_4(u) = f_0(0, 2-u) = \overline{f_0(2-u, 0)} = \overline{f_{-1}(0, 2-u)} = \overline{w_n^{-1} f_0(0, 2-u)} = w_n \overline{\gamma_4(u)}.$$

This is possible only if $\gamma_4(u)$ lies on a straight line with angle either $\arg\gamma_4(u) = \pi/n$ or $\arg\gamma_4(u) = \pi + \pi/n$ for all $u \in U$. As before, one shows that $|\gamma_4(u)|$ is strictly monotone, and that $\gamma_4(u) \neq 0$ for all $u \in U$. Hence, $\gamma_4(U)$ does not contain real points. By assumption, $\gamma_1(U) \cap \mathbb{R}_0^- = \emptyset$, and the same is true for $\gamma_3(U) = \gamma_1(U)/\lambda$. Together, we have shown that all real points on the image of the boundary of $\mathbf{\Sigma}_\mathrm{h}$ are positive, i.e., $f_\mathrm{h}(\partial\mathbf{\Sigma}_\mathrm{h}) \cap \mathbb{R}_0^- = \emptyset$. Since, by the inverse function theorem, for a regular map the boundary of the image is a subset of the image of the boundary, we conclude that $f_\mathrm{h}(\mathbf{\Sigma}_\mathrm{h}) \cap \mathbb{R}_0^- = \emptyset$.

Second, we show that the minimum and maximum of the angle $\arg f_\mathrm{h}(\boldsymbol{\sigma})$ are attained if and only if $\boldsymbol{\sigma} \in \partial_2\mathbf{\Sigma}$ and $\partial_4\mathbf{\Sigma}$, respectively. Using the findings of the first part of the proof, we see that $\arg f_\mathrm{h} \in (-\pi, \pi)$ is well defined so that the search for extrema actually makes sense. Now, let $\boldsymbol{\sigma} \in \operatorname{argmin} \arg f_\mathrm{h}$ be an argument corresponding to the minimal angle. As before, we conclude that $\boldsymbol{\sigma}$ has to lie on the boundary of $\mathbf{\Sigma}_\mathrm{h}$. If $\boldsymbol{\sigma}$ does not lie in $\partial_2\mathbf{\Sigma} \cup \partial_4\mathbf{\Sigma}$ then there exists $u \in (0,1)$ such that $\arg\gamma_1(u) = \arg\gamma_3(1-u)$ is minimal. Moreover, the tangents satisfy $\gamma_1'(u) = -\lambda\gamma_3'(1-u) \neq 0$. That is, the tangents point into opposite directions, while the image of f_h always lies on the left hand side of the boundary curve. Hence, there exists a point $\tilde{\boldsymbol{\sigma}}$ in the neighborhood either of $[u,1]$ or of $[2-2u,2]$ with $\arg f_\mathrm{h}(\tilde{\boldsymbol{\sigma}}) < \arg f_\mathrm{h}(\boldsymbol{\sigma})$ contradicting the minimality assumption. Analogously, one shows that $\operatorname{argmax} \arg f_\mathrm{h} \subset \partial_2\mathbf{\Sigma} \cup \partial_4\mathbf{\Sigma}$. Since the image of f_h lies on the left hand side of the curve γ_2, which is oriented from left to right on the positive real axis, we see that there exist positive angles so that $\arg\gamma_2 = 0$ cannot be maximal. Hence, $\min \arg f_\mathrm{h} = \arg\gamma_2 = 0$, and $\max \arg f_\mathrm{h} = \arg\gamma_4 = \pi/n$.

Third, we show that f_h is injective. Since the image $f_\mathrm{h}(\mathbf{\Sigma}_\mathrm{h})$ does not contain the origin, there exists $r \in \mathbb{N}$ such that $\lambda^r < |f_\mathrm{h}(\boldsymbol{\sigma})| \leq \lambda^{-r}$ for all $\boldsymbol{\sigma} \in \mathbf{\Sigma}_\mathrm{h}$. We define the domain $\mathbf{\Sigma}_\mathrm{h}^r := \cup_{|m|<r} 2^m \mathbf{\Sigma}_\mathrm{h}$ and the map

$$f_\mathrm{h}^r : \mathbf{\Sigma}_\mathrm{h}^r \ni \boldsymbol{\sigma} \mapsto \lambda^m f_\mathrm{h}(2^m\boldsymbol{\sigma}) \quad \text{if} \quad 2^m\boldsymbol{\sigma} \in \mathbf{\Sigma}_\mathrm{h}$$

consisting of scaled copies of f_h. This map is smooth and regular since consecutive parts satisfy contact conditions up to order k analogously to (13). The four boundary curves $\gamma_1^r, \ldots, \gamma_4^r$ of f_h^r are defined as shown in Figure 8,*right*. Now, we consider the function

$$\nu : \mathbb{C} \ni z \mapsto |\{\boldsymbol{\sigma} \in \mathbf{\Sigma}_\mathrm{h}^r : f_\mathrm{h}^r(\boldsymbol{\sigma}) = z\}| \in \mathbb{N}_0$$

assigning the number of pre-images to points in the complex plane. Since f_h is continuous, ν is upper semi-continuous. Since f_h is regular, ν is lower semi-continuous at all points not contained in the image $f_\mathrm{h}^r(\mathbf{\Sigma}_\mathrm{h}^r)$ of the boundary. From the results above it follows immediately that $\nu(\gamma_2(0)) = 1$. Further, by definition of r, the curves γ_1^r or γ_3^r do not intersect $f_\mathrm{h}(\mathbf{\Sigma}_\mathrm{h})$. Hence, ν is continuous and equal to one on $f_\mathrm{h}(\mathbf{\Sigma}_\mathrm{h})$.

Fourth, we show that f is injective. From

$$f(s,t,j) = \begin{cases} w_n^j f_\mathrm{h}(s,t) & \text{if } s \leq t \\ w_n^j f_\mathrm{h}(t,s) & \text{if } t \leq s \end{cases}$$

it follows

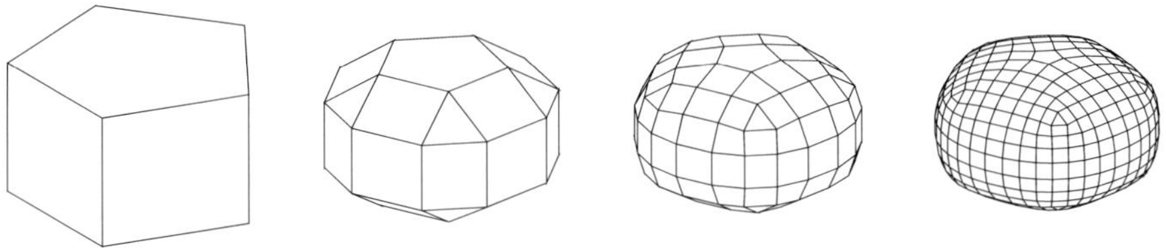

Fig. 9. Mesh refinement of the Doo-Sabin algorithm.

$$\arg f(s,t,j) \in \begin{cases} [0,\pi/n] + 2j\pi/n & \text{if } s \le t \\ [-\pi/n, 0] + 2j\pi/n & \text{if } t \le s. \end{cases}$$

For $(s,t,j) \neq (s',t',j')$, equal angles and moduli are only possible if $j' = j+1$ and $(s,t) = (t',s') \in \partial_4\Sigma_h$, or $j' = j$ and $(s,t) = (s',t') \in \partial_2\Sigma_h$, but these points are identified in $\mathbf{S}^0$. □

Typically, the check for intersections of the curve γ_1 with the non-positive real line is easy to accomplish since, in case of injectivity, the angle of γ_1 varies between 0 and π/n, so that the two sets are not even close. We demonstrate the procedure in the next section at hand of the Doo-Sabin algorithm.

Bibliographic notes

- The relevance of the Discrete Fourier Transform for the analysis of subdivision matrices was already recognized in [24].
- Symmetry properties of the characteristic map and their relation to its Fourier index are discussed in [63].
- In [99], an alternate criterion for injectivity of a regular characteristic map is given. It is based on the winding number of its boundary curve.

7. An Example: The Doo-Sabin Algorithm

The Doo-Sabin algorithm generalizes subdivision (uniform knot insertion) of biquadratic tensor-product B-splines. For each n-gon of the original mesh, a new, smaller n-gon is created and connected with its neighbors as depicted in Figure 9. Figure 10 shows the mask for generating a new n-gon from an old one for the regular case $n = 4$ (*left*) and the general case (*middle*). The standard weights suggested by Doo and Sabin in [24] are

$$a_j := \frac{\delta_{j,0}}{4} + \frac{3 + 2\cos(2\pi j/n)}{4n}. \tag{17}$$

Each of the n segments $\mathbf{x}_j^m, j \in \mathbb{Z}_n$, of the mth spline ring generated by the Doo-Sabin algorithm consists of three biquadratic B-spline patches. Accordingly, we can split the control points $\mathbf{Q}^m$ into n groups of nine control points, each, ordered as

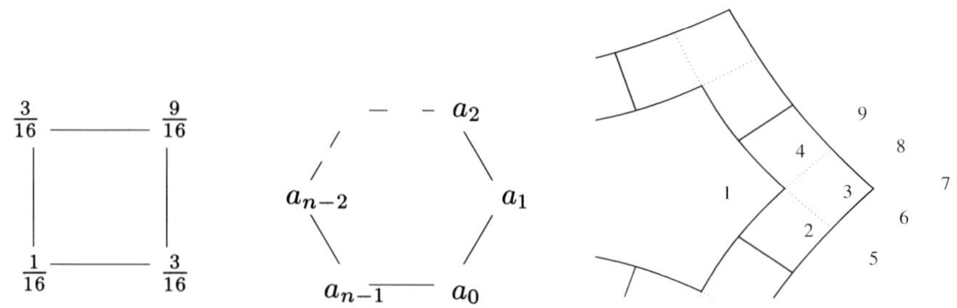

Fig. 10. Regular refinement rule (*left*), general refinement rule (*middle*), and labeling of the control points (*right*) of the Doo-Sabin algorithm.

shown in Figure 10 (*right*). Since the scheme is symmetric, we can apply DFT as in Section 6 to obtain

$$\hat{A}_k = \begin{pmatrix} \hat{a}_k & 0 & 0 \\ \hat{A}_k^{1,0} & \hat{A}_k^{1,1} & 0 \\ \hat{A}_k^{2,0} & \hat{A}_k^{2,1} & 0 \end{pmatrix},$$

where $\hat{a}_k = \sum_j w_n^{-jk} a_j$ are the entries of the DFT of the vector $[a_0, \ldots, a_{n-1}]$ of inner weights. The sub-matrices $\hat{A}_k^{i,j}$ do not depend on the special choice of weights or on the index k. Rather, they contain information on the subdivision rules in the tensor product case. For instance,

$$\hat{A}_k^{1,1} = \frac{1}{16} \begin{bmatrix} 3 & 0 & \overline{w} \\ 3 & 1 & 3 \\ w & 0 & 3 \end{bmatrix}, \quad w := 16 w_n^k.$$

The weights a_j are assumed to sum to one, i.e., $\hat{a}_0 = 1$, and to be symmetric according to $a_j = a_{n-j}$, i.e., $\hat{a}_k = \hat{a}_{n-k}$ is real. The eigenvalues of $\hat{A}_k$ are $\hat{a}_k$, $1/4$, $1/8$, $1/16$, 0 so that the subdominant eigenvalue must be $\lambda := \hat{a}_1 = \hat{a}_{n-1} \in (1/4, 1)$. Using a computer algebra system, one can determine the complex eigenvector $\hat{v}$ of $\hat{A}_1$ corresponding to λ explicitly:

$$\hat{v} = \begin{bmatrix} 2\lambda(16\lambda-1)(8\lambda-1)(4\lambda-1) \\ 6\lambda(16\lambda-1)(6\lambda-1+2\overline{w_n}\lambda) \\ 18\lambda(32\lambda^2-1+4c_n\lambda) \\ 6\lambda(16\lambda-1)(6\lambda-1+2w_n\lambda) \\ (16\lambda-1)\left(12\lambda^2+18\lambda-3+\overline{w_n}(4\lambda^2+12\lambda-1)\right) \\ 6\lambda\left(32\lambda^2+64\lambda-12+c_n(20\lambda+1)-is_n(16\lambda-1)\right) \\ 64\lambda^3+512\lambda^2-46\lambda-8+36c_n\lambda(2\lambda+1) \\ 6\lambda\left(32\lambda^2+64\lambda-12+c_n(20\lambda+1)+is_n(16\lambda-1)\right) \\ (16\lambda-1)\left(12\lambda^2+18\lambda-3+w_n(4\lambda^2+12\lambda-1)\right) \end{bmatrix},$$

where $w_n = c_n + is_n$. In particular, for the original Doo-Sabin weights in (17), we have $\lambda = 1/2$ and, rearranging the entries of $\hat{v}$ in a (3×3)-matrix according to

Figure 10, *right*,

$$\begin{bmatrix} \hat{v}_5 & \hat{v}_6 & \hat{v}_7 \\ \hat{v}_2 & \hat{v}_3 & \hat{v}_8 \\ \hat{v}_1 & \hat{v}_4 & \hat{v}_9 \end{bmatrix} = 3 \begin{bmatrix} 21 + 14\overline{w_n} & 28 + 2w_n + 9\overline{w_n} & 35 + 12c_n \\ 14 + 7\overline{w_n} & 21 + 6c_n & 28 + 2\overline{w_n} + 9w_n \\ 7 & 14 + 7w_n & 21 + 14w_n \end{bmatrix}.$$

By elementary computations, one can determine the Bernstein-Bézier-form of all three biquadratic patches forming the first segment of the complex characteristic map f. For $\lambda \in (1/4, 1)$, the minimum of the real parts of all Bernstein-Bézier coefficients is positive. Hence, by the convex hull property, the condition $\gamma_1(U) \cap \mathbb{R}_0^- = \emptyset$ is always satisfied. In particular, for $\lambda = 1/2$, we obtain the minimal value $\min \operatorname{Re} \gamma_1(U) = \operatorname{Re} f(0,1,0) = 21 + 21c_n > 0$. It remains to show regularity of the characteristic map. The Jacobian determinant ${}^{\times}\mathcal{D}f$ consists of three bicubic patches, which can also be expressed explicitly in Bernstein-Bézier-form. A careful analysis shows that all coefficients are positive if

$$p(\lambda) := 128\lambda^2(1-\lambda) - 7\lambda - 2 + 9\lambda c_n > 0.$$

Again by the convex hull property, we conclude ${}^{\times}\mathcal{D}f > 0$ if $p(\lambda) > 0$. In particular, for $\lambda = 1/2$, we obtain $p(1/2) = 3/2\,(7 + 3c_n) > 0$ proving that the Doo-Sabin in its standard form is a C_1^1-scheme. Surprisingly, there is an upper bound $\lambda_{\mathrm{sup}}(n)$ with $p(\lambda) < 0$ for $1 > \lambda > \lambda_{\mathrm{sup}}(n)$. For such λ, ${}^{\times}\mathcal{D}f$ actually reveals a change of sign, and the corresponding algorithm cannot be C_1^1. Fortunately, the upper bounds are quite close to 1, so that they do not impose severe restrictions when designing variants on the standard Doo-Sabin algorithm. More precisely, the lowest upper bound occurs for $n = 3$,

$$\lambda_{\mathrm{sup}}(n) \geq \lambda_{\mathrm{sup}}(3) = \frac{\sqrt{187}}{24} \cos\left(\frac{1}{3} \arctan\left(\frac{27\sqrt{5563}}{1576}\right)\right) + \frac{1}{3} \approx 0.8773.$$

The asymptotic behavior for $n \to \infty$ is

$$\lambda_{\mathrm{sup}}(n) \doteq 1 - \frac{\pi^2}{7n^2}.$$

Summarizing, we have shown the following:

Theorem 22. *Let $\hat{a}_0, \ldots, \hat{a}_{n-1}$ be the Fourier coefficients of a symmetric set of weights for the generalized Doo-Sabin algorithm. Then a standard scheme is obtained if $\lambda := \hat{a}_1 = \hat{a}_{n-1}$ satisfies the condition*

$$1 > \lambda > \max\{1/4, |\hat{a}_2|, \ldots, |\hat{a}_{n-2}|\}.$$

The scheme is C_1^k if $p(\lambda) > 0$, and not C_1^k if $p(\lambda) < 0$. In particular, the scheme is C_1^k when choosing the standard weights.

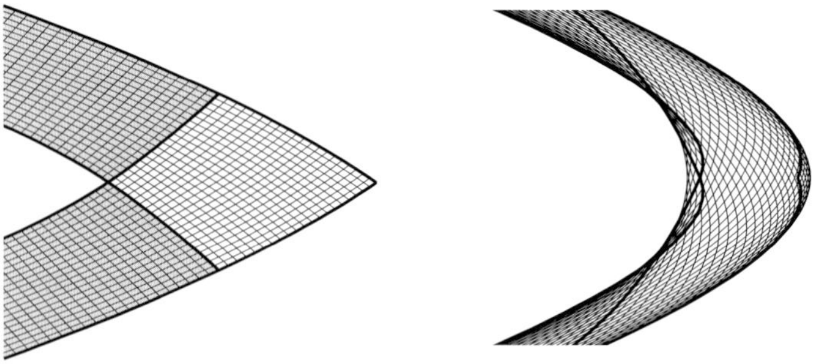

Fig. 11. Characteristic map for $n = 3$ and subdominant eigenvalue (*left*) $\lambda = 0.5$ and (*right*) $\lambda = 0.95$.

Bibliographic notes

- Details of the analysis of the Doo-Sabin algorithms can be found in [63]. In the same paper, also the Catmull-Clark algorithm and possible generalizations are analyzed.
- Today, all relevant algorithms have been thoroughly analyzed. For instance, simplest subdivision in [62], Loop's scheme in [93], $\sqrt{3}$-subdivision in [44], and the butterfly scheme in [100].
- A quite universal algorithm for numerical verification has been provided in [99].

8. Conditions for C_2^k-Schemes

As we have shown, C_1^k-subdivision schemes are well understood and relatively easy to construct. By contrast, the development of schemes providing regularity of higher order is subject to very restrictive conditions on the eigenvalues and eigenfunctions. In this section, we derive these conditions and discuss some of their consequences.

While the central point $\mathbf{x}^c$ and the central normal $\mathbf{n}^c$ are determined by the eigencoefficients $\mathbf{p}_0$ and $\mathbf{p}_1$, $\mathbf{p}_2$ related to the dominant and subdominant eigenvalue $\lambda_0 = 1$ and $\lambda_1 = \lambda_2 = \lambda$, curvature properties rely on the next smaller eigenvalue and the corresponding eigenfunctions. Let us assume that (A, G) is a standard C_2^2-scheme with eigenvalues

$$1 > \lambda > \mu := |\lambda_3| = \cdots = |\lambda_{\bar{q}}| > |\lambda_{\bar{q}+1}|$$

and, for the sake of simplicity, a full set of eigenvectors $v_3, \ldots, v_{\bar{q}}$ to the eigenvalues with modulus μ. The general case of non-trivial Jordan blocks is slightly more complicated from a purely technical point of view without providing further insight. Then the second order expansion of $\mathbf{x}^m$ reads

$$\mathbf{x}^m = GVJ^mV^{-1}\mathbf{Q} = FJ^m\mathbf{P} \doteq \mathbf{x}^c + \lambda^m\psi[\mathbf{p}_1; \mathbf{p}_2] + \mu^m\mathbf{r}^m \tag{18}$$

with

$$\mathbf{r}^m := \sum_{q=3}^{\bar{q}} a_q^m f_q \mathbf{p}_q, \quad a_q := \lambda_q/\mu. \tag{19}$$

The sequences a_q^m have modulus 1 and account for the oscillating behavior of λ_q^m in the case when λ_q is negative or complex.

For the rest of this section, the initial data are always assumed to be generic. Then, by definition of a C_2^2-scheme, the central height function h is twice differentiable at the origin. With H the symmetric (2×2)-matrix of second derivatives of h at the origin, also called the *central Hessian*, we obtain, using (4), the Taylor expansion

$$h(\boldsymbol{\xi}) = \frac{1}{2}\boldsymbol{\xi} H \boldsymbol{\xi}^{\mathrm{T}} + o(|\boldsymbol{\xi}|^2).$$

The mth spline ring of the projection $\boldsymbol{\xi}$ according to (2) is

$$\boldsymbol{\xi}^m(\mathbf{s}) = (\mathbf{x}^m(\mathbf{s}) - \mathbf{x}^{\mathrm{c}}) \cdot \mathbf{T}^{\mathrm{c}} \doteq \lambda^m \psi L, \quad L := [\mathbf{p}_1; \mathbf{p}_2] \cdot \mathbf{T}^{\mathrm{c}}, \tag{20}$$

where the (2×2)-matrix L has full rank. Since $[\mathbf{p}_1; \mathbf{p}_2] \cdot \mathbf{n}^{\mathrm{c}} = \mathbf{0}$, the central height function according to (3) is $h(\boldsymbol{\xi}^m(\mathbf{s})) \doteq \mu^m \mathbf{r}^m(\mathbf{s}) \cdot \mathbf{n}^{\mathrm{c}}$. Hence, after dividing (20) by λ^{2m}, we obtain the condition

$$\varrho^m \mathbf{r}^m \cdot \mathbf{n}^{\mathrm{c}} = \psi L H L^{\mathrm{T}} \psi^{\mathrm{T}} + o(1 + \varrho^m), \quad \varrho := \mu/\lambda^2, \tag{21}$$

which has to be satisfied for an appropriate choice of H. By Lemma 7, the eigenfunctions f_q are non-zero implying that also $\mathbf{r}^m \cdot \mathbf{n}^{\mathrm{c}}$ is in general non-zero. We distinguish three cases for the crucial ratio ϱ:

- If $\varrho < 1$, then $H = 0$ follows independent of the chosen data. This case of an enforced flat spot at the center will be referred to as ***trivial curvature continuity***.
- If $\varrho = 1$, then the right-hand side of (21) converges to a constant, and a simple argument shows that $\mathbf{r}^m \cdot \mathbf{n}^{\mathrm{c}}$ can be constant only if $a_3 = \cdots = a_{\bar{q}} = 1$.
- If $\varrho > 1$, then the left-hand side of (21) diverges faster than the right-hand side, and asymptotic equality of the sequences is impossible.

Together, it follows that *non-trivial curvature continuity* is possible, although by no means guaranteed, only if

$$\lambda^2 = \mu = \lambda_3 = \cdots = \lambda_{\bar{q}}.$$

We call μ the ***subsubdominant eigenvalue***, and elaborate on that case now. Passing to the limit, (21) reduces to

$$\sum_{q=3}^{\bar{q}} f_q p_q = \psi L H L^{\mathrm{T}} \psi^{\mathrm{T}}, \quad p_q := \mathbf{p}_q \cdot \mathbf{n}^{\mathrm{c}}. \tag{22}$$

One can easily show that for any set of coefficients $p_3, \ldots, p_{\bar{q}}$ and arbitrary $\varepsilon > 0$ there exist generic initial data $\mathbf{Q}^\varepsilon$ such that the corresponding coefficients $p_3^\varepsilon, \ldots, p_{\bar{q}}^\varepsilon$ differ by less than ε. By continuity, we conclude that (22) holds for any choice of coefficients. In particular, for any $\ell = 3, \ldots, \bar{q}$, we can set $p_q := \delta_{q,\ell}$ to obtain

$$f_\ell = \psi L H L^{\mathrm{T}} \psi^{\mathrm{T}} = a f_1^2 + b f_2^2 + c f_1 f_2$$

for certain constants a, b, c. This means that all subsubdominant eigenfunctions must lie in the space of homogeneous quadratic polynomials in the two subdominant eigenfunctions. We summarize our findings as follows:

Theorem 23. *Let (A, G) be a standard C_2^2-scheme which is non-trivial in the sense that the central Hessian does not necessarily vanish. Then the subsubdominant eigenvalue μ satisfies*

$$\lambda^2 = \mu = \lambda_3 = \cdots = \lambda_{\bar{q}} > |\lambda_{\bar{q}+1}|,$$

and the subsubdominant eigenfunctions $f_3, \ldots, f_{\bar{q}}$ satisfy

$$f_\ell \in \operatorname{span}\{f_1^2, f_2^2, f_1 f_2\}. \tag{23}$$

It is not too difficult to construct subdivision schemes that have the desirable distribution of eigenvalues. By contrast, the conditions on the eigenfunctions are extremely restrictive and the major obstacle to remove when trying to construct C_2^k-schemes. Let us consider an implication on the important special case of piecewise polynomial subdivision surfaces. We define the bi-degree $\deg f$ of a spline ring $f \in C^k(\mathbf{S}^0, \mathbb{R}, G)$ as the maximal bi-degree of the polynomial pieces of f. For vector-valued spline rings, the bi-degree is the maximum over all components.

Theorem 24. *Let (A, G) be a standard C_2^k-scheme, and assume that the generating system consists of piecewise polynomials. If the scheme is flexible in the sense that there exist generic initial data such that the corresponding subdivision surface has positive Gaussian curvature at the center, then either $n = 4$ or*

$$\deg G \geq 2k + 2. \tag{24}$$

Proof. First, we show $\deg \psi > k$ if $n \neq 4$. Suppose that $\deg \psi \leq k$. Then patches ψ_j of the characteristic map are not piecewise polynomial functions, but in fact single polynomials, which we now consider to be extended from $\mathbf{\Sigma}^0$ to all of $\mathbb{R}^2$. By means of the conditions (6), each patch ψ_{j+1} is completely determined by its predecessor ψ_j, and we have

$$\psi_{j+1}(s, t) = \psi_j(-t, s), \quad j \in \mathbb{Z}_n.$$

Repeated use of this equation yields $\psi_{j+4} = \psi_j$. For the regular case $n = 4$ this is just fine, but otherwise it implies that the characteristic map is not injective contradicting Definition 15.

Positive Gaussian curvature means that $\det H > 0$. Hence, H is positive or negative definite. For generic initial data, L has full rank implying that LHL^{T} is positive or negative definite as well. We easily conclude for the degree of the scheme

$$\deg G = \deg(\psi L H L^{\mathrm{T}} \psi^{\mathrm{T}}) = 2 \deg \psi \geq 2k + 2$$

since cancellation of the leading coefficients is impossible. □

Since $k \geq 2$ for a scheme generating curvature continuous surfaces, we see that the simplest C_2^k-scheme has degree 6. Further, no scheme generalizing uniform B-spline subdivision, like the Catmull-Clark-scheme, comes into question because here $\deg G = k + 1 < 2k + 2$.

Bibliographic notes

- The case $\mu < \lambda^2$, which yields vanishing principal curvatures at the center, is discussed in [68]
- The importance of $\mu = \lambda^2$ for C_2^k has already been observed in [24].
- Necessity of (23) is proven in [74], and in [67], it could be shown that it is also sufficient. In the same paper, similar conditions for C_r^k-schemes are specified.
- In [74], it is pointed out that the degree estimate (24) relies on the piecewise polynomial structure of the surfaces rather than on properties of the subdivision algorithm. Thus, it applies even to non-stationary or non-linear schemes as long as they live up to certain symmetry properties.
- Generalizations of the degree estimate based on the concept of flexibility can be found in [69].
- C_2^k-algorithms, and even general C_r^k-algorithms are constructed in [76,66]. However, they are less elegant than the popular C_1^k-schemes and rarely used in practice.

9. Curvature Analysis

In this section we discuss the limit behavior of curvature at the center. The results are useful to understand certain artifacts in subdivision surfaces, and provide insight for the construction of ameliorated schemes.

To conveniently analyze a subdivision surface $\mathbf{x}$ with eigencoefficients $\mathbf{P} = V^{-1}\mathbf{Q}$, we introduce a *local coordinate system* in $\mathbb{R}^3$ such that the center $\mathbf{x}^c =: \mathbf{0}$ is the origin, and the unit vectors are

$$\mathbf{e}_1 := \mathbf{p}_1/|\mathbf{p}_1|, \quad \mathbf{e}_2 := \mathbf{n}^c \times \mathbf{e}_1, \quad \mathbf{e}_3 := \mathbf{n}^c.$$

It is chosen such that the central tangent plane is spanned by $\mathbf{e}_1$ and $\mathbf{e}_2$. That is, the matrix $\mathbf{T}^c$ according to (2) is given by $\mathbf{T}^c := [\mathbf{e}_1; \mathbf{e}_2]$. As in Section 8, we assume

$$1 > \lambda > \mu := |\lambda_3| = \cdots = |\lambda_{\bar{q}}| > |\lambda_{\bar{q}+1}|$$

and a full set of eigenvectors $v_3, \ldots, v_{\bar{q}}$. The second order expansion of the spline rings according to (18) reads,

$$\mathbf{x}^m \doteq \lambda^m \psi[\mathbf{p}_1; \mathbf{p}_2] + \mu^m \mathbf{r}^m$$

where $\mathbf{r}^m$ is defined by (19). With α the angle between $\mathbf{p}_1$ and $\mathbf{p}_2$, the first two components of $\mathbf{x}^m$ are asymptotically given by

$$\mathbf{x}^m \cdot \mathbf{T}^c \doteq \lambda^m \psi L, \quad L := [\mathbf{p}_1; \mathbf{p}_2] \cdot \mathbf{T}^c = \begin{bmatrix} |\mathbf{p}_1| & 0 \\ |\mathbf{p}_2| \cos\alpha & |\mathbf{p}_2| \sin\alpha \end{bmatrix}, \tag{25}$$

while the third component is

$$\mathbf{x} \cdot \mathbf{n}^{\mathrm{c}} \doteq \mu^m \mathbf{r}^m \cdot \mathbf{n}^{\mathrm{c}}.$$

Assume that an eigenvalue λ_{q_0} with $|\lambda_{q_0}| = \mu$ is not positive. Then the sign of $\mathbf{r}^m \cdot \mathbf{n}^{\mathrm{c}}$ is incessantly changing as $m \to \infty$ if $\mathbf{p}_{q_0} \cdot \mathbf{n}^{\mathrm{c}}$ is large compared with the other coefficients $\mathbf{p}_q \cdot \mathbf{n}^{\mathrm{c}}, q \in \{3, \ldots, \bar{q}\} \backslash \{q_0\}$. Schemes revealing such an oscillating behavior should be discarded, so that we focus now on the case of coinciding positive subsubdominant eigenvalues.

Definition 25. A standard C_1^k-scheme with $k \geq 2$ and eigenvalues

$$\lambda > \mu = \lambda_3 = \cdots = \lambda_{\bar{q}} > |\lambda_{\bar{q}+1}| \tag{26}$$

is called a *(λ, μ)-scheme.*

For a (λ, μ)-scheme, the real-valued spline rings $\mathbf{r}^m \cdot \mathbf{n}^{\mathrm{c}}$ are independent of m, and we define

$$h^{\mathrm{c}} := \mathbf{r}^m \cdot \mathbf{n}^{\mathrm{c}} = \sum_{q=3}^{\bar{q}} f_q \mathbf{p}_q \cdot \mathbf{n}^{\mathrm{c}} \in C^k(\mathbf{S}^0, \mathbb{R}). \tag{27}$$

Together, we find the expansion

$$\mathbf{x}^m \doteq [\lambda^m \psi L, \mu^m h^{\mathrm{c}}] = [\psi L, h^{\mathrm{c}}] \operatorname{diag}(\lambda^m, \lambda^m, \mu^m). \tag{28}$$

This means that, asymptotically, the spline rings $\mathbf{x}^m$ are just scaled copies of the surface $[\psi L, h^{\mathrm{c}}]$. For the forthcoming investigation of curvature and shape properties, this surface plays a most important role.

Definition 26. The *central surface* $\mathbf{z}^{\mathrm{c}}$ corresponding to the spline $\mathbf{x} = B\mathbf{Q}$ generated by a (λ, μ)-scheme is defined by

$$\mathbf{z}^{\mathrm{c}} := [\psi^{\mathrm{c}}, h^{\mathrm{c}}] \in C^k(\mathbf{S}^0, \mathbb{R}^3), \quad \psi^{\mathrm{c}} := \psi L,$$

where the (2×2)-matrix L and the real-valued spline ring h^{c} are given by (25) and (27), respectively.

It is important to notice that unlike the characteristic map, the central surface depends on the initial data. Using $\partial_1 := \partial_s, \partial_2 := \partial_t$, and the differential operators

$$D := \begin{bmatrix} \partial_1 \\ \partial_2 \end{bmatrix}, \quad D_{i,j} := \begin{bmatrix} D \\ \partial_i \partial_j \end{bmatrix},$$

the fundamental forms of a subdivision surface can be expressed conveniently.

Theorem 27. *For a (λ, μ)-scheme and generic initial data, the first fundamental form of the spline ring $\mathbf{x}^m$ and its inverse are given by*

$$I^m \doteq \lambda^{2m} I, \quad (I^m)^{-1} \doteq \lambda^{-2m} I^{-1}, \quad I := D\psi^{\mathrm{c}} \cdot D\psi^{\mathrm{c}}. \tag{29}$$

With I^c and II^c the first and second fundamental form of the central surface $\mathbf{z}^c$, the second fundamental form of $\mathbf{x}^m$ is

$$II^m \doteq \mu^m II, \quad II := \sqrt{\frac{\det I^c}{\det I}}\, II^c. \tag{30}$$

Proof. The first formula follows immediately from $I^m = D\mathbf{x}^m \cdot D\mathbf{x}^m$, $D\mathbf{x}^m \doteq \lambda^m D\psi^c \mathbf{T}^c$, and $\mathbf{T}^c \cdot \mathbf{T}^c = \mathrm{Id}$. To compute the inverse, we note that

$$\det I = ({}^{\times}\!D\psi^c)^2 = ({}^{\times}\!D\psi)^2 (\det L)^2 = ({}^{\times}\!D\psi)^2\, |\mathbf{p}_1 \times \mathbf{p}_2|^2.$$

For generic initial data, the cross product does not vanish, while $({}^{\times}\!D\psi)^2 \geq c > 0$ for some constant c by regularity of ψ, compactness of the domain, and continuity of ${}^{\times}\!D\psi$. Hence,

$$(\det I^m)^{-1} \doteq \lambda^{-4m} (\det I)^{-1}$$

and the formula for $(I^m)^{-1}$ follows easily.

From (28), we conclude

$$\det D_{i,j}\mathbf{x}^m \doteq \lambda^{2m} \mu^m \det D_{i,j}\mathbf{z}^c,$$

and (30) is obtained by comparing the definitions

$$II^m_{i,j} = \frac{\det D_{i,j}\mathbf{x}^m}{\sqrt{\det I^m}}, \quad II^c_{i,j} = \frac{\det D_{i,j}\mathbf{z}^c}{\sqrt{\det I^c}}.$$

□

It is important to notice that the second fundamental form of $\mathbf{x}^m$ and the central surface $\mathbf{z}^c$ differ only be a scalar factor. For that reason, their shape properties are closely related. With the help of the fundamental forms we can compute the *Weingarten map*, which in turn determines the asymptotic behavior of the principal curvatures and directions. We recall that the Weingarten map (also known as the *shape operator*) is defined as the differential of the normal map. Its eigenvalues are the principal curvatures, while its left eigenvectors in the parameter domain are mapped to the principal directions by the Jacobian of the surface parametrization.

Theorem 28. *For a (λ, μ)-scheme and generic initial data, the Weingarten map W^m of $\mathbf{x}^m$ is*

$$W^m \doteq \varrho^m W, \quad W := II\, I^{-1}, \quad \varrho := \frac{\mu}{\lambda^2}. \tag{31}$$

Let Q be the matrix of normalized left eigenvectors and K the diagonal matrix of eigenvalues of W,

$$W = Q^{-1} K Q. \tag{32}$$

The diagonal matrix of principal curvatures $K^m := \mathrm{diag}(\kappa_1^m, \kappa_2^m)$ of $\mathbf{x}^m$ is given by

$$K^m \doteq \varrho^m K, \tag{33}$$

while the principal directions $\mathbf{P}^m := [\mathbf{p}_1^m; \mathbf{p}_2^m]$ converge to the $\mathbf{e}_1\mathbf{e}_2$-plane:

$$\mathbf{P}^m \doteq \mathbf{P}, \quad \mathbf{P} := Q\, D\psi^c\, \mathbf{T}^c. \tag{34}$$

Proof. With respect to the parameter domain, the Weingarten map is given by $W^m := \mathit{II}^m\,(I^m)^{-1}$, and (31) follows from (29) and (30). The principal curvatures κ_1^m, κ_2^m are the eigenvalues of W^m, and that implies (33). The matrix Q^m of normalized left eigenvectors of W^m converges according to $Q^m \doteq Q$. Further, by (28), $D\mathbf{x}^m \doteq \lambda^m\, D\psi^c\, \mathbf{T}^c$. Hence, using appropriate scaling, the principal directions of $\mathbf{x}^m$ are $\mathbf{P}^m = \lambda^{-m} Q D\mathbf{x}_m \doteq Q\, D\psi\, \mathbf{T}^c$, as stated. □

As in the preceding section, we see that the ratio ϱ plays a central role for the limit behavior of curvature.

Theorem 29. *For a (λ,μ)-scheme and generic initial data, the principal curvatures near the center behave according to the ratio ϱ.*
- *If $\varrho < 1$, then both principal curvatures converge to 0.*
- *If $\varrho = 1$, then both principal curvatures are bounded and at least one of them does not converge to 0.*
- *If $\varrho > 1$, then at least one principal curvature diverges.*

Proof. In view of (33), it remains to show that $K \neq 0$ for generic initial data. If $K = 0$, then $W = 0$ and $\mathit{II} = \mathit{II}^c = 0$. The second fundamental form II^c of $\mathbf{z}^c$ vanishes only if $\mathbf{z}^c$ is planar. This is the case if and only if h^c and ψ are linearly dependent, i.e., if there are constants $a, b, c \in \mathbb{R}$ which do not vanish simultaneously, such that

$$a h^c + \psi[b; c] = 0.$$

Let $\mathbf{s} = (s, t, j)$ be an arbitrary point on the outer boundary of the domain $\mathbf{S}^0$, i.e., $\max\{s, t\} = 2$. Then, by (12), $\psi(2^{-1}\mathbf{s}) = \lambda\psi(\mathbf{s})$. Analogously, since h^c is an eigenfunction to μ, one can show $h^c(2^{-1}\mathbf{s}) = \mu h^c(\mathbf{s})$. Hence,

$$\begin{aligned} a h^c(\mathbf{s}) + \psi(\mathbf{s})[b; c] &= 0 \\ a\mu h^c(\mathbf{s}) + \lambda\psi(\mathbf{s})[b; c] &= 0. \end{aligned}$$

This implies $\psi(\mathbf{s})[b; c] = 0$. Let us assume that $[b; c] = [0; 0]$. Then $a \neq 0$ and $h^c = 0$. By Lemma 7, the eigenfunctions $f_3, \ldots, f_{\bar{q}}$ to μ are linearly independent so that all coefficients $\mathbf{p}_q \cdot \mathbf{n}^c$ in the definition (27) of h^c must vanish. This contradicts the assumption that the initial data are generic. Now, we assume $[b; c] \neq [0; 0]$. In this case, all outer boundary points $\psi(\mathbf{s})$ lie on the straight line $xb + yc = 0$. Since $\psi(2^{-1}\mathbf{s}) = \lambda\psi(\mathbf{s})$, also all inner boundary points lie on the same straight line. Since ψ is regular, the boundary of the image is a subset of the image of the boundary, which is part of a straight line. Hence, the complete image of ψ must be part of a straight line; but this is impossible for a regular map. □

In the previous section, we have derived a necessary condition for C_2^k-schemes. Now, we are able to show that this condition is also sufficient.

Theorem 30. *A (λ,μ)-scheme is C_2^k if and only if $\mu = \lambda^2$ and the subsubdominant eigenfunctions $f_3, \ldots, f_{\bar{q}}$ satisfy*

$$f_\ell \in \operatorname{span}\{f_1^2, f_2^2, f_1 f_2\}.$$

Proof. In Theorem 23, we have shown that the given conditions are necessary. Now, let us assume that they are satisfied. Then there exists a symmetric (2×2)-matrix S with constant entries such that

$$h^{\mathrm{c}} = \frac{1}{2}\psi^{\mathrm{c}} S \cdot \psi^{\mathrm{c}}.$$

It is easily verified by inspection that

$$I = D\psi^{\mathrm{c}} \cdot D\psi^{\mathrm{c}}, \quad I\!I = D\psi^{\mathrm{c}} S \cdot D\psi^{\mathrm{c}}.$$

Hence, by (31), $W = I\!I\, I^{-1} = D\psi^{\mathrm{c}}\, S\, (D\psi^{\mathrm{c}})^{-1}$. That is, the eigenvalues of S and W coincide and are constant. More precisely, if $S = RKR^{-1}$ for a diagonal matrix K, then

$$W = Q^{-1} K Q, \quad Q := (D\psi\, R)^{-1}.$$

Comparison with (32) shows that the principal curvatures converge according to $K^m \doteq K$. By (34), also the principal directions converge to a constant limit:

$$\mathbf{P}^m \doteq Q\, D\psi\, \mathbf{T}^{\mathrm{c}} = R^{-1}\, \mathbf{T}^{\mathrm{c}}.$$

□

We conclude our discussion of the limit behavior of curvature by specifying limit exponents for L^p-integrability. More precisely, for $1 \leq p \leq \infty$, we say that a subdivision surface is $H^k_{2,p}$, if it is C^k_1, and if the principal curvatures are L^p-integrable when restricted to a sufficiently small neighborhood of the center.

Theorem 31. *For a (λ, μ)-scheme and generic initial data, the generated subdivision surface is*
- $H^k_{2,\infty}$, *if* $\varrho \leq 1$.
- $H^k_{2,p}$ *for all* $p < -2\log\lambda/\log\varrho$, *if* $\varrho > 1$.

In particular, for any ϱ, the surface is $H^k_{2,2}$.

Proof. For $\varrho \leq 1$, the principal curvatures are bounded, as stated. For $\varrho > 1$, we choose m_0 so large that I^m is regular for all $m \geq m_0$. Then, with the surface element $d\mathbf{x}^m = \sqrt{\det I^m}\, dsdt \doteq \lambda^{2m}\sqrt{\det I} dsdt$, the surface integral of the pth power of the principal curvatures of the mth spline ring is

$$\int_{\mathbf{x}^m} |K^m|^p d\mathbf{x}^m \doteq \varrho^{mp}\lambda^{2m}\, \bar{K}, \quad \bar{K} := \sum_{j \in \mathbf{Z}_n} \int_{\mathbf{\Sigma}^0} K_j \sqrt{\det I_j}\, dsdt,$$

where K_j and I_j denote the jth segment of K and I, respectively. Summing over all $m \geq m_0$, we obtain

$$\sum_{m=m_0}^{\infty} \int_{\mathbf{x}^m} |K^m|^p d\mathbf{x}^m \doteq \frac{(\varrho^p\lambda^2)^{m_0}\bar{K}}{1 - \varrho^p\lambda^2},$$

which is finite for $p < -2\log\lambda/\log\varrho$. Since $\varrho < \lambda^{-1}$, the upper bound is always ≥ 2. □

Bibliographic notes

- A first careful analysis of curvature in a vicinity of the center was given in [59], and bounds on their oscillation were specified in [60].
- The concept of the central surface and its relation to the limit behavior of curvature was introduced in [64]. Applications of the theory are discussed in [39].
- The basic limit behavior of principal curvatures according to Theorem 29 was observed in [24,52,64].
- L^p-regularity of principal curvatures was investigated in [79]. The results of Theorem 31 are crucial for using subdivision surfaces in the finite element analysis of higher order problems as in [13,12].
- The central surface provides further information on the local shape of a subdivision surface near the center. The analysis in [64] shows that the subsubdominant eigenvalue μ must be at least triple with Fourier index $\{0, 2, n-2\} \subset \mathcal{F}(\mu)$ in order to avoid severe restrictions on what type of shapes can be modeled.
- If $\mathcal{F}(\mu) = \{0, 2, n-2\}$, [39] defines a chart that characterizes, for a subdivision algorithm and for the full gamut of input data, the shape of resulting surfaces.

10. Conclusion

Subdivision surfaces are remarkably similar to spline surfaces. Their distinct character reveals itself in the neighborhood of extraordinary points where $n \neq 4$ quadrilateral patches join. This paper summarizes the structure of subdivision surfaces near extraordinary points. It adds two new building blocks to the foundations by clarifying the role of linearly dependent generating systems and simplifying the test for injectivity of the characteristic map.

Acknowledgements

We would like to thank Malcolm Sabin for many fruitful discussions and for providing his list of references.

References

1. Alexa, M., Refinement operators for triangle meshes, *Computer Aided Geometric Design* **19** (2002), 169–172.
2. Bajaj, C., Schaefer, S., Warren, J. and Xu, G., A subdivision scheme for hexahedral meshes, *The Visual Computer* **18** (2002), 343–356.
3. Ball, A. A. and Storry, D. J. T., A matrix approach to the analysis of recursively generated B-spline surfaces, *Computer Aided Design* **18** (1986), 437–442.

4. Ball, A. A. and Storry, D. J. T., Conditions for tangent plane continuity over recursively generated B-spline surfaces, *ACM Trans. on Graphics* **7** (1988), 83–102.
5. Ball, A. A. and Storry, D. J. T., An investigation of curvature variations over recursively generated B-spline surfaces, Technical report, Loughborough University of Technology, 1990.
6. Barthe, L., Gérot, C., Sabin, M. and Kobbelt, L., Simple computation of the eigencomponents of a subdivision matrix in the Fourier domain, in: *Advances in Multiresolution for Geometric Modelling* (N. A. Dodgson, M. S. Floater and M. A. Sabin, Eds.), Springer, Berlin, 2005, pp. 245–258.
7. Biermann, H., Levin, A. and Zorin, D., Piecewise smooth subdivision surfaces with normal control, in: *SIGGRAPH 2000 Conference Proceedings: Computer Graphics Annual Conference Series* (K. Akeley, Ed.), ACM Press, New York, 2000, pp. 113–120.
8. Böhm, W., Subdividing multivariate splines, *Computer Aided Design* **15** (1983), 345–352.
9. Catmull, E. and Clark, J., Recursively generated B-spline surfaces on arbitrary topological meshes, *Computer Aided Design* **10** (1978), 350–355.
10. Cavaretta, A. S., Dahmen, W. and Micchelli, C. A., *Stationary Subdivision*, Memoirs of the Amer. Math. Soc., Vol. 93(453), AMS, Providence, 1991.
11. Chaikin, G. M., An algorithm for high-speed curve generation, *Computer Graphics and Image Processing* **3** (1974), 346–349.
12. Cirak, F., Scott, M. J., Antonsson, E. K., Ortiz, M. and Schröder, P., Integrated modeling, finite element analysis and engineering design for thin-shell structures using subdivision, *Computer Aided Design* **34** (2002), 137–148.
13. Cirak, F., Ortiz, M. and Schröder, P., Subdivision surfaces: A new paradigm for thin-shell finite element analysis, *Int. J. Numer. Meth. Eng.* **47**(2000), 2039–2072.
14. Cohen-Steiner, D., Alliez, P. and Desbrun, M., Variational shape approximation, *ACM Trans. on Graphics* **23** (2004), 905–914.
15. Conti, C. and Zimmermann, G., Interpolatory rank-1 vector subdivision schemes, *Computer Aided Geometric Design* **21** (2004), 341–351.
16. Dahmen, W., Dyn, N. and Levin, D., On the convergence rates of subdivision algorithms for box splines, *Constr. Approximation* **1** (1985), 305–322.
17. Dahmen, W. and Micchelli, C. A., Subdivision algorithms for the generation of box splines, *Computer Aided Geometric Design* **1** (1984), 77–85.
18. van Damme, R., Bivariate Hermite subdivision, *Computer Aided Geometric Design* **14** (1997), 847–875.
19. DeRose, T., Halstead, M. and Kass, M., Efficient, fair interpolation using Catmull-Clark surfaces, in: *SIGGRAPH '93 Proceedings*, ACM Press, New York, 1993, pp. 35–44.
20. DeRose, T., Kass, M. and Truong, T., Subdivision surfaces in character animation, in: *SIGGRAPH 1998 Conference Proceedings: Computer Graphics Annual Conference Series* (M. Cohen, Ed.), ACM Press, New York, 1998, pp. 85–94.

21. Dodgson, N., Ivrissimtzis, I. and Sabin, M., Characteristics of dual triangular $\sqrt{3}$ subdivision, in: *Curve and Surface Fitting*, Proc. Saint Malo 2002 (A. Cohen, J.-L. Merrien and L. L. Schumaker, Eds.), Nashboro Press, Brentwood, 2003, pp. 119–128.
22. Dodgson, N., Sabin, M., Barthe, L. and Hassan, M., Towards a ternary interpolating scheme for the triangular mesh, Technical Report UCAM-CL-TR-539, Computer Laboratory, University of Cambridge, 2002.
23. Doo, D., A subdivision algorithm for smoothing down irregularly shaped polyhedrons, in: *Proc. International Conf. on Interactive Techniques in Computer Aided Design*, IEEE Computer Soc., 1978, pp. 157–165.
24. Doo, D. and Sabin, M. A., Behaviour of recursive subdivision surfaces near extraordinary points, *Computer Aided Design* **10** (1978), 356–360.
25. Dubuc, S., Interpolation through an iterative scheme, *J. Math. Anal. Appl.* **114** (1986), 185–204.
26. Dyn, N., Gregory, J. A. and Levin, D., A four-point interpolatory subdivision scheme for curve design, *Computer Aided Geometric Design* **4**, 257–268.
27. Dyn, N., Gregory, J. A. and Levin, D., Analysis of uniform binary subdivision schemes for curve design, *Constr. Approximation* **7** (1991), 127–147.
28. Dyn, N. and Levin, D., Analysis of asymptotically equivalent binary subdivision schemes, *J. Math. Anal. Appl.* **193** (1995), 594–621.
29. Dyn, N. and Levin, D., Subdivision schemes in geometric modelling, *Acta Numerica* **11** (2002), 73–144.
30. Dyn, N., Levin, D. and Gregory, J. A., A butterfly subdivision scheme for surface interpolation with tension control, *ACM Transactions on Graphics* **9** (1990), 160–169.
31. Dyn, N., Levin, D. and Liu, D., Interpolatory convexity-preserving subdivision schemes for curves and surfaces, *Computer Aided Design* **24** (1992), 211–216.
32. Gérot, C., Barthe, L., Dodgson, N. and Sabin, M., Subdivision as a sequence of sampled C^p surfaces, in: *Advances in Multiresolution for Geometric Modelling* (N. A. Dodgson, M. S. Floater and M. A. Sabin, Eds.), Springer, Berlin, 2005, pp. 259–270.
33. Han, B., Computing the smoothness exponent of a symmetric multivariate refinable function, *SIAM J. Matrix Anal. Appl.* **24** (2003), 693–714.
34. Hassan, M., Ivrissimtzis, I. P., Dodgson, N. and Sabin, M., An interpolating 4-point C^2 ternary stationary subdivision scheme, *Computer Aided Geometric Design* **19** (2002), 1–18.
35. Holt, F., Towards a curvature-continuous stationary subdivision algorithm, *Z. Angew. Math. Mech.* **76** (1995), 423–424.
36. Ivrissimtzis, I. P., Dodgson, N. and Sabin, M., $\sqrt{5}$ subdivision, in: *Advances in Multiresolution for Geometric Modelling* (N. A. Dodgson, M. S. Floater and M. A. Sabin, Eds.), Springer, Berlin, 2005, pp. 285–299.
37. Ivrissimtzis, I. P., Shrivastava, K. and Seidel, H.-P., Subdivision rules for general meshes, in: *Curve and Surface Fitting*, Proc. Saint Malo 2002 (A. Cohen, J. L. Merrien and L. L. Schumaker, Eds.), Nashboro Press, Brentwood, 2003, pp. 229–237.

38. Ivrissimtzis, I. P., Sabin, M. and Dodgson, N., On the support of recursive subdivision surfaces, Technical Report UCAM-CL-TR-544, Computer Laboratory, University of Cambridge, 2002.
39. Karciauskas, K., Peters, J. and Reif, U., Shape characterization of subdivision surfaces – case studies, *Computer Aided Geometric Design* **21** (2004), 601–614.
40. Kobbelt, L., Interpolatory refinement by variational methods, in: *Approximation Theory VIII, Vol. 2: Wavelets and Multilevel Approximation* (C. K. Chui and L. L. Schumaker, Eds.), World Scientific, Singapore, 1995, pp. 217–224.
41. Kobbelt, L., Interpolatory subdivision on open quadrilateral nets with arbitrary topology, *Computer Graphics Forum* **15** (1996), 409–420.
42. Kobbelt, L., A variational approach to subdivision, *Computer Aided Geometric Design* **13** (1996), 743–761.
43. Kobbelt, L., Using the discrete Fourier-transform to analyze the convergence of subdivision schemes, *Appl. Comp. Harmonic Analysis* **5** (1998), 68–91.
44. Kobbelt, L., $\sqrt{3}$ subdivision, in: *Proceedings of ACM SIGGRAPH 2000* (K. Akeley, Ed.), ACM Press, New York, 2000, pp. 103–112.
45. Lane, J. M. and Riesenfeld, R. F., A theoretical development for the computer generation of piecewise polynomial surfaces, *IEEE Trans. Pattern Anal. Machine Intell.* **2** (1980), 35–46.
46. Leber, M., *Interpolierende Unterteilungsalgorithmen*, Master's thesis, Universität Stuttgart, 1994.
47. Levin, A., *Combined Subdivision Schemes*, PhD thesis, Tel-Aviv University, 1999.
48. Levin, A., Combined subdivision schemes for the design of surfaces satisfying boundary conditions, *Computer Aided Geometric Design* **16** (1999), 345–354.
49. Levin, D., Using Laurent polynomial representation for the analysis of non-uniform binary subdivision schemes, *Adv. Comput. Math.* **11** (1999), 41–54.
50. Litke, N., Levin, A. and Schröder, P., Trimming for subdivision surfaces, *Computer Aided Geometric Design* **18** (2001), 463–481.
51. Loop, C., Bounded curvature triangle mesh subdivision with the convex hull property, *The Visual Computer* **18** (2002), 316–325.
52. Loop, C. T., *Smooth Subdivision for Surfaces Based on Triangles*, Master's thesis, University of Utah, 1987.
53. Micchelli, C. A. and Prautzsch, H., Computing curves invariant under halving, *Computer Aided Geometric Design* **4** (1987), 133–140.
54. Micchelli, C. A. and Prautzsch, H., Computing surfaces invariant under subdivision, *Computer Aided Geometric Design* **4** (1987), 321–328.
55. Morin, G., Warren, J. and Weimer, H., A subdivision scheme for surfaces of revolution, *Computer Aided Geometric Design* **18** (2001), 483–502.
56. Nasri, A. and Sabin, M., Taxonomy of interpolation constraints on recursive subdivision curves, *The Visual Computer* **18** (2002), 259–272.
57. Oswald, P. and Schröder, P., Composite primal/dual $\sqrt{3}$ subdivision schemes, *Computer Aided Geometric Design* **20** (2003), 135–164.

58. Peters, J., Geometric continuity, in: *Handbook of Computer Aided Geometric Design* (G. Farin, J. Hoschek and M. Kim, Eds.), Elsevier, Amsterdam, 2002, pp. 193–229.
59. Peters, J. and Umlauf, G., Gaussian and mean curvature of subdivision surfaces, in: *The Mathematics of Surfaces IX* (R. Cipolla and R. Martin, Eds.), Springer, Berlin, 2000, pp. 59–69.
60. Peters, J. and Umlauf, G., Computing curvature bounds for bounded curvature subdivision, *Computer Aided Geometric Design* **18** (2001), 455–461.
61. Peters, J. and Shiue, L.-J., Combining 4- and 3-direction subdivision, *ACM Transactions on Graphics* **23** (2004), 980–1003.
62. Peters, J. and Reif, U., The simplest subdivision scheme for smoothing polyhedra, *ACM Trans. on Graphics* **16** (1997), 420–431.
63. Peters, J. and Reif, U., Analysis of algorithms generalizing B-spline subdivision, *SIAM J. Numer. Anal.* **35** (1998), 728–748.
64. Peters, J. and Reif, U., Shape characterization of subdivision surfaces: basic principles, *Computer Aided Geometric Design* **21** (2004), 585–599.
65. Peters, J. and Wu, X., On the local linear independence of generalized subdivision functions, submitted.
66. Prautzsch, H., Freeform splines, *Computer Aided Geometric Design* **14** (1997), 201–206.
67. Prautzsch, H., Smoothness of subdivision surfaces at extraordinary points, *Adv. Comput. Math.* **9** (1998), 377–390.
68. Prautzsch, H. and Umlauf, G., Improved triangular subdivision schemes, in: *Computer Graphics International*, Proc. Hannover 1998 (F. E. Wolter and N. M. Patrikalakis, Eds.), pp. 626–632.
69. Prautzsch, H. and Reif, U., Necessary conditions for subdivision surfaces, *Adv. Comput. Math.* **10** (1999), 209–217.
70. Qin, K. and Wang, H., Eigenanalysis and continuity of non-uniform Doo-Sabin surfaces, in: *Proc. Pacific Graphics 99*, 1999, pp. 179–186.
71. Reif, U., *Neue Aspekte in der Theorie der Freiformflächen beliebiger Topologie*, PhD thesis, Universität Stuttgart, 1993.
72. Reif, U., Some new results on subdivision algorithms for meshes of arbitrary topology, in: *Approximation Theory VIII, Vol. 2, Wavelets and Multilevel Approximation* (C. K. Chui and L. L. Schumaker, Eds.), World Scientific, Singapore, 1995, pp. 367–374.
73. Reif, U., A unified approach to subdivision algorithms near extraordinary vertices, *Computer Aided Geometric Design* **12** (1995), 153–174.
74. Reif, U., A degree estimate for subdivision surfaces of higher regularity. *Proc. Amer. Math. Soc.* **124** (1996), 2167–2174.
75. Reif, U., A refinable space of smooth spline surfaces of arbitrary topological genus, *J. Approximation Theory* **90** (1997) 174–199.
76. Reif, U., TURBS – Topologically unrestricted rational B-splines, *Constr. Approximation* **14** (1998), 57–77.
77. Reif, U., *Analyse und Konstruktion von Subdivisionsalgorithmen für Freiformflächen beliebiger Topologie*, Habilitationsschrift, Shaker Verlag, Aachen, 1999.

78. Reif, U. and Peters, J., *Structural Analysis of Subdivision Surfaces*, in preparation.
79. Reif, U. and Schröder, P., Curvature smoothness of subdivision surfaces, *Adv. Comput. Math.* **14** (2001), 157–174.
80. Sabin, M., Recursive division, in: *The Mathematics of Surfaces* (J. A. Gregory, Ed.), Clarendon Press, Oxford, 1986, pp. 269–282.
81. Sabin, M., Cubic recursive division with bounded curvature, in: *Curves and Surfaces* (P.-J. Laurent, A. Le Méhauté and L. L. Schumaker, Eds.), Academic Press, San Diego, 1991, pp. 411–414.
82. Sabin, M., Eigenanalysis and artifacts of subdivision curves and surfaces, chapter 4 in: *Tutorials on Multiresolution in Geometric Modelling* (A. Iske, E. Quak and M. S. Floater, Eds.), Springer, Berlin, 2002, pp. 69–92.
83. Sabin, M., Subdivision surfaces, in: *Handbook of Computer Aided Design* (G. Farin, J. Hoschek and M. Kim, Eds.), 2002, pp. 309–326.
84. Sabin, M., Dodgson, N., Hassan, M. and Ivrissimtzis, I., Curvature behaviours at extraordinary points of subdivision surfaces, *Computer Aided Design* **35** (2003), 1047–1051.
85. Sabin, M. and Barthe, L., Artifacts in recursive subdivision surfaces, in: *Curve and Surface Fitting*, Proc. Saint Malo 2002 (A. Cohen, J.-L. Merrien and L. L. Schumaker, Eds.), Nashboro Press, Brentwood, 2003, pp. 353–362.
86. Schaefer, S. and Warren, J., On C^2 triangle/quad subdivision, *ACM Transaction on Graphics*, to appear.
87. Schröder, P., Subdivision as a fundamental building block of digital geometry processing algorithms, *J. Comput. Appl. Math.* **149** (2002), 207–219.
88. Shiue, L. and Peters, J., Mesh refinement based on Euler encoding, in: *Proceedings of The International Conference on Shape Modeling and Applications* (N. Patrikalakis, Ed.), 2005, pp. 343–348.
89. Shiue, L., Jones, I. and Peters, J., A realtime GPU subdivision kernel, in: *ACM Siggraph 2005, Computer Graphics Proceedings*, (M. Gross, Ed.), to appear.
90. Stam, J., Exact evaluation of Catmull-Clark subdivision surfaces at arbitrary parameter values, in: *Proceedings of SIGGRAPH 1998* (M. Cohen, Ed.), ACM Press, New York, 1998, pp. 395–404.
91. Stam, J., On subdivision schemes generalizing uniform B-spline surfaces of arbitrary degree, *Computer Aided Geometric Design* **18** (2001), 383–396.
92. Storry, D. J. T., *B-Spline Surfaces Over an Irregular Topology by Recursive Subdivision.* PhD thesis, Loughborough University of Technology, 1985.
93. Umlauf, G., Analyzing the characteristic map of triangular subdivision schemes, *Constr. Approximation* **16** (2000), 145–155.
94. Velho, V., Quasi 4-8 subdivision, *Computer Aided Geometric Design* **18** (2001), 345–358.
95. Velho, L. and Zorin, D., 4-8 subdivision, *Computer Aided Geometric Design* **18** (2001), 397–427.
96. Warren, J. and Weimer, H., *Subdivision Methods for Geometric design*, Morgan Kaufmann, Los Altos, 2002.

97. Warren, J. and Schäfer, S., A factored approach to subdivision surfaces, *Computer Graphics and Applications, IEEE* **24** (2004), 74–81.
98. Zorin, D., *Subdivision and Multiresolution Surface Representations*, PhD thesis, Caltech, Pasadena, 1997.
99. Zorin, D., Smoothness of stationary subdivision on irregular meshes, *Constr. Approximation* **16** (2000), 359–397.
100. Zorin, D., Schröder, P. and Sweldens, W., Interpolatory subdivision for meshes with arbitrary topology, in: *Proc ACM SIGGRAPH 1996* (H. Rushmeier, Ed.), ACM Press, New York, 1996, pp. 189–192.
101. Zorin, D. and Schröder, P., A unified framework for primal/dual quadrilateral subdivision schemes, *Computer Aided Geometric Design* **18** (2001), 429–454.

K. Jetter et al., Editors

Polynomial Interpolation in Several Variables: Lattices, Differences, and Ideals

Tomas Sauer

Lehrstuhl für Numerische Mathematik, Justus–Liebig–Universität Gießen, Heinrich–Buff–Ring 44, D–35192 Gießen, Germany

Abstract

When passing from one to several variables, the nature and structure of polynomial interpolation changes completely: the solvability of the interpolation problem with respect to a given finite dimensional polynomial space, like all polynomials of at most a certain total degree, depends not only on the number, but significantly on the geometry of the nodes. Thus the construction of interpolation sites suitable for a given space of polynomials or of appropriate interpolation spaces for a given set of nodes become challenging and nontrivial problems. The paper will review some of the basic constructions of interpolation lattices which emerge from the geometric characterization due to Chung and Yao. Depending on the structure of the interpolation problem, there are different representations of the interpolation polynomial and several formulas for the error of interpolation, reflecting the underlying point geometry and employing different types of differences. In addition, we point out the close relationship with constructive ideal theory and degree reducing interpolation, whose most prominent representer is the least interpolant, introduced by de Boor et al.

Key words: Polynomial interpolation, geometric characterization, finite difference, ideal basis

2000 MSC: 41A05, 41A63, 13P10

Email address: tomas.sauer@math.uni-giessen.de (Tomas Sauer).

1. Introduction

One of the basic but fundamental tasks in the process of trying to understand the nature of multivariate polynomial interpolation is to provide extensions of the well–known and classical theory for univariate interpolation. As tautologic as this may sound, finding such extensions is often highly nontrivial, and in many cases it is not even agreed upon what the natural or even a reasonable counterpart of a (more or less) related univariate concept or result is. This makes polynomial interpolation a much wider field in several variables than in one variable with contributions ranging from analysis to algebraic geometry and commutative algebra and there are "generalizations" that do not even have any counterpart in one variable. In this paper, I will try to summarize some old and derive some new facts on three specific aspects of multivariate polynomial interpolation: the question of lattice generation, some forms of finite and divided differences that occur as coefficients in the interpolation polynomials and error formulas, and some algebraic aspects, especially the close connection between degree reducing interpolation and the "algorithmic" ideal bases like Gröbner and H–bases. Such bases are essential for computational methods in ideal theory based on computing a remainder of division by an ideal and thus are the basis on which algorithms are built that solve ideal theoretic problems like the question whether a polynomial is contained in a given ideal or not.

Due to its selective scope, this paper cannot be a survey and has no ambition to be one. Readers interested in surveys on multivariate polynomial interpolation can still find a lot of information in [33,31,44,45]. The selection of topics in this paper, however, is unbiased, incomplete and guided entirely by personal preferences. In particular, a lot of valuable and interesting work, like interpolation on spheres and other manifolds, applications in the spline or finite element context or the connection to polynomial system solving will be mentioned not at all or just briefly in passing.

On the other hand, this is also no research paper, though it contains some new results together with full proofs, as quite some part of it just lists (more or less widely) known facts and tries to put them into a context. Once in a while I will even mix in a short proof into these reviews – mainly when I feel that it could be illustrative or give an idea of the type of arguments used at that point. The chapter on algebraic concepts, on the other hand, will be equipped more substantially with proofs, some for the aforementioned reason, but mainly because it contains new material that simply needs to be proved. Also the list of references does not claim completeness. Though it is long and, to some extent, exotic, it is by no means exhaustive and only contains items which are referenced somewhere in the text. I apologize to everyone whose work has not been mentioned here, but "*... every story one chooses to tell is a kind of censorship, it prevents the telling of other tales ...*" [64].

According to [3], the name *interpolation* was introduced by Wallis in 1655. At this time, the goal of interpolation was to estimate the value of a function at a certain point based on the value of the function at other points; the typical application was to fill in values of functions available only in tabulated form like the logarithm. In fact, even the later papers by Aitken [1] and Neville [58] are still mainly driven by this application of interpolation. But also in several variables polynomials still play a fundamental role in the ***local*** approximation of smooth functions, and interpolation is one of the simplest projectors to polynomial spaces, though it is significantly more intricate than in the univariate case. The "modern" point of view of interpolation as means of reconstructing curves or surfaces from "measurements", on the other hand, has only been taken much more recently in the time of CAGD, and it is a fair question whether interpolation by polynomials can play a reasonable role at all in applications as the polynomials' oscillating nature and the sensitivity of polynomial interpolation to the node configuration really tend to accumulate in the multivariate case. In my opinion, it is more the mathematical beauty of polynomials and the interplay with other fields of mathematics, in particular with computer algebra via constructive ideal theory, that still makes interpolation by algebraic polynomials in several variables a worthwhile topic to study.

While polynomial interpolation in one variable is classical material in almost any textbook on Numerical Analysis, facts about interpolation in several variables are hard to find, and, if at all, it is more likely to find them in the "classics" like Steffensen's book [76] or in [36]. And though the oldest paper on polynomial interpolation in several variables, due to Kronecker [38], considers nodes that are ***not*** on a tensor product grid, most of the later textbook material only covers ***tensor product*** interpolation or the structurally similar interpolation on the triangular grid. In view of that, one might be tempted to conjecture that there is something peculiar about interpolation by polynomials in two and more variables, even if this was explicitly denied in [2], cf. [33]. As a general disclaimer that warns of some of the side effects of polynomial interpolation, we will begin by pointing out some of these peculiarities in Section 2. A more positive approach will be taken in Section 3 on lattice generation by means of intersection of hyperplanes; the classical paper by Chung and Yao reviewed there is still the main source for the explicit construction of point configurations that allow for unique interpolation by a total degree space. Section 4 deals with errors, not in the sense of mathematical flaws or roundoff errors in numerical computations, but with the global deviation of an interpolation polynomial from the function it interpolates. This concept is related to generalizations of the concept of divided differences and some of them are given and put into relation with each other. Section 5 simply makes use of the fact that polynomials can also be multiplied to provide an algebraic background in terms of graded rings and polynomial ideals for degree reducing interpolation spaces, pointing out their intimate relationship with normal forms modulo Γ–bases, a concept that simultaneously generalizes Gröbner bases and H–bases. The final Section 6 uses some more special though still sufficiently general types of gradings to arrive at the conclusion that tensor product data is best interpolated by tensor product spaces and to give a brief ideal theoretic view at error formulas. Interested? So let us start ...

2. Some Basics on Multivariate Polynomial Interpolation

The general form of the *polynomial interpolation problem* is as follows: given a finite linearly independent set Θ of functionals and an associated vector $Y = (y_\theta \; : \; \theta \in \Theta)$ of prescribed values, find a polynomial f such that

$$\Theta f = Y, \qquad \text{i.e.} \qquad \theta f = y_\theta, \quad \theta \in \Theta. \tag{1}$$

Here we consider polynomials in d variables, $d \geq 1$, involving, for some field $\mathbb{K}$, the *ring* $\Pi := \mathbb{K}[x] = \mathbb{K}[x_1, \ldots, x_d]$ of polynomials in d variables, writing $x = (x_1, \ldots, x_d)$. It is important to make two comments on the underlying algebra here: first, the field $\mathbb{K}$ should be infinite, with the three "role models" $\mathbb{K} = \mathbb{Q}, \mathbb{R}, \mathbb{C}$; interpolation on finite fields is a very interesting topic with important applications in coding theory, but structurally it is significantly different from what we will consider here. Second, we will make use of the *ring* structure of polynomials when employing concepts from ideal theory. The polynomial interpolation problem by itself can easily be written as a linear algebra problem with respect to the vector space of polynomials, but the additional multiplicative structure of the ring will allow us to draw further conclusions on the behavior of interpolation.

Clearly, the interpolation problem as stated in (1) can be expected to have many solutions as without further restrictions we will be permitted to choose from the infinite dimensional vector space Π an interpolation polynomial f that must only match a finite number of conditions. In view of this it makes perfect sense to consider spaces that allow for *unique* interpolation, i.e., finite dimensional subspaces of Π whose dimension coincides with $\#\Theta$.

Definition 1. Let $\mathcal{P}$ be a subspace of Π. The polynomial interpolation problem with respect to Θ is said to be *poised for* $\mathcal{P}$ or *correct for* $\mathcal{P}$, if for any $Y \in \mathbb{K}^\Theta$ there exists a *unique* $f \in \mathcal{P}$ such that $\Theta f = Y$.

There is a somewhat tautologic characterization of poisedness in terms of algebraic geometry and linear algebra. To that end, we will denote by $P \subset \mathcal{P}$ a basis for the finite dimensional space $\mathcal{P}$.

Theorem 2. *For $\mathcal{P} \subset \Pi$ and a finite set $\Theta \subset \Pi'$ the following statements are equivalent:*

(i) The polynomial interpolation problem with respect to Θ is poised for $\mathcal{P}$.

(ii) $\dim \mathcal{P} = \#\Theta$ *and the matrix*

$$\Theta P = [\theta p \; : \; \theta \in \Theta, p \in P] \in \mathbb{K}^{\Theta \times P} \tag{2}$$

satisfies $\det \Theta P \neq 0$ *for any basis P of $\mathcal{P}$.*

(iii) $\ker \Theta \cap \mathcal{P} = \{0\}$.

Usually, condition (iii) of the above theorem is phrased as *there is no algebraic hypersurface in $\mathcal{P}$ that contains Θ*, at least when Θ consists of point evaluation functionals or as points with multiplicities – we will get to this issue in the next chapter. Unfortunately, none of the two criteria above is easier to verify than the uniqueness of polynomial interpolation, at least not in general.

One remark on the notation in (2) which will be used throughout the paper: Θ and P are sets which can be used to index the elements of vectors or matrices like in (2). Moreover, each set gives rise to a natural vector, e.g., $P = (p : p \in P)$, which can be multiplied to such matrices. This notation allows, for example, to conveniently write

$$L = (\Theta P)^{-1} \, P = \sum_{p \in P} (\Theta P)_p^{-1} \, p \qquad \Rightarrow \qquad \Theta L = (\Theta P)^{-1} \, \Theta P = I,$$

for the dual basis L of Θ without having to state explicitly the cardinalities of Θ and P – except that, in order have an inverse, the matrix ΘP must be a square one, that is, $\#\Theta = \#P$.

2.1. Lagrange and the trouble with Hermite

It is common practice to classify interpolation problems according to the type of the functionals $\theta \in \Theta$ that have to be matched by the polynomial. If those functionals are point evaluations, i.e., $\theta = \delta_\xi$, $\xi \in \mathbb{K}^d$, the interpolation problem is called a *Lagrange interpolation problem* and the points are called *nodes* or *(interpolation) sites*. If functionals are consecutive derivatives at some points, one usually speaks of a *Hermite interpolation problem* while the remaining problems are usually named *Hermite–Birkhoff interpolation problems*. At least, this is the terminology for the univariate case. In several variables it is not even agreed upon what a Hermite interpolation problem should be. The most convincing generalization requires that Θ is an *ideal interpolation scheme* in the sense of Birkhoff [5] which means that the set $\ker \Theta = \{f \in \Pi : \Theta f = 0\}$ of all polynomials annihilated by all the functionals in Θ forms an ideal in Π. Substantial investigations of Hermite interpolation in connection with ideal schemes have been performed even earlier by Möller [52,53] who already proposed ideal schemes as the natural generalization of Hermite interpolation problem. And indeed the concept of ideal interpolation schemes provides precisely what one would expect, namely interpolation of derivatives at different nodes. More precisely, it was shown in [11,48] that a set Θ of functionals defines an ideal interpolation scheme if and only if there exists a finite subset $\Xi \subset \mathbb{K}^d$ and finite dimensional polynomial subspaces $\mathcal{Q}_\xi \subset \Pi$, $\xi \in \Xi$, each of them closed under differentiation, such that

$$\operatorname{span} \Theta = \operatorname{span} \{\delta_\xi \circ q(D) \, : \, q \in \mathcal{Q}_\xi, \xi \in \Xi\}, \qquad \bigcup_{j=1}^{d} \frac{\partial}{\partial x_j} \mathcal{Q}_\xi \subseteq \mathcal{Q}_\xi, \quad \xi \in \Xi,$$

where δ_ξ denotes the point evaluation functional at ξ. In fact, this is nothing else than interpolation *with multiplicities* at the finite set Ξ! To that end, one has to take into account that the multiplicity of a zero of a function or a common zero of finitely many functions is no more a number, i.e., a matter of counting, in two and more variables: it is a ***structural*** quantity, most conveniently expressed as a ***D–invariant*** space of polynomials, that is a finite dimensional subspace $\mathcal{Q} \subset \Pi$ closed under differentiation.

In the univariate case, Hermite interpolation problems are introduced rather conveniently as the limit of Lagrange interpolation problems with coalescing nodes; alternatively, this could also be seen as starting with the interpolation of (divided or at least reasonably normalized) differences which then converge to derivatives in the dual of the interpolation space. But the main point in the univariate case is that the Lagrange interpolation problem with respect to $n+1$ distinct sites is poised for Π_n, the polynomials of degree up to n, and this also remains a valid interpolation space upon coalescence. This property holds true no more in two or more variables as a very simple example shows: consider, for $h \in \mathbb{R} \setminus \{0\}$, the six point evaluation functionals that map f to

$$f(0,0), \; f(h,0), \; f(0,h), \; f(1,1), \; f(1+h,1), \; f(1,1+h)$$

which give rise to a poised Lagrange interpolation problem for

$$\Pi_2 = \text{span}\left\{1, x, y, x^2, xy, y^2\right\}$$

as long as $h \neq 0$. More precisely, the Vandermonde determinant of this interpolation problem can be easily computed to be $-4h^5$ which, on the other hand, already indicates that there may be some trouble with the limit problem that has $\mathcal{Q}_{(0,0)} = \mathcal{Q}_{(1,1)} = \Pi_1$, interpolating point values and first derivatives at the two points. And indeed, this Hermite interpolation problem is not poised any more for Π_2 which can be verified by direct computations, but also follows from a slightly more general principle due to [73]: interpolating all derivatives of order k and k' at $\xi, \xi' \in \Xi$ by means of Π_n requires that $n > k + k'$. But it can even be seen directly from the discretization of the limit functionals in terms of the directional divided differences

$$\begin{array}{lll} f(0,0), & \dfrac{f(h,0)-f(0,0)}{h}, & \dfrac{f(0,h)-f(0,0)}{h}, \\ f(1,1), & \dfrac{f(1+h,1)-f(1,1)}{h}, & \dfrac{f(1,1+h)-f(1,1)}{h}, \end{array}$$

that trouble is to be expected: the Vandermonde determinant associated to the interpolation problem with respect to these functionals is $-4h$ then and this still converges to zero for $h \to 0$.

To make it clear: if the Hermite interpolation problem is poised for some subspace $\mathcal{P}$ of Π, then so are almost all Lagrange interpolation problems where partial derivatives are replaced by the associated divided differences; this is a simple consequence of the continuity of the determinant. However, there is no converse any more to the above statement as even the poisedness of *all* Lagrange interpolation problems does not guarantee the poisedness of the limit Hermite problem, and this is an exclusively multivariate phenomenon.

2.2. The loss of Haar

Depending on the point of view, the most scary or the most challenging aspect of multivariate polynomial interpolation is the fact that geometry really begins to matter now. In one variable, the Lagrange interpolation problem with respect to $n+1$ nodes is poised for Π_n, so all that has to be done was to match the number of interpolation sites and the dimension of the space. Univariate polynomials up to a certain degree form a *Haar space*: they allow for unique interpolation at an appropriate *number* of distinct nodes, regardless of their *position*. This is no more true in two and more variables, cf. [40], and generally Mairhuber's theorem [47] makes it clear that Haar spaces exist essentially for two different topologies: the interval and the unit circle which can be seen as the periodic relative of the interval. Consequently, the loss of Haar (and I do not claim to have invented this terminology) is unavoidable in $\mathbb{K}^d$, $d > 1$.

In particular, this means that for no polynomial subspace $\mathcal{P}$ of Π the interpolation problem can be poised with respect to *all* interpolation conditions Θ, not even if Θ consists only of point evaluation functionals. On the other hand, if $\Theta = \{\delta_\xi \, : \, \xi \in \Xi\}$ defines a Lagrange interpolation problem and P is a basis of an interpolation space $\mathcal{P}$, then the matrix

$$P(\Xi) := \Theta P = [p(\xi) = \delta_\xi p \, : \, \xi \in \Xi, \, p \in P]$$

is a nonzero polynomial in the $d \times \#\Xi$ variables $(\xi \, : \, \xi \in \Xi)$ and therefore vanishes at most on a set of measure zero. This fact we record in the following statement.

Proposition 3. *If for a subspace $\mathcal{P}$ of Π there exists a set Ξ of* $\dim \mathcal{P}$ *interpolation sites in $\mathbb{K}^d$ such that the interpolation problem with respect to Ξ is poised for $\mathcal{P}$ then* $\det P(\Xi) \neq 0$ *for any basis P of $\mathcal{P}$ and almost any $\Xi \subset \mathbb{K}^d$, $\#\Xi = \dim \mathcal{P}$.*

This property of Lagrange interpolation problems is called *almost poisedness* of the interpolation problem for $\mathcal{P} \subset \Pi$ and is the "proper" counterpart for the Haar space property. In particular, almost poisedness again simply requires the dimension of the interpolation space to match the cardinality of the set of interpolation nodes.

Proposition 4. *For every finite dimensional subspace $\mathcal{P}$ of Π the* Lagrange interpolation problem *is poised with respect to almost any $\Xi \subset \mathbb{K}^d$ as long as $\#\Xi = \dim \mathcal{P}$.*

Proof. We first show by induction on $N := \#P = \dim \mathcal{P}$ that there always exists $\Xi \subset \mathbb{K}^d$ such that $\det P(\Xi) \neq 0$ from which the claim follows immediately by applying Proposition 3. If $N = 1$ then P consists of a single nonzero polynomial which clearly does not vanish at some $\xi \in \mathbb{K}^d$. To advance the induction hypothesis from N to $N+1$, we choose $p \in P$, set

$$P' := P \setminus \{p\}$$

and note that by the induction hypothesis there exists $\Xi' \subset \mathbb{K}^d$, $\#\Xi' = \dim \mathcal{P} - 1$, such that $\det P'(\Xi') \neq 0$. Therefore, the polynomial

$$q := p - p(\Xi')^T \, P'(\Xi')^{-1} \, P'$$

vanishes at Ξ', but is nonzero by linear independence, thus $q(\xi) \neq 0$ for some point $\xi \in \mathbb{K}^d \setminus \Xi'$. With $\Xi = \Xi' \cup \{\xi\}$ we then get that

$$\det P(\Xi) = \det \{q, P'\} (\{\xi, \Xi'\}) = q(\xi) \det P'(\Xi') \neq 0,$$

which completes the proof. □

So it seems as if almost poisedness is the property to go for and therefore all the difference between the univariate and the multivariate case is something that happens just on a set of measure zero? Unfortunately, this is not the case, as there is also the issue of ***never poisedness*** when derivatives enter the scene: there are *Hermite* interpolation problems that will never be poised for a given interpolation space of fitting dimension, regardless of how we choose the nodes. And we already know an example: the innocent interpolation of first derivatives at two points in $\mathbb{R}^2$ from Π_2. This issue, that definitely adds a different flavor to Birkhoff interpolation problems in several variables, has been addressed to some extent in [73] for Hermite problems, but the much more systematic and substantial treatment has been done by Lorentz and Lorentz in [41–43] and is mostly summarized in [44] to which the reader is referred for additional information.

2.3. Let us keep it simple

We have seen that there is a difference between Lagrange and Hermite interpolation problems in two and more variables and that the simple and intuitive way to relate them by means of coalescence cannot be followed so easily – and we haven't even mentioned the question of points coalescing along lines or curves and similar issues yet. In fact, though many of the facts and concepts presented in what follows also continue to persist for Hermite interpolation (in Möller's, Birkhoff's and de Boor's sense of ideal interpolation; and I consider it justified to vote with this majority myself), the main focus will be on Lagrange interpolation and potential or possible extensions to Hermite problems will only be mentioned in passing. Nevertheless, I can assure the ambitious reader that Lagrange interpolation in several variables is by itself a sufficiently intricate issue.

3. Lattice Generation – The Legacy of Chung and Yao

One of the consequences of the "loss of Haar" is that there is no a priori certainty that a given node configuration (or a given set of ***data sites***) really admits a poised interpolation problem for a fixed polynomial subspace. This problem is frequently "resolved" for "applications" by making use of the handwaving argument that essentially all this fuss is just about a set of measure zero only, but unfortunately any interpolation problem "close" to such configurations becomes terribly ill–conditioned and thus useless. Consequently, it makes sense to ask for explicit constructions of data sites with respect to which the interpolation problem is poised for a given subspace $\mathcal{P}$. The most common choice for $\mathcal{P}$ is Π_n, the space of all polynomials of total degree at most n. Recall that a polynomial f is a *finite* sum of the form

$$f(x) = \sum_{\alpha \in \mathbb{N}_0^d} f_\alpha x^\alpha, \qquad f_\alpha \in \mathbb{K}, \qquad \#\{\alpha \,:\, f_\alpha \neq 0\} < \infty,$$

for which the *total degree* $\deg f$ is defined as

$$\deg f := \max\{|\alpha| \,:\, f_\alpha \neq 0\}, \qquad |\alpha| := \sum_{j=1}^d \alpha_j.$$

Then $\Pi_n := \{f \in \Pi \,:\, \deg f \leq n\}$ is a finite dimensional $\mathbb{K}$–vector space of dimension $\binom{n+d}{n}$. At present, I just want to mention that the total degree is a very natural, but by far not the only way to extend the notion of degree to polynomials in several variables; we will encounter the more general concept of *graded rings* later in Section 5. For the time being, however, we will focus on the problem to construct sets $\Xi \subset \mathbb{K}^d$ of $\binom{n+d}{n}$ distinct nodes such that the interpolation problem with respect to Ξ is poised for Π_n.

3.1. The geometric characterization

The geometric characterization due to Chung and Yao [22] is based on the intersection of hyperplanes. Recall that a hyperplane $H \subset \mathbb{K}^d$ is the zero set of a nondegenerate affine function $h \in \Pi_1$, i.e.,

$$H = \left\{x \in \mathbb{K}^d \,:\, 0 = h(x) = v^T x + c\right\}, \qquad v \in \mathbb{K}^d \setminus \{0\},\, c \in \mathbb{K}.$$

A set $\Xi \subset \mathbb{K}^d$, $\#\Xi = \binom{n+d}{d}$, satisfies the *geometric characterization* of degree n if for any $\xi \in \Xi$ there exist hyperplanes $H_{\xi,1}, \dots, H_{\xi,n}$ such that

$$\xi \notin \bigcup_{j=1}^n H_{\xi,j}, \qquad \Xi \setminus \{\xi\} \subset \bigcup_{j=1}^n H_{\xi,j}. \tag{3}$$

The geometric characterization is clearly a sufficient condition for the poisedness of the interpolation problem with respect to Ξ for Π_n, as it implies that the *Lagrange fundamental polynomials*

$$\ell_\xi := \prod_{j=1}^n \frac{h_{\xi,j}}{h_{\xi,j}(\xi)}, \qquad \xi \in \Xi, \tag{4}$$

are well defined and satisfy $\ell_\xi(\xi') = \delta_{\xi,\xi'}$ for $\xi, \xi' \in \Xi$. However, the fundamental polynomials have an additional property: they are *factorizable* into affine factors, a property quite "normal" for univariate polynomials but very rare for multivariate ones. On the other hand, if the fundamental polynomial ℓ_ξ which must be of total degree precisely n, can be factored into n affine factors then the zero set of this polynomial is a union of n hyperplanes not containing ξ while the remaining sites in $\Xi \setminus \{\xi\}$ must be distributed over these hyperplanes. This, however, is nothing else than (3) and so we have proved the following result.

Theorem 5. *Let $\Xi \subset \mathbb{K}^d$ be a set of $\binom{n+d}{d}$ distinct points. The Lagrange fundamental polynomials $\ell_\xi \in \Pi_n$, $\xi \in \Xi$, defined by $\ell_\xi(\xi') = \delta_{\xi,\xi'}$ can be factorized into affine factors if and only if Ξ satisfies the geometric characterization.*

Though condition (3) even characterizes a certain type of point configurations for interpolation, it mainly serves the purpose of being a ***sufficient*** condition for poisedness or as device for constructing node configurations which lead to poised interpolation problems. However, this appears paradoxical in some sense: on the one hand, the Lagrange interpolation problem with respect to Π_n is poised for almost any choice Ξ of cardinality $\binom{n+d}{d}$, on the other hand, the most popular constructions of such interpolation sets place the sites appropriately on (intersections of) finitely many hyperplanes, i.e., on sets of measure zero. Nevertheless, these point configurations offer a lot of appealing properties that have been discovered in the course of a substantial amount of research, most of which is summarized and explained in [26].

3.2. Natural lattices

The first specific method in [22] to construct point configurations that satisfy the geometric characterization is the approach of *natural lattices*. Here, one chooses, depending of the order $n \in \mathbb{N}_0$ of the lattice, a set $\mathcal{H}$ of $n+d$ hyperplanes *in general position*. The latter means that any d of the hyperplanes intersect in a point (which can be assured by requiring that all the normal vectors to the hyperplanes are in general position, i.e., that any d of them are linearly independent) and that no $d+1$ hyperplanes have a point in common. That is, any set $\mathcal{K} \subset \mathcal{H}$ of cardinality d, or, in the convenient shorthand notation from [8], any $\mathcal{K} \in \binom{\mathcal{H}}{d}$, defines a unique point $\xi_\mathcal{K}$ and for $\mathcal{K}, \mathcal{K}' \in \binom{\mathcal{H}}{d}$, $\mathcal{K} \neq \mathcal{K}'$, we have that $\xi_\mathcal{K} \neq \xi_{\mathcal{K}'}$. Since there are $\binom{n+d}{d}$ such subsets $\mathcal{K} \in \binom{\mathcal{H}}{d}$, this process defines a set Ξ of $\binom{n+d}{d}$ distinct nodes ξ_K which satisfy the geometric characterization.

There has been a lot of work on natural lattices, in particular by Mariano Gasca and his collaborators; for a fairly complete list of references and a more detailed description see [26,31].

3.3. Principal lattices

Although the very first explicit work on multivariate polynomial interpolation [38], cf. [33], that I could track down so far is concerned with what is called the *complete intersection case* nowadays, the "classical", "early" point configurations for bivariate and multivariate interpolation have been the tensor product and the structurally similar triangular grid. These grids can be seen as the intersection of a set of lines, for example those parallel to the y–axis, with "transversal" lines, for example those parallel to the x–axis. This is the geometric idea behind the concept of ***principal lattices*** that start with d families of $n+1$ distinct hyperplanes, $H_{j,k}$, $j = 1,\ldots,d$, $k = 0,\ldots,n$. For $|\alpha| \le n$, the interpolation sites are now chosen as the intersections

$$\xi_\alpha := \bigcap_{j=1}^{d} H_{j,\alpha_j},$$

with the *assumption* that these points are distinct. Since the polynomials

$$p_\alpha := \prod_{j=1}^{d} \prod_{k=0}^{\alpha_j - 1} \frac{h_{j,k}}{h_{j,k}(\xi_\alpha)}, \qquad |\alpha| \le n, \tag{5}$$

satisfy $p_\alpha(\xi_\beta) = \delta_{\alpha,\beta}$, $|\beta| \le |\alpha| \le n$, if follows that for $P = \{p_\alpha : |\alpha| \le n\}$ and $\Xi = \{\xi_\alpha : |\alpha| \le n\}$ the matrix $[p_\beta(\xi_\alpha) : |\alpha|, |\beta| \le n]$, being a rearrangement of $P(\Xi)$, is block upper triangular with identities on the diagonal and thus invertible.

Geometrically, principal lattices define the nodes by intersection of their duals, namely the hyperplanes, which are the same as the interpolation sites in one variable. In this respect principal lattices form a very natural generalization of univariate polynomial interpolation, and indeed some univariate features can be recovered in principal lattices. For example, as shown in [32] for the bivariate case, coalescence of the hyperplanes always leads to a poised Hermite interpolation problem, but then several sites become "Hermite sites" simultaneously. It also turns out that a geometric variant of the Aitken–Neville method (by means of barycentric coordinates and affine combinations) works only if the interpolation nodes form a principal lattice, see [75].

There is a multitude of constructions for principal lattices, many using ***pencils of hyperplanes*** as introduced by Lee and Phillips [39] and elaborated in a series of papers by Gasca and Mühlbach [28–30]. This approach nicely connects to projective geometry and again much of this work is summarized and referenced in [31]. However, it is worthwhile to also look at the recent [20] which employs the algebraic concept of ***addition of points on a cubic***.

3.4. There are still open questions

Let us consider natural and principal lattices in $d = 2$ for the moment, that is, the hyperplanes simply become lines in the plane. In the first case of natural lattices, any line intersects with any of the $n+1$ other lines and so even ***any*** line contains $n+1$ interpolation sites. In the case of principal lattices, on the other hand, the lines $H_{1,0}$ and $H_{2,0}$ also have $n+1$ interpolation sites lying on them, namely $(0,k)$ and $(k,0)$, $k = 0,\ldots,n$. This evidence is a first suggestion of the fascinating observation is that ***all*** lattices satisfying the geometric characterization in $d = 2$ seem to have this particularly nice structure which has been conjectured in [27].

Conjecture 6 (Gasca). *If $\Xi \subset \mathbb{R}^2$ satisfies the geometric characterization of degree n in $d = 2$, then there always exists a line H that contains $n+1$ sites.*

This conjecture, which still remains unproven at the time of writing, would allow for a recursive "Newton" approach to interpolation on natural lattices: if H is any line containing $n+1$ nodes then the removal of H and the $n+1$ sites on it leaves a set Ξ' that satisfies the geometric characterization of degree $n-1$ and can be further decomposed by the same argument. The conjecture is trivial for $n = 1$, easily proved by means of a counting argument for $n = 2$ and there also exists a quite complicated proof for $n = 3$, see [17]. The most recent work on Conjecture 6, to be found in [18,19], does not yet prove the conjecture but gives a classification of node sets that satisfy the geometric condition and actually indicates that probably an even stronger version of Conjecture 6 may hold.

4. Error, Formulas, and Differences

Clearly, interpolation polynomials coincide with the function f to be interpolated at the nodes. At least they should, even if they usually refuse to do so in numerical computations once the degree exceeds certain bounds, cf. [65], but also [59,60] for some further aspects of roundoff error analysis for multivariate polynomials. In addition to the unavoidable numerical difficulties, it is important to understand the approximation behavior of the interpolation polynomials, i.e., the behavior of the error of interpolation $f - L_n f$. Here, for simplicity, f is a sufficiently smooth function defined on some compact set $\Omega \subset \mathbb{R}^d$ (we neglect the more general fields for the sake of analysis), and L_n is the interpolation operator with respect to some set $\Xi \subset \Omega$ such that the interpolation problem with respect to Ξ is poised for Π_n.

4.1. The Newton approach and the finite difference

The first step towards the error formula was originally motivated by a very simple observation: the Vandermonde matrix

$$[\xi^\alpha \ : \ \xi \in \Xi, \, |\alpha| \le n]$$

with respect to the canonical *monomial basis* of Π_n indexes the nodes and the basis elements in two different ways. So why not re–index the nodes as $\Xi = \{\xi_\alpha \ : \ |\alpha| \le n\}$? Clearly, this is always possible in many ways, so that we make the additional requirement that the interpolation problem with respect to the node subsets $\Xi_k := \{\xi_\alpha \ : \ |\alpha| \le k\} \subseteq \Xi$, $k = 0, \dots, n$, is poised for Π_k. The "dual" polynomials for this arrangement of nodes are the *Newton fundamental polynomials* p_α, $|\alpha| \le n$, defined by the requirement that

$$p_\alpha \in \Pi_{|\alpha|}, \qquad p_\alpha(\xi_\beta) = \delta_{\alpha,\beta}, \quad |\beta| \le |\alpha| \le n. \tag{6}$$

The *Newton basis* $P = \{p_\alpha : |\alpha| \leq n\}$ of Π_n has the property that the matrix $[p_\beta(\xi_\alpha) : |\alpha|, |\beta| \leq n]$ is block upper triangular with identity matrices on the diagonal. Moreover, the polynomials in (6), which generalize the ones from (5), can be computed as a by–product of the indexing process of Ξ, cf. [6,72], either by means of Gauss elimination or a Gram–Schmidt orthogonalization process; mathematically, both approaches are equivalent, but they differ in implementation details. Even the polynomials p_α do not depend uniquely on Ξ: in the generic case, even *any* $\xi \in \Xi$ can be chosen as ξ_α and the polynomials have to adapt accordingly. This is the point where some subtle problems arise with Hermite interpolation: for a reasonable generalization of the Newton basis it would be necessary to index Θ as $\{\theta_\alpha : |\alpha| \leq n\}$ in such a way that, for $k = 0, \ldots, n$,

(i) the interpolation problem with respect to Θ_k is poised for Π_k *and*

(ii) ker Θ_k is also an ideal.

It is not difficult to ensure each of these properties *separately*, which corresponds to row or column permutations in Gaussian elimination, but how to satisfy them simultaneously is a tricky issue. Recall that the above conditions permit "putting the interpolation problem into block" [73], hold trivially for Lagrange interpolation problems (since condition ii is always satisfied then) and were essential for extending the Newton approach from Lagrange to Hermite interpolation problems in [73]. At the present it is not clear whether or not this can be done for a general set of functionals, but evidence fortunately is more on the optimist's than on the pessimist's side.

Conjecture 7. *Any finite set $\Theta \subset \Pi'$ of linear functionals which admits a poised ideal interpolation problem for Π_n can be graded as $\Theta_0 \subset \Theta_1 \subset \cdots \subset \Theta_n = \Theta$ such that for $k = 0, \ldots, n$ the interpolation with respect to Θ_k is poised for Π_k and* ker Θ_k *is an ideal in* Π.

As shown in [72], the interpolant of degree n to f can now be written in terms of the Newton basis as

$$L_n f = \sum_{|\alpha| \leq n} \lambda\left[\Xi_{|\alpha|-1}, \xi_\alpha\right] f \cdot p_\alpha, \tag{7}$$

where the *finite differences* $\lambda\left[\Xi_k, x\right] f$ satisfy the recurrence relation

$$\lambda[x] f = f(x), \tag{8}$$

$$\lambda\left[\Xi_k, x\right] f = \lambda\left[\Xi_{k-1}, x\right] - \sum_{|\alpha|=k} \lambda\left[\Xi_{k-1}, \xi_\alpha\right] f \cdot p_\alpha(x), \qquad k = 0, \ldots, n, \tag{9}$$

and

$$\lambda\left[\Xi_k, x\right] = \left(f - L_k f\right)(x), \qquad x \in \mathbb{R}^d. \tag{10}$$

Substituting (10) into (7) shows that the Newton approach to polynomial interpolation is repeated interpolation of errors:

$$L_n f = L_{n-1} + \sum_{|\alpha|=n} (f - L_{n-1})(\xi_\alpha) \cdot p_\alpha.$$

The difference $\lambda[\Xi_k, x] f$ is called *finite* difference for the simple reason that it is **not** a *divided* difference. This is most easily confirmed by looking at (10) from which it follows that

$$\lim_{x \to \Xi_k} \lambda[\Xi_k, x] = 0,$$

where $x \to \Xi_k$ means that $x \to \xi_\alpha$ for some $|\alpha| \le k$. Another, even more instructive, example is to consider $\xi_\alpha = h\alpha$, $h > 0$, as then

$$\lambda[\Xi_{k-1}, \xi_\alpha] f = \Delta_h^\alpha f(0), \qquad |\alpha| \le n,$$

with the standard forward difference Δ_h^α with step width h. The associated *divided* difference, on the other hand, would be $h^{-|\alpha|}\Delta_h^\alpha$ in this case, hence, it is precisely the division that is lacking in $\lambda\left[\Xi_{|\alpha|-1}, \xi_\alpha\right] f$. Also in the univariate case there is a big difference between the differences: denoting by $[x_0, \dots, x_n] f$ the divided difference, we get that

$$\lambda[x_0, \dots, x_n] f = (x_n - x_0) \cdots (x_n - x_{n-1}) [x_0, \dots, x_n] f.$$

4.2. Error formulas

The error formulas we consider here are describing the error of interpolation $f - L_n f$ in terms of integrating $(n+1)$–st derivatives of the function f against linear combinations of simplex splines as introduced by Micchelli in [50,51]. Recall that the *simplex spline* for the finite knot set $T \subset \mathbb{R}^d$, most conveniently written as a matrix $\mathrm{T} = [\tau : \tau \in \mathrm{T}] \subset \mathbb{R}^{d \times \#\mathrm{T}}$, is the density of the functional

$$f \mapsto \int_{[T]} f = \int_{\mathbb{S}_{\#T}} f(\mathrm{T}\, u)\, du, \tag{11}$$

where

$$\mathbb{S}_m := \left\{ u = (u_0, \dots, u_m) \in \mathbb{R}^{m+1} \;:\; u_j \ge 0, \sum_{j=0}^{m} u_j = 1 \right\}$$

denotes the standard m–simplex. Based on simplex splines, de Boor introduced in [7] a concept of a ***multivariate divided difference*** for $\mathrm{T} = [\tau_j : j = 0, \dots, n] \in \mathbb{R}^{d \times n+1}$ and $\mathrm{H} = [\eta_j : j = 1, \dots, n] \in \mathbb{R}^{d \times n}$ whose columns are interpreted as nodes and associated directions, respectively; the difference is then defined as

$$[\mathrm{T}; \mathrm{H}] f = \int_{[\mathrm{T}]} D_{\eta_1} \cdots D_{\eta_n} f.$$

Recalling the classical Hermite–Genocchi formula, which expresses the univariate divided difference in terms of a B–spline integral and is the origin from which many property of simplex splines were derived, cf. [51], one uses as "directions" differences of successive nodes, and so we specialize de Boor's divided difference to

$$[\mathrm{T}]\, f := [\mathrm{T}; \mathrm{T}D]\, f, \qquad D = \begin{bmatrix} -1 & & \\ 1 & \ddots & \\ & \ddots & -1 \\ & & 1 \end{bmatrix} \in \mathbb{R}^{n+1 \times n},$$

which will become one essential building block for the remainder formula. The other is the concept of a *path* of length n which is a vector $\mu = (\mu^0, \dots, \mu^n)$ of multiindices such that $|\mu^k| = k$, $k = 0, \dots, n$. The set of all paths of length n will be denoted by M_n. Associated to a path $\mu \in \mathrm{M}_n$ and a properly indexed set $\Xi = \{\xi_\alpha \ : \ |\alpha| \le n\}$ we obtain a matrix $\Xi_\mu := [\xi_{\mu^j} \ : \ j = 0, \dots, n]$ of sites visited along the path as well as the number

$$\pi_\mu := \prod_{j=0}^{n-1} p_{\mu^j} \left(\xi_{\mu^{j+1}}\right).$$

With this notation at hand, the error formula from [72] takes the convenient form

$$(f - L_n f)(x) = \sum_{\mu \in \mathrm{M}_n} p_{\mu^n}(x)\, \pi_\mu \, [\Xi_\mu, x]\, f. \tag{12}$$

This formula illustrates how significantly things change when passing from univariate to multivariate polynomial interpolation. In one variable, there is precisely one path and (12) is the well–known error formula with B–splines as Peano kernels that can be encountered in many textbooks as for example in [25,36]. The multivariate version of this formula, however, contains

$$\prod_{j=0}^{n} \binom{j+d-1}{d-1}$$

terms in the summation which already becomes $(n+1)!$ for $d = 2$ and grows much faster for higher values of d. In particular, these numbers by far exceed even the "dimension curse" factor d^n which is widely accepted as an unavoidable growth rate in multivariate problems. But the number of terms in (12) appears even more peculiar when comparing it to another formula for the error of multivariate polynomial interpolation which is due to Ciarlet and Raviart [23] and uses the Lagrange fundamental polynomials ℓ_α, $|\alpha| \le n$, to express the error as

$$(f - L_n f)(x) = \sum_{|\alpha| \le n} \ell_\alpha(x)\ [\xi_\alpha^n, x; (x - \xi_\alpha)^n]\, f, \tag{13}$$

where exponentiation indicates n–fold repetition of a column in the respective matrix. This formula, based on a multipoint Taylor expansion, has become an important tool in the analysis of finite elements and it has the apparent advantage that the number of terms in the sum equals the number of interpolation sites, a number much smaller than the number of paths – except, of course, in the univariate case where it becomes a formula due to Kowalewski, [37, eq. (24), p. 24], but not the "standard" error formula. In this respect, (13) is *no* (direct) generalization of the univariate formula, while (12) definitely is. Another fundamental difference is that the simplex spline integrals in (12) normally run over nondegenerate convex sets in $\mathbb{R}^d$, while the one in (13) only integrates along lines. The surprising fact that these two expressions nevertheless describe precisely the same quantity, namely the error of interpolation, shows that there is still quite a bit of magic going on among simplex splines.

4.3. The error on lattices

For interpolation sites located on natural or principal lattices, one can also hope for the associated error formulas to take a very special form. We begin with the error by interpolation on natural lattices which is expressed in a beautiful and very compact way in the formula

$$(f - L_n f)(x) = \sum_{\mathcal{K} \in \binom{\mathcal{H}}{d}} p_{\mathcal{K}}(x) \left[\Xi_{\mathcal{K}}, x; D_{\eta_{\mathcal{K}}}^{n+1}\right] f \tag{14}$$

from [8]. Apparently, some of the ingredients in this formula we haven't encountered yet, so let us figure out what they are. In fact, the notation $\mathcal{K} \in \binom{\mathcal{H}}{d}$ has appeared in Section 3.2 and it stands for choosing a subset $\mathcal{K}$ of $\mathcal{H}$ consisting of d hyperplanes; since they are in general position, they intersect in a line with normalized directional vector $\eta_{\mathcal{K}}$. On the other hand, the complement $\mathcal{H} \setminus \mathcal{K}$ consists of $n+d-d=n$ hyperplanes whose associated normalized affine polynomials multiply to $p_{\mathcal{K}}$, and each of these hyperplanes intersects with the line defined by $\mathcal{K}$ in a different point – these n nodes are collected in $\Xi_{\mathcal{K}}$. Note that this is quite much in the spirit of the formula in (12): since they are all lining up along the line in direction $\eta_{\mathcal{K}}$, the difference of any two nodes in $\Xi_{\mathcal{K}}$ is clearly a multiple of η_k, only the directional derivative in the direction $x - \xi$ for some $\xi \in \Xi_{\mathcal{K}}$ is missing relative to (12); apparently it is compensated in the $p_{\mathcal{K}}$.

Also for principal lattices there exists a special form for the error formula which has been developed for the bivariate case in [32], the general situation will be considered in the forthcoming [21]. The expression takes the form

$$(f - L_n f)(x) = \sum_{\mu \in \mathrm{M}_n^*} \frac{p_{\mu^n}(x)}{\theta_\mu} \left[\Xi_\mu, x; \mathrm{H}_\mu, x - \xi_{\mu^n}\right] f, \tag{15}$$

where M_n^* denotes the (much smaller) set of all paths μ such that $\mu^{k+1} - \mu^k \in \mathbb{N}_0^d$, H_μ is the set of *normalized* directions that are followed along this path while θ_μ – the "new" quantity – is the product of the "angles" encountered along the path. Without going into the intricate details, cf. [32], it is just worthwhile to mention here that the identity (15) highlights the crucial geometric parameters in this type of interpolation, namely the angle of intersection between the hyperplanes that define the lattice.

Another common property of the formulas (14) and (15) is that, in accordance with line geometry, the formulas are invariant under scaling – the distance between the lines is irrelevant in the error of interpolation. This, in turn, is the major tool that allows us for such special configurations to pass from Lagrange to Hermite interpolation problems by means of coalescing lines (not points!), see [32] for the hairy details.

4.4. Different differences

We have made the point before that the finite difference from (8) is *not* a divided difference, simply because it lacks division. So the question remains what a *divided difference* is in the context of multivariate polynomial interpolation. One way to generalize the notion of divided difference is to retreat to saying that it is the leading coefficient of the interpolation polynomial, i.e., $L_n f(x) = [x_0, \dots, x_n] f\, x^n + q(x)$, $q \in \Pi_{n-1}$. This notion of a divided difference has been investigated in [62] and also, for special configurations, in [75]. On the other hand, until we will know "better" in the next section, the leading coefficient of a polynomial is now a *vector* of coefficients. More precisely, setting $X^k = (X_\alpha = (\cdot)^\alpha \; : \; |\alpha| = k)$, we can write a polynomial of degree n as

$$f = \sum_{k=0}^{n} f_k^T X^k, \qquad f_k = (f_\alpha \; : \; |\alpha| = k).$$

With $P^k = (p_\alpha \; : \; |\alpha| = k)$ and $\lambda[\Xi_k] f = (\lambda[\Xi_{k-1}, \xi_\alpha] f \; : \; |\alpha| = k)$ we thus get from the Newton formula (7) that

$$L_n f = \sum_{k=0} \lambda[\Xi_k] f^T P^k = \sum_{k=0}^{n} c_k^T X^k,$$

where c_n would be the desired divided difference. If M_α denotes the *monomial Newton basis polynomial*, defined by the requirements

$$M_\alpha - (\cdot)^\alpha \in \Pi_{|\alpha|-1}, \qquad M_\alpha(\xi_\beta) = 0, \quad |\beta| < |\alpha|,$$

then there exist invertible matrices A_k such that $P^k = A_k M^k$, $k = 0, \dots, n$, and therefore

$$L_n f = \sum_{k=0} \lambda[\Xi_k] f^T P^k = \sum_{k=0} \lambda[\Xi_k] f^T A_k M^k = \sum_{k=0} \left(A_k^T \lambda[\Xi_k] f\right)^T M^k,$$

so that $\Delta[\Xi_k] f := A_k^T \lambda[\Xi_k] f$ gives the divided difference of order k:

$$L_n f = \Delta[\Xi_k] f^T X^k + q, \qquad q \in \Pi_{n-1}. \tag{16}$$

A simple recursive application of this argument immediately gives another Newton formula, namely

$$L_n f = \sum_{k=0}^{n} \Delta[\Xi_k] f^T M^k. \tag{17}$$

Taking into account (12) and the fact that $\lambda\,[\Xi_n]\,f = ((f - L_{n-1}f)\,(\xi_\alpha)\;:\;|\alpha| = n)$ we thus obtain for $|\alpha| = n$ the spline integral representation

$$\begin{aligned}\Delta\,[\Xi_n]_\alpha\,f &= \sum_{|\beta|=n} (A_k)_{\beta,\alpha} \sum_{\mu\in\mathrm{M}_{n-1}} p_{\mu^{n-1}}\,(\xi_\beta)\,\pi_\mu\,[\Xi_\mu,\xi_\beta]\,f \\ &= \sum_{\mu\in\mathrm{M}_n} (A_k)_{\mu^n,\alpha}\,\pi_\mu\,[\Xi_\mu]\,f = \sum_{\mu\in\mathrm{M}_n} \frac{1}{\alpha!}\frac{\partial^n p_{\mu^n}}{\partial x^\alpha}\,\pi_\mu\,[\Xi_\mu]\,f,\end{aligned}$$

that is,

$$\Delta\,[\Xi_n]\,f = D^n \sum_{\mu\in\mathrm{M}_n} p_{\mu^n}\,\pi_\mu\,[\Xi_\mu]\,f, \qquad D^n = \left[\frac{1}{\alpha!}\frac{\partial^n}{\partial x^\alpha}\;:\;|\alpha| = n\right]. \tag{18}$$

This, of course, is mainly a starting point for a study of the divided difference, understood as leading homogeneous coefficients of interpolation polynomials; at this point, however, we will not pursue this issue any further and just refer to [62], where the interested reader can find recurrences, for example.

5. The Algebraic Approach

This will become the longest part of this article and the one that contains the largest amount of new material, thus also the largest amount of proofs. While so far all we considered dealt exclusively with the vector space structure of polynomials by regarding interpolation as a linear algebra problem, we will now make more and more use of the ring structure of polynomials, taking into account multiplication and ideals.

The starting point here is the question of finding appropriate interpolation spaces for a given set Θ of functionals; though I will try to keep the presentation as general as possible here, it is not wrong to think of Lagrange interpolation problems, but definitely Θ is now supposed to be an ideal interpolation problem, i.e., $\ker\Theta$ is an ideal in $\Pi = \mathbb{K}[x]$. Of course, as mentioned before, there are still many subspaces $\mathcal{P}$ of Π such that the interpolation problem with respect to Θ is poised for $\mathcal{P}$. In order to find "good" interpolation spaces among this manifold, it seems to make very good sense to keep the degree low ...

5.1. Degree reducing interpolation

First, recall that polynomials of high degree are not very desirable, especially not from a numerical point of view: the memory consumption for storing them is higher, they tend to be more critical and unstable in all computations and the higher the degree, the more the polynomials feel free to follow their desire to oscillate. The first requirement to be made on a reasonable interpolation $\mathcal{P}$ space is that it should be of ***minimal degree***, i.e., that any other interpolation space $\mathcal{Q}$ should satisfy $\deg\mathcal{Q} \geq \deg\mathcal{P}$, where $\deg\mathcal{P} := \max\{\deg p\;:\;p\in\mathcal{P}\}$. Spaces of minimal degree, however, can still provide unwanted effects as the following simple example shows: the Lagrange interpolation problem with respect to $\Xi = \{(1,0),(0,1)\}$ is poised for $\mathcal{P} = \operatorname{span}\{x,y\}$, and since $\deg \mathcal{P} = 1$, this is an interpolation space of minimal degree. On the other hand, when interpolating the constant polynomial that has degree 0, one obtains the polynomial $x+y$ which is obviously of degree 1. In this sense, the interpolation operator $L_{\mathcal{P}} = L_{\mathcal{P},\Theta}$ with respect to Θ and $\mathcal{P}$, defined as

$$\Theta L_{\mathcal{P}} f = \Theta f, \qquad f \in \Pi,$$

does not behave "well" for polynomials. In response to this undesirable phenomenon, we will restrict ourselves to interpolation spaces that treat the degree of the polynomial to be interpolated in a more respectful way.

Definition 8. A subspace $\mathcal{P}$ of Π is called a *degree reducing interpolation space* with respect to $\Theta \subset \Pi'$ if the interpolation problem with respect to Θ is poised for $\mathcal{P}$ and if $\deg L_{\mathcal{P}} f \le \deg f$ for any $f \in \Pi$.

Degree reduction is a key property of the *least interpolant* introduced by de Boor and Ron in [10] and further elaborated in [12,13]. Motivated by their investigations which also included more general (total) degree reducing interpolation spaces sharing some (but not all) of the properties of the least interpolant as well as a comparison between minimal degree and degree reducing interpolation, a "direct" approach to degree reducing interpolation was pursued later in [66], deriving some properties directly from the requirement of degree reduction.

Degree reduction is really a stronger property than being of minimal degree. Let us formally state this and recall one simple proof for the reader's convenience, and keep in mind that the above example shows that the inclusion is indeed a strict one.

Proposition 9. *Any degree reducing interpolation space $\mathcal{P}$ is also of minimal degree.*

Proof. Suppose that $\mathcal{Q} \subset \mathcal{P}$ were an interpolation space with $\deg \mathcal{Q} < \deg \mathcal{P}$ and let $\ell_\theta \in \mathcal{Q}$, $\theta \in \Theta$, denote the fundamental polynomials: $\theta \ell_{\theta'} = \delta_{\theta,\theta'}$, $\theta, \theta' \in \Theta$. Then $\{L_{\mathcal{P}} \ell_\theta \ : \ \theta \in \Theta\}$ is a basis of $\mathcal{P}$ and by degree reduction it follows that $\deg L_{\mathcal{P}} \ell_\theta \le \deg \ell_\theta \le \deg \mathcal{Q} < \deg \mathcal{P}$, which is a contradiction since $\deg \mathcal{P} \ge \max_\theta L_{\mathcal{P}} \ell_\theta$. □

So far we have always made the implicit assumption that **the** degree of a multivariate polynomial is its total degree; this is natural from the analytic and even the numerical point of view (as [59] shows, for example, that the roundoff error in evaluating a polynomial depends linearly on the total degree), but in no way the only possible choice. Quite the contrary, the notion of *multidegree* based on term orders, cf. [24], plays an important role in Computer Algebra ever since the introduction [15] of Gröbner bases. Moreover, it will turn out that there is a very close relationship between degree reducing interpolation and good ideal bases which persists for a much more general concept of degree and therefore it will be worthwhile to assume a more generous point of view of to what degree we consider something the degree of a polynomial.

5.2. Grading rings

The most natural generalization of the concept of degree, not only to multivariate polynomials, but to arbitrary rings, is that of a *graded ring*. To comprehend the appealing and intuitive idea behind this concept, it is instructive to begin with a look at the total degree. First, note that the degree of a polynomial essentially depends on a *homogeneous* polynomial, namely its leading part. Indeed, if $f \in \Pi_n$ is written as

$$f(x) = \sum_{|\alpha|=n} f_\alpha x^\alpha + \sum_{|\alpha|<n} f_\alpha x^\alpha =: f_k(x) + f_{<n}(x), \qquad x \in \mathbb{K}^d,$$

then the polynomial $f_{<k}$ does not influence the degree of f which depends entirely of the homogeneous part f_k. Moreover, adding two homogeneous polynomials of the same degree either gives the zero polynomial or another homogeneous polynomial of ***precisely*** the same (total) degree, and, finally, the multiplication of two homogeneous polynomials (and thus of two arbitrary polynomials) corresponds to adding the degrees. In other words: the degree should be an additive structure that defines "homogeneous" polynomials and the multiplication of such objects should correspond to the addition of degrees.

And this is precisely the concept of a graded ring. The additive structure is a semigroup Γ with neutral element 0, that is, a *monoid*, and the structure of Γ is connected to our ring $\Pi = \mathbb{K}[x]$ via the *direct sum decomposition*

$$\Pi = \bigoplus_{\gamma\in\Gamma} \Pi_\gamma^0, \qquad \text{i.e.,} \qquad f = \sum_{\gamma\in\Gamma} f_\gamma, \quad f \in \Pi,\ f_\gamma \in \Pi_\gamma^0, \tag{19}$$

where each Π_γ^0 is an abelian group under addition (one can add and subtract homogeneous polynomials) and multiplication of homogeneous polynomials is reflected by addition in Γ:

$$\Pi_\gamma^0 \cdot \Pi_{\gamma'}^0 \subseteq \Pi_{\gamma+\gamma'}^0, \qquad \gamma, \gamma' \in \Gamma. \tag{20}$$

In fact, we can even retreat to more familiar concepts in the case of Π: the homogenous spaces Π_γ^0 must be $\mathbb{K}$ vector spaces. This follows by first noticing that (20) yields that $1 \in \Pi_0^0$ and then realizing that any *unit* in $f \in \Pi^*$, i.e., any element such that $f^{-1} \in \Pi$, must have degree 0 again, as follows from considering

$$\Pi_\gamma^0 = 1 \cdot \Pi_\gamma^0 = \left(f\, f^{-1}\right) \Pi_\gamma^0 = \sum_{\eta,\eta'\in\Gamma} f_\eta\, f_{\eta'}^{-1} \Pi_{\gamma+\eta+\eta'}^0 .$$

It is worthwhile to mention that this is a general property of graded rings: units have degree zero. As innocent and natural as this sounds, it has the somewhat surprising consequence that Laurent polynomials (which are analytically very similar to polynomials, but are spanned by units – all the monomials are invertible) only admit the *trivial grading* $\Gamma = \{0\}$; this makes the construction of Gröbner- and H–bases more intricate, cf. [57].

We also denote by

$$\Pi_\gamma := \bigoplus_{\eta \le \gamma} \Pi^0_\gamma, \qquad \gamma \in \Gamma,$$

the vector space of all polynomials of (Γ–)degree at most γ. Note that, depending on the underlying grading, these spaces can be finite dimensional (like with the grading by total degree) or infinite dimensional (the grading by a purely lexicographic term order serves as the standard example here).

To step forward from grading to degree, we need some additional structure of Γ, namely a *well–ordering* "$<$" that is compatible with the semigroup structure. Recall that "well–ordering" means that any strictly descending chain $\gamma > \gamma' > \cdots$ must be finite, i.e., any subset of Γ has a smallest element and that "compatibility" is equivalent to $\gamma < \gamma'$ implying $\gamma + \gamma'' < \gamma' + \gamma''$ for all $\gamma, \gamma', \gamma'' \in \Gamma$. It is not hard to guess, and not much harder to prove, that this property implies $0 \le \Gamma$. To any polynomial $f \in \Pi$ which can be written as a *finite* sum (19) of its homogeneous components, we can now associate the *degree*

$$\delta(f) := \max\{\gamma \in \Gamma \ : \ f_\gamma \ne 0\}, \qquad \delta : \Pi \to \Gamma,$$

and the *leading term*

$$\Lambda(f) := f_{\delta(f)} \ne 0, \qquad \Lambda : \Pi \to \Pi^0 := \bigcup_{\gamma \in \Gamma} \Pi^0_\gamma,$$

which leaves, as usually $\delta(0)$ and $\Lambda(0)$ undefined.

So the degree of a polynomial now depends on two ingredients: the monoid Γ and the well ordering "$<$". In the case of total degree, the monoid is $\mathbb{N}_0$, the well–ordering is the canonical one and the homogenous spaces are

$$\Pi^0_n = \operatorname{span}_{\mathbb{K}} \{x^\alpha \ : \ |\alpha| = n\}, \qquad k \in \mathbb{N}_0.$$

A well–ordering on $\mathbb{N}_0^d$ is called a *term order* where classical examples are the *lexicographic* term order

$$\alpha \prec_\ell \beta \qquad \Leftrightarrow \qquad \alpha_j = \beta_j,\ j = 1, \dots, k-1, \quad \alpha_k < \beta_k,$$

and the *graded lexicographical* term order

$$\alpha \prec_g \beta \qquad \Leftrightarrow \qquad |\alpha| < |\beta| \text{ or } |\alpha| = |\beta|,\ \alpha \prec_\ell \beta.$$

In fact, the graded lexicographical term order induces a grading that can be seen as a refinement of the grading by total degree. In the case of term orders, we have "homogeneous" spaces

$$\Pi^0_\alpha = \operatorname{span}_{\mathbb{K}} \{x^\alpha\}, \qquad \alpha \in \mathbb{N}_0^d,$$

which are of dimension 1. Some more esoteric gradings can be defined by $\Gamma = \mathbb{N}_0$ and

$$\Pi^0_k := \operatorname{span}_{\mathbb{K}} \left\{x^\alpha \ : \ \omega^T \alpha = k\right\}, \qquad k \in \mathbb{N}_0,\ 0 \ne \omega \in \mathbb{N}_0^d, \tag{21}$$

or, driving this idea even further, $\Gamma = \mathbb{N}_0^m$, equipped with some term order, and

$$\Pi_\beta := \operatorname{span}_{\mathbb{K}} \{x^\alpha \,:\, M\alpha = \beta\}, \qquad \beta \in \mathbb{N}_0^m, 0 \neq M \in \mathbb{N}_0^{m\times d}.$$

The latter, however, is much more than academic eccentricity and leads to the concepts like the "Gröbner fan" and the "Gröbner walk", see [63] as a starting point. Finally, we single out two classes of gradings which will be useful in the sequel.

Definition 10. A grading is called *strict* if $\Pi_0^0 = \mathbb{K}$ and *monomial* if all homogeneous spaces Π_γ^0 are spanned by monomials.

Recall that we always have $\mathbb{K} \subseteq \Pi_0^0$ and so strict gradings are those where *exactly* the constants have degree zero; an example for a non–strict grading is obtained by setting

$$\Gamma = \mathbb{N}_0, \qquad \Pi_k^0 = \operatorname{span}_{\mathbb{K}} \{x^\alpha \,:\, \alpha_1 = k\}, \quad k \in \mathbb{N}_0, \tag{22}$$

choosing $\omega = (1, 0, \ldots, 0)$ in (21). This yields a grading that only cares for the first variable x_1. Non–monomial gradings are a little bit more obscure and not really to be found in the literature, but nevertheless do exist. It has to be noted, however, that all the "standard" gradings are strict monomial ones.

With this terminology at hand, we can literally repeat the definition of degree reducing interpolation. To that end, we introduce the shorthand notation

$$f \preceq g \qquad \Leftrightarrow \qquad \delta(f) \leq \delta(g), \tag{23}$$

which determines a well–ordering on Π.

Definition 11. A subspace $\mathcal{P} \subset \Pi$ is called a *degree reducing interpolation space* for $\Theta \subset \Pi'$ if the interpolation problem with respect to Θ is poised for $\mathcal{P}$ and if the interpolation operator satisfies $L_\mathcal{P} f \preceq f$, $f \in \Pi$.

Since the proof of Proposition 9 works for any arbitrary grading, it follows immediately that degree reducing interpolation spaces are always of minimal degree, i.e., $\mathcal{P} \preceq \mathcal{Q}$ for any interpolation space $\mathcal{Q} \subset \Pi$. total

5.3. A good basis for idealism

Finally, now arises the time when ideals really matter. An *ideal* $\mathcal{I} \subseteq \Pi$ is a subset, in fact a vector subspace, of Π that is closed under addition, $\mathcal{I} + \mathcal{I} = \mathcal{I}$, and closed under multiplication by arbitrary polynomials, $\mathcal{I} \cdot \Pi = \mathcal{I}$. Consequently, if $F \subseteq \mathcal{I}$ then also $\langle F \rangle \subseteq \mathcal{I}$, where

$$\langle F \rangle := \left\{ \sum_{f \in F} g_f\, f \,:\, g_f \in \Pi \right\},$$

is the *ideal generated by* F. A set $F \subset \Pi$ is called a *basis* of $\mathcal{I}$ if $\mathcal{I} = \langle F \rangle$. Recall that Hilbert's famous *Basissatz* [35] states that every polynomial ideal, i.e., any ideal in Π, has a *finite* basis. This is the background and the theoretical justification for the way in which polynomial ideals are represented in computer algebra, namely as finite (unordered) sets of polynomials that generate the ideal. Note, however, that though the word "basis" is used, there is in general nothing like independence or uniqueness of representation as one would expect from linear algebra – in this respect an ideal basis is only a generating system for the ideal.

A finite set G is called a Γ–basis for the ideal $\langle G \rangle$ if

$$f \in \langle G \rangle \qquad \Leftrightarrow \qquad f = \sum_{g \in G} f_g \, g, \quad f_g \, g \preceq f, \, g \in G. \tag{24}$$

The representation of f on the right hand side of (24) – relevant here is the degree restriction on the terms in the sum – is called a *Γ–representation* and can be understood as a *non–redundant* representation: if there were some g such that $f_g \, g \succ f$, then all terms in $f_g \, g$ whose degree exceeds that of f must be canceled by the rest of the sum, thus is redundant and therefore should better not appear from the beginning. As pointed out in [70], Γ–bases exist for any grading Γ and can be constructed (more or less) efficiently. At least two instances of Γ–bases are classical: when the underlying grading is the one by total degree, they become *H–bases*, due to Macaulay [46], in the case of term orders, on the other hand, they are *Gröbner bases*. These two notions are not mutually exclusive as there exist Gröbner bases that are H–bases as well, but there are also H–bases which are no Gröbner bases and vice versa.

Next, there is another characterization of Γ–bases that will become useful in the sequel. To that end, we need the notion of the *homogeneous ideal*

$$\langle F \rangle^0 = \bigcup_{\gamma \in \Gamma} \left\{ \sum_{f \in F} p_f \, f \; : \; p_f \in \Pi_{\gamma - \delta(f)} \right\} \subset \Pi^0$$

generated by a finite set $F \subset \Pi^0$ of *homogeneous* polynomials. In the case of term orders, the homogeneous ideals are *monomial ideals* which are the essence of Dickson's Lemma, cf. [24]. Homogeneous ideals in connection with dehomogeneization, on the other hand, play a fundamental role in the original definition of H–bases, cf. [34]. The characterization of Γ–bases in terms of homogeneous ideals is now as follows.

Proposition 12. *A set $G \subset \Pi$ is a Γ–basis of an ideal $\mathcal{I} \subseteq \Pi$ if and only if $\Lambda(\mathcal{I}) = \langle \Lambda(G) \rangle^0$.*

Proof. If G is a Γ–basis then define, for a Γ–representation (24), $G^* \subset G$ as $G^* = \{ g \in G \; : \; \delta(f_g \, g) = \delta(f) \}$ and obtain

$$\Lambda(f) = \sum_{g \in G^*} \Lambda(f_g \, g) = \sum_{g \in G^*} \Lambda(f_g) \, \Lambda(g) \in \langle \Lambda(G^*) \rangle^0 \subseteq \langle \Lambda(G) \rangle^0 .$$

Conversely, if $\Lambda(\mathcal{I}) = \langle \Lambda(G) \rangle^0$ then we represent, for $f \in \mathcal{I}$,

$$\Lambda(f) = \sum_{g \in G} f_g \, \Lambda(g), \qquad f_g \in \Pi^0_{\delta(f)-\delta(g)}, \; g \in G,$$

and replace f by

$$f' = f - \sum_{g \in G} f_g \, g, \qquad \text{i.e.} \qquad f' \prec f.$$

Since "$<$" is a well–ordering on Γ we can rest assured that finitely many applications of this process lead to a Γ–representation of f with respect to G. □

The main computational advantage of Γ–bases, however, and as stated in Buchberger's thesis [14], the main intention of Gröbner when addressing the research topic for Buchberger's thesis, rests in the fact that they enable us to efficiently *compute* remainders modulo an ideal. As a first step towards the notion of a *reduced* polynomial, we define for $F \subset \Pi$ and $\gamma \in \Gamma$ the vector space

$$V_\gamma(F) := \langle \Lambda(F) \rangle^0 \cap \Pi^0_\gamma.$$

Moreover, let $(\cdot,\cdot) : \Pi \to \mathbb{K}$ denote an inner product on Π such that $\left(\Pi^0_\gamma, \Pi^0_{\gamma'}\right) = 0$ for $\gamma \neq \gamma' \in \Gamma$ and write $f \perp g$ if $(f,g) = 0$. If Γ is a monomial grading, then canonical choices for

$$f(x) = \sum_{\alpha \in \mathbb{N}_0^d} f_\alpha \, x^\alpha, \qquad g(x) = \sum_{\alpha \in \mathbb{N}_0^d} g_\alpha \, x^\alpha,$$

are

$$(f,g) = \sum_{\alpha \in \mathbb{N}_0^d} f_\alpha \, g_\alpha \qquad \text{or} \qquad (f,g) = (f(D)g)(0) = \sum_{\alpha \in \mathbb{N}_0^d} \alpha! \, f_\alpha \, g_\alpha.$$

The latter inner product can be found in the literature under various names, but in the context of polynomial interpolation it has been established by de Boor and Ron [10–13] as a useful tool to characterize the *least interpolant*. Also observe that in the case of a term order the inner product is irrelevant as the orthogonality between monomial spaces of different degree already defines it up to the positive constants (x^α, x^α), $\alpha \in \mathbb{N}_0^d$.

A polynomial $g \in \Pi$ is called *reduced* with respect to $F \subset \Pi$ if any of its homogeneous components g_γ is in the orthogonal complement of the respective $V_\gamma(F)$, that is

$$g \in W(F) := \bigoplus_{\gamma \in \Gamma} W_\gamma(F), \qquad \Pi^0_\gamma = V_\gamma(F) \oplus W_\gamma(F), \quad V_\gamma(F) \perp W_\gamma(F). \tag{25}$$

To illustrate what this property means, we have a brief look at the univariate case. Since univariate polynomials form a principal ideal ring, we can assume that $F = \{f\}$ and get that $V_k(F)$ is either $\mathbb{K} \cdot (\cdot)^k$ if $k \geq \deg f$ or $\{0\}$ otherwise. Consequently, a *univariate* polynomial is reduced with respect to f, $\deg f = k$, if and only if it belongs to $\Pi_{k-1} \simeq \Pi / \langle F \rangle$.

A polynomial $r \in \Pi$ is called a ***remainder of division*** of f with respect to $G \subset \Pi$ if it is a reduced remainder of a Γ–representation of f, that is,

$$f = \sum_{g \in G} f_g \, g + r, \qquad f_g \, g \preceq f, \quad r \in W(G). \tag{26}$$

The representation in (26) can be computed more or less efficiently by means of ***reduction***: suppose that $\delta(f) = \gamma$, compute the orthogonal projection of $\Lambda(f)$ to $V_\gamma(G)$, yielding

$$\Lambda(f) = \sum_{g \in G} f_{g,\gamma} \, \Lambda(g) + r_\gamma, \qquad r_\gamma \in W_\gamma(G), \tag{27}$$

and continue this process with

$$f' := f - \sum_{g \in G} f_{g,\gamma} \, g - r_\gamma, \qquad \text{hence,} \qquad f' \prec f. \tag{28}$$

This process terminates after finitely many steps and yields the terms in (26) as

$$f_g = \sum_{\gamma \leq \delta(f)} f_{g,\gamma}, \quad g \in G, \qquad r = \sum_{\gamma \leq \delta(f)} r_\gamma.$$

The use of orthogonal projections, introduced in [68] for the grading by total degree, provides the "gradual" notion of divisibility needed for non–term–order gradings: while a monomial x^α either does divide another monomial x^β or is does not, homogeneous terms are much more indifferent towards each other in terms of divisibility. For example, the bivariate polynomial $x^2 + y^2$ does not divide $x^3 + y^3$, but nevertheless they have something "in common" as well. From a computational point of view, it should be clear that division of monomials is definitely much simpler than to determine orthogonal projection though the latter is "only" a problem from linear algebra; an efficient method to compute this projection for the inner product $f(D)g(0)$ has been given recently in [61]. On the other hand, H–bases can be very helpful in avoiding ***representation singularities*** in polynomial system solving, cf. [56].

To finish the "short course" on Γ–bases, note that the reduction algorithm looks very similar to what was done in the proof of Proposition 12, so Γ–bases can be expected to play a fundamental role in reduction. In fact, they turn out to be absolutely necessary for reduction to work at all. This is due to the subtle fact that the representation of $\Lambda(f)$ in (27) need not be unique and, in general, will not be because of the ubiquitous syzygies. And while this does not affect the leading term $r_{\delta(f)}$ of the remainder, different choices of this representation can lead to different polynomials f' in (28) which may very well affect lower order terms of r. Consequently, the remainder of division can depend on how the orthogonal projections are expanded – except in the case when we divide by a Γ–basis. Specifically, we have the following result, cf. [70].

Theorem 13. *A finite set $G \subset \Pi$ is a Γ–basis if and only if for any $f \in \Pi$ and any pair of Γ–representations*

$$f = \sum_{g \in G} f_g\, g + r = \sum_{g \in G} f'_g\, g + r', \qquad r, r' \in W(G),$$

one has $r = r'$.

In other words, if G is a Γ–basis, we can speak of **the** remainder of the division of f by G, or of the *normal form* $\nu_G(f) := r$ of f modulo G. This immediately implies that $\nu_G(f) = 0$ if and only $f \in \langle G \rangle$ giving an algorithmic criterion to decide the *ideal membership problem.* But Γ–bases offer even more: the normal form depends entirely on the ideal (and of course on the grading and the inner product), but not on the actual Γ–basis!

Proposition 14. *If $G, G' \subset \Pi$ are two Γ–bases for the same ideal $\langle G \rangle = \langle G' \rangle$ then $\nu_G = \nu_{G'}$.*

Proof. We expand $G' \ni g' = \sum_g p_{g',g}\, g$ in terms of G and substitute this into a Γ–representation of $f \in \Pi$ with respect to G',

$$f = \sum_{g' \in G'} f_{g'}\, g' + r' = \sum_{g \in G} \left(\sum_{g' \in G'} f_{g'} p_{g',g} \right) g + r',$$

and a comparison of degrees yields that the term on the right hand side is a Γ–representation of f with respect to G. Consequently, Theorem 13 implies that $r' = r$. □

5.4. What remains is interpolation

Proposition 14 states that for a given grading and a given inner product (which can be seen as a part of the grading), there is a *unique* normal form $\nu_{\mathcal{I}}(f)$ for a polynomial f modulo an ideal $\mathcal{I}$, and the polynomial $\nu_{\mathcal{I}}(f)$ is the canonical representer of the equivalence class $[f] = f + \mathcal{I}$ in $\Pi / \mathcal{I}$. This finally returns us to interpolation as it says that normal forms are natural interpolants for ideal interpolation schemes.

Theorem 15. *If $\Theta \subset \Pi'$ has the property that $\mathcal{I}(\Theta) := \ker \Theta$ is an ideal in Π, then $\mathcal{P}^* := \nu_{\mathcal{I}(\Theta)}(\Pi)$ is a degree reducing interpolation space for Θ and the interpolation operator takes the form $L_{\mathcal{P}^*} = \nu_{\mathcal{I}(\Theta)}$.*

Proof. Since $f - \nu_{\mathcal{I}(\Theta)}(f) \in \mathcal{I}(\Theta)$, the mapping $\nu_{\mathcal{I}(\Theta)}$ is an interpolation operator whose linearity follows from the uniqueness of remainders and the fact that $f + f'$ has the Γ–representation

$$f+f'=\sum_{g\in G} h_g\, g+r+r', \qquad r+r'\in W(G), \quad h_g\in\Pi.$$

Degree reduction of the normal form operator, on the other hand, is seen directly from the reduction algorithm. □

Hence, normal forms are canonical degree reducing interpolants and $\mathcal{P}^*=\nu_{\mathcal{I}(\Theta)}(\Pi)$ is the canonical degree reducing interpolation space. Moreover, reduction even allows us to compute an interpolant to a polynomial f if the interpolation problem is only known in a *implicit* way, namely in terms of a basis of $\mathcal{I}(\Theta)$. On the other hand, in the case of a Lagrange interpolation problem, that is, if a finite node set $\Xi\subset\mathbb{K}^d$ is given, we can easily give the Lagrange fundamental polynomials *explicitly*. To that end, let $v\in\mathbb{K}^d$ be any vector such that $v^T(\xi-\xi')\neq 0$, $\xi,\xi'\in\Xi$ – such vectors exist in abundance, in fact there are only finitely many (normalized) vectors that do not have this property. Now simply define for $\xi\in\Xi$ the polynomials

$$\widehat{\ell}_\xi := \prod_{\xi'\in\Xi\setminus\{\xi\}} \frac{v^T(\cdot-\xi')}{v^T(\xi-\xi')}, \qquad \ell_\xi:=\nu_{\mathcal{I}(\Xi)}(\widehat{\ell}_\xi), \tag{29}$$

and ℓ_ξ, $\xi\in\Xi$, is the Lagrange basis of $\mathcal{P}^*=\nu_{\mathcal{I}(\Xi)}(\Pi)$, so that the interpolant takes the form

$$L_{\mathcal{P}^*}f=\sum_{\xi\in\Xi} f(\xi)\,\nu_{\mathcal{I}(\Xi)}\left(\prod_{\xi'\in\Xi\setminus\{\xi\}} \frac{v^T(\cdot-\xi')}{v^T(\xi-\xi')}\right).$$

But this tells us even more: the basis is obtained by taking the normal forms of polynomials of total degree $\#\Xi-1$, leading to the following observation.

Corollary 16. *For any finite node set $\Xi\subset\mathbb{K}^d$ the normal form interpolation space $\mathcal{P}^*$ satisfies $\mathcal{P}^*=\nu_{\mathcal{I}(\Xi)}(\Pi_{\#\Xi-1})$.*

In other words, the minimal degree interpolation space is spanned by the polynomials $p_\alpha=\nu_{\mathcal{I}(\Xi)}((\cdot)^\alpha)$ and the finite matrix

$$P(\Xi)=[p_\alpha(\xi)\ :\ \xi\in\Xi,\ |\alpha|\le\#\Xi-1]$$

has the maximal rank $\#\Xi$. In contrast to the approach in [9], where infinite matrices are considered, the linear system associated to a Lagrange interpolation problem can therefore always be assumed to be finite, even if the number of polynomials exceeds the number of points significantly.

Normal forms are closely tied to the *Hilbert function* of the ideal $\mathcal{I}(\Xi)$. Though the Hilbert function of an ideal is usually defined and considered for the homogeneous grading, cf. [34,54,55], it can be extended in quite a straightforward way to arbitrary gradings, which we will describe next. For an ideal $\mathcal{I}\subset\Pi$ we set $W_\gamma(\mathcal{I})=W_\gamma(G)$, G any Γ–basis of $\mathcal{I}$, and define the *homogeneous* Hilbert function of $\mathcal{I}$ as

$$H^0_{\mathcal{I}}(\gamma)=\dim W_\gamma(\mathcal{I})=\dim\Pi^0_\gamma-\dim\left(\Lambda(\mathcal{I})\cap\Pi^0_\gamma\right), \qquad \gamma\in\Gamma, \tag{30}$$

Moreover, the "summarized" function

$$H_{\mathcal{I}}(\gamma) = \sum_{\gamma' \le \gamma} H^0_{\mathcal{I}}(\gamma'), \qquad \gamma \in \Gamma,$$

is called the *affine* Hilbert function of $\mathcal{I}$. The latter one is monotonically increasing, i.e., $H_{\mathcal{I}}(\gamma) \le H_{\mathcal{I}}(\gamma')$ if $\gamma \le \gamma'$. Also note that, strictly speaking the second identity in (30) is only valid if Π^0_γ is of finite dimension which is the case, for example, for the grading by total degree, but not for the one from (22). The affine Hilbert function describes the dimensions of the nested interpolation spaces $\mathcal{P}^*_\gamma = \mathcal{P}^* \cap \Pi_\gamma$, $\gamma \in \Gamma$.

Proposition 17. *For $\gamma \in \Gamma$ we have that*

$$H^0_{\mathcal{I}(\Xi)}(\gamma) = \dim\left(\mathcal{P}^* \cap \Pi^0_\gamma\right), \qquad H_{\mathcal{I}(\Xi)}(\gamma) = \dim \mathcal{P}^*_\gamma. \tag{31}$$

Proof. We first note that $\mathcal{P}^* = \nu_{\mathcal{I}(\Xi)}(\Pi)$ is spanned by homogeneous polynomials: any polynomial $f \in W_\gamma(\mathcal{I}) \subset \Pi^0_\gamma$ has the property that $f = \nu_{\mathcal{I}(\Xi)}(f)$. Hence $H^0_{\mathcal{I}(\Xi)}(\gamma) \le \dim\left(\mathcal{P} \cap \Pi^0_\gamma\right)$. On the other hand, any homogeneous $p \in \mathcal{P}^* \cap \Pi^0_\gamma$ also satisfies $p = \nu_{\mathcal{I}(\Xi)}(p) \in W_\gamma(\mathcal{I})$ and therefore $H^0_{\mathcal{I}(\Xi)}(\gamma) \ge \dim\left(\mathcal{P} \cap \Pi^0_\gamma\right)$ as well. □

The understanding obtained so far of degree reducing interpolation for graded rings and the prominent role that normal forms play, finally allows us to define the notion of a *Newton basis* for arbitrary gradings, and even to construct such bases. To that end, we start off with the ***homogeneous bases*** $P^0_\gamma \subset \mathcal{P}^*$, $\gamma \le \delta(\mathcal{P}^*)$, which are defined by the requirement that

$$\operatorname{span} P^0_\gamma = \mathcal{P}^*_\gamma \cap \Pi^0_\gamma = W_\gamma(\mathcal{I}(\Xi)). \tag{32}$$

Since $\dim \mathcal{P}^* = \#\Xi < \infty$ there is a finite set $\Gamma^* \subset \Gamma$ such that P^0_γ, $\gamma \in \Gamma^*$, still form a homogeneous basis for $\mathcal{P}^*$. Moreover, we can assume that $0 \in \Gamma^*$ as $W_0(\mathcal{I}(\Xi)) = \{0\}$ implies that $V_0(\mathcal{I}(\Xi)) = \Pi^0_0$ and thus $\mathcal{I}(\Xi) = \Pi$, i.e., $\Xi = \emptyset$.

The construction of the Newton basis proceeds as follows: beginning with $0 = \min \Gamma^*$ and taking into account that the $P^0_0(\Xi)$ has rank $H_{\mathcal{I}(\Xi)}(0)$, we can find $\Xi_0 \subset \Xi$ such that $\det P^0_0(\Xi_0) \ne 0$, and so $N_0 := P^0_0(\Xi_0)^{-1} P^0_0$ satisfies $N_0(\Xi_0) = I$.

Suppose now that for the k smallest elements $0 = \gamma^0 < \cdots < \gamma^{k-1}$ of Γ^* we have constructed $\Xi_0, \dots, \Xi_{\gamma^{k-1}}$ such that

$$\det P_{k-1}(\Xi_{k-1}) \ne 0, \qquad P_{k-1} := \bigcup_{j=0}^{k-1} N_{\gamma^j}, \qquad \Xi_{k-1} := \bigcup_{j=0}^{k-1} \Xi_{\gamma^j},$$

and set $\gamma = \gamma^k = \min\left\{\gamma' \in \Gamma \;:\; \gamma' > \gamma^{k-1}\right\}$. Then the polynomials in the vector

$$P' := P^0_\gamma - P^0_\gamma(\Xi_{k-1})^T \, P_{k-1}(\Xi_{k-1})^{-T} \, P_{k-1}$$

are linearly independent, vanish on Ξ_{k-1}, and their leading terms $\Lambda(P')$ span $W_\gamma(\mathcal{I}(\Xi))$. In addition, the rank of the matrix $P'(\Xi \setminus \Xi_{k-1})$ coincides with the cardinality of P'. Consequently, there is a subset Ξ_γ of $\Xi \setminus \Xi_{k-1}$ such that $\det P'(\Xi_\gamma) \ne$

0, and once we define the polynomial vector $N_\gamma = P'(\Xi_\gamma)^{-1} P'$ it follows that $N_\gamma(\Xi_\gamma) = I$. This process, first used in [16] in the case of monomial gradings based on term orders, is nothing but the Γ–equivalent of ***Gauß elimination by segments***, see [6], or the Gram–Schmidt process in [66], and creates a decomposition of Ξ into Ξ_γ and a Newton basis N_γ, $\gamma \in \Gamma^*$, such that

$$N_\gamma(\Xi_{\gamma'}) = \delta_{\gamma',\gamma} I, \quad \gamma' \le \gamma, \qquad \operatorname{span} \Lambda(N_\gamma) = \operatorname{span} \Lambda\left(P_\gamma^0\right) = W_\gamma(\mathcal{I}(\Xi)).$$

Once more the normal form space $\mathcal{P}^*$ serves as a prototype for degree reducing interpolation spaces in the sense that we can even *characterize* degree reducing interpolation in terms of the existence of a Newton basis. For the grading by total degree, this result has been given in [66], the presentation here, however, is strongly influenced by discussions with Carl de Boor on [67] and his comments made in [9].

Theorem 18. *A polynomial subspace $\mathcal{P} \subset \Pi$ is a degree reducing interpolation space for a finite node set Ξ if and only if there exists a* Newton basis $N_\gamma \subset \mathcal{P}$, *$\gamma \in \Gamma^* \subset \Gamma$, and a decomposition of Ξ into Ξ_γ, $\gamma \in \Gamma^*$, such that*

$$N_\gamma(\Xi_{\gamma'}) = \delta_{\gamma,\gamma'}, \quad \gamma' \le \gamma, \qquad \Pi_\gamma^0 = \operatorname{span} \Lambda(N_\gamma) + V_\gamma(\mathcal{I}(\Xi)), \quad \gamma \in \Gamma. \tag{33}$$

Proof. Suppose that $\mathcal{P}$ is a degree reducing interpolation space. Together with $L_\mathcal{P}\mathcal{I}(\Xi) = \nu_{\mathcal{I}(\Xi)}(\mathcal{I}(\Xi)) = 0$ and Theorem 15 this implies for any $f \in \Pi$ that

$$\delta(L_\mathcal{P} f) = \delta\left(L_\mathcal{P} \nu_{\mathcal{I}(\Xi)}(f)\right) \le \delta\left(\nu_{\mathcal{I}(\Xi)}(f)\right) = \delta\left(\nu_{\mathcal{I}(\Xi)}(L_\mathcal{P} f)\right) \le \delta(L_\mathcal{P} f),$$

i.e.,

$$\delta(L_\mathcal{P} f) = \delta\left(\nu_{\mathcal{I}(\Xi)}(f)\right) = \delta(L_{\mathcal{P}^*} f). \tag{34}$$

Let N_γ^*, $\gamma \in \Gamma^*$, denote a Newton basis of $\mathcal{P}^*$, constructed as above, and set $N_\gamma = L_\mathcal{P} N_\gamma^*$, $\gamma \in \Gamma^*$. The interpolation property of $L_\mathcal{P}$ then yields that

$$N_\gamma(\Xi_{\gamma'}) = N_\gamma^*(\Xi_{\gamma'}) = \delta_{\gamma,\gamma'} I, \qquad \gamma' \le \gamma.$$

Hence, for any $\xi \in \Xi_\gamma$ there exist $N_\xi \in N_\gamma$ and $N_\xi^* \in N_\gamma^*$ such that $N_\xi - N_\xi^* \in \mathcal{I}(\Xi)$ which is improved by (34) into

$$N_\xi - N_\xi^* \in \mathcal{I}(\Xi) \cap \Pi_\gamma \qquad \text{i.e.,} \qquad \Lambda(N_\xi) - \Lambda\left(N_\xi^*\right) \in V_\gamma(\mathcal{I}(\Xi)).$$

Since $\Pi_\gamma^0 = \operatorname{span} \Lambda\left(N_\gamma^*\right) + V_\gamma(\mathcal{I}(\Xi))$ for all $\gamma \in \Gamma$, this implies (33).

Conversely, suppose that $\mathcal{P}$ has a basis that satisfies (33), hence, $\mathcal{P}$ is an interpolation space. Moreover, the sets $N_\gamma^* \subset \mathcal{P}_\gamma^*$, $\gamma \in \Gamma^*$, defined by $N_\gamma^* = \nu_{I(\Xi)}(N_\gamma)$ also satisfy $N_\gamma^*(\Xi_{\gamma'}) = \delta_{\gamma,\gamma'} I$, $\gamma, \gamma' \in \Gamma^*$, and thus form a Newton basis for the normal form interpolation space $\mathcal{P}^*$. Now, for $f \in \Pi$,

$$L_{\mathcal{P}^*} f = \sum_{\gamma \in \Gamma^*} f_\gamma^T N_\gamma^* \qquad \Leftrightarrow \qquad L_\mathcal{P} f = \sum_{\gamma \in \Gamma^*} f_\gamma^T N_\gamma$$

and since the leading terms of both sets, N_γ and N_γ^*, each span Π_γ^0 jointly with $V_\gamma(\mathcal{I}(\Xi))$, they consist of $H_{\mathcal{I}(\Xi)}^0(\gamma)$ linearly independent elements, which leads to the conclusion that $\delta(L_\mathcal{P}(f)) = \delta(L_{\mathcal{P}^*}(f)) \le \delta(f)$; in other words, $\mathcal{P}$ is a degree reducing interpolation space. □

From the proof of Theorem 18 we can draw another interesting conclusion which further justifies the notation "Γ^*".

Corollary 19. *Let $\Xi \subset \mathbb{K}^d$. Then the index set γ^* for the Newton basis is the same for all degree reducing interpolation spaces and takes the form*

$$\Gamma^* = \left\{ \gamma \in \Gamma \; : \; H^0_{\mathcal{I}(\Xi)}(\gamma) \neq 0 \right\}.$$

Allright, the existence of a Newton basis characterizes degree reducing interpolation spaces and of course those will depend on the parameters involved with the grading, that is, the monoid with its total order and the inner product that is used for the orthogonal projection. We should first note that the normal form interpolant is singled out by requiring that its Newton basis makes the decomposition in (33) an *orthogonal* one. But even then there is in general a freedom of choice as different decompositions of Ξ into Ξ_γ, $\gamma \in \Gamma^*$, lead to different bases. This is, however, not a multivariate phenomenon as even the univariate Newton polynomials

$$\frac{(x-\xi_0)\cdots(x-\xi_{k-1})}{(\xi_k-\xi_0)\cdots(\xi_k-\xi_{k-1})}, \qquad k = 0,\ldots,n, \qquad \Xi = \{\xi_0,\ldots,\xi_n\},$$

depend on the ordering of the points in Ξ which need not be in any "natural", for example ascending, order.

It is clear that if N_γ is a Newton basis for an interpolation space $\mathcal{P}$, then so is

$$N'_\gamma = N_\gamma + \sum_{\gamma' \leq \gamma} V_\gamma\left(\mathcal{I}(\Xi)\right), \qquad \gamma \in \Gamma^*,$$

that is, even after fixing the decomposition Ξ_γ, the elements of a Newton basis are only defined up to addition of an ideal element of at most the same degree – and this *is* a truly multivariate feature. On the other hand, the above modification is also the only way to pass from one Newton basis to another so that they are normal forms up to addition of ideal elements of restricted degree.

Corollary 20. *If $\delta(\mathcal{P}) < \min\{\delta(g) \; : \; g \in G\}$ for some Γ–basis G of $\mathcal{I}(\Xi)$ then the degree reducing interpolation space is unique and so is its Newton basis up to indexing of nodes.*

5.5. When Newton goes Γ ...

We close this section with a brief look at an appealing algebraic property of the Newton basis. To that end we fix $\gamma \in \Gamma^*$ and consider the node set

$$\Xi \supseteq \Xi' := \Xi'_\gamma = \bigcup_{\gamma' \leq \gamma} \Xi_\gamma$$

and the associated ideal $\mathcal{I}(\Xi') \supseteq \mathcal{I}(\Xi)$. How can we obtain a Γ–basis G' for $\mathcal{I}(\Xi')$? By the inclusion relation between the ideals we can consider this as a basis extension problem, i.e., $G' = G + H$, $H \subset \Pi$, but what is H? Let $\mathcal{P}$ be a degree reducing interpolation space and consider $f \in \mathcal{I}(\Xi')$: since $f(\Xi') = 0$ the interpolant $L_{\mathcal{P}} f$ lies in the space spanned by $N_{\gamma'}$, $\gamma' > \gamma$, and so it follows that

$$f = L_{\mathcal{P}} f + q = \sum_{\gamma' > \gamma} f_\gamma^T N_\gamma + \sum_{g \in G} q_g \, g, \qquad q = \sum_{g \in G} q_g \, g \in \mathcal{I}(\Xi).$$

Since $L_{\mathcal{P}} f \preceq f$, all degrees appearing in the sums on the right hand side of this equation are $\leq \delta(f)$, hence this equation is in fact a Γ–representation. In other words: the set

$$G \cup \bigcup_{\gamma' > \gamma} N_\gamma$$

is a Γ–basis for $\mathcal{I}(\Xi')$. This observation is the motivation behind Conjecture 7. In order to make a Newton basis work as a supplement to a Γ–basis, it is necessary that one is dealing with an ideal structure on all levels: the decomposition of $\Theta \subset \Pi'$ into Θ_γ, $\gamma \in \Gamma^*$, must have the property that ker (Θ'_γ) is an ideal for any $\gamma \in \Gamma^*$, where $\Theta'_\gamma = \bigcup_{\gamma' \leq \gamma} \Theta_{\gamma'}$. Whether or not such a decomposition exists is not known at the present moment, but it would definitely be useful in deriving remainder formulas, see [66,73,72]. Anyway, the generalization of Conjecture 7 is as follows.

Conjecture 21. *Any finite set $\Theta \subset \Pi'$ of linear functionals which admits an ideal interpolation scheme can be graded as*

$$\Theta = \bigcup_{\gamma \in \Gamma^*} \Theta_\gamma, \qquad \Theta'_\gamma := \bigcup_{\gamma' \leq \gamma} \Theta_{\gamma'}, \quad \gamma \in \Gamma^*$$

such that for $\gamma \in \Gamma^*$

(i) the interpolation problem with respect to Θ'_γ is poised for $\mathcal{P}^*_\gamma = \mathcal{P}^* \cap \Pi_\gamma$,
(ii) ker Θ'_γ *is an ideal.*

6. More Special General Gradings

The approach by means of arbitrary gradings that we pursued in the preceding section provides the advantage of great generality and allows us to comprehend the general principle behind degree reducing interpolation. On the other hand, it is a very simple piece of wisdom that imposing further structure allows us to obtain more detailed results. To that end, we will now say goodbye to our general gradings and assume for this section that our grading is a ***strict monomial*** one, the concept that includes all the "popular" notions of degree and thus still offer sufficient generality. In particular, it covers the grading by total degree, by term orders and the gradings supported by the Gröbner fan.

6.1. Kernels from the right product line

So far we have not really put emphasis on the inner product used in the reduction process which nevertheless affects the behavior of the normal forms and thus the interpolation space and process quite significantly. It seems as if "the" inner product is the one first used for interpolation by de Boor and Ron, which is defined as

$$(f,g) = (f(D)g)(0) = \sum_{\alpha \in \mathbb{N}_0^d} f_\alpha \, g_\alpha \, \alpha!, \qquad f,g \in \Pi. \tag{35}$$

To be precise: this is only the *real* version of the inner product, in the case $\mathbb{K} = \mathbb{C}$, complex conjugation has to be added, but for simplicity I prefer being a realist here. The inner product in (35) seems to carry many names: Bombieri inner product or Fisher inner product, I have even been told by Ch. Dunkl that it has been used by Calderon in the context of harmonic polynomials. But whatever name it bears, this inner product turns out to be quite useful, not only in the practical problem of computing reductions [61], but also for an explicit description of the normal form interpolation space that has been given in [70], extending results e.g. from [13].

Theorem 22. *Suppose that $\Theta \subset \Pi'$ defines an ideal interpolation scheme and let G be a Γ–basis of $\mathcal{I}(\Theta)$, where the grading Γ is based on the inner product from (35). Then the canonical normal form interpolation space*

$$\mathcal{P}^* = \nu_{I(\Theta)}(\Pi) = \bigcap_{g \in G} \ker \Lambda(g)(D) \tag{36}$$

is closed under differentiation.

The "explicit" form (36) of the interpolation space $\mathcal{P}^*$ is well–known for the grading by total degree: it is precisely the *least interpolation* scheme from [10–13] which is thus the equivalence of normal form interpolation for any strict monomial grading as long as the associated inner product is chosen as in (35). Actually, there is even more behind this relationship. As has been shown in [70] for arbitrary gradings and in [67] in the Gröbner context, one can replay the whole formalism of leading terms for polynomials and least terms in power series for general gradings to obtain (36). In this respect, the argument is a purely algebraic one to identify normal forms as kernel elements of the differential operators $\Lambda(g)(D)$ and this, of course, has to take into account the nature of the inner product. On the other hand, the kernels of the differential operators can even be used to characterize Γ–bases, a fact proven in [69] as a tool to study quotient ideals that arise in connection with refinable functions.

Theorem 23. *A finite set $G \subset \Pi$ of polynomials is a Γ–basis for an ideal $\mathcal{I}$ if and only if*

$$\dim \Pi/\mathcal{I} = \dim \bigcap_{g \in G} \ker \Lambda(g)(D).$$

6.2. Lower sets and tensor product data

The fact that the total ordering "<" on Γ is compatible with addition makes it immediate that $\gamma \geq \gamma - \Gamma$ where, of course, $\gamma - \Gamma := \{\gamma' \in \Gamma \,:\, \gamma \in \gamma' + \Gamma\}$. Therefore, the span or direct sum of $\Pi_{\gamma'}$, $\gamma' \in \gamma - \Gamma$, is a natural subset of polynomials whose degree is bounded by γ. We now call a set $L \subset \Gamma$ a *lower set* if $\gamma \in L$ implies that $\gamma - \Gamma \subseteq L$. If, specifically, $\Gamma = \mathbb{N}_0^d$, then the lower sets take the form

$$\alpha - \Gamma = \alpha - \mathbb{N}_0^d = \left\{ \beta \in \mathbb{N}_0^d \ : \ \beta_j \le \alpha_j, \, j = 1, \dots, d \right\}, \qquad \alpha \in \mathbb{N}_0^d.$$

Note that being a lower set is independent of the ordering, it is exclusively a property of the additive structure of the monoid Γ, the fact that lower sets interact properly with the ordering was an extra assumption that we had to make.

The "classical" gradings, i.e., the ones by total degree and by term orders, have an appealing property: the set Γ^* of degrees appearing in $\mathcal{P}^*$ (or, for the readers who prefer it in "algebraic": the degrees of the homogeneous components whose direct sum is $\mathcal{P}^*$) form a lower set, cf. [66,67]. For arbitrary strict monomial gradings the problem is more intricate: the simple example of the bivariate gradings based on

$$\Gamma = \mathbb{N}_0, \qquad \delta \left(x^\alpha \right) = \alpha_1 + n \, \alpha_2, \qquad n \in \mathbb{N},$$

have, for $\Xi = \{(0,0), (1,0), (0,1)\}$ their normal form spaces spanned by $\{1, x, y\}$ and these are homogeneous contributions are of degree $0, 1, n$ which is no lower set as soon as $n \ge 2$. However, this example clearly shows what really goes wrong here: there is a coordinate polynomial, namely y, that belongs to Π_n and this $n > 1$ can be decomposed into summands in $\mathbb{N}$; in other words,

$$n - \mathbb{N}_0 \ne \{n\} \qquad \text{or} \qquad n - \mathbb{N} = n - (\mathbb{N}_0 \setminus \{0\}) \ne \emptyset.$$

This can be overcome by requiring that for $j = 1, \dots, d$

$$x_j \in \Pi_\gamma \qquad \Rightarrow \qquad \gamma - \Gamma = \{\gamma\}, \tag{37}$$

and indeed it can be shown that (37) even characterizes the situation where Γ^* is a lower set – provided that Γ is a strict monomial grading and that the monomials are all orthogonal with respect to the inner product, which is obviously the case for the one from (35). In a way, this observation and the above example can be understood in the sense that the grading should be "democratic" with respect to coordinate polynomials. Just note that in the example the coordinate polynomial y had the same degree as an nth power of the other coordinate polynomial x.

But let us continue with ***tensor product sites*** now, that is, with a very special, very classical, but also very organized and structural way of choosing the nodes. To that end, let $L \subset \mathbb{N}_0^d$ a finite lower set of multiindices and suppose that abscissae

$$\xi_{j,k} \in \mathbb{K}, \qquad j = 1, \dots, d, \, k \in \mathbb{N}_0,$$

are given; we assume no ordering among them, and that k ranges over $\mathbb{N}_0$ is for convenience only. The tensor product sites are then given as

$$\Xi = \left\{ \xi_\alpha = \left(\xi_{1,\alpha_1}, \dots, \xi_{d,\alpha_d} \right) \ : \ \alpha \in L \right\}. \tag{38}$$

This is slightly more general than the usual notion of tensor product sites that corresponds to choosing $L = \{\alpha \ : \ \alpha_j \le n_j\}$, $n_j \in \mathbb{N}$, $j = 1, \dots, d$. On the other hand, this concept also includes "triangular" data by means of $L = \{\alpha \ : \ |\alpha| \le n\}$, $n \in \mathbb{N}_0$. This latter configuration, however, could also be interpreted as the principal lattice with the hyperplanes $H_{j,k} = x_j - \xi_{j,k}$.

There is a canonical interpolation space for node configurations as in (38) which is the associated space of "tensor product polynomials"

$$\mathcal{P} = \operatorname{span}_{\mathbb{K}} \{x^\alpha \ : \ \alpha \in L\}, \tag{39}$$

and this connection has made is possible to investigate the "triangular grid" mainly by means of univariate methods more than 100 years ago in [4], also leading remainder formulas by means of mean values, cf. [31,36,76]. The extension to lower sets was also studied by Werner [77] in the context of arbitrary Chebyshev systems. Deriving a remainder formula of the type (12) in terms of integrals against splines, on the other hand, was left to [74]. The splines appearing there were actually box splines and on the way of obtaining them the simplex spline decomposition of cube splines, cf. [49, p. 192ff], appeared in a very natural way.

An interesting property of the specific minimal degree interpolation procedures, in particular the ***least interpolant*** for which this fact is pointed out in [12], is that when applied to tensor product sites they are what one expects them to be, namely the tensor product space from (39). And so it is fair to ask whether this natural behavior is a specific aspect of particularly well–chosen interpolation spaces or whether this is a more general fact that holds for "any" choice of interpolation space. It turns out that the latter is true; more precisely, the following result holds true [71].

Theorem 24. *If Γ is a strict monomial grading and Ξ is of the form (38) then*

$$\nu_{\mathcal{I}(\Xi)}(\Pi) = \operatorname{span}\{x^\alpha \ : \ \alpha \in L\}.$$

In other words, any normal form interpolant with respect to a strict monomial grading – and this includes the least interpolant – ***must*** be a tensor product space. Of course, there always exist non–tensor interpolants as well, but in view of the preceding section those are more unnatural and the examples of such spaces always are very artificial in appearance. The proof of Theorem 24 is based on the fact that for tensor product sites Ξ from (38) the ideal $\mathcal{I}(\Xi)$ has a ***universal*** Γ***–basis***, that is, a set G which is a Γ–basis for all strict monomial gradings (and any inner product) simultaneously.

6.3. H–bases and the error formula revisited

Getting close to the end of our travel through some parts of the world of multivariate polynomial interpolation, we will enjoy a brief look on the ideal theoretic interpretation of error formulas. The starting point is the simple observation that for $f \in \Pi$ and an interpolation space $\mathcal{P}$ we clearly have that $\Theta(f - L_{\mathcal{P}} f) = 0$, i.e., $f - L_{\mathcal{P}} f \in \mathcal{I}(\Theta)$, regardless of whether we deal with Lagrange or Hermite interpolation schemes. Hence, as long as f is a polynomial and G is a Γ–basis of $\mathcal{I}(\Theta)$, there is a representation

$$f - L_{\mathcal{P}} f = \sum_{g \in G} P_g(f)\, g,$$

where any P_g is an operator that maps a polynomial f to a polynomial of degree at most $\delta(f) - \delta(g)$. Note that it is not all clear how to find such coefficient operators, and in general, they are in no way unique. Once an explicit formula for such operators P_g has been given – usually in terms of a simplex spline integral over certain derivatives of f or linear combinations thereof – we can then hope to extend this operator from polynomials to sufficiently smooth functions. In fact, smoothness would not even be needed for the error formulas from Section 4 as the formulas always hold in the ***distributional*** sense. On the other hand, note that unfortunately the temptation to assume duality, $P_g(g') = 0$, $g, g' \in G$, has to be resisted or at least been considered carefully: the representation with respect to a Γ–basis, not even with respect to a minimal Gröbner or H–basis, is usually ***not*** unique due to the unavoidable appearance of syzygies and so there is no a priori justification that the "good" operators have to be of this type. On the other hand, lack of uniqueness is not by itself a disadvantage as it allows for certain degrees of flexibility. Be it good or bad, the task to determine P_g a really hard one, and in full generality this is still an unsolved problem, though it is conjectured in [8] what a decent remainder formula should look like, albeit supported only by evidence for configurations that satisfy the very restrictive geometric characterization.

To interpret the error formulas from Section 4 in algebraic terms, we first note that the notion of degree employed there was that of ***total degree*** and so the Γ–bases will become H–bases. An almost perfect formula from the algebraic point of view is de Boor's formula (14). The $\binom{n+d}{d}$ polynomials p_K there even form a ***minimal reduced H–basis*** for $\mathcal{I}(\Xi)$ and the representation operators P_K are given explicitly as simplex spline integrals over a $n+1$–fold directional derivative which takes care of reducing the degree of polynomials by a proper amount.

But also (12) is quite easily converted into an expression with H–bases in it by observing that

$$[\Xi_\mu, x]\, f = \sum_{j=1}^{d} (x_j - \xi_{\mu^n, j}) \left[\Xi_\mu, x; \Xi_\mu D, \frac{\partial}{\partial x_j}\right] f$$

and so, with $\mathrm{M}_n(\alpha) = \{\mu \in \mathrm{M}_n \ : \ \mu^n = \alpha\}$,

$$(f - L_n f)(x) = \sum_{|\alpha|=n} \sum_{j=1}^{d} (x_j - \xi_{\alpha,j})\, p_\alpha(x) \sum_{\mu \in \mathrm{M}_n(\alpha)} \pi_\mu \left[\Xi_\mu, x; \Xi_\mu D, \frac{\partial}{\partial x_j}\right] f.$$

The polynomials $(x_j - \xi_{\alpha,j})\, p_\alpha(x)$, $|\alpha| = n$, $j = 1, \ldots, d$, form an H–basis for $\mathcal{I}(\Theta)$, but while the above representation has the "proper" algebraic form, it definitely lacks the simplicity of (14): the H–basis is by no means minimal and there is a loss of symmetry due to different ways of indexing. But, on the other hand, the formula holds true for *any* node set Ξ with respect to which the interpolation problem is poised for Π_n and thus is valid in a sufficiently general context. If there are simpler, more symmetric formulas for arbitrary point configurations, or even something of the form conjectured in equation (5.1) of [8], remains an open question at the moment.

Acknowledgements

A lot of people have contributed to this paper, consciously or not, willingly or not, by discussions, encouragement, collaboration or simply by listening patiently. There are, however, four persons to whom I am particularly indebted: Yuan Xu, for starting the work on polynomial interpolation with me, Michael Möller for guiding me into the wonderful world of algebra, Mariano Gasca for all his help and encouragement and Carl de Boor who spent an enormous effort on controversial discussion from which I benefitted very much. Thanks to all of them!

In addition, Michael Möller and Carl de Boor provided me with a lot of helpful comments and suggestions for this present paper which helped a lot to improve content and presentation and which I am very grateful for.

References

1. Aitken, A. G., On interpolation by iteration of proportional parts, without the use of differences, *Proc. Edinburgh Math. Soc.* **3** (1932), 56–76.
2. Andoyer, H., Interpolation, in: *Encyclopédie de Sciences Mathématiques, Tôme I, Vol. 4*, Gauthier–Villars, Paris, 1906.
3. Bauschinger, J., Interpolation, in: *Encyklopädie der Mathematischen Wissenschaften, Bd. I, Teil 2*, Teubner, Leipzig, 1900, pp. 800–821.
4. Biermann, O., Über näherungsweise Kubaturen, *Monatsh. Math. Phys.* **14** (1903), 211–225.
5. Birkhoff, G., The algebra of multivariate interpolation, in: *Constructive Approaches to Mathematical Models* (C. Coffman and G. Fix, Eds.), Academic Press, Boston, 1979, pp. 345–363.
6. de Boor, C., Gauss elimination by segments and multivariate polynomial interpolation, in: *Approximation and Computation: A Festschrift in Honor of Walter Gautschi* (R. V. M. Zahar, Ed.), Birkhäuser, Basel, 1994, pp. 87–96.
7. de Boor, C., A multivariate divided difference, in: *Approximation Theory VIII, Vol. 1: Approximation and Interpolation* (C. K. Chui and L. L. Schumaker, Eds.), World Scientific, Singapore, 1995, pp. 87–96.
8. de Boor, C., The error in polynomial tensor-product, and in Chung-Yao, interpolation, in: *Surface Fitting and Multiresolution Methods* (A. Le Méhauté, C. Rabut and L. L. Schumaker, Eds.), Vanderbilt University Press, Nashville, 1997, pp. 35–50.
9. de Boor, C., Interpolation from spaces spanned by monomials, *Adv. Comput. Math.*, to appear.
10. de Boor, C. and Ron, A., On multivariate polynomial interpolation, *Constr. Approximation* **6** (1990), 287–302.
11. de Boor, C. and Ron, A., On polynomial ideals of finite codimension with applications to box spline theory, *J. Math. Anal. Appl.* **158** (1991), 168–193.
12. de Boor, C. and Ron, A., Computational aspects of polynomial interpolation in several variables, *Math. Comp.* **58** (1992), 705–727.

13. de Boor, C. and Ron, A., The least solution for the polynomial interpolation problem, *Math. Z.* **210** (1992), 347–378.
14. Buchberger, B., *Ein Algorithmus zum Auffinden der Basiselemente des Restklassenrings nach einem nulldimensionalen Polynomideal*, PhD thesis, Innsbruck, 1965.
15. Buchberger, B., Ein algorithmisches Kriterium für die Lösbarkeit eines algebraischen Gleichungssystems, *Aequationes Math.* **4** (3) (1970), 374–383.
16. Buchberger, B. and Möller, H. M., The construction of multivariate polynomials with preassigned zeros, in: *Computer Algebra*, Proc. EUROCAM 1982 (G. Goos and J. Hartmanis, Eds.), Lecture Notes in Computer Science, Vol. 144, Springer, Berlin, 1982, pp. 24–31.
17. Busch, J. R., A note on Lagrange interpolation in $\mathbb{R}^2$, *Rev. Union Matem. Argent.* **36** (1990), 33–38.
18. Carnicer, J. M. and Gasca, M., A conjecture on multivariate polynomial interpolation, Rev. R. Acad. Cien., Serie A. Mat. **95** (2001), 145–153.
19. Carnicer, J. M. and Gasca, M., Classification of bivariate GC configurations for interpolation, *Adv. Comput. Math.* **20** (2004), 5–16.
20. Carnicer, J. M. and Gasca, M., Interpolation on lattices generated by cubic pencils, *Adv. Comput. Math.*, to appear.
21. Carnicer, J., Gasca, M. and Sauer, T., Interpolation lattices in several variables, preprint (2005).
22. Chung, K. C. and Yao, T. H., On lattices admitting unique Lagrange interpolation, *SIAM J. Numer. Anal.* **14** (1977), 735–743.
23. Ciarlet, P. G. and Raviart, P. A., General Lagrange and Hermite interpolation in $\mathbb{R}^n$ with applications to finite element methods, *Arch. Rational Mech. Anal.* **46** (1972), 178–199.
24. Cox, D., Little, J. and O'Shea, D., *Ideals, Varieties and Algorithms*, 2nd Edition, Springer, Berlin. 1996.
25. Davis, P. J., *Interpolation and Approximation*, Dover, New York, 1975.
26. Gasca, M., Multivariate polynomial interpolation, in: *Computation of Curves and Surfaces* (W. Dahmen, M. Gasca and C. A. Micchelli, Eds.), Kluver Academic Publ., Dordrecht, 1990, pp. 215–236.
27. Gasca, M. and Maeztu, J. I., On Lagrange and Hermite interpolation in $\mathbb{R}^k$, *Numer. Math.* **39** (1982), 1–14.
28. Gasca, M. and Mühlbach, G., Multivariate polynomial interpolation under projectivities. Part I: Lagrange and Newton interpolation formulas, *Numer. Algorithms* **1** (1991), 375–400.
29. Gasca, M. and Mühlbach, G., Multivariate polynomial interpolation under projectivities II: Neville–Aitken formulas, *Numer. Algorithms* **2** (1992), 255–278.
30. Gasca, M. and Mühlbach, G., Multivariate polynomial interpolation under projectivities III: Remainder formulas, *Numer. Algorithms* **8** (1994), 103–109.
31. Gasca, M. and Sauer, T., Polynomial interpolation in several variables, *Adv. Comput. Math.* **12** (2000), 377–410.

32. Gasca, M. and Sauer, T., On bivariate Hermite interpolation with minimal degree polynomials, *SIAM J. Numer. Anal.* **37** (2000), 772–798.
33. Gasca, M. and Sauer, T., On the history of multivariate polynomial interpolation, *J. Comput. Appl. Math.* **122** (2000), 23–35.
34. Gröbner, W., *Algebraische Geometrie II*, B.I–Hochschultaschenbücher, Vol. 737, Bibliographisches Institut Mannheim, 1970.
35. Hilbert, D., Über die Theorie von algebraischen Formen, *Math. Ann.* **36** (1890), 473–534.
36. Isaacson, E. and Keller, H. B., *Analysis of Numerical Methods*, John Wiley & Sons, New York, 1966.
37. Kowalewski, G., *Interpolation und genäherte Quadratur*, Teubner, Berlin, 1932.
38. Kronecker, L., Über einige Interpolationsformeln für ganze Funktionen mehrerer Variabeln, in: *L. Kroneckers Werke* (H. Hensel, Ed.), Chelsea Publ. Company, New York 1966, pp. 133–141.
39. Lee, S. L. and Phillips, G. M., Polynomial interpolation at points of a geometric mesh on a triangle, *Proc. Roy. Soc. Edinburgh* **108A** (1988), 75–87.
40. Lorentz, G. G., *Approximation of Functions*, 2nd ed., Chelsea Publ. Company, New York, 1986.
41. Lorentz, G. and Lorentz, R., Solvability problems of bivariate interpolation. I, *Constr. Approximation* **2** (1986), 153–170.
42. Lorentz, G. and Lorentz, R., Solvability problems of bivariate interpolation. II. Applications, *Approx. Theory Appl.* **3** (1987), 79–97.
43. Lorentz, G. and Lorentz, R., Bivariate Hermite interpolation and applications to algebraic geometry, *Numer. Math.* **57** (1990), 669–680.
44. Lorentz, R. A., *Multivariate Birkhoff Interpolation*, Lecture Notes in Mathematics, Vol. 1516, Springer, Berlin, 1992.
45. Lorentz, R. A., Multivariate Hermite interpolation by algebraic polynomials: a survey, *J. Comput. Appl. Math.* **122** (2000), 167–201.
46. Macaulay, F. S., *The Algebraic Theory of Modular Systems*, Cambridge Tracts in Math. and Math. Physics, Vol. 19, Cambridge Univ. Press, Cambridge, 1916.
47. Mairhuber, J. C., On Haar's theorem concerning Chebychev approximation problems having unique solutions, *Proc. Amer. Math. Soc.* **7** (1956), 609–615.
48. Marinari, M. G., Möller, H. M. and Mora, T., Gröbner bases of ideals given by dual bases, in: *Proc. ISAAC*, 1991.
49. Micchelli, C. A., *Mathematical Aspects of Geometric Modeling*, CBMS-NSF Reg. Conf. Series in Applied Mathematics, Vol. 65, SIAM, Philadelphia, 1995.
50. Micchelli, C. A., On a numerically efficient method of computing multivariate B–splines, in: *Multivariate Approximation Theory* (W. Schempp and K. Zeller, Eds.), Birkhäuser, Basel, 1979, pp. 211–248.
51. Micchelli, C. A., A constructive approach to Kergin interpolation in $\mathbb{R}^k$: multivariate B–splines and Lagrange interpolation, *Rocky Mountain J. Math.* **10** (1979), 485–497.
52. Möller, H. M., Mehrdimensionale Hermite–Interpolation und numerische Integration, *Math. Z.* **148** (1976), 107–118.

53. Möller, H. M., Hermite interpolation in several variables using ideal–theoretic methods, in: *Constructive Theory of Functions of Several Variables* (W. Schempp and K. Zeller, Eds.), Lecture Notes in Mathematics, Vol. 571, Springer, Berlin, 1977, pp. 155–163.
54. Möller, H. M. and Sauer, T., H–bases for polynomial interpolation and system solving, *Adv. Comput. Math.* **12** (2000), 335–362.
55. Möller, H. M. and Sauer, T., H–bases I: The foundation, in: *Curve and Surface Fitting*, Proc. Saint–Malo 1999 (A. Cohen, C. Rabut and L. L. Schumaker, Eds.), Vanderbilt University Press, Nashville, 2000, pp. 325–332.
56. Möller, H. M. and Sauer, T., H–bases II: Applications to numerical problems, in: *Curve and Surface Fitting*, Proc. Saint–Malo 1999 (A. Cohen, C. Rabut and L. L. Schumaker, Eds.), Vanderbilt University Press, Nashville, 2000, pp. 333–342.
57. Möller, H. M. and Sauer, T., Multivariate refinable functions of high approximation order via quotient ideals of Laurent polynomials, *Adv. Comput. Math.* **20** (2004), 205–228.
58. Neville, E. H., Iterative interpolation, *J. Indian Math. Soc.* **20** (1934), 87–120.
59. Peña, J. M. and Sauer, T., On the multivariate Horner scheme, *SIAM J. Numer. Anal.* **37** (2000), 1186–1197.
60. Peña, J. M. and Sauer, T., On the multivariate Horner scheme II: Running error analysis, *Computing* **65** (2000), 313–322.
61. Peña, J. M. and Sauer, T., Efficient polynomial reduction, *Adv. Comput. Math.*, to appear.
62. Rabut, C., Generalized divided difference with simple knots, *SIAM J. Numer. Anal.* **38** (2000), 1294–1311.
63. Robbiano, L., On the theory of graded structures, *J. Symbolic Computation* **2** (1986), 139–170.
64. Rushdie, S., *Shame*, Vintage Books, London, 1983.
65. Sauer, T., Computational aspects of multivariate polynomial interpolation, *Adv. Comput. Math.* **3** (3) (1995), 219–238.
66. Sauer, T., Polynomial interpolation of minimal degree, *Numer. Math.* **78** (1) (1997), 59–85.
67. Sauer, T., Polynomial interpolation of minimal degree and Gröbner bases, in: *Groebner Bases and Applications*, Proc. Conf. 33 Years of Groebner Bases (B. Buchberger and F. Winkler, Eds.), London Math. Soc. Lecture Notes, Vol. 251, Cambridge University Press, Cambridge, 1998, pp. 483–494.
68. Sauer, T., Gröbner bases, H–bases and interpolation, *Trans. Amer. Math. Soc.* **353** (2001), 2293–2308.
69. Sauer, T., Polynomial interpolation, ideals and approximation order of refinable functions, *Proc. Amer. Math. Soc.* **130** (2002), 3335–3347.
70. Sauer, T., Ideal bases for graded polynomial rings and applications to interpolation, in: *Multivariate Approximation and Interpolation with Applications* (M. Gasca, Ed.), Monograph. Academia de Ciencias de Zaragoza, Vol. 20, Academia de Ciencias Zaragoza, Zaragoza, 2002, pp. 97–110.

71. Sauer, T., Lagrange interpolation on subgrids of tensor product grids, *Math. Comp.* **73** (2004), 181–190.
72. Sauer, T. and Xu, Y., On multivariate Lagrange interpolation, *Math. Comp.* **64** (1995), 1147–1170.
73. Sauer, T. and Xu, Y., On multivariate Hermite interpolation, *Adv. Comput. Math.* **4** (4) (1995), 207–259.
74. Sauer, T. and Xu, Y., A case study in multivariate Lagrange interpolation, in: *Approximation Theory, Wavelets and Applications*, NATO–ASI Proceedings (S. Singh, Ed.), Kluwer Academic Publ., Dordrecht, 1995, pp. 443–452.
75. Sauer, T. and Xu, Y., The Aitken–Neville scheme in several variables, in: *Approximation Theory X* (C. K. Chui, L. L. Schumaker and J. Stöckler, Eds.), Vanderbilt University Press, Nashville, 2002, pp. 353–366.
76. Steffensen, I. F., *Interpolation*, Chelsea, New York, 1927.
77. Werner, H., Remarks on Newton type multivariate interpolation for subsets of grids, *Computing* **25** (1980), 181–191.

Topics in Multivariate Approximation and Interpolation
K. Jetter et al., Editors

Computational Aspects of Radial Basis Function Approximation

Holger Wendland

Institut für Numerische und Angewandte Mathematik, Universität Göttingen
Lotzestr. 16-18, D–37083 Göttingen, Germany

Abstract

This paper gives a recent overview on numerical aspects of multivariate interpolation and approximation by radial basis functions. It comments on the correct choice of the basis function. It discusses the reduction of complexity by different methods as well as the problem of ill-conditioning. It is aimed to be a user's guide to an efficient employment of radial basis functions for the reconstruction of multivariate functions.

Key words: multivariate approximation theory, scattered data approximation, positive definite kernel, domain decomposition, multilevel, multipole expansion, partition of unity

2000 MSC: 65D05, 65D10, 65D15

1. Introduction

Over the past decades radial basis functions or, more generally, (conditionally) positive definite kernels have very successfully been used for reconstructing multivariate functions from scattered data. This success is mainly based upon the following facts:

Email address: wendland@math.uni-goettingen.de (Holger Wendland).

(i) Radial basis functions can be used in *any* space dimension.
(ii) They work for arbitrarily ***scattered data***, bearing no regularity at all.
(iii) They allow interpolants of ***arbitrary*** smoothness.
(iv) The interpolants have a simple structure, which makes RBFs in particular interesting to users outside mathematics.

However, these positive properties do not come for free. For example, building a smooth interpolant using a smooth basis function leads also to an ill-conditioned linear system that has to be solved. Moreover, since most basis functions are globally supported, a large number of interpolation points leads to an unacceptable complexity concerning both space and time.

For these reasons recent research concentrated on resolving these problems. Fast methods for evaluating and computing an RBF interpolant have been developed and thoroughly investigated. Smoothing techniques have been employed to regularize ill-conditioned systems and to smooth out measurement errors.

It is the goal of this paper to describe the most successful approaches in this research direction. Naturally, the selection is biased by this author's point of view and, even if the attempt is made to cover as many methods as possible, there is no guarantee that some important ones are missing.

Nonetheless, this paper should help the reader to understand which basis function and which efficient method she or he should employ for her or his particular reconstruction problem. It should also encourage the reader to consult the literature pointed out in the bibliography for further study.

This paper is organized as follows. Section 2 will give a short review on interpolation and approximation by (conditionally) positive definite kernels. Section 3 is devoted to efficient methods, while the last section deals with stabilizing and regularization.

2. Scattered Data Interpolation and Approximation

A general scattered data interpolation problem can be described as follows. Suppose we are given N distinct ***data sites*** $X = \{x_1, \ldots, x_N\}$ situated in a bounded region $\Omega \subseteq \mathbb{R}^d$. Attached to each data site x_j comes a ***data value*** f_j, which might represent a measured quantity at site x_j. Finally, suppose that it is known, or at least assumed, that the function values f_j are generated by an unknown function f, i.e. $f_j = f(x_j)$. It might also be possible that the data values are generated by functionals other than point evaluations. Examples include derivative or cell average information, but we will not pursue this here, although, to a large extent, the theory can be developed in a more general context.

The information given so far (the data sites, the data values and the fact that the data values are generated by a function) is not enough to built a trustworthy interpolant or approximant to the unknown function, since it provides only very ***localized*** information. To have a decent reconstruction of the unknown function on all of Ω and not only at X also certain ***global*** information is necessary. This global information can often be provided by certain *smoothness* assumptions on the unknown target function f. This is equivalent to assuming the function to belong to a certain, normed function space $\mathcal{H}$. With this information at hand, it is possible and natural to look for the solution of

$$\min \{\|s\|_{\mathcal{H}} : s \in \mathcal{H} \text{ with } s|X = f|X\}, \tag{1}$$

or, if the data values are known to contain noise, of

$$\min \left\{ \sum_{j=1}^{N} [s(x_j) - f(x_j)]^2 + \lambda \|s\|_{\mathcal{H}}^2 : s \in \mathcal{H} \right\}, \tag{2}$$

where $\lambda > 0$ is a certain *smoothing parameter*, which has to be chosen carefully, to balance between interpolation and approximation.

2.1. Reproducing kernel Hilbert spaces

To discuss the solutions to both problems we have to make two more assumptions on the function space. The first assumption is a natural one. Since we want to work with point evaluation functionals, it is reasonable to assume that point evaluation functionals are continuous on $\mathcal{H}$, i.e. that for every $x \in \Omega$ there exists a constant $C_x > 0$ with

$$|f(x)| \le C_x \|f\|_{\mathcal{H}}, \qquad \text{for all } f \in \mathcal{H}.$$

Our second assumption is not that natural but it will simplify the theory dramatically and provides no severe restriction in applications. We will assume that our function space $\mathcal{H}$ is a *Hilbert space* of functions.

A Hilbert space $\mathcal{H}$ of functions $f : \Omega \to \mathbb{R}$ with continuous point evaluation functionals is known to be a *reproducing kernel Hilbert space* (RKHS) (see e.g. [1]).

Definition 1. A *reproducing kernel Hilbert space* $\mathcal{H}$ is a Hilbert space of functions $f : \Omega \to \mathbb{R}$, which has a unique kernel $\Phi : \Omega \times \Omega \to \mathbb{R}$, satisfying

(i) $\Phi(\cdot, x) \in \mathcal{H}$ for all $x \in \Omega$,
(ii) $f(x) = (f, \Phi(\cdot, x))_{\mathcal{H}}$ for all $x \in \Omega$ and all $f \in \mathcal{H}$.

In a RKHS the reproducing kernel Φ is always symmetric and positive semi-definite, i.e. for arbitrary point sets $X = \{x_1, \dots, x_N\}$, the matrices

$$A = A_{\Phi,X} = (\Phi(x_i, x_j))_{i,j} \tag{3}$$

are symmetric and positive semi-definite. This follows immediately from the reproduction property applied to the kernel itself:

$$\Phi(x, y) = (\Phi(\cdot, y), \Phi(\cdot, x))_{\mathcal{H}},$$

and this already ensures that (2) has a unique solution for $\lambda > 0$. However, for guaranteeing a unique solution to (1) as well, we have to make the additional assumption that point evaluation functionals are linearly independent on $\mathcal{H}$. This is equivalent to the fact that all matrices A of the form (3) are positive definite.

Theorem 2. *Let $\mathcal{H} \subseteq C(\Omega)$ be a reproducing kernel Hilbert space with reproducing kernel Φ. Assume that point evaluation functionals are linearly independent on $\mathcal{H}$. Then, for every $f \in \mathcal{H}$, both problems (1) and (2) have a unique solution s_0 and s_λ, respectively. This solution has a representation*

$$s_\lambda = \sum_{j=1}^{N} \alpha_j \Phi(\cdot, x_j),$$

where the coefficient vector $\alpha \in \mathbb{R}^N$ can be determined by solving the linear system

$$(A + \lambda I)\alpha = f|X.$$

Here, A is the matrix (3) and I denotes the identity matrix.

For a proof, we refer the reader to [38] for problem (1) and to [36] for problem (2). From this theorem one can read off, that the solutions s_λ, $\lambda > 0$, of (2) converge to the solution $s_0 = I_X f$ of (1) with λ approaching zero.

Our main interest will be in kernels that are radial and hence defined on all of $\mathbb{R}^d$.

Definition 3. A kernel $\Phi : \mathbb{R}^d \times \mathbb{R}^d \to \mathbb{R}$ is said to be *translation invariant*, if it can be written in the form $\Phi(x, y) = \varphi(x - y)$ for all $x, y \in \mathbb{R}^d$ with an even function $\varphi : \mathbb{R}^d \to \mathbb{R}$.

The kernel is said to be *radial* if it can be written as $\Phi(x, y) = \phi(\|x - y\|_2)$, $x, y \in \mathbb{R}^d$, with a function $\phi : [0, \infty) \to \mathbb{R}$.

Typical examples of positive definite and radial kernels are Gaussians $\phi(r) = e^{-r^2}$, inverse multiquadrics $\phi(r) = 1/\sqrt{1 + r^2}$, and compactly supported functions like $\phi(r) = (1 - r)_+^4(4r + 1)$. While the associated Hilbert spaces for the first two kernels are rather small and consist mainly of analytic functions, the RKHS to the latter kernel is a classical Sobolev space.

This is due to the fact that in all practical situations, the RKHS to a positive definite, translation invariant kernel can be characterized by Fourier transformation. Recall that the *Fourier transform* of $f \in L_1(\mathbb{R}^d)$ is defined by

$$\widehat{f}(\omega) := (2\pi)^{-d/2} \int_{\mathbb{R}^d} f(x) e^{-ix^T\omega} dx, \qquad \omega \in \mathbb{R}^d.$$

Then, the RKHS to $\Phi(x, y) = \varphi(x - y) = \phi(\|x - y\|_2)$ can be characterized as

$$\{f \in L_2(\mathbb{R}^d) \cap C(\mathbb{R}^d) : \widehat{f}/\sqrt{\widehat{\varphi}} \in L_2(\mathbb{R}^d)\},$$

which shows that for Gaussians and inverse multiquadrics we are confronted with rather small spaces. The spaces we are mainly interested in are *Sobolev spaces*

$$H^\tau(\mathbb{R}^d) := \{f \in L_2(\mathbb{R}^d) : \widehat{f}(\cdot)(1 + \|\cdot\|_2^2)^{\tau/2} \in L_2(\mathbb{R}^d)\},$$

they possess reproducing kernels of finite smoothness. The Sobolev space $H^\tau(\mathbb{R}^d)$ is known to be a subset of $C(\mathbb{R}^d)$ provided that $\tau > d/2$, which will be our general assumption.

Note that a kernel $\Phi(x,y) = \varphi(x-y)$ with a Fourier transform satisfying

$$c_1(1+\|\omega\|_2^2)^{-\tau} \le \widehat{\varphi}(\omega) \le c_2(1+\|\omega\|_2^2)^{-\tau}$$

has $H^\tau(\mathbb{R}^d)$ as its RKHS and the *native* inner product

$$(f,g)_{\mathcal{H}} := (2\pi)^{-d/2}\int_{\mathbb{R}^d} \widehat{f}(\omega)\overline{\widehat{g}(\omega)}\widehat{\Phi}(\omega)^{-1}d\omega$$

leads to a norm which is equivalent to the Sobolev space norm. In this sense, we will call Φ also a reproducing kernel of $H^\tau(\mathbb{R}^d)$.

2.2. From thin-plate splines to conditionally positive definite kernels

If $\tau = m$ is a nonnegative integer, the norm on $H^m(\mathbb{R}^d)$ can equivalently be expressed by

$$\|f\|_{H^m(\mathbb{R}^d)}^2 = \sum_{k=0}^{m}\sum_{|\alpha|=k}\frac{k!}{\alpha!}\|D^\alpha f\|_{L_2(\mathbb{R}^d)}^2$$

and it is interesting to see that problems (1) and (2) also have a solution if the full norm is replaced by the semi-norm

$$|f|_{\mathrm{BL}_m(\mathbb{R}^d)}^2 := \sum_{|\alpha|=m}\frac{m!}{\alpha!}\|D^\alpha f\|_{L_2(\mathbb{R}^d)}^2$$

and minimization is done over all functions from the *Beppo-Levi space*

$$\mathrm{BL}_m := \{f \in C(\mathbb{R}^n) : D^\alpha f \in L_2(\mathbb{R}^n) \text{ for all } |\alpha| = m\}.$$

The solution can be expressed in terms of the so-called *surface* or *thin-plate splines* which are defined as

$$\phi_{d,m}(r) := \begin{cases} \dfrac{\Gamma(\frac{d}{2}-m)}{2^{2m}\pi^{d/2}(m-1)!}r^{2m-d}, & \text{for } d \text{ odd},\\[2ex] \dfrac{(-1)^{m+\frac{d-2}{2}}}{2^{2m-1}\pi^{d/2}(m-1)!(m-d/2)!}r^{2m-d}\log r, & \text{for } d \text{ even}, \end{cases}$$

and form a Green's function to the m-th iterated Laplacian operator. To formulate the result, we introduce $\pi_{m-1}(\mathbb{R}^d)$ as the space of d-variate polynomials of degree at most $m-1$ and let $p_1,\dots,p_Q$ with $Q = \dim(\pi_{m-1}(\mathbb{R}^d))$ be a basis of $\pi_{m-1}(\mathbb{R}^d)$. Moreover, we say that a set X is $\pi_{m-1}(\mathbb{R}^d)$ *unisolvent*, if the only polynomial $p \in \pi_{m-1}(\mathbb{R}^d)$ with $p|X = 0$ is the zero polynomial. The proof of the following theorem can again be found in [36,38].

Theorem 4. *Suppose that $m > d/2$ and that $X \subseteq \mathbb{R}^d$ is $\pi_{m-1}(\mathbb{R}^d)$ unisolvent. Let $f \in \mathrm{BL}_m(\mathbb{R}^d)$ be the unknown target function. Then, the solution s_λ to*

$$\min\left\{\sum_{j=1}^{N}[f(x_j)-s(x_j)]^2 + \lambda|s|_{\mathrm{BL}_m(\mathbb{R}^n)}^2 : s \in \mathrm{BL}_m(\mathbb{R}^d)\right\}$$

as well as the solution s_0 *to*

$$\min\left\{|s|_{\mathrm{BL}_m(\mathbb{R}^n)} : s \in \mathrm{BL}_m(\mathbb{R}^n) \text{ with } s|X = f|X\right\}$$

is given by

$$s_\lambda(x) = \sum_{j=1}^{N} a_j \phi_{d,m}(\|x - x_j\|_2) + \sum_{k=1}^{Q} b_j p_j(x).$$

The coefficients $a \in \mathbb{R}^N$ *and* $b \in \mathbb{R}^Q$ *can be determined as follows. Let* $A = (\phi_{d,m}(\|x_i - x_j\|_2)) \in \mathbb{R}^{N\times N}$ *and* $P = (p_j(x_i)) \in \mathbb{R}^{N\times Q}$. *Then* a *and* b *are the solutions of*

$$\begin{aligned}(A + \lambda I)a + Pb &= f|X \\ P^T b &= 0,\end{aligned}$$

where I *denotes the identity matrix.*

Thin-plate splines form a particular class of *conditionally positive definite* functions. In general, a function $\varphi : \mathbb{R}^d \to \mathbb{R}$ is said to be conditionally positive definite of order m, if the associated matrices $A_{\varphi,X} \in \mathbb{R}^{N\times N}$ are positive definite on the set

$$\left\{\alpha \in \mathbb{R}^N : \sum_{j=1}^{N} \alpha_j p(x_j) = 0 \text{ for all } p \in \pi_{m-1}(\mathbb{R}^d)\right\}.$$

Besides thin-plate splines, multiquadrics, defined by $\phi(r) = \sqrt{1 + r^2}$, are the most prominent examples of conditionally positive definite functions. A formal, more general definition for conditionally positive definite kernels is the following one:

Definition 5. Suppose $\mathcal{P}$ is a finite dimensional subspace of $C(\Omega)$, $\Omega \subseteq \mathbb{R}^d$. A continuous symmetric kernel $\Phi : \Omega \times \Omega \to \mathbb{R}$ is said to be conditionally positive definite on Ω with respect to $\mathcal{P}$ if for any N pairwise distinct centers $x_1, \ldots, x_N \in \Omega$ and all $\alpha \in \mathbb{R}^N \setminus \{0\}$ with $\sum_{j=1}^N \alpha_j p(x_j) = 0$ for all $p \in \mathcal{P}$, the quadratic form

$$\sum_{j=1}^{N}\sum_{k=1}^{N} \alpha_j \alpha_k \Phi(x_j, x_k)$$

is positive.

The results of Theorem 4 remain true in the situation of a conditionally positive definite kernel, if the Beppo-Levi space is replaced by an adequate, *natural* function space, the so called *native space* $\mathcal{H} = \mathcal{N}_\Phi(\Omega)$ of the kernel Φ. For a thorough discussion of native spaces see [34,35] and Chapter 11 of [38]. Here, only the following information is important. The native space $\mathcal{H}$ is a function space carrying a semi-inner product $(\cdot,\cdot)_\mathcal{H}$ which has the finite dimensional space $\mathcal{P}$ as its kernel. Moreover, the quotient space $\mathcal{H}/\mathcal{P}$ is a Hilbert space. This can be used to equip $\mathcal{H}$ with a new inner product and a new kernel so that $\mathcal{H}$ becomes a reproducing kernel

Hilbert space itself. Since this will play a role in a later section we will be more precise here. Suppose $\Xi = \{\xi_1, \ldots, \xi_Q\} \subseteq \Omega$ with $Q = \dim(\mathcal{P})$ is $\mathcal{P}$-unisolvent, meaning that the zero function is the only function from $\mathcal{P}$ that vanishes on Ξ. Suppose further that $p_1, \ldots, p_Q$ form a Lagrange basis for $\mathcal{P}$, meaning in particular $p_j(\xi_i) = \delta_{ij}$. Then, we can define a new inner product

$$(f, g) := (f, g)_{\mathcal{H}} + \sum_{\ell=1}^{Q} f(\xi_\ell) g(\xi_\ell)$$

and a new kernel

$$\begin{aligned} K(x, y) = \Phi(x, y) &- \sum_{k=1}^{Q} p_k(x) \Phi(\xi_k, y) - \sum_{\ell=1}^{Q} p_\ell(y) \Phi(x, \xi_\ell) \\ &+ \sum_{\ell=1}^{Q} \sum_{k=1}^{Q} p_k(x) p_\ell(y) \Phi(\xi_k, \xi_\ell) + \sum_{\ell=1}^{Q} p_\ell(x) p_\ell(y). \end{aligned}$$

Proposition 6. *With this kernel and this inner product the native space $\mathcal{H}$ becomes a RKHS with kernel K.*

Hence, in many situations one can reduce the conditionally positive definite case to the positive definite one.

2.3. Error estimates and condition numbers

To understand the numerical behavior of the interpolant or approximant it is essential to have bounds on the approximation error and on the condition number of the interpolation matrix. These bounds are usually expressed employing two different geometric measures. For the approximation error, it is crucial to know how well the data sites X fill the region Ω. This can be measured by the ***fill distance***

$$h_{X,\Omega} := \sup_{x \in \Omega} \min_{x_j \in X} \|x - x_j\|_2,$$

which gives the radius of the largest "data-site free ball" in Ω. The condition number, however, will obviously only depend on the data sites X and not on the region Ω. Moreover, if two data sites tend to coalesce then the corresponding interpolation matrix has two rows which are almost identical. Hence, it is reasonable to measure the condition number in terms of the ***separation distance***

$$q_X = \frac{1}{2} \min_{j \neq k} \|x_j - x_k\|_2.$$

In the case of interpolation, i.e., for the solution of (1), interpolation error and condition number are well investigated. See for example [18–22,27–29,32,33,37,40] for error estimates and [2,3,13,23–26,32,38]) for investigations on the condition number. Both lists are far from being complete.

In our situation, when working in $H^\tau(\Omega)$ or $\mathrm{BL}_\tau(\Omega)$, respectively, error estimates for the solutions $s_0 = I_X f$ of (1) are of the form

$$\|D^\alpha(f - I_X f)\|_{L_p(\Omega)} \le C h_{X,\Omega}^{\tau-|\alpha|-d(1/2-1/p)_+} \|f\|_{H^\tau(\Omega)}, \tag{4}$$

provided that $f \in H^\tau(\Omega)$, Ω is a bounded Lipschitz domain, and $\tau > |\alpha| + d/2$.

The condition number of the interpolation process is mainly determined by the smallest eigenvalue of the matrix A, i.e., by

$$\lambda_{\min}(A) = \inf\left\{\frac{\alpha^T A\alpha}{\alpha^T\alpha} : \alpha \in \mathbb{R}^N \setminus \{0\} \text{ with } P^T\alpha = 0\right\},$$

where the condition $P^T\alpha = 0$ can be neglected in the case of positive definite kernels. The asymptotic behavior of $\lambda_{\min}(A)$ has been studied extensively over the past years. It is now well known that $\lambda_{\min}(A)$ behaves like

$$\lambda_{\min}(A) \ge C q_X^{2\tau-d}. \tag{5}$$

The interpolation error (4) and the eigenvalue estimate (5) indicate that interpolation with smooth kernels will lead to a good approximation behavior at the price of an ill-conditioned problem.

In the case of the smoothing spline solution of (2) there exist also estimates on the approximation error and on the condition number, but we will delay this discussion until later.

2.4. Choosing an appropriate kernel

By now it should be apparent, that the knowledge on the assumed smoothness of the unknown target function f leads automatically to a specific interpolation problem and hence to a specific kernel. The user should use this smoothness information, if at hand, to set up the correct interpolation problem.

If the precise smoothness of the target function is unknown, it is better to overestimate the smoothness, as long as no ill-conditioning problems arise, and to work with a kernel that might be smoother than necessary.

It can be shown, that the resulting norm-minimal interpolant still produces optimal approximation rates, which are now determined by the rougher smoothness of f, as long as the data sets are quasi-uniform (see [14,27,30]).

3. Reducing the Complexity

The naive approach to solve the linear systems associated to the interpolation or smoothing problems leads to a complexity of $\mathcal{O}(N^3)$ time and $\mathcal{O}(N^2)$ space. Furthermore, every evaluation needs another $\mathcal{O}(N)$ time. For a large number N of data sites this is unacceptable. Thus more efficient methods are necessary. We will review some of the most promising approaches in this section.

3.1. The multipole expansion

In the case of a globally supported basis function and a large number of data sites it is impossible to use direct methods for solving the resulting linear equations. Instead, iterative methods have to be employed. For the time being, we do not want to discuss a specific iterative method here. It is only important to realize that the main operation in such an iterative method is a matrix by vector multiplication which takes generally $\mathcal{O}(N^2)$ operations. In our situation, the matrix vector multiplication reduces mainly to the evaluation of N sums of the form

$$s(x) = \sum_{j=1}^{N} \alpha_j \Phi(x, x_j) \tag{6}$$

neglecting the low order polynomial in the case of conditionally positive definite kernels.

Hence, it is crucial to find efficient algorithms that are able to perform this evaluation in less than $\mathcal{O}(N)$ time, preferably in $\mathcal{O}(\log N)$ or even constant time. Obviously we cannot achieve this goal if we want to reproduce the exact value of s at x. But since s is already an approximation to an unknown function, an additional error might be acceptable. Hence, we only try to approximate s up to a certain accuracy $\epsilon > 0$.

In the following, we will call t in $\Phi(x,t)$ a *source point* and x an *evaluation point*. The idea of *multipole expansions* is based on a *far field expansion* of Φ. Suppose all source points are situated in a certain region, also called *panel*, which is centered at a point t_0. Suppose further that we want to evaluate the function (6) at a point x that is sufficiently far away from the source panel. If we can expand Φ in the form

$$\Phi(x,t) = \sum_{k=1}^{p} \phi_k(x)\psi_k(t) + R(x,t) \tag{7}$$

with a remainder R that tends to zero for $\|x - t_0\|_2 \to \infty$ or for $p \to \infty$ if $\|x - t_0\|_2$ is sufficiently large, then we call (7) a *far field expansion* for Φ around the source t_0. Using (7) to evaluate (6) yields

$$\begin{aligned}
s(x) &= \sum_{j=1}^{N} \alpha_j \Phi(x, x_j) \\
&= \sum_{j=1}^{N} \alpha_j \sum_{k=1}^{p} \phi_k(x)\psi_k(x_j) + \sum_{j=1}^{N} \alpha_j R(x, x_j) \\
&= \sum_{k=1}^{p} \phi_k(x) \sum_{j=1}^{N} \alpha_j \psi_k(x_j) + \sum_{j=1}^{N} \alpha_j R(x, x_j) \\
&=: \sum_{k=1}^{p} \beta_k \phi_k(x) + \sum_{j=1}^{N} \alpha_j R(x, x_j).
\end{aligned}$$

Hence, if we use the approximation $\widetilde{s}(x) = \sum_{k=1}^{p} \beta_k \phi_k(x)$ we have an error bound

$$|s(x) - \widetilde{s}(x)| \le \|\alpha\|_1 \max_{1 \le j \le N} |R(x, x_j)|,$$

which is small if x is far enough away from the sources x_j. Moreover, each coefficient β_k can be computed in advance in linear time. Thus, if p is much smaller than N, we can consider it as constant and we need therefore $\mathcal{O}(N)$ time to compute the coefficients $\{\beta_k\}$ and constant time for each evaluation of $\widetilde{s}$.

One could say that we have averaged the information given at the sources x_j to one information given at the center t_0 of the panel. In this sense, (7) is a unipole expansion of Φ. Note that in the case of translation invariant kernels it is easy to get the far field expansion of Φ around any t_0 if the far field expansion around 0 is known. To see this, suppose (7) is the far field expansion of $\Phi(x,t) = \Phi(x-t)$ around zero. Then we can write

$$\Phi(x-t) = \Phi((x-t_0) - (t-t_0)) = \sum_{k=1}^{p} \phi_k(x-t_0)\psi_k(t-t_0) + R(x-t_0, t-t_0)$$

to derive the far field expansion around t_0.

In general, the evaluation points are close to at least a few of the data sites. To cope with this situation, we have to refine our approach. The solution is a hierarchical subdivision of the region Ω of interest into panels or cells of sources. This is generally achieved using tree-based algorithms. Let us explain the idea by an example. Suppose all data points are contained in the interval $[0,1]$ and suppose we hierarchically divide this interval into panels as shown in Figure 1.

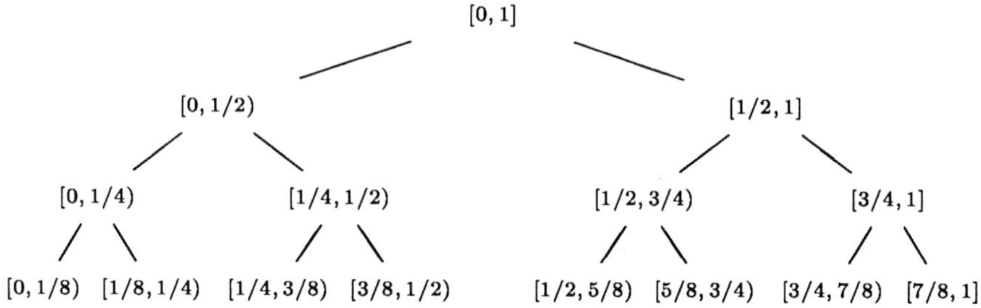

Fig. 1. Hierarchical decomposition of $[0,1]$.

With every source panel T we associate the part of the function s that corresponds to the sources in that panel by setting

$$s_T = \sum_{x_j \in T} \alpha_j \Phi(\cdot, x_j).$$

Moreover, we also assign the far field expansion $\widetilde{s}_T$ of s_T to the panel T.

Then, the approximation $\widetilde{s}$ to the function s at evaluation point x is computed by adding the associated functions s_T for the panel that contains x itself and all neighboring panels. Those panels that are well-separated from x contribute only by their far field expansion $\widetilde{s}_T$ to $\widetilde{s}$. Here, we say that a point x is *well-separated* from a panel T if it has at least the distance diam(T) from T. A panel U is called well-separated from a panel T if all points in U are well-separated from T. Since we want to save floating point operations we always use the largest possible source panel for the far field series.

Let us return to the example given in Figure 1. If we want to approximate $s(x)$ for x in the panel $[0, 1/8)$ we form

$$\widetilde{s}(x) = s_{[0,\frac{1}{8})}(x) + s_{[\frac{1}{8},\frac{1}{4})}(x) + \widetilde{s}_{[\frac{1}{4},\frac{3}{8})}(x) + \widetilde{s}_{[\frac{3}{8},\frac{1}{2})}(x) + \widetilde{s}_{[\frac{1}{2},\frac{3}{4})}(x) + \widetilde{s}_{[\frac{3}{4},1]}(x).$$

Note, that we use the two level 2 approximants $\widetilde{s}_{[\frac{1}{2},\frac{3}{4})}$ and $\widetilde{s}_{[\frac{3}{4},1]}(x)$ instead of the four level 3 approximants $\widetilde{s}_{[\frac{1}{2},\frac{5}{8})}, \dots, \widetilde{s}_{[\frac{7}{8},1]}$. This halves the computational complexity in this case. We can do this because the panels $[1/2, 3/4)$ and $[3/4, 1]$ are well-separated from $[0, 1/4)$, the panel on the same level that contains x. On the other hand, we could not use the approximant $\widetilde{s}_{[\frac{1}{2},1]}$ because its panel $[1/2, 1]$ is not well-separated from the panel $[0, 1/2]$, which is the panel on the same level that contains x.

Similarly, to approximately evaluate $s(x)$ in the panel $[\frac{3}{8}, \frac{1}{2})$ we would use

$$\begin{aligned}\widetilde{s}(x) = s_{[\frac{3}{8},\frac{1}{2})}(x) + s_{[\frac{1}{4},\frac{3}{8})}(x) + s_{[\frac{1}{2},\frac{5}{8})}(x) \\ + \widetilde{s}_{[0,\frac{1}{8})}(x) + \widetilde{s}_{[\frac{1}{8},\frac{1}{4})}(x) + \widetilde{s}_{[\frac{5}{8},\frac{3}{4})}(x) + \widetilde{s}_{[\frac{3}{4},1]}(x).\end{aligned}$$

This finishes the description of the general idea of the multipole technique to speed up the evaluation of sums like (6). What still has to be done is to find a far field expansion for the underlying kernel. Since this, in general, has to be done for every kernel separately, it would go beyond the scope of this text. Hence, we refer the reader to the constructions, for example, in [4–9,11,12,31].

3.2. The domain decomposition method

Now, we want to discuss one particular iterative method for interpolation, which has been introduced in [10]. The idea of this method is to subdivide the original data set into several smaller data sets and to iteratively solve the interpolation equations and to form residuals. To be more precise, let us decompose X in subsets $X_1, \dots, X_k$. These subsets need not be disjoint but their union must be X. Then the algorithm starts to interpolate on the first set X_1, forms the residual, interpolates this on X_2 and so on. After k steps one cycle of the algorithm is complete and it starts over again. A more formal description is

(i) Set $f_0 = f$, $s_0 = 0$.

(ii) For $n = 0, 1, 2, \dots$

For $r = 1, \dots, k$

$$\begin{aligned} f_{nk+r} &= f_{nk+r-1} - I_{X_r} f_{nk+r-1} \\ s_{nk+r} &= s_{nk+r-1} + I_{X_r} f_{nk+r-1} \end{aligned}$$

If $\|f_{(n+1)k}\|_{L_\infty(X)} < \epsilon$ stop.

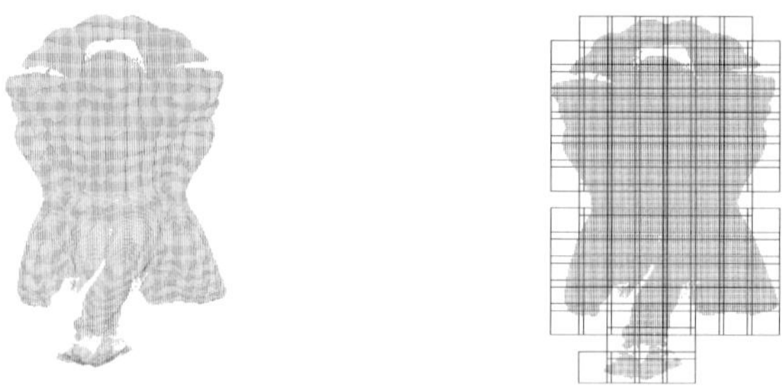

Fig. 2. The test data set consisting of 23574 points (left) and the decomposition (right).

This algorithm approximates the interpolant $I_X f$ up to the specified accuracy. The convergence result is based upon the fact that the interpolant $s_0 = I_X f$ is also the *best approximant* to f from the subspace

$$V_X := \left\{ \sum_{j=1}^{N} \alpha_j \Phi(x, x_j) : P^T \alpha = 0 \right\} + \mathcal{P}.$$

Convergence is achieved under very mild assumptions on the decomposition. The data sets X_j have to be *weakly disjoint* meaning that $X_j \neq Y_j$ and $Y_{j+1} \neq Y_j$ for each $1 \le j \le k-1$, where $Y_j = \cup_{i=j}^{k} X_i$, $1 \le j \le k$. This is, for example, satisfied, if each X_j contains at least one data site, which is not contained in any other X_i.

Theorem 7. *Let $f \in \mathcal{H}$ be given. Suppose $X_1, \dots, X_k$ are weakly distinct subsets of $\Omega \subseteq \mathbb{R}^d$. Set $Y_j = \cup_{i=j}^{k} X_i$, $1 \le j \le k$. Denote with $s^{(j)}$ the approximant after j completed cycles. Then there exists a constant $c \in (0,1)$ so that*

$$\|s_{f,Y_1} - s^{(n)}\|_{\mathcal{H}} \le c^n \|f\|_{\mathcal{H}}.$$

For a proof of this theorem and for a more thorough discussion on how the subsets X_j have to be chosen we refer the reader to [10,38].

We want to demonstrate the behavior of this method as well as that of the ones still coming. To this end we applied all methods to the data set represented in the left part of Figure 2.

This data set consists of a moderate number of points; most of them are arranged in a grid like manner. However, as one can clearly see the data set contains regions without points and the reader should compare the behavior of the numerical methods discussed in this paper in particular in these regions.

The results for the alternating-projection algorithm are shown in Figure 3. The point sets X_k have been chosen as axes parallel boxes containing at most 300 points, see the right part of Figure 2. The choice is not optimal and can be improved by an adaptive domain decomposition algorithm as we will employ it in the next subsection.

The first picture from the left in Figure 3 shows the reconstruction after one half of the first outer cycle. The second and the third picture show the reconstruction after one completed outer cycle, viewed from different perspectives. While the reconstruction seems accurate from above, the side view reveals that the solution is not yet decent. The final two pictures present the reconstruction after 88 cycles where the residual has dropped below 10^{-6}.

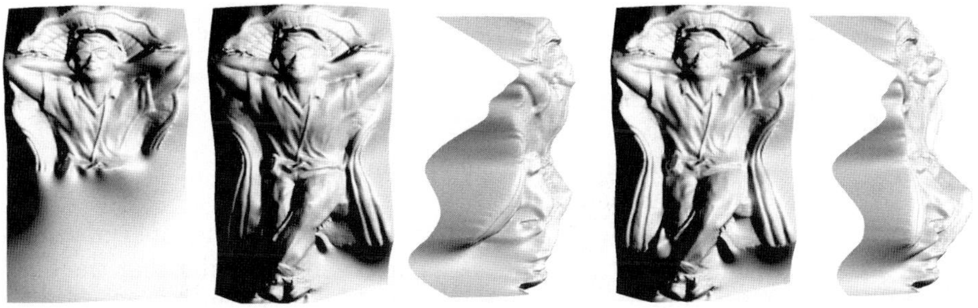

Fig. 3. Alternating-projection reconstruction.

3.3. The partition of unity idea

The idea of a partition of unity method is comparable to the idea of the domain decomposition method. In the context of radial basis function interpolation, a partition of unity approach has first been introduced in [15,16].

This time not only the set of data sites is decomposed but the entire region Ω, i.e. we choose an overlapping covering $\{\Omega_j\}_{j=1}^M$ of Ω. Along with this covering we need a partition of unity, i.e. a family of compactly supported, nonnegative, continuous functions $\{w_j\}$ with $\mathrm{supp}(w_j) \subseteq \Omega_j$ and

$$\sum_{j=1}^{M} w_j(x) = 1, \qquad x \in \Omega.$$

Moreover, we choose for every cell Ω_j an approximation space V_j. Then, a function f is approximated on each cell by a local approximant $s_j \in V_j$ and the global approximant is formed by the weighted sum of local approximants:

$$s_f = \sum_{j=1}^{M} s_j w_j. \tag{8}$$

Note that this process respects an interpolation property, i.e. if all s_j are interpolants at $X \cap \Omega_j$ then s_f is an interpolant at the entire set X.

Moreover, even in this general situation, the simple estimate

$$\begin{aligned}|f(x) - s_f(x)| &= \left|\sum_{j=1}^{M}[f(x) - s_j(x)]w_j(x)\right| \\ &\leq \sum_{j=1}^{M} |f(x) - s_j(x)|w_j(x) \leq \max_{1\leq j\leq M} \|f - s_j\|_{L_\infty(\Omega_j)}\end{aligned}$$

shows, that the global approximation error is governed by the worst local error. In other words, if the local approximants provide good approximations so will the global one. In our situation, the local approximants are formed as solutions of (1) or (2) where the global data set is replaced by the local data sets $X_j := \Omega_j \cap X$.

The complexity of this approach is governed by the following assumptions.

(i) We need a data structure for our point set such that we can find the data sites in each cell efficiently.

(ii) We need a cell structure such that

- each cell contains only a small number of points,
- each $x \in \Omega$ is contained only in a small number of cells,
- these cells can be found efficiently.

Note that these assumptions lead to the requirement that the number of cells M is proportional to the number N of points. This means in particular that the $\mathcal{O}(N)$ interpolation/approximation problems can be solved in $\mathcal{O}(N)$ time.

Again, tree-like decompositions lead to good data structures in both cases and, employing such data structures, one can show that the operations can be performed in $\mathcal{O}(\log N)$ time with an additional preprocessing step to build the data structure in $\mathcal{O}(N \log N)$ time.

The results for our test data sets are shown in the left part of Figure 4. The right part of that figure indicates how the overlapping boxes are chosen this time.

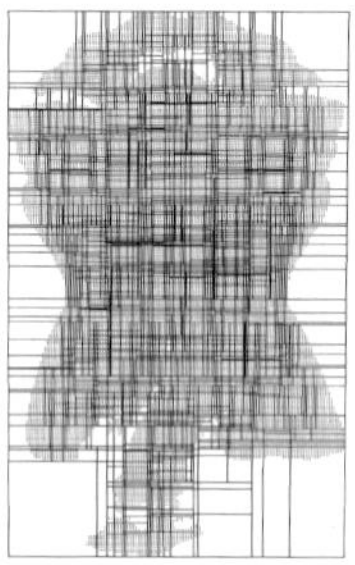

Fig. 4. Partition-of-unity reconstruction.

3.4. Multilevel and compactly supported RBFs

The numerical methods discussed so far are tailored for globally supported radial basis functions. In particular, the domain decomposition method necessarily needs a multipole expansion for an efficient implementation. Even if compactly supported basis functions can be used as globally supported ones, this does not take the local character of these functions into account. Hence, it is time to discuss ideas for efficiently using compactly supported functions.

One possibility is to adjust the support radius as a function of the data density. Instead of using the function Φ one better employs the function $\Phi_\delta = \Phi(\cdot/\delta)$ and chooses δ as a function of the fill distance $h_{X,\Omega}$. The choice $\delta = ch_{X,\Omega}$ with a sufficiently small $c > 0$ will always lead to a sparse matrix. Moreover, since the interpolation matrix using Φ_δ and X coincides with the interpolation matrix using Φ and X/δ, we know from (5) that our scaled problem has a smallest eigenvalue, which behaves like

$$\lambda_{\min} \geq C\left(\frac{q_X}{\delta}\right)^{2\tau-d}.$$

Hence, in the case of quasi-uniform data, the choice $\delta = ch_{X,\Omega}$ leads to a stable interpolation problem.

Unfortunately, it is also possible to show that the error scales in a similar way. To see this, we cannot directly use (4), since the constant there includes a norm equivalence constant, which now depends also on δ. Nonetheless, it is possible to show the following result:

Theorem 8. *Let $\Omega \subseteq \mathbb{R}^d$ be a bounded Lipschitz domain satisfying an interior cone condition. If Φ is the reproducing kernel of $H^\tau(\Omega)$, $\tau = k + \sigma$ with $k > d/2$ and $0 < \sigma \leq 1$, then interpolation with the scaled kernel Φ_δ, for $0 < \delta \leq 1$, leads to*

$$\|f - I_{X,\Phi_\delta} f\|_{L_2(\Omega)} \leq C\left(\frac{h_{X,\Omega}}{\delta}\right)^\tau \|f\|_{H^\tau(\Omega)},$$

with a constant $C > 0$ independent of $h_{X,\Omega}$, δ, and f.

Proof. Since the function $u := f - I_{X,\Phi_\delta} f \in W_2^\tau(\Omega)$ vanishes on X, a general result from [29] gives

$$\|f - I_{X,\Phi_\delta} f\|_{L_2(\Omega)} \leq C h_{X,\Omega}^\tau |f - I_{X,\Phi_\delta} f|_{H^\tau(\Omega)}$$

It remains to show that the expression $|f - I_{X,\Phi_\delta} f|_{H^\tau(\Omega)}$ can be bounded by a constant times $\delta^{-\tau}\|f\|_{H^\tau(\Omega)}$. To this end, we assume that the interpolant is given by

$$s := I_{X,\Phi_\delta} f = \sum_{j=1}^N c_j \Phi((\cdot - x_j)/\delta).$$

Moreover, since Ω is a Lipschitz domain, we can extend $f \in H^\tau(\Omega)$ continuously to a function from $H^\tau(\mathbb{R}^d)$. Next, the fractional semi-norm can, up to a constant, be expressed as

$$|s|_{H^\tau(\Omega)}^2 = \sum_{|\alpha|=k} \int_\Omega \int_\Omega \frac{|D^\alpha s(x) - D^\alpha s(y)|^2}{\|x - y\|_2^{d+2\sigma}} dx dy,$$

which leads, with $\widetilde{s} = \sum_j c_j \Phi(\cdot - y_j)$, $y_j = x_j/\delta$, to

$$\begin{aligned}
|s|^2_{H^\tau(\Omega)} &\le |s|^2_{H^\tau(\mathbb{R}^d)} \\
&= \delta^{-2k} \sum_{|\alpha|=k} \int_{\mathbb{R}^d}\int_{\mathbb{R}^d} \frac{|D^\alpha \widetilde{s}(x/\delta) - D^\alpha \widetilde{s}(y/\delta)|^2}{\|x-y\|_2^{d+2\sigma}} dxdy \\
&= \delta^{d-2\tau} \sum_{|\alpha|=k} \int_{\mathbb{R}^d}\int_{\mathbb{R}^d} \frac{|D^\alpha \widetilde{s}(x) - D^\alpha \widetilde{s}(y)|^2}{\|x-y\|_2^{d+2\sigma}} dxdy \\
&= \delta^{d-2\tau} |I_{X/\delta,\Phi} f(\delta\cdot)|^2_{H^\tau(\mathbb{R}^d)} \\
&\le \delta^{d-2\tau} \|f(\delta\cdot)\|_{H^\tau(\mathbb{R}^d)}
\end{aligned}$$

using the fact that $\widetilde{s} = I_{X/\delta,\Phi} f(\delta\cdot)$ is the $\|\cdot\|_{H^\tau(\mathbb{R}^d)}$-norm minimal interpolant to $f(\delta\cdot)$. Finally, the latter norm can, up to a constant, be bounded by

$$\begin{aligned}
\|f(\delta\cdot)\|^2_{H^\tau(\mathbb{R}^d)} &= \int_{\mathbb{R}^d} |\widehat{f(\delta\cdot)}(\omega)|^2 (1+\|\omega\|_2^2)^\tau d\omega \\
&= \delta^{-2d} \int_{\mathbb{R}^d} |\widehat{f}(\omega/\delta)|^2 (1+\|\omega\|_2^2)^\tau d\omega. \\
&= \delta^{-d} \int_{\mathbb{R}^d} |\widehat{f}(\omega)|^2 (1+\|\omega\|_2^2 \delta^2)^\tau d\omega \\
&\le \delta^{-d} \|f\|^2_{H^\tau(\mathbb{R}^d)}
\end{aligned}$$

provided that $\delta \le 1$. Putting things together leads to

$$|f-s|_{H^\tau(\Omega)} \le |f|_{H^\tau(\Omega)} + |s|_{H^\tau(\Omega)} \le (1+C\delta^{-\tau})\|f\|_{H^\tau(\Omega)} \le C\delta^{-\tau}\|f\|_{H^\tau(\Omega)},$$

which finally gives the stated result. □

Thus, the choice $\delta = ch_{X,\Omega}$ leads only to an acceptable error if c is sufficiently large, which stands in contrast to choosing c as small as possible. This is also reflected in the so called ***uncertainty*** or ***trade-off principle***, see [32].

However, in practical applications generally only *one* data set is given and it is often possible to find a compromise in choosing an appropriate support radius. Nonetheless, our goal now is to describe a possible approach so that the application benefits from both a stable interpolation matrix and a good approximation property.

The idea of a multilevel method is again based on a decomposition of the set of data sites X, but this time in a nested sequence of subsets,

$$X_1 \subseteq X_2 \subseteq \ldots \subseteq X_k = X. \tag{9}$$

If X is quasi-uniform meaning that q_X has comparable size to $h_{X,\Omega}$, then the subsets X_j should also be quasi-uniform. Moreover, they should satisfy $q_{X_{j+1}} \approx c_a q_{X_j}$ and $h_{X_{j+1},\Omega} \approx c_a h_{X_j,\Omega}$ with a fixed constant c_a. A good choice for c_a would be $c_a = 1/2$.

Now, the ***multilevel method***, introduced in [17], is simply one cycle of the domain decomposition method discussed in Section 3.2. But this time we use compactly supported basis functions with a different support radius at each level. We could even use different basis functions at different levels. Hence, a general formulation goes as follows. For every $1 \le j \le k$ we choose a basis function Φ_j and form the interpolant

$$I_j f := I_{X,\Phi_j} f = \sum_{x_j \in X_j} c_{x_j}(f)\Phi_j(\cdot - x_j),$$

but using now the basis function Φ_j on level j. We have in mind to take Φ_j as $\Phi(\cdot/\delta_j)$ with a compactly supported basis function Φ and scaling parameter δ_j proportional to $h_{X_j,\Omega}$. The idea behind this algorithm is that one starts with a very thin, widely spread set of points and uses a smooth basis function to recover the global behavior of the function f. In the next level a finer set of points is used and a less smooth function possibly with a smaller support is employed to resolve more details and so on.

As said before, the algorithm performs one cycle of the domain decomposition algorithm. This means

(i) set $f_0 = f$ and $s_0 = 0$.

(ii) for $1 \le j \le k$:

$s_j = s_{j-1} + I_j f_{j-1}$,

$f_j = f_{j-1} - I_j f_{j-1}$.

Even if the multilevel algorithm resembles the alternating projection algorithm, the idea behind it is completely different. The most obvious differences are that we use different basis functions at each level and that we perform only one cycle. The latter is reasonable since any further cycle would not change our interpolant because the data sets are nested.

The results of our test data sets are represented in Figure 5. The data set has been subdivided into a sequence of 5 nested subsets. The first row of Figure 5 shows the accumulated interpolants while the second row contains the level interpolants $I_j f_{j-1}$ for $2 \le j \le 5$. Note that the first image in the first row is also $I_1 f_0$. The interpretation of this sequence of images is as follows. The ith image of the first row and the ith image of the second row give together the $(i+1)$th image of the first row.

3.5. Which method to choose?

As mentioned earlier, the right choice of the basis function depends mainly upon additional information on the target function such as smoothness. The right choice of the particular reconstruction method depends also on the application. For example, if exact interpolation is necessary, any method that is based on a far field expansion is only of limited use, since such a method lives from the approximate character in the far field. On the other hand, if the data set has holes which have to be filled (e.g. for mesh repair) any purely local method is the wrong choice, even the partition of unity approach has to be handled with care here. Better suited is a global method such as domain decomposition or the multilevel method. In all other cases, it seems that, in particular for huge data sets, the partition of unity method is the fastest one.

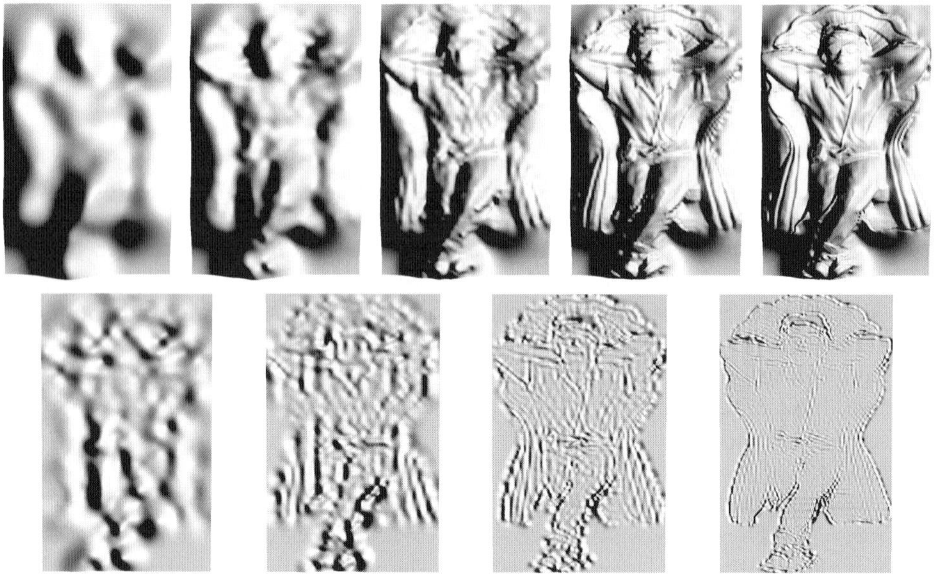

Fig. 5. Multilevel-reconstruction.

4. Stabilizing the Interpolation Process and Dealing with Noise

4.1. Error estimates and condition numbers

Our discussion of error estimates and condition numbers in Section 2.3 has in particular shown that, when working in $H^\tau(\Omega)$ or $\mathrm{BL}_\tau(\Omega)$, we are confronted with an error and a smallest eigenvalue behaving like

$$\begin{aligned} \|f - I_X f\|_{L_\infty(\Omega)} &\le C h_{X,\Omega}^{\tau-d/2} \|f\|_{W_2^\tau(\Omega)} &&=: C_f F(h_{X,\Omega}) \ , \\ \lambda_{\min}(A) &\ge c q_X^{2\tau-d} &&=: cG(q_X) \ , \end{aligned}$$

and the reader should be aware of the fact that the exponent $2\tau - d$ of q_X in the lower bound of $\lambda_{\min}$ is precisely twice the exponent of $h_{X,\Omega}$ in the error estimate, or, in other words, $F(h^2) = G(h)$, using the above introduced functions. This is indeed an intrinsic relation between error estimates and smallest eigenvalues. To be more precise, the *uncertainty relation* (cf. [32]) states roughly that

$$cG(q_X) \le \lambda_{\min}(A) \le CF(h_{X,\Omega})^2.$$

This has the unfortunate side-effect that in the case of quasi-uniform data sets, i.e. data sets where q_X and $h_{X,\Omega}$ are of comparable size, the smallest eigenvalue indeed behaves like

$$ch_{X,\Omega}^{2\tau-d} \le \lambda_{\min}(A) \le Ch_{X,\Omega}^{2\tau-d},$$

so that this is the best rate that we can expect for deriving a prescribed error of size $h_{X,\Omega}^{\tau-d/2}$.

Proposition 9. *The interpolation process derives an error of size $\mathcal{O}(h_{X,\Omega}^{\tau-d/2})$ with a smallest eigenvalue of size $\mathcal{O}(q_X^{2\tau-d})$. Moreover, the most stable interpolation process, where q_X is comparable to $h_{X,\Omega}$, still has a smallest eigenvalue of size $\mathcal{O}(h_{X,\Omega}^{2\tau-d})$.*

This proposition has direct consequences for any numerical strategy for solving the interpolation equations. In particular, we have to deal, at least, with the following two situations:

(i) The data set has a moderate sized $h_{X,\Omega}$, but a small q_X.

(ii) Both quantities $h_{X,\Omega}$ and q_X are small, but of comparable size.

If $h_{X,\Omega}$ is rather small, it might still happen that q_X is still essentially smaller than $h_{X,\Omega}$. However, the solution for the first case will show that we can avoid such a situation, and that it suffices to restrict ourselves then to the situation described in the second case.

In the first case, our point set is actually sparsely distributed over Ω with some almost identical points. Proposition 9 tells us that we achieve the accuracy at a cost that is too high, namely $\mathcal{O}(q_X^{2\tau-d})$, while a more "intelligent" chosen data set would lead to the same accuracy but with a smallest eigenvalue behaving like $\mathcal{O}(h_{X,\Omega}^{2\tau-d})$, which is acceptable, for a not too small $h_{X,\Omega}$.

One strategy to avoid the situation described in the first case is to sort out nearly identical points in a preprocessing step. This, assuming a good data structure, can be done efficiently, as long as the number of coalescing points is small when compared to the number of all points.

However, sorting out coalescing points only makes sense if the information carried by two such points is also almost identical. But if, for example, the points x_j and x_k are almost identical and if the associated function values f_j and f_k differ significantly, additional information from the user is required to specify that point that is to be sorted out. Unfortunately, such a situation can and will quite often appear in real-world applications, where the data represents measured quantities that contain noise.

Fortunately, the solution of (2) gives another possibility for stabilizing the interpolation process by regularization. We will discuss this in the next subsection, while the final subsection deals with the situation in the second case.

4.2. Regularization and stabilization by smoothing

Suppose we still work in the spaces $H^\tau(\Omega)$ or $\mathrm{BL}_\tau(\Omega)$. Let s_λ denote the solution of the smoothing problem (2). From Theorems 2 and 4 we see that s_λ can be computed by inverting a slightly modified interpolation matrix. The modification simply consists in adding a λ to the diagonal of the main part A of the interpolation matrix. This already gives the first part of the following theorem. The second part has recently been proved in [39].

Theorem 10. *Suppose the data are generated by a function $f \in H^\tau(\Omega)$ or $f \in \mathrm{BL}_\tau(\Omega)$, respectively. Let s_λ be the solution of (2). Then the smallest eigenvalue of the associated system matrices behaves like*

$$\lambda_{\min}(A+\lambda I) = \lambda_{\min}(A) + \lambda \geq C q_X^{2\tau-d} + \lambda,$$

where I denotes the $N \times N$ identity matrix. Moreover, the error between f and s_λ can be bounded by

$$\|D^\alpha(f-s_\lambda)\|_{L_p(\Omega)} \leq C\left(h_{X,\Omega}^{\tau-|\alpha|-d(1/2-1/p)_+} + h_{X,\Omega}^{-|\alpha|}\sqrt{\lambda}\right)\|f\|_{H^\tau(\Omega)},$$

provided that Ω is a bounded Lipschitz domain satisfying an interior cone condition, and $\tau = k + \sigma$ with $k > |\alpha| + d/2$.

If we look at the particular situation of pure $L_\infty(\Omega)$ error estimates we see that

$$\|f-s_\lambda\|_{L_\infty(\Omega)} \leq C\left(h_{X,\Omega}^{\tau-d/2} + \sqrt{\lambda}\right)\|f\|_{H^\tau(\Omega)}\ ,$$
$$\lambda_{\min}(A+\lambda I) \geq c q_X^{2\tau-d} + \lambda.$$

Hence, choosing $\lambda = \tilde{c} h_{X,\Omega}^{2\tau-d}$ leads to

$$\|f-s_\lambda\|_{L_\infty(\Omega)} \leq C h_{X,\Omega}^{\tau-d/2}\|f\|_{H^\tau(\Omega)}\ ,$$
$$\lambda_{\min}(A+\lambda I) \geq c q_X^{2\tau-d} + c h_X^{2\tau-d} = c h_X^{2\tau-d}.$$

Corollary 11. *If the smoothing parameter $\lambda > 0$ is determined in this way then the best possible approximation order is achieved in the most stable way, meaning that the "interpolation matrix" has a largest possible smallest eigenvalue.*

Here, we have restricted ourselves to the most important case of L_∞-error estimates. But Theorem 10 also indicates how the smoothing parameter has to be chosen if one is interested in estimates employing different L_p-norms and derivatives.

Theorem 10 can also be interpreted in a slightly different, more practical way. To guarantee stability, the user can simply add an a priori λ to the diagonal entries of A, for example something like $\lambda = 10^{-6}$. Then, Theorem 10 implies stability and an error which is of size $\mathcal{O}(10^{-3})$, which, in many applications, is sufficient. As a matter of fact, since this is a worst case upper bound, the user can, in general, expect a much better behavior.

4.3. Change of basis

So far, we have learned that smoothing is an adequate choice in the situation of highly non-uniform data sets. It also helps in the case of quasi-uniform data sets and infinitely smooth basis functions, like Gaussians and (inverse) multiquadrics, since their associated interpolation matrices are already in that particular situation for moderate separation distances highly ill-conditioned. Unfortunately there exists no theoretical coverage of error estimates in that situation, even if numerical tests (cf. [39]) show promising results.

Our final task, for basis functions of finite smoothness, is to deal with the case of really dense data sets. Due to the last section we can restrict ourselves to quasi-uniform data sets.

In the first place, the bad conditioning in this situation is due to the naturally chosen basis, namely $\Phi(\cdot, x_1), \ldots, \Phi(\cdot, x_N)$ (plus a basis for $\mathcal{P}$), provided Φ is conditionally positive definite with respect to $\mathcal{P}$. Thus, it seems to be natural to investigate and to search for better suited bases for the subspace

$$V_X := \left\{ \sum_{j=1}^{N} \alpha_j \Phi(\cdot, x_j) : P^T \alpha = 0 \right\} + \mathcal{P},$$

where $P \in \mathbb{R}^{N \times Q}$ is the usual matrix with entries $p_j(x_i)$. The ideas we are going to describe now are in particular fitted to the situation of thin-plate splines, while they might also work for other conditionally positive definite kernels. They mainly come from [10].

To introduce a different basis of V_X, we first select a $\mathcal{P}$-unisolvent subset $\Xi = \{\xi_1, \ldots, \xi_Q\}$ of $X \subset \Omega$ and a cardinal basis $p_1, \ldots, p_Q$ for $\mathcal{P}$ satisfying $p_\ell(\xi_k) = \delta_{\ell,k}$. Then we define the kernels

$$\begin{aligned} \kappa(x, y) :=& \Phi(x, y) - \sum_{k=1}^{Q} p_k(x) \Phi(\xi_k, y) - \sum_{\ell=1}^{Q} p_\ell(y) \Phi(x, \xi_\ell) \\ &+ \sum_{\ell=1}^{Q} \sum_{k=1}^{Q} p_k(x) p_\ell(y) \Phi(\xi_k, \xi_\ell) \end{aligned} \tag{10}$$

and

$$K(x, y) = \kappa(x, y) + \sum_{\ell=1}^{Q} p_\ell(x) p_\ell(y).$$

It is easily verified that the resulting functions $\kappa(\cdot, x_j)$ and $K(\cdot, x_j)$ belong to the space V_X defined above. However, κ has the additional feature $\kappa(\cdot, \xi_\ell) = 0$, $1 \le \ell \le Q$. The following results can be found in [10], see also [38].

Theorem 12. *The kernel $K : \Omega \times \Omega \to \mathbb{R}$ is positive definite on Ω. Moreover, if $\widetilde{\Omega} = \Omega \setminus \Xi$ then $\kappa : \widetilde{\Omega} \times \widetilde{\Omega} \to \mathbb{R}$ is positive definite on $\widetilde{\Omega}$.*

Thus we can restate our initial interpolation problem in two new ways. Let us start with the simpler one.

Corollary 13. *If $\Xi \subseteq X$ then the interpolant $I_X f$ can be written as*

$$I_X f = \sum_{j=1}^{N} \alpha_j K(\cdot, x_j),$$

where the coefficients are determined by $I_X f(x_j) = f_j$, $1 \le j \le N$.

When using κ we have to be more careful, since $\Xi \subseteq X$ does not lead to linearly independent functions $\kappa(\cdot, x_j)$. But we need $\Xi \subseteq X$ to ensure that we get the same interpolant. So assume that $x_j = \xi_j$ for $1 \le j \le Q$. Then we know at least that the matrix

$$C = (\kappa(x_i, x_j))_{Q+1 \le i,j \le N}$$

is positive definite. Or with other words the family $\{\kappa(\cdot, x_j) : Q+1 \le j \le N\}$ is linearly independent. Since $\kappa(\cdot, x_j) = 0$ for $1 \le j \le Q$, we can immediately conclude that $\{\kappa(\cdot, x_j) : Q+1 \le j \le N\} \cup \{p_k : 1 \le k \le Q\}$ is a basis for V_X.

Thus we can restate the interpolation problem using this basis.

Corollary 14. *If $\Xi \subseteq X$ satisfies $x_j = \xi_j$ for $1 \le j \le Q$ then the interpolant can be written as*

$$I_X f(x) = \sum_{j=1}^{Q} \beta_k p_k(x) + \sum_{j=Q+1}^{N} \alpha_j \kappa(x, x_j)$$

and the coefficients are again determined by $I_X f(x_j) = f(x_j)$, $1 \le j \le N$.

Since the $\{p_\ell\}$ form a Lagrangian basis for Ξ and since κ vanishes if one of its arguments is an element from Ξ the interpolation conditions lead to the matrix equation

$$\begin{pmatrix} I & O \\ \widetilde{P} & C \end{pmatrix} \begin{pmatrix} \beta \\ \alpha \end{pmatrix} = f|X$$

with the $Q \times Q$ identity matrix I and the $(N-Q) \times Q$ matrix $\widetilde{P} = (p_j(x_i))$ where i runs over the last indices starting with $i = Q+1$. The first Q rows in this system indicate $\beta = (f(x_1), \dots, f(x_Q))^T$, so that we can reduce the system to solving

$$C\alpha = \widetilde{f} - \widetilde{P}\beta$$

where $\widetilde{f} = (f(x_{Q+1}), \dots, f(x_N))$.

Hence, we now have three possibilities to solve the interpolation equations. We can either use the conventional approach using the $(N+Q) \times (N+Q)$-interpolation matrix $\widetilde{A}$ from Theorem 4, or one of our new approaches.

Table 1 shows the results for all three approaches using a regular grid of 25 points with grid size h. This reflects in particular the situation in the partition of unity approach, where the number of data sites is small but the fill and separation distance can become arbitrarily small. The results show that both new methods are

Table 1. Condition numbers for a fixed number of centers.

Fill distance h	Conventional matrix $\widetilde{A}$	Reproducing kernel matrix K	Homogeneous matrix C
0.001	2.4349×10^8	8.4635×10^8	5.4938×10^3
0.01	2.4364×10^6	8.4640×10^6	5.4938×10^3
0.1	2.5179×10^4	8.5134×10^4	5.4938×10^3
1.0	3.6458×10^2	1.3660×10^3	5.4938×10^3
10	1.8742×10^6	1.2609×10^3	5.4938×10^3
100	1.1520×10^{11}	1.1396×10^5	5.4938×10^3
1000	3.5478×10^{15}	$1.1386e \times 10^7$	5.4938×10^3

superior to the conventional one, and that the system involving the matrix C even leads to a condition number which is independent of the parameter h.

The theoretical background for such a behavior is given in the next theorem. It states that for a specific class of basis functions, which includes thin-plate splines, the associated kernel κ is invariant under scaling. This means, in particular, that the eigenvalues of the matrix C scale in the same way showing that the condition number of C is independent of the scaling parameter.

Theorem 15. *Suppose the symmetric kernel $\Phi \in C(\mathbb{R}^d \times \mathbb{R}^d)$ satisfies $\Phi(hx, hy) = h^\lambda \Phi(x, y) + q_h(x-y)$ for all $h > 0$ and $x, y \in \mathbb{R}^d$, where $\lambda \in \mathbb{R}$ and $q_h \in \pi_{2m-1}(\mathbb{R}^d)$. Let $\Xi = \{\xi_1, \ldots, \xi_Q\}$ be unisolvent for $\pi_{m-1}(\mathbb{R}^d)$ with associated Lagrange basis $p_1, \ldots p_Q$. Let κ be the kernel (10) and κ^h for $h > 0$ be the kernel κ for the set $h\Xi = \{h\xi_1, \ldots, h\xi_Q\}$ and the Lagrange functions $p_1^h, \ldots, p_Q^h$ associated to this set. Then $\kappa^h(hx, hy) = h^\lambda \kappa(x, y)$ for all $x, y \in \mathbb{R}$.*

References

1. Aronszajn, N., Theory of reproducing kernels, *Trans. Amer. Math. Soc.* **68** (1950), 337–404.
2. Ball, K., Eigenvalues of Euclidean distance matrices, *J. Approx. Theory* **68** (1992), 74–82.
3. Ball, K., Sivakumar, N. and Ward, J. D., On the sensitivity of radial basis interpolation to minimal data separation distance, *Constr. Approximation* **8** (1992), 401–426.
4. Beatson, R. K. and Chacko, E., Fast evaluation of radial basis functions: A multivariate momentary evaluation scheme, in: *Curve and Surface Fitting*, Proc. Saint-Malo 1999 (A. Cohen, C. Rabut and L. L. Schumaker, Eds.), Vanderbilt University Press, Nashville, 2000, pp. 37–46.
5. Beatson, R. K., Cherrie, J. B. and Ragozin, D. L., Polyharmonic splines in $\mathbb{R}^d$: Tools for fast evaluation, in: *Curve and Surface Fitting*, Proc. Saint-Malo 1999

(A. Cohen, C. Rabut and L. L. Schumaker, Eds.), Vanderbilt University Press, Nashville, 2000, pp. 47–56.

6. Beatson, R. K., Cherrie, J. B. and Ragozin, D. L. Fast evaluation of radial basis functions: Methods for four-dimensional polyharmonic splines, *SIAM J. Math. Anal.* **32** (2001), 1272–1310.
7. Beatson, R. K., Goodsell, G. and Powell, M. J. D., On multigrid techniques for thin plate spline interpolation in two dimensions, *Lect. Appl. Math.* **32** (1996), 77–97.
8. Beatson, R. K. and Greengard, L., A short course on fast multipole methods, in: *Wavelets, Multilevel Methods and Elliptic PDEs*, 7th EPSRC Numerical Analysis Summer School, University of Leicester 1996 (M. Ainsworth, J. Levesley, W. Light and M. Marletta, Eds.), Clarendon Press, Oxford, 1997, pp. 1–37.
9. Beatson, R. K. and Light, W. A., Fast evaluation of radial basis functions: Methods for two-dimensional polyharmonic splines, *IMA J. Numer. Anal.* **17** (1997), 343–372.
10. Beatson, R. K., Light, W. A. and Billings, S., Fast solution of the radial basis function interpolation equations: Domain decomposition methods, *SIAM J. Sci. Comput.* **22** (2000), 1717–1740.
11. Beatson, R. K. and Newsam, G. N., Fast evaluation of radial basis functions: I, *Comput. Math. Appl.* **24** (1992), 7–19.
12. Beatson, R. K. and Newsam, G. N., Fast evaluation of radial basis functions: Moment-based methods, *SIAM J. Sci. Comput.* **19** (1998), 1428–1449.
13. Binev, P. and Jetter, K., Estimating the condition number for multivariate interpolation problems, in: *Numerical Methods in Approximation Theory, Vol. 9* (D. Braess and L. L. Schumaker, Eds.), ISNM, Vol. 105, Birkhäuser, Basel, 1992, pp. 41–52.
14. Brownlee, R. and Light, W., Approximation orders for interpolation by surface splines to rough functions, *IMA J. Numer. Anal.* **24** (2002), 179–192.
15. Franke, R., Locally determined smooth interpolation at irregularly spaced points in several variables, *J. Inst. Math. Applications* **19** (1977), 471–482.
16. Franke, R., Smooth interpolation of scattered data by local thin plate splines, *Comput. Math. Appl.* **8** (1982), 273–281.
17. Floater, M. S. and Iske, A., Multistep scattered data interpolation using compactly supported radial basis functions, *J. Comput. Appl. Math.* **73** (1996), 65–78.
18. Light, W. A. and Wayne, H., Error estimates for approximation by radial basis functions, in: *Wavelet Analysis and Approximation Theory* (S. P. Singh and A. Carbone, Eds.), Kluver Academic Publ., Dordrecht, 1995, pp. 215–246.
19. Light, W. A. and Wayne, H., On power functions and error estimates for radial basis function interpolation, *J. Approx. Theory* **92** (1998), 245–266.
20. Madych, W. R., Error estimates for interpolation by generalized splines, in: *Curves and Surfaces* (P.-J. Laurent, A. Le Méhauté and L. L. Schumaker, Eds.), Academic Press, Boston, 1991, pp. 297–306.

21. Madych, W. R. and Nelson, S. A., Error bounds for multiquadric interpolation, in: *Approximation Theory VI, Vol. 2* (C. K. Chui, L. L. Schumaker and J. D. Ward, Eds.), Academic Press, New York, 1989, pp. 413–416.
22. Madych, W. R. and Nelson, S. A., Bounds on multivariate polynomials and exponential error estimates for multiquadric interpolation, *J. Approx. Theory* **70** (1992), 94–114.
23. Narcowich, F. J. and Ward, J. D., Norms of inverses and condition numbers for matrices associated with scattered data, *J. Approx. Theory* **64** (1991), 69–94.
24. Narcowich, F. J. and Ward, J. D., Norms of inverses for matrices associated with scattered data, in: *Curves and Surfaces* (P.-J. Laurent, A. Le Méhauté and L. L. Schumaker, Eds.), Academic Press, Boston, 1991, pp. 341–348.
25. Narcowich, F. J. and Ward, J. D., Norm estimates for the inverse of a general class of scattered-data radial-function interpolation matrices, *J. Approx. Theory* **69** (1992), 84–109.
26. Narcowich, F. J. and Ward, J. D., On condition numbers associated with radial-function interpolation, *J. Math. Anal. Appl.* **186** (1994), 457–485.
27. Narcowich, F. J. and Ward, J. D., Scattered-data interpolation on $\mathbb{R}^n$: Error estimates for radial basis and band-limited functions, *SIAM J. Math. Anal.* **36** (2004), 284–300.
28. Narcowich, F. J., Ward, J. D. and Wendland, H., Refined error estimates for radial basis function interpolation, *Constr. Approximation* **19** (2003), 541–564.
29. Narcowich, F. J., Ward, J. D. and Wendland, H., Sobolev bounds on functions with scattered zeros, with applications to radial basis function surface fitting, *Math. Comp.* **74** (2005), 743–763.
30. Narcowich, F. J., Ward, J. D. and Wendland, H., Sobolev error estimates and a Bernstein inequality for scattered data interpolation via radial basis functions, preprint (2004).
31. Roussos, G., *Computation with Radial Basis Functions*, PhD thesis, Imperial College of Science Technology and Medicine, University of London, 1999.
32. Schaback, R., Error estimates and condition number for radial basis function interpolation, *Adv. Comput. Math.* **3** (1995), 251–264.
33. Schaback, R., Improved error bounds for scattered data interpolation by radial basis functions, *Math. Comp.* **68** (1999), 201–216.
34. Schaback, R., Native Hilbert spaces for radial basis functions I, in: *New Developments in Approximation Theory*, Proc. IDoMAT 1998 (M. D. Buhmann, M. Felten, D. H. Mache and M. W. Müller, Eds.), ISNM, Vol. 132, Birkhäuser, Basel, 1999, pp. 255–282.
35. Schaback, R., A unified theory of radial basis functions: Native Hilbert spaces for radial basis functions II, *J. Comput. Appl. Math.* **121** (2000), 165–177.
36. Wahba, G., *Spline Models for Observational Data*, CBMS-NSF Reg. Conf. Series in Applied Mathematics, Vol. 59, Siam, Philadelphia, 1990.
37. Wendland, H., Error estimates for interpolation by compactly supported radial basis functions of minimal degree, *J. Approx. Theory* **93** (1998), 258–272.
38. Wendland, H., *Scattered Data Approximation*, Cambridge Monographs on Applied and Computational Mathematics, Cambridge University Press,

Cambridge, 2005.

39. Wendland, H. and Rieger, C., Approximate interpolation with applications to selecting smoothing parameters, report, Universität Göttingen (2004).
40. Wu, Z. and Schaback, R., Local error estimates for radial basis function interpolation of scattered data, *IMA J. Numer. Anal.* **13** (1993), 13–27.

K. Jetter et al., Editors

Learning Theory: From Regression to Classification

Qiang Wu, Yiming Ying, Ding-Xuan Zhou [1]

Department of Mathematics, City University of Hong Kong,
83 Tat Chee Avenue, Kowloon, Hong Kong, P. R. China

Abstract

We give a brief survey of regularization schemes in learning theory for the purposes of regression and classification, from an approximation theory point of view. First, the classical method of empirical risk minimization is reviewed for regression with a general convex loss function. Next, we explain ideas and methods for the error analysis of regression algorithms generated by Tikhonov regularization schemes associated with reproducing kernel Hilbert spaces. Then binary classification algorithms given by regularization schemes are described with emphasis on support vector machines and noise conditions for distributions. Finally, we mention further topics and some open problems in learning theory.

Key words: Learning theory, regression, classification, regularization scheme, reproducing kernel Hilbert space, error analysis

2000 MSC: 68T05, 62J02

Email addresses: 50004455@student.cityu.edu.hk (Qiang Wu),
ymying@cityu.edu.hk (Yiming Ying),
mazhou@math.cityu.edu.hk (Ding-Xuan Zhou).

[1] Corresponding author

1. Introduction

Learning theory started with the classical work on statistical learning theory and support vector machines [75]. It studies learning of function relations from samples and provides powerful learning algorithms for the purpose of regression, classification, feature selection, and many other applications [2,26,76]. These learning algorithms are mostly reduced to well studied convex optimization procedures and are very effective for practical problems of large sample size and huge dimensions [31,51,59]. One example is support vector machine algorithms which are implemented by convex quadratic programming optimizations and can handle large size samples [75,20]. Compared with the vast applications of learning algorithms due to fast developed computing tools, the theoretical analysis of machine learning procedures is still in its starting stage. The purpose of this paper is to give a brief survey on recent developments of learning theory including regression and classification, from an approximation theory point of view.

2. Empirical Risk Minimization for Regression

The main idea of the classical statistical learning theory is Empirical Risk Minimization (ERM). We shall describe this method for regression [62,31,21].

The goal of the regression problem is to find a desired function $f^* : X \to Y$ or its approximation from samples. Here X is a complete metric space (e.g. a subset of $\mathbb{R}^n$) and $Y = \mathbb{R}$. The sample set $\mathbf{z} = \{z_i\}_{i=1}^m = \{(x_i, y_i)\}_{i=1}^m$ is produced by the desired (unknown) function f^*. To allow noise, we would not expect the sample function value y_i to be exactly $f^*(x_i)$, but its approximation $y_i \approx f^*(x_i)$. The most adopted model in learning theory is to assume that the samples are drawn according to a probability distribution ρ on the product space $Z := X \times Y$, and that the desired function f^* is its regression function f_ρ. If we denote ρ_X as the marginal distribution of ρ on X and $\rho(\cdot|x)$ its conditional distribution at $x \in X$, then the *regression function* is defined as

$$f_\rho(x) = \int_Y y d\rho(y|x), \qquad x \in X.$$

That is, for a fixed point $x \in X$, the conditional probability distribution $\rho(\cdot|x)$ on $Y = \mathbb{R}$ reflects the noise while its mean is exactly the value of the desired function at x: $f^*(x) = f_\rho(x)$.

To learn the function f_ρ, we draw the set $\mathbf{z} = \{(x_i, y_i)\}_{i=1}^m$ of m random samples according to ρ. Various learning algorithms can be applied to the samples and produce approximations of the regression function. ERM forms a rich family of learning algorithms.

The idea of ERM learning algorithms is to minimize empirical errors with respect to a loss function ψ. The loss function measures how the sample value y approximates the function value $f(x)$ by evaluating $\psi(y - f(x))$.

Definition 1. We say that $\psi : \mathbb{R} \to \mathbb{R}_+$ is a *regressing loss function* if it is even, convex, and continuous with $\psi(0) = 0$.

For $(x, y) \in Z$, the value $\psi(y - f(x))$ is the local error suffered from the use of f as a model for the process producing y at x. The condition $\psi(0) = 0$ ensures a zero error when $y = f(x)$. Examples of loss functions for regression include the least-square loss and Vapnik's ϵ-insensitive norm.

Example 2. The least-square loss corresponds to the loss function $\psi(t) = t^2$. For $\epsilon > 0$, the *ϵ-insensitive norm* is the loss function defined by

$$\psi(t) = \psi_\epsilon(t) = \begin{cases} |t| - \epsilon, & \text{if } |t| > \epsilon, \\ 0, & \text{otherwise.} \end{cases}$$

The *error* (generalization error) associated with the loss function ψ is defined by

$$\mathcal{E}(f) = \int_Z \psi(y - f(x))d\rho. \tag{1}$$

2.1. Example: Regression with the least-square loss

To show the basic idea of ERM, consider the least-square method. Here the error or risk $\mathcal{E}(f)$ for a function $f : X \to Y$ takes the form

$$\mathcal{E}(f) = \int_Z \psi(y - f(x))d\rho = \int_Z (y - f(x))^2 d\rho. \tag{2}$$

Separating ρ into the marginal distribution and conditional distributions, we see from $\int_Y \{y - f_\rho(x)\}d\rho(y|x) = 0$ that

$$\begin{aligned} \mathcal{E}(f) &= \int_X \left\{ \int_Y (y - f_\rho(x))^2 d\rho(y|x) + (f_\rho(x) - f(x))^2 \right\} d\rho_X(x). \\ &= \mathcal{E}(f_\rho) + \int_X (f(x) - f_\rho(x))^2 d\rho_X(x) \geq \mathcal{E}(f_\rho). \end{aligned}$$

If we denote the weighted L^2 norm as $||f||_{L^2_{\rho_X}} = (\int_X |f(x)|^2 d\rho_X)^{1/2}$, then

$$\mathcal{E}(f) - \mathcal{E}(f_\rho) = ||f - f_\rho||^2_{L^2_{\rho_X}} \geq 0. \tag{3}$$

It means that the regression function f_ρ minimizes the error.

Since the samples are drawn according to ρ, the law of large numbers tells us that for a fixed function $f : X \to Y$, the random variable $\xi(z) = (y - f(x))^2$ on (Z, ρ) satisfies $\frac{1}{m}\sum_{i=1}^m \xi(z_i) \to E(\xi)$ in probability. But $E(\xi) = \int_Z \xi(z)d\rho$ is exactly $\mathcal{E}(f)$, the error of f, while $\frac{1}{m}\sum_{i=1}^m \xi(z_i) = \frac{1}{m}\sum_{i=1}^m (y_i - f(x_i))^2$. We denote this as $\mathcal{E}_\mathbf{z}(f) = \frac{1}{m}\sum_{i=1}^m (y_i - f(x_i))^2$ and call it the empirical error of f at $\mathbf{z}$. Since $\mathcal{E}_\mathbf{z}(f) \to \mathcal{E}(f)$, we expect that a minimizer of the empirical error $\mathcal{E}_\mathbf{z}(f)$ is a good approximation of the minimizer f_ρ of the error $\mathcal{E}(f)$. This is the idea of ERM. If one allows the minimizer to be taken over the set of all measurable functions, he can choose a function with $f(x_i) = y_i$ making $\mathcal{E}_{\mathbf{z}}(f) = 0$, leading to an overfitting. To avoid this, we choose a set $\mathcal{H}$ of functions on X and take the minimization of the empirical error only over $\mathcal{H}$ to get a minimizer

$$f_{\mathbf{z}} := \arg\min_{f\in\mathcal{H}} \mathcal{E}_{\mathbf{z}}(f) = \arg\min_{f\in\mathcal{H}} \frac{1}{m}\sum_{i=1}^{m}(y_i - f(x_i))^2. \tag{4}$$

It turns out that $f_{\mathbf{z}}$ is a good approximation of f_ρ when $\mathcal{H}$ is suitably chosen.

2.2. ERM regression with a general loss function

Now we can define the ERM learning algorithm for the regression problem associated with a general regressing loss function ψ.

Definition 3. Let $\mathcal{H}$ be a set of uniformly bounded measurable functions on X and $\psi : \mathbb{R} \to \mathbb{R}_+$ be a regressing loss function. The *empirical target function* $f_{\mathbf{z}}$ associated with the sample $\mathbf{z} = \{(x_i, y_i)\}_{i=1}^m \in Z^m$ is defined as

$$f_{\mathbf{z}} := \arg\min_{f\in\mathcal{H}} \frac{1}{m}\sum_{i=1}^{m}\psi(y_i - f(x_i)). \tag{5}$$

We call $\mathcal{H}$ the *hypothesis space* for the regression problem.

If we define the *empirical error* or *risk* associated with the loss function ψ as

$$\mathcal{E}_{\mathbf{z}}(f) = \frac{1}{m}\sum_{i=1}^{m}\psi(y_i - f(x_i)),$$

then $f_{\mathbf{z}} = \arg\min_{f\in\mathcal{H}} \mathcal{E}_{\mathbf{z}}(f)$.

The theory of ERM studies how $f_{\mathbf{z}}$ approximates (with respect to the error) the *target function* defined as

$$f_{\mathcal{H}} := \arg\min_{f\in\mathcal{H}} \mathcal{E}(f) = \arg\min_{f\in\mathcal{H}} \int_Z \psi(y - f(x))d\rho. \tag{6}$$

To describe this approximation, we decompose the *sample error*

$$\mathcal{E}(f_{\mathbf{z}}) - \mathcal{E}(f_{\mathcal{H}}) = \mathcal{E}(f_{\mathbf{z}}) - \mathcal{E}_{\mathbf{z}}(f_{\mathbf{z}}) + \mathcal{E}_{\mathbf{z}}(f_{\mathbf{z}}) - \mathcal{E}_{\mathbf{z}}(f_{\mathcal{H}}) + \mathcal{E}_{\mathbf{z}}(f_{\mathcal{H}}) - \mathcal{E}(f_{\mathcal{H}}).$$

Since $f_{\mathbf{z}}$ minimizes the empirical error over $\mathcal{H}$, there holds $\mathcal{E}_{\mathbf{z}}(f_{\mathbf{z}}) - \mathcal{E}_{\mathbf{z}}(f_{\mathcal{H}}) \le 0$. Therefore, we have an *ERM error decomposition*:

$$\mathcal{E}(f_{\mathbf{z}}) - \mathcal{E}(f_{\mathcal{H}}) \le \{\mathcal{E}(f_{\mathbf{z}}) - \mathcal{E}_{\mathbf{z}}(f_{\mathbf{z}})\} + \{\mathcal{E}_{\mathbf{z}}(f_{\mathcal{H}}) - \mathcal{E}(f_{\mathcal{H}})\}. \tag{7}$$

Set a random variable $\xi(z) = \psi(y - f_{\mathcal{H}}(x))$ on (Z, ρ) corresponding to the target function $f_{\mathcal{H}}$, we find that in probability $\mathcal{E}_{\mathbf{z}}(f_{\mathcal{H}}) - \mathcal{E}(f_{\mathcal{H}}) = \frac{1}{m}\sum_{i=1}^m \xi(z_i) - E(\xi) \to 0$ (as $m \to \infty$). Thus, the second term in (7) converges to zero and its convergence rate can be estimated by some well-known probability inequalities such as Bernstein and Hoeffding inequalities.

Lemma 4. *Let ξ be a random variable on a probability space Z with mean μ and variance σ^2 satisfying $|\xi - \mu| \leq M$ almost surely. Then, for every $\epsilon > 0$, we have*

$$\text{[Bernstein]} \quad \mathrm{Prob}_{\mathbf{z}\in Z^m}\left\{\frac{1}{m}\sum_{i=1}^{m}\xi(z_i) - \mu > \epsilon\right\} \leq \exp\left\{-\frac{m\epsilon^2}{2(\sigma^2 + \frac{1}{3}M\epsilon)}\right\},$$

$$\text{[Hoeffding]} \quad \mathrm{Prob}_{\mathbf{z}\in Z^m}\left\{\left|\frac{1}{m}\sum_{i=1}^{m}\xi(z_i) - \mu\right| > \epsilon\right\} \leq 2\exp\left\{-\frac{m\epsilon^2}{2M^2}\right\}.$$

The Bernstein inequality in the above form is called a one-side inequality.

The first term in (7) is more involved. If one writes $\mathcal{E}(f_{\mathbf{z}}) - \mathcal{E}_{\mathbf{z}}(f_{\mathbf{z}})$ as $\int_Z \xi(z)d\rho - \frac{1}{m}\sum_{i=1}^m \xi(z_i)$ with $\xi(z) = \psi(y - f_{\mathbf{z}}(x))$, then ξ is not a single random variable, it depends on the sample $\mathbf{z}$. Thus the usual law of large numbers does not guarantee the convergence of the first term of (7) to zero. The main part of the classical statistical learning theory deals with this error term. It is called the *theory of uniform convergence*, playing the role of some uniform law of large numbers. To see this, we consider the quantity

$$\sup_{f\in\mathcal{H}} |\mathcal{E}_{\mathbf{z}}(f) - \mathcal{E}(f)| \tag{8}$$

which bounds the first term of (7), hence estimates the error $\mathcal{E}(f_{\mathbf{z}}) - \mathcal{E}(f_{\mathcal{H}})$. The theory of uniform convergence studies the convergence of this quantity to zero.

2.3. Sample error bounds using covering numbers

To get convergence rates for the ERM scheme, quantitative behaviors of capacity of the hypothesis space $\mathcal{H}$ are needed. There is a rich literature concerning the error analysis (7) by means of the capacity of the hypothesis space $\mathcal{H}$. In particular, when $\mathcal{H}$ is a compact subset of $C(X)$, the error can be estimated by covering numbers [39,21,9,58].

Definition 5. For a subset $\mathcal{F}$ of a metric space and $\eta > 0$, the *covering number* $\mathcal{N}(\mathcal{F}, \eta)$ is defined to be the minimal integer $\ell \in \mathbb{N}$ such that there exist ℓ disks with radius η covering $\mathcal{F}$. For a set $\mathcal{H}$ of continuous functions on X, we denote $\mathcal{N}(\mathcal{H}, \eta)$ the covering number of $\mathcal{H}$, as a subset of $C(X)$.

For the estimates of covering numbers, see [89,90]. The following result for the least-square loss was provided in [21].

Proposition 6. *Let $\psi(t) = t^2$ and $\mathcal{H}$ be a compact subset of $C(X)$. Assume that, for all $f \in \mathcal{H}$, $|f(x) - y| \leq M$ almost everywhere. Let $\sigma^2 = \sup_{f\in\mathcal{H}} \sigma^2((y - f(x))^2)$ where $\sigma^2((y - f(x))^2)$ is the variance of the random variable $\xi(z) = (y - f(x))^2$ on (Z, ρ). Then, for all $\epsilon > 0$,*

$$Prob_{\mathbf{z}\in Z^m}\left\{\mathcal{E}(f_{\mathbf{z}})-\mathcal{E}(f_{\mathcal{H}})\leq\epsilon\right\}\geq 1-\mathcal{N}\left(\mathcal{H},\frac{\epsilon}{16M}\right)2\exp\left\{-\frac{m\epsilon^2}{8(4\sigma^2+\frac{1}{3}M^2\epsilon)}\right\}.$$

To show the idea, we give a similar result with a general loss function. We say that ψ is *Lipschitz s* $(0 < s \leq 1)$ on $[-M, M]$ if there is some constant $C \geq 0$ such that

$$|\psi(t)-\psi(t')|\leq C|t-t'|^s, \qquad \forall t,t'\in[-M,M]. \tag{9}$$

Theorem 7. *Let $\mathcal{H}$ be a subset of $C(X)$ such that for every $f \in \mathcal{H}$, there holds $|f(x) - y| \leq M$ almost surely. If $\psi : \mathbb{R} \to \mathbb{R}_+$ is a regressing loss function satisfying (9), then for every $\epsilon > 0$ there holds*

$$Prob_{\mathbf{z}\in Z^m}\left\{\mathcal{E}(f_{\mathbf{z}})-\mathcal{E}(f_{\mathcal{H}})\leq\epsilon\right\}\geq 1-\mathcal{N}\left(\mathcal{H},\left(\frac{\epsilon}{8C}\right)^{1/s}\right)2\exp\left\{-\frac{m\epsilon^2}{128C^2M^{2s}}\right\}.$$

Proof. Let $\{f_j\}_{j=1}^N \subset \mathcal{H}$ with $N = \mathcal{N}(\mathcal{H}, \eta)$ and $\eta := \big(\epsilon/(4C)\big)^{1/s}$ such that for every $f \in \mathcal{H}$ there is some $j \in \{1, \dots, N\}$ satisfying $\|f - f_j\|_\infty \leq \eta$. Then according to the Lipschitz condition given in our assumption, we see from $|y - f(x)| \leq M$ that

$$|\mathcal{E}(f)-\mathcal{E}(f_j)|\leq\int_Z|\psi(y-f(x))-\psi(y-f_j(x))|d\rho\leq\int_Z C|f(x)-f_j(x)|^s d\rho\leq C\eta^s$$

and that almost surely

$$|\mathcal{E}_{\mathbf{z}}(f)-\mathcal{E}_{\mathbf{z}}(f_j)|\leq\frac{1}{m}\sum_{i=1}^m|\psi(y_i-f(x_i))-\psi(y_i-f_j(x_i))|\leq C\eta^s.$$

Thus

$$|\mathcal{E}(f)-\mathcal{E}_{\mathbf{z}}(f)|\leq|\mathcal{E}(f_j)-\mathcal{E}_{\mathbf{z}}(f_j)|+2C\eta^s.$$

By the choice of η, we find $2C\eta^s = \epsilon/2$. Therefore, the event $\{\mathbf{z} \in Z^m : |\mathcal{E}(f) - \mathcal{E}_{\mathbf{z}}(f)| > \epsilon$ for some $f \in \mathcal{H}\}$ is contained in the event $\cup_{j=1}^N\{\mathbf{z} \in Z^m : |\mathcal{E}(f_j) - \mathcal{E}_{\mathbf{z}}(f_j)| > \epsilon/2\}$. It follows that

$$\mathrm{Prob}_{\mathbf{z}\in Z^m}\left\{\sup_{f\in\mathcal{H}}|\mathcal{E}(f)-\mathcal{E}_{\mathbf{z}}(f)|>\epsilon\right\}\leq\sum_{j=1}^N\mathrm{Prob}_{\mathbf{z}\in Z^m}\left\{|\mathcal{E}(f_j)-\mathcal{E}_{\mathbf{z}}(f_j)|>\epsilon/2\right\}.$$

For each j, we apply Lemma 4 to the random variable $\xi(z) = \psi(y - f_j(x))$ on (Z, ρ). It satisfies $\mu = \mathcal{E}(f_j)$ and $\frac{1}{m}\sum_{i=1}^m \xi(z_i) = \mathcal{E}_{\mathbf{z}}(f_j)$. Moreover, $|\xi(z) - \mu|$ can be expressed as

$$|\psi(y-f_j(x))-\int\psi(y'-f_j(x'))d\rho(z')|=|\int\psi(y-f_j(x))-\psi(y'-f_j(x'))d\rho(z')|$$

which is bounded by $\int C|y - f_j(x) - (y' - f_j(x'))|^s d\rho(z') \leq 2CM^s$. Then we have from the Hoeffding inequality in Lemma 4 that

$$\mathrm{Prob}_{\mathbf{z}\in Z^m}\left\{|\mathcal{E}(f_j)-\mathcal{E}_{\mathbf{z}}(f_j)|>\epsilon/2\right\}\leq 2\exp\left\{-\frac{m(\epsilon/2)^2}{2(2CM^s)^2}\right\}.$$

It follows that

$$\text{Prob}_{\mathbf{z}\in Z^m}\left\{\sup_{f\in\mathcal{H}}|\mathcal{E}(f)-\mathcal{E}_{\mathbf{z}}(f)|>\epsilon\right\}\le\mathcal{N}(\mathcal{H},\eta)2\exp\left\{-\frac{m\epsilon^2}{32C^2M^{2s}}\right\}.$$

Since $f_{\mathbf{z}}$ and $f_{\mathcal{H}}$ are both in $\mathcal{H}$, by (7) we have $\mathcal{E}(f_{\mathbf{z}})-\mathcal{E}(f_{\mathcal{H}})\le 2\sup_{f\in\mathcal{H}}|\mathcal{E}(f)-\mathcal{E}_{\mathbf{z}}(f)|\le 2\epsilon$ with the above confidence. Then our conclusion follows after replacing ϵ by $\epsilon/2$. □

By setting the confidence to be $1-\delta$, we can get convergence rates from Theorem 7. Recall the elementary fact [22] that the positive solution ϵ^* to the equation $\epsilon^p - c_1\epsilon^q - c_2 = 0$ with $c_1, c_2 > 0$ and $p > q > 0$ satisfies $\epsilon^* \le \max\{(2c_1)^{1/(p-q)}, (2c_2)^{1/p}\}$.

Corollary 8. *If the assumption of Theorem 7 holds, and with some $r, C_0 > 0$ the covering number satisfies $\log\mathcal{N}(\mathcal{H},\eta)\le C_0\eta^{-r}$, then for every $0<\delta<1$, we have with confidence $1-\delta$,*

$$\mathcal{E}(f_{\mathbf{z}})-\mathcal{E}(f_{\mathcal{H}})\le\widetilde{C}\max\left\{\sqrt{\frac{\log(2/\delta)}{m}}, m^{\frac{1}{2+r/s}}\right\},$$

where $\widetilde{C}:=\max\{16CM^s,(256C^2M^{2s}C_0(8C)^{r/s})^{1/(2+r/s)}\}$.

The condition for the covering number is satisfied when the hypothesis space is a ball of Sobolev spaces embedded in $C(X)$. When r is small enough (the smoothness is high enough for the case of Sobolev spaces), the bound in Corollary 8 is almost of the form $O(1/\sqrt{m})$. This can be improved to a bound of the form $O(1/m^{1-\epsilon})$ when the hypothesis space is convex and the loss function satisfies some high convexity condition, see [39,4,21] and the discussion in Section 3.3 by means of ratio probability inequalities.

2.4. Theory of uniform convergence

The theory of uniform convergence also characterizes those function sets $\mathcal{H}$ such that the quantity (8) tends to zero in probability as $m\to\infty$. This is required for all Borel probability distributions μ, which leads to the following definition.

Definition 9. We say that a set $\mathcal{H}$ of real-valued functions on a metric space X is *uniform Glivenko-Cantelli* (UGC) if for every $\epsilon>0$,

$$\lim_{\ell\to+\infty}\sup_{\mu}\text{Prob}\left\{\sup_{m\ge\ell}\sup_{f\in\mathcal{H}}\left|\frac{1}{m}\sum_{i=1}^{m}f(x_i)-\int_X f(x)d\mu\right|\ge\epsilon\right\}=0$$

where the supremum is taken with respect to all Borel probability distributions μ on X, and Prob denotes the probability with respect to the samples $x_1, x_2, \ldots$ independently drawn according to such a distribution μ.

When $\mathcal{H}$ consists of indicator functions (characteristic functions), the UGC property can be characterized by VC-dimensions [75]. For general sets of real-valued functions, it can be characterized by V_γ-dimensions [1].

Definition 10. Let $\mathcal{H}$ be a set of functions from X to $[0, 1]$. We say that $A \subset X$ is V_γ shattered by $\mathcal{H}$ if there is a number $\alpha \in \mathbb{R}$ with the following property: for every subset E of A there exists some function $f_E \in \mathcal{H}$ such that $f_E(x) \leq \alpha - \gamma$ for every $x \in A \setminus E$, and $f_E(x) \geq \alpha + \gamma$ for every $x \in E$. The V_γ dimension of $\mathcal{H}$, $V_\gamma(\mathcal{H})$, is the maximal cardinality of a set $A \subset X$ which is V_γ shattered by $\mathcal{H}$.

The concept of V_γ-dimension is related to many others about capacity of function sets in approximation theory or functional analysis: covering number, entropy number, packing number, metric entropy and more. See [1,26,52,49,55].

The following characterization of the UGC property was given in [1].

Theorem 11. *Let $\mathcal{H}$ be a set of functions from X to $[0, 1]$. Then it is UGC if and only if the V_γ dimension of $\mathcal{H}$ is finite for every $\gamma > 0$.*

Theorem 11 may be used to verify the convergence of ERM schemes when the hypothesis space $\mathcal{H}$ is a UGC set, but is not compact with respect to the norm $\|\cdot\|_\infty$ in $C(X)$. In [83] we consider the union of unit balls of reproducing kernel Hilbert spaces $(\mathcal{H}_K, \|\cdot\|_K)$ (to be defined below) associated with a set of Mercer kernels K, and find conditions for such a hypothesis space to be UGC. In particular, for the Gaussian kernels, we verify the UGC property.

Example 12. Let X be a compact subset of $\mathbb{R}^n$ and $\{K_\sigma\}_{0<\sigma<\infty}$ be the set of Gaussian kernels with $K_\sigma(x, y) = \exp(-\frac{|x-y|^2}{\sigma^2})$. Then the set

$$\mathcal{H} = \cup_{0<\sigma<\infty} \{f \in \mathcal{H}_{K_\sigma} : \|f\|_{K_\sigma} \leq 1\}$$

is UGC, but its closure in $C(X)$ is not compact.

Quantitative behavior of the convergence of the ERM scheme with the hypothesis space in Example 12 is not fully understood, though we have some partial estimates [83,48] by means of local Rademacher averages.

There are still many open fundamental problems about the UGC property. As an example, let us consider the empirical covering numbers.

Definition 13. For $\mathbf{x} = (x_i)_{i=1}^m \in X^m$, $1 \leq p \leq \infty$, and $\mathcal{H} \subset C(X)$, the ℓ^p-*empirical covering number* $\mathcal{N}_p(\mathcal{H}, \mathbf{x}, \eta)$ is the covering number of $\mathcal{H}|_{\mathbf{x}} = \{(f(x_i))_{i=1}^m : f \in \mathcal{H}\}$, as a subset of $\mathbb{R}^m$ under the metric defined for $f, g \in C(X)$ as $d_{p,\mathbf{x}}(f, g) = \left\{\frac{1}{m}\sum_{i=1}^m |f(x_i) - g(x_i)|^p\right\}^{1/p}$. The *metric entropy* of $\mathcal{H}$ is

$$H_m(\mathcal{H}, \eta) = \sup_{\mathbf{x} \in X^m} \log \mathcal{N}_\infty(\mathcal{H}, \mathbf{x}, \eta), \qquad m \in \mathbb{N}, \eta > 0.$$

It is known [28] that a function set $\mathcal{H}$ from X to $[0,1]$ is UGC if and only if $\lim_{m\to\infty} H_m(\mathcal{H},\eta)/m = 0$ for every $\eta > 0$. In this case, there holds $H_m(\mathcal{H},\eta) = O(\log^2 m)$ for every $\eta > 0$. It is conjectured in [1] that $H_m(\mathcal{H},\eta) = O(\log m)$ is true for every $\eta > 0$. The following problem is a weak form.

Problem 14. Is it true that for some $\alpha \in [1,2)$, every UGC set $\mathcal{H}$ satisfies

$$H_m(\mathcal{H},\eta) = O(\log^\alpha m), \qquad \forall \eta > 0?$$

2.5. Approximation error and the ψ-regression function

The target function $f_{\mathcal{H}}$ defined by (6) is the one minimizing the error over $\mathcal{H}$. If we allow the set of all measurable functions, the minimizer is the function we desire with respect to the regressing loss function ψ.

Definition 15. Given the regressing loss function ψ and a probability distribution ρ on Z, the *ψ-regression function* is the one minimizing the error, given by

$$f_\rho^\psi(x) = \arg\min_{t\in\mathbb{R}} \int_Y \psi(y-t)d\rho(y|x), \qquad x \in X.$$

The *approximation error* associated with the hypothesis space $\mathcal{H}$ is defined as

$$\mathcal{E}(f_{\mathcal{H}}) - \mathcal{E}(f_\rho^\psi) = \inf_{f\in\mathcal{H}} \mathcal{E}(f) - \mathcal{E}(f_\rho^\psi).$$

The convexity of ψ tells us some information about f_ρ^ψ. For $x \in X$, denote

$$\Psi(t) = \Psi_x(t) = \int_Y \psi(y-t)d\rho(y|x) = \int_Y \psi(t-y)d\rho(y|x), \qquad t \in \mathbb{R}. \tag{10}$$

Then Ψ is also a convex function on $\mathbb{R}$. We know that $\Psi''(t) \geq 0$ almost everywhere. The left and right one-side derivatives of Ψ, $\Psi'_-(t)$ and $\Psi'_+(t)$, exist, are nondecreasing, and satisfy $\Psi'_-(t) \leq \Psi'_+(t)$ for every $t \in \mathbb{R}$. Hence the following fact holds.

Theorem 16. *Let ψ be a regressing loss function and $x \in X$. Then:*

(a) *The univariate function Ψ given by (10) is strictly decreasing on $(-\infty, f_\rho^-(x)]$, strictly increasing on $[f_\rho^+(x), +\infty)$, and is constant on $[f_\rho^-(x), f_\rho^+(x)]$, where*

$$f_\rho^-(x) := \sup\{t\in\mathbb{R} : \Psi'_-(t) < 0\}, \quad f_\rho^+(x) := \inf\{t\in\mathbb{R} : \Psi'_+(t) > 0\}.$$

(b) *$f_\rho^\psi(x)$ is a minimizer of Ψ, that is, $f_\rho^-(x) \leq f_\rho^\psi(x) \leq f_\rho^+(x)$.*

(c) *If $\epsilon \geq 0$ is the maximal zero of ψ and $|y| \leq M$ almost surely, then $f_\rho^\psi(x) \in [-M-\epsilon, M+\epsilon]$.*

Now we turn to the approximation error estimates. Denote a weighted L^p norm as $\|f\|_{L^p_{\rho_X}} = (\int_X |f(x)|^p d\rho_X)^{1/p}$ for $p \geq 1$.

Theorem 17. *Assume $|y - f(x)| \le M$ and $|y - f_\rho^\psi(x)| \le M$ almost surely.*

(a) *If ψ is a Lipschitz s regressing loss function satisfying (9), then*

$$\mathcal{E}(f) - \mathcal{E}(f_\rho^\psi) \le C \int_X |f(x) - f_\rho^\psi(x)|^s d\rho_X \le C\|f - f_\rho^\psi\|_{L^1_{\rho_X}}^s \le C\|f - f_\rho^\psi\|_{L^2_{\rho_X}}^{s/2}.$$

(b) *If ψ is C^1 on $[-M, M]$, and its derivative is Lipschitz s satisfying (9), then*

$$\mathcal{E}(f) - \mathcal{E}(f_\rho^\psi) \le C\|f - f_\rho^\psi\|_{L^{1+s}_{\rho_X}}^{1+s} \le C\|f - f_\rho^\psi\|_{L^2_{\rho_X}}^{(1+s)/2}.$$

(c) *If $\psi''(u) \ge c > 0$ for every $u \in [-M, M]$, then*

$$\mathcal{E}(f) - \mathcal{E}(f_\rho^\psi) \ge \frac{c}{2}\|f - f_\rho^\psi\|_{L^2_{\rho_X}}^2.$$

Proof. By the definition of the error, we have

$$\mathcal{E}(f) - \mathcal{E}(f_\rho^\psi) = \int_X \left\{ \int_Y \psi(y - f(x)) - \psi(y - f_\rho^\psi(x)) d\rho(y|x) \right\} d\rho_X.$$

Then the statement in (a) follows from the Lipschitz condition (9) with $t = y - f(x), t' = y - f_\rho^\psi(x) \in [-M, M]$.

If ψ is C^1 on $[-M, M]$, and ψ' satisfies (9), then for each $x \in X$, the function $\Psi = \Psi_x$ is also C^1 and $\Psi'(f_\rho^\psi(x)) = 0$ since $f_\rho^\psi(x)$ is a minimum. Moreover, $\Psi'(t) = \int_Y \psi'(t - y) d\rho(y|x)$. It implies from the Lipschitz condition (9) for ψ' that for t with $|t - y| \le M$,

$$|\Psi'(t) - \Psi'(f_\rho^\psi(x))| = \left| \int_Y \psi'(t - y) - \psi'(f_\rho^\psi(x) - y) d\rho(y|x) \right| \le C|t - f_\rho^\psi(x)|^s.$$

Then

$$\Psi(f(x)) - \Psi(f_\rho^\psi(x)) = \int_{f_\rho^\psi(x)}^{f(x)} \Psi'(t) - \Psi'(f_\rho^\psi(x)) dt \le C|f(x) - f_\rho^\psi(x)|^{1+s}.$$

Hence

$$\mathcal{E}(f) - \mathcal{E}(f_\rho^\psi) = \int_X \Psi(f(x)) - \Psi(f_\rho^\psi(x)) d\rho_X \le C \int_X |f(x) - f_\rho^\psi(x)|^{1+s} d\rho_X.$$

This proves the statement in (b).

Now assume $\psi''(u) \ge c > 0$ for every $u \in [-M, M]$. If $f(x) > f_\rho^\psi(x)$, for any $t \in [f_\rho^\psi(x), f(x)]$ we have $\Psi''(t) = \int_Y \psi''(t - y) d\rho(y|x) \ge c$. Hence $\Psi'(t) - \Psi'(f_\rho^\psi(x)) = \int_{f_\rho^\psi(x)}^t \Psi''(u) du \ge c(t - f_\rho^\psi(x))$. Then

$$\Psi(f(x)) - \Psi(f_\rho^\psi(x)) \ge \int_{f_\rho^\psi(x)}^{f(x)} c(t - f_\rho^\psi(x)) dt = \frac{c}{2}(f(x) - f_\rho^\psi(x))^2.$$

In the same way, if $f(x) < f_\rho^\psi(x)$, we have $\Psi'(t) - \Psi'(f_\rho^\psi(x)) \le c(t - f_\rho^\psi(x))$ for any $t \in [f(x), f_\rho^\psi(x)]$. Hence $\Psi(f(x)) - \Psi(f_\rho^\psi(x)) \ge \frac{c}{2}(f(x) - f_\rho^\psi(x))^2$. In both cases, we get

$$\mathcal{E}(f) - \mathcal{E}(f_\rho^\psi) = \int_X \Psi(f(x)) - \Psi(f_\rho^\psi(x)) d\rho_X \ge \frac{c}{2} \int_X |f(x) - f_\rho^\psi(x)|^2 d\rho_X.$$

This verifies the result. □

To see the approximation of f_ρ^ψ by $f_{\mathbf{z}}$, we need to get the error between $f_{\mathbf{z}}$ and f_ρ^ψ in some function spaces, from the excess generalization error $\mathcal{E}(f_{\mathbf{z}}) - \mathcal{E}(f_\rho^\psi)$. Part (c) of Theorem 17 ensures this in the case when ψ'' is bounded from below.

Problem 18. It would be interesting for a general loss function ψ to derive lower bounds of the following form with $c, p > 0$:

$$\mathcal{E}(f) - \mathcal{E}(f_\rho^\psi) \geq c \int_X |f(x) - f_\rho^\psi(x)|^p d\rho_X.$$

When ψ is the ϵ-insensitive norm, it is Lipschitz 1 with $C = 1$ and $M = \infty$. It follows from (a) of Theorem 17 that $\mathcal{E}(f) - \mathcal{E}(f_\rho^\psi) \leq \|f - f_\rho^\psi\|_{L^1_{\rho_X}}$.

When ψ is the least-square loss, it is C^1 and its derivative satisfies (9) with $s = 1$, $C = 2$ and $M = \infty$. Hence $\mathcal{E}(f) - \mathcal{E}(f_\rho^\psi) \leq \|f - f_\rho^\psi\|^2_{L^2_{\rho_X}}$ (in fact, equality holds).

Theorem 17 (parts (a) and (b)) gives a way to estimate the approximation error by means of results from approximation theory [91].

For many ERM schemes, the hypothesis is a ball of a dense subspace $(H, \|\cdot\|_H)$ of a Banach space $(B, \|\cdot\|_B)$ of functions on X. For example, B may be some weighted L^p space $L^p_{\rho_X}$ and H is a Sobolev space or a reproducing kernel Hilbert space. For such a setting, the approximation error can be characterized by interpolation spaces. Recall that the $\mathcal{K}$-functional of the couple (B, H) is defined for $f \in B$ as

$$\mathcal{K}(f,t) = \inf_{g \in B} \{\|f - g\|_B + t\|g\|_H\}, \qquad t > 0.$$

For $0 < \theta < 1$, the interpolation space $(B, H)_{\theta,\infty}$ consists of those elements $f \in B$ such that the following norm is finite

$$\|f\|_{\theta,\infty} = \sup_{t>0} \left\{\mathcal{K}(f,t)/t^\theta\right\}.$$

The next result can be found in [64].

Theorem 19. *Let $(B, \|\cdot\|_B)$ be a Banach space and $(H, \|\cdot\|_H)$ be a dense subspace with $\|g\|_B \leq \|g\|_H$ for $g \in H$. Let $0 < \theta < 1$. If $f \in (B, H)_{\theta,\infty}$, then*

$$I(f,R) := \inf\{\|f - g\|_B : g \in H, \|g\|_H \leq R\} \leq \|f\|_{\theta,\infty}^{\frac{1}{1-\theta}} \left(\frac{1}{R}\right)^{\frac{\theta}{1-\theta}}.$$

Conversely, if $I(f,R) \leq C\left(\frac{1}{R}\right)^{\frac{\theta}{1-\theta}}$ for all $R > 0$, then $f \in (B,H)_{\theta,\infty}$ and $\|f\|_{\theta,\infty} \leq 2C^{1-\theta}$.

The most prominent function spaces used in learning theory are reproducing kernel Hilbert spaces. The importance of these function spaces can be seen from regularization schemes.

3. Regularization Schemes for Regression

3.1. Regularization schemes and convex optimization

Another approach in learning theory is to replace the ERM by the minimization of a penalized empirical error over a function space. Reproducing kernel Hilbert spaces generated by Mercer kernels [3] are often used for machine learning.

Definition 20. We say that $K : X \times X \to \mathbb{R}$ is a *Mercer kernel* if it is continuous, symmetric and positive semidefinite, that is, for any finite set of distinct points $\{x_1, \cdots, x_\ell\} \subset X$, the matrix $(K(x_i, x_j))_{i,j=1}^{\ell}$ is positive semidefinite.

The *Reproducing Kernel Hilbert Space* (RKHS) $\mathcal{H}_K$ associated with the kernel K is defined to be the closure of the linear span of the set of functions $\{K_x := K(x, \cdot) : x \in X\}$ with the inner product $\langle \cdot, \cdot \rangle_K$ given by $\langle K_x, K_y \rangle_K = K(x, y)$.

The reproducing property takes the form

$$\langle K_x, f \rangle_K = f(x), \qquad \forall\, x \in X,\ f \in \mathcal{H}_K. \tag{11}$$

Denote $\kappa = \sup_{x \in X} \sqrt{K(x,x)}$. Then the above reproducing property tells us that

$$\|f\|_\infty \le \kappa \|f\|_K, \qquad \forall\, f \in \mathcal{H}_K.$$

The *regularization scheme* for the regression problem associated with the RKHS $\mathcal{H}_K$ takes the form

$$f_{\mathbf{z},\lambda} := \arg\min_{f \in \mathcal{H}_K} \left\{ \frac{1}{m} \sum_{i=1}^{m} \psi(y_i - f(x_i)) + \lambda \|f\|_K^2 \right\}. \tag{12}$$

Here $\lambda > 0$ is a constant called the *regularization parameter*. Usually it is chosen to depend on m, $\lambda = \lambda(m)$, and $\lim_{m\to\infty} \lambda(m) = 0$. It plays the role of R in ERM.

The Hilbert space structure of $\mathcal{H}_K$ and the convexity of ψ reduce the scheme (12) to a convex optimization problem over $\mathbb{R}^m$. First, if we denote the projection in $\mathcal{H}_K$ onto the finite dimensional space $\mathcal{H}_{K,\mathbf{x}} = \operatorname{span}\{K_{x_i}\}_{i=1}^m$ as P, then for any $f \in \mathcal{H}_K$ and $i = 1, \ldots, m$, there holds $\langle Pf, K_{x_i} \rangle_K = \langle f, K_{x_i} \rangle_K$, meaning that $(Pf)(x_i) = f(x_i)$ and hence $\mathcal{E}_\mathbf{z}(Pf) = \mathcal{E}_\mathbf{z}(f)$. This tells us that $f_{\mathbf{z},\lambda}$ must lie in $\mathcal{H}_{K,\mathbf{x}}$. Second, if we write $f_{\mathbf{z},\lambda} = \sum_{i=1}^m c_{i,\mathbf{z}} K_{x_i}$, then $c_\mathbf{z} := (c_{i,\mathbf{z}})_{i=1}^m$ is a minimizer of a convex function over $\mathbb{R}^m$:

$$c_\mathbf{z} = \arg\min_{c \in \mathbb{R}^m} \frac{1}{m} \sum_{i=1}^{m} \psi\Big(y_i - \sum_{j=1}^{m} c_j K(x_i, x_j) \Big) + \lambda \sum_{i,j=1}^{m} c_i K(x_i, x_j) c_j.$$

This is a convex optimization problem in $\mathbb{R}^m$, which can be solved by many efficient algorithms from optimization theory. If ψ is a convex quadratic spline such as the least-square loss or the ϵ-insensitive loss, then the convex optimization problem is a quadratic programming, and simpler algorithms are available.

3.2. Error decomposition and regularization error

The convergence of the regularization scheme (12) can be studied by the *excess generalization error* $\mathcal{E}(f_{\mathbf{z},\lambda}) - \mathcal{E}(f_\rho^\psi)$.

Let $\widetilde{f}_\lambda$ be an arbitrary function from $\mathcal{H}_K$. The penalized excess generalization error $\mathcal{E}(f_{\mathbf{z},\lambda}) - \mathcal{E}(f_\rho^\psi) + \lambda\|f_{\mathbf{z},\lambda}\|_K^2$ can be decomposed as

$$\{\mathcal{E}(f_{\mathbf{z},\lambda}) - \mathcal{E}_{\mathbf{z}}(f_{\mathbf{z},\lambda})\} + \left\{\mathcal{E}_{\mathbf{z}}(\widetilde{f}_\lambda) - \mathcal{E}(\widetilde{f}_\lambda)\right\} + \left\{\mathcal{E}(\widetilde{f}_\lambda) - \mathcal{E}(f_\rho^\psi) + \lambda\|\widetilde{f}_\lambda\|_K^2\right\}$$
$$+ \left\{\mathcal{E}_{\mathbf{z}}(f_{\mathbf{z},\lambda}) + \lambda\|f_{\mathbf{z},\lambda}\|_K^2 - \left(\mathcal{E}_{\mathbf{z}}(\widetilde{f}_\lambda) + \lambda\|\widetilde{f}_\lambda\|_K^2\right)\right\}.$$

The last term is at most 0, according to the definition of $f_{\mathbf{z},\lambda}$. Thus, $\mathcal{E}(f_{\mathbf{z},\lambda}) - \mathcal{E}(f_\rho^\psi) + \lambda\|f_{\mathbf{z},\lambda}\|_K^2$ can be bounded by

$$\{\mathcal{E}(f_{\mathbf{z},\lambda}) - \mathcal{E}_{\mathbf{z}}(f_{\mathbf{z},\lambda})\} + \left\{\mathcal{E}_{\mathbf{z}}(\widetilde{f}_\lambda) - \mathcal{E}(\widetilde{f}_\lambda)\right\} + \left\{\mathcal{E}(\widetilde{f}_\lambda) - \mathcal{E}(f_\rho^\psi) + \lambda\|\widetilde{f}_\lambda\|_K^2\right\}.$$

The last term above is called the regularization error [65]. It does not depend on the sample.

Definition 21. The *regularization error* for a *regularizing function* $\widetilde{f}_\lambda \in \mathcal{H}_K$ is defined as

$$\widetilde{\mathcal{D}}(\lambda) = \mathcal{E}(\widetilde{f}_\lambda) - \mathcal{E}(f_\rho^\psi) + \lambda\|\widetilde{f}_\lambda\|_K^2.$$

It is called the *regularization error of the scheme* (12), denoted as $\mathcal{D}(\lambda)$, when $\widetilde{f}_\lambda = f_\lambda$:

$$f_\lambda = \arg\inf_{f \in \mathcal{H}_K} \left\{\mathcal{E}(f) - \mathcal{E}(f_\rho^\psi) + \lambda\|f\|_K^2\right\}. \tag{13}$$

The regularization error can be estimated by Theorem 17 and the rich knowledge of approximation theory in function spaces. When ψ is the least-square loss, $f_\rho^\psi = f_\rho$ and $\mathcal{E}(f) - \mathcal{E}(f_\rho^\psi) = \|f - f_\rho\|_{L^2_{\rho_X}}^2$. In this case, the regularization error can be easily bounded by the integral operator $L_K : L^2_{\rho_X} \to L^2_{\rho_X}$ defined as

$$L_K(f)(x) = \int_X K(x,y) f(y) d\rho_X(y), \qquad x \in X, f \in L^2_{\rho_X}.$$

Note that the range of L_K is in $\mathcal{H}_K$, hence L_K can also be regarded [71] as an operator from $L^2_{\rho_X}$ to $\mathcal{H}_K$. The following result can be found in [66]. The condition $L_K^{-r} f_\rho \in L^2_{\rho_X}$ means $f_\rho = L_K^r g$ for some $g \in L^2_{\rho_X}$ with $\|g\|_{L^2_{\rho_X}} = \|L_K^{-r} f_\rho\|_{L^2_{\rho_X}}$. When $0 < r < 1/2$, this means that $f_\rho \in (L^2_{\rho_X}, \mathcal{H}_K)_{r,\infty}$.

Theorem 22. *Define f_λ by (13). If $L_K^{-r} f_\rho \in L^2_{\rho_X}$, then*

$$\mathcal{D}(\lambda) = \|f_\lambda - f_\rho\|_{L^2_{\rho_X}}^2 + \lambda\|f_\lambda\|_K^2 \le \lambda^{2r}\|L_K^{-r} f_\rho\|_{L^2_{\rho_X}}^2, \qquad \text{if } 0 < r \le \frac{1}{2}$$

and

$$\|f_\lambda - f_\rho\|_{L^2_{\rho_X}} \le \lambda^r \|L_K^{-r} f_\rho\|_{L^2_{\rho_X}}, \qquad \text{if } 0 < r \le 1.$$

When $\frac{1}{2} < r \le 1$, we have

$$\|f_\lambda - f_\rho\|_K \le \lambda^{r-\frac{1}{2}} \|L_K^{-r} f_\rho\|_{L^2_{\rho_X}}.$$

Notice that for $\frac{1}{2} < r \le 1$, Theorem 22 yields $\|f_\lambda - f_\rho\|^2_{L^2_{\rho_X}} = O(\lambda^{2r})$ while $\mathcal{D}(\lambda)$ is at most $O(\lambda)$.

Turn back to the error decomposition. One can write the sample error as (7) and then derive convergence rates of some weak form (of type $O(1/\sqrt{m})$) by Hoeffding's inequality, as in Corollary 8. But we can do better with the Bernstein inequality by considering the excess random variable $\psi(y-f(x)) - \psi(y - f_\rho^\psi(x))$, not $\psi(y-f(x))$. Then we have a ***regularization scheme error decomposition***:

$$\begin{aligned}&\mathcal{E}(f_{\mathbf{z},\lambda}) - \mathcal{E}(f_\rho^\psi) + \lambda \|f_{\mathbf{z},\lambda}\|_K^2 \\ &\quad \le \{E(\xi_1) - E_{\mathbf{z}}(\xi_1)\} + \{E_{\mathbf{z}}(\xi_2) - E(\xi_2)\} + \widetilde{\mathcal{D}}(\lambda).\end{aligned} \tag{14}$$

Here

$$\xi_1 := \psi(y - f_{\mathbf{z},\lambda}(x)) - \psi(y - f_\rho^\psi(x)), \qquad \xi_2 := \psi(y - \widetilde{f}_\lambda(x)) - \psi(y - f_\rho^\psi(x))$$

and for $\xi = \xi(x,y)$, we denote $E(\xi) = \int_Z \xi(x,y) d\rho$ and $E_{\mathbf{z}}(\xi) = \frac{1}{m}\sum_{i=1}^m \xi(x_i, y_i)$. We remark that $E(\xi_1)$ is not the expected value of a single random variable since it depends on $\mathbf{z}$.

To show how (14) improves the analysis of the error decomposition of the form (7), we first look at the middle term $E_{\mathbf{z}}(\xi_2) - E(\xi_2)$.

Lemma 23. *Assume $|\psi(y - \widetilde{f}_\lambda(x)) - \psi(y - f_\rho^\psi(x))| \le M$ almost surely. If for some $\tau \in [0,1]$ and $C_\psi > 0$ there holds*

$$\int \left(\psi(y - \widetilde{f}_\lambda(x)) - \psi(y - f_\rho^\psi(x))\right)^2 d\rho \le C_\psi \left\{\mathcal{E}(\widetilde{f}_\lambda) - \mathcal{E}(f_\rho^\psi)\right\}^\tau,$$

then for any $0 < \delta < 1$, with confidence at least $1 - \delta$, we have

$$E_{\mathbf{z}}(\xi_2) - E(\xi_2) \le \frac{2M \log(1/\delta)}{3m} + \left(\frac{2C_\psi \log(1/\delta)}{m}\right)^{\frac{1}{2-\tau}} + \frac{1}{2}\widetilde{\mathcal{D}}(\lambda).$$

Proof. Consider the random variable ξ_2 on (Z, ρ). It satisfies $|\xi_2| \le M$ and

$$\sigma^2 \le E(\xi_2^2) \le C_\psi \left\{\mathcal{E}(\widetilde{f}_\lambda) - \mathcal{E}(f_\rho^\psi)\right\}^\tau \le C_\psi \left(\widetilde{\mathcal{D}}(\lambda)\right)^\tau.$$

Applying the one-side Bernstein inequality from lemma 4 to ξ_2, we know that for any $\epsilon > 0$, with confidence at least

$$1 - \exp\left\{-\frac{m\epsilon^2}{2(\sigma^2 + \frac{1}{3}M\epsilon)}\right\}$$

there holds $E_{\mathbf{z}}(\xi_2) - E(\xi_2) \le \epsilon$. Take $\epsilon = \epsilon^*$ to be the positive solution of the quadratic equation

$$-\frac{m\epsilon^2}{2(\sigma^2 + \frac{1}{3}M\epsilon)} = \log \delta.$$

That is, $m\epsilon^2 - \frac{2M}{3}\log(1/\delta)\epsilon - 2\sigma^2\log(1/\delta) = 0$. Then

$$\epsilon^* = \frac{\frac{M}{3}\log(1/\delta) + \sqrt{(\frac{M}{3}\log(1/\delta))^2 + 2m\sigma^2\log(1/\delta)}}{m}$$
$$\leq \frac{2M}{3m}\log(1/\delta) + \sqrt{\frac{2\log(1/\delta)}{m}C_\psi\left(\widetilde{\mathcal{D}}(\lambda)\right)^\tau}.$$

Recall the elementary inequality

$$\frac{1}{q} + \frac{1}{q^*} = 1 \text{ with } q, q^* > 1 \Longrightarrow a \cdot b \leq \frac{1}{q}a^q + \frac{1}{q^*}b^{q^*}, \qquad \forall a, b \geq 0.$$

Using this with $q = \frac{2}{2-\tau}, q^* = \frac{2}{\tau}$ and $a = \sqrt{\frac{2C_\psi \log(1/\delta)}{m}}, b = \left(\widetilde{\mathcal{D}}(\lambda)\right)^{\tau/2}$, we see that with confidence $1-\delta$, $E_{\mathbf{z}}(\xi_2) - E(\xi_2) \leq \epsilon^*$ can be bounded as stated. □

Definition 24. We say that ψ has *variancing exponent* $\tau \in [0, 1]$ if for any $M > 0$ there is a constant $C_{\psi,M}$ such that

$$\int \left(\psi(y - f(x)) - \psi(y - f_\rho^\psi(x))\right)^2 d\rho \leq C_{\psi,M}\left\{\mathcal{E}(f) - \mathcal{E}(f_\rho^\psi)\right\}^\tau$$

for any measurable function f with $|y - f(x)|, |y - f_\rho^\psi(x)| \leq M$ almost surely.

The least-square loss has variancing exponent $\tau = 1$ with $C_{\psi,M} = 4M^2$.

3.3. Error analysis by ratio probability inequalities

Now we show the error analysis for the regularization scheme (12). The error rate has the form $O((\frac{1}{m})^{1-\epsilon})$ with ϵ depending on the regularization error and capacity of $\mathcal{H}_K$. In particular, when K is C^∞ and the regularization error is $O(\lambda)$, ϵ can be arbitrarily small.

Ratio probability inequalities such as Talagrand's inequalities using empirical covering numbers [10,11,5,6] play an essential role in deriving good sample error estimates. Here, to illustrate the idea, we give the following simple form [80].

Proposition 25. *Let ξ be a random variable on Z satisfying $\mu = E(\xi) \geq 0$, $|\xi - \mu| \leq M$ almost surely, and $\sigma^2 \leq C\mu^\tau$ for some $0 \leq \tau \leq 2$. Then for every $\varepsilon > 0$*

$$Prob_{\mathbf{z}\in Z^m}\left\{\frac{\mu - \frac{1}{m}\sum_{i=1}^m \xi(z_i)}{(\mu^\tau + \varepsilon^\tau)^{\frac{1}{2}}} > \varepsilon^{1-\frac{\tau}{2}}\right\} \leq \exp\left\{-\frac{m\varepsilon^{2-\tau}}{2(C + \frac{1}{3}M\varepsilon^{1-\tau})}\right\}.$$

This proposition is an easy consequence of the one-side Bernstein inequality applied to the event $\mu - \frac{1}{m}\sum_{i=1}^m \xi(z_i) > \varepsilon^{1-\frac{\tau}{2}}(\mu^\tau + \varepsilon^\tau)^{\frac{1}{2}}$.

Note that the function $f_{\mathbf{z},\lambda}$ depends on $\mathbf{z}$ and thus runs over a set of functions as $\mathbf{z}$ changes. So we need a probability inequality concerning the uniform convergence. Denote $E(f) = \int_Z f(z)d\rho$.

Proposition 26. *Let $0 \leq \tau \leq 1$, $M > 0$, $C \geq 0$, and $\mathcal{F}$ be a set of functions on Z such that for every $f \in \mathcal{F}$, $E(f) \geq 0$, $|f - E(f)| \leq M$ and $E(f^2) \leq C(E(f))^\tau$. Then for $\varepsilon > 0$, there holds*

$$\begin{aligned} &Prob_{\mathbf{z}\in Z^m}\left\{\sup_{f\in\mathcal{F}} \frac{E(f) - \frac{1}{m}\sum_{i=1}^m f(z_i)}{((E(f))^\tau + \varepsilon^\tau)^{\frac{1}{2}}} > 4\varepsilon^{1-\frac{\tau}{2}}\right\} \\ &\qquad \leq \mathcal{N}(\mathcal{F},\varepsilon)\exp\left\{\frac{-m\varepsilon^{2-\tau}}{2(C + \frac{1}{3}M\varepsilon^{1-\tau})}\right\}. \end{aligned}$$

The proof of this proposition is the same as the procedure in Theorem 7, by means of covering numbers.

The error estimates for (12) follow from Proposition 26 and the regularization error. As an example, consider the scheme associated with the least-square loss

$$f_{\mathbf{z},\lambda} := \arg\min_{f\in\mathcal{H}_K}\left\{\frac{1}{m}\sum_{i=1}^m (y_i - f(x_i))^2 + \lambda\|f\|_K^2\right\}. \tag{15}$$

We have the following result presented in [79].

Theorem 27. *Let $f_{\mathbf{z},\lambda}$ be defined by (15). Assume $|y| \leq M$ almost surely and that f_ρ is not identically zero. If for some $s, C_0 > 0$, $\log\mathcal{N}(\eta) \leq C_0\,(1/\eta)^s$, then for any $0 < \zeta < \frac{1}{1+s}$, $0 < \delta < 1$ and $m \geq m_{\delta,\zeta}$, with confidence $1-\delta$ there holds*

$$\|f_{\mathbf{z},\lambda} - f_\rho\|^2_{L^2_{\rho_X}} \leq \frac{\widetilde{C}}{\|f_\rho\|^2_{L^2_{\rho_X}}}\log\left(\frac{2}{\delta}\right)\mathcal{D}\left(m^{-\zeta}\right) \qquad \text{by taking } \lambda = m^{-\zeta}$$

where $m_{\delta,\zeta}$ and $\widetilde{C}$ are constants depending on C_0, s, ζ, κ, M, and $m_{\delta,\zeta}$ also on δ.

In particular, if K is C^∞ on $X \subset \mathbb{R}^n$ and $f_\rho \in \mathcal{H}_K$, then by taking $\lambda = \lambda(m) = m^{2\epsilon-1}$ with an arbitrary $\epsilon > 0$, for any $0 < \delta < 1$, with confidence $1-\delta$ there holds

$$\|f_{\mathbf{z},\lambda} - f_\rho\|^2_{L^2_{\rho_X}} \leq \widetilde{C}\log\left(\frac{2}{\delta}\right)\left(\frac{1}{m}\right)^{1-\epsilon}$$

for $m \geq m_{\delta,\epsilon}$, where the constants $m_{\delta,\epsilon}$ and $\widetilde{C}$ are independent of m.

Let us briefly mention the idea of deriving the error bounds stated in Theorem 27. For details, see [79]. Take $\widetilde{f_\lambda} = f_\lambda$. Denote $B_R = \{f \in \mathcal{H}_K : \|f\|_K \leq R\}$. The main point is to apply Proposition 26 with $\tau = 1$ to the function set

$$\mathcal{F}_R = \left\{\left(f(x) - y\right)^2 - \left(f_\rho(x) - y\right)^2 : f \in B_R\right\} \tag{16}$$

for various $R>0$. Proposition 26 says that for all $\varepsilon>0$ and $R\geq M$,

$$\mathrm{Prob}_{\mathbf{z}\in Z^m}\left\{\sup_{f\in B_R}\frac{\mathcal{E}(f)-\mathcal{E}(f_\rho)-\big(\mathcal{E}_{\mathbf{z}}(f)-\mathcal{E}_{\mathbf{z}}(f_\rho)\big)}{\sqrt{\mathcal{E}(f)-\mathcal{E}(f_\rho)+\varepsilon}}\leq\sqrt{\varepsilon}\right\}$$

$$\geq 1-\mathcal{N}\left(B_1,\frac{3\varepsilon}{4(\kappa+3)^2R^2}\right)\exp\left\{-\frac{3m\varepsilon}{160(\kappa+3)^2R^2}\right\}.$$

Denote $v^*(m,\delta)$ as the unique positive solution to the equation

$$\log\mathcal{N}(B_1,\eta)-\frac{m\eta}{40}=\log\delta.$$

Then the confidence becomes $1-\delta/2$ if $\varepsilon=\frac{4(\kappa+3)^2R^2}{3}v^*(m,\delta/2)$. It follows that there is a set $V_R'\subseteq Z^m$ of measure at most $\delta/2$ such that for all $f\in B_R$ and $\mathbf{z}\in Z^m\setminus V_R'$,

$$\mathcal{E}(f)-\mathcal{E}(f_\rho)-\big(\mathcal{E}_{\mathbf{z}}(f)-\mathcal{E}_{\mathbf{z}}(f_\rho)\big)\leq\frac{1}{2}\big(\mathcal{E}(f)-\mathcal{E}(f_\rho)\big)+\frac{4}{3}(\kappa+3)^2R^2v^*(m,\delta/2).$$

Thus, if we denote

$$\mathcal{W}(R)=\{\mathbf{z}\in Z^m:\|f_{\mathbf{z},\lambda}\|_K\leq R\},\tag{17}$$

then for $\mathbf{z}\in\mathcal{W}(R)\setminus V_R'$, $f_{\mathbf{z},\lambda}\in B_R$ and $\mathcal{E}(f_{\mathbf{z},\lambda})-\mathcal{E}(f_\rho)-\big(\mathcal{E}_{\mathbf{z}}(f_{\mathbf{z},\lambda})-\mathcal{E}_{\mathbf{z}}(f_\rho)\big)$ becomes

$$E(\xi_1)-\frac{1}{m}\sum_{i=1}^m\xi_1(z_i)\leq\frac{1}{2}\big(\mathcal{E}(f_{\mathbf{z},\lambda})-\mathcal{E}(f_\rho)\big)+\frac{4}{3}(\kappa+3)^2R^2v^*(m,\delta/2).$$

On the other hand, since ξ_2 satisfies $|\xi_2-E(\xi_2)|\leq 2\big(\kappa\sqrt{\mathcal{D}(\lambda)/\lambda}+3M\big)^2$, Lemma 23 tells us that there is another subset $V_R''\subset Z^m$ of measure at most $\delta/2$ such that for all $\mathbf{z}\in Z^m\setminus V_R''$,

$$\frac{1}{m}\sum_{i=1}^m\xi_2(z_i)-E(\xi_2)\leq\mathcal{D}(\lambda)\left(1+\frac{4\kappa^2\log(2/\delta)}{m\lambda}\right)+\frac{36M^2\log(2/\delta)}{m}.$$

Thus for all $\mathbf{z}\in\mathcal{W}(R)\setminus(V_R'\cup V_R'')$,

$$\begin{aligned}\mathcal{E}(f_{\mathbf{z},\lambda})-\mathcal{E}(f_\rho)+\lambda\|f_{\mathbf{z},\lambda}\|_K^2&\leq 3(\kappa+3)^2R^2v^*(m,\delta/2)\\&+\mathcal{D}(\lambda)\left(4+\frac{8\kappa^2\log(2/\delta)}{m\lambda}\right)+\frac{72M^2\log(2/\delta)}{m}.\end{aligned}\tag{18}$$

The key point to derive satisfactory error bounds is to use (18) iteratively. First, by taking $f=0$ in the the definition (15) of the scheme, we see that $\|f_{\mathbf{z},\lambda}\|_K\leq R^{[0]}:=M/\sqrt{\lambda}$. So $\mathcal{W}(R^{[0]})=Z^m$. Then we apply (18) to sharpen the norm and get $Z^m=\mathcal{W}(R^{[0]})\subseteq\mathcal{W}(R^{[1]})\cup V^{[1]}$ where $V^{[1]}=V_{R^{[0]}}'\cup V_{R^{[0]}}''$ and

$$R^{[1]}=(\kappa+3)R\sqrt{3v^*(m,\delta/2)/\lambda}+\sqrt{\frac{\mathcal{D}(\lambda)}{\lambda}}\left(4+\frac{8\kappa^2\log(2/\delta)}{m\lambda}\right)^{1/2}+M\sqrt{\frac{72\log(2/\delta)}{m\lambda}}.$$

By the choice of λ, $R^{[1]}<R^{[0]}$ (for sufficiently large m). That is, the bound $R^{[0]}$ for $\|f_{\mathbf{z},\lambda}\|_K$ has been sharpened to $R^{[1]}$ (with confidence $1-\delta$). Iterating this process

J times with $\frac{1}{1-\zeta(1+s)} - 1 \leq J \leq \frac{1}{1-\zeta(1+s)}$, we can show that with confidence $1 - \frac{\delta}{1-\zeta(1+s)}$ there holds

$$\|f_{\mathbf{z},\lambda}\|_K \leq R^{[J]} = \widetilde{c}\sqrt{\log(2/\delta)}\left(\sqrt{\mathcal{D}(\lambda)/\lambda} + 1\right).$$

Then (18) with $R = R^{[J]}$ yields the desired error bound for $\mathcal{E}(f_{\mathbf{z},\lambda}) - \mathcal{E}(f_\rho)$ stated in Theorem 27.

3.4. Leave-one-out capacity independent error bounds

Another interesting approach for the error analysis of learning schemes is the so-called leave-one-out techniques [75]. They derive error bounds from the behavior of local error changes when only one sample point is removed or replaced. Let us state some results for the regularization scheme (15) associated with $\psi(t) = t^2$.

In [14,86], leave-one-out error and stability analysis were used to derive the expected value of general learning schemes. For the scheme (15), the result in [86] can be expressed as

$$E_{\mathbf{z}\in Z^m}\left(\mathcal{E}(f_{\mathbf{z},\lambda})\right) \leq \left(1 + \frac{2\kappa^2}{m\lambda}\right)^2 \inf_{f\in\mathcal{H}_K}\left\{\mathcal{E}(f) + \lambda\|f\|_K^2\right\}.$$

In terms of the regularization error, it can be restated as

$$E_{\mathbf{z}\in Z^m}\left(\|f_{\mathbf{z},\lambda} - f_\rho\|_{L^2_{\rho_X}}^2\right) \leq \mathcal{D}(\lambda) + \left(\mathcal{E}(f_\rho) + \mathcal{D}(\lambda)\right)\left\{\frac{2\kappa^2}{m\lambda} + \left(\frac{\kappa^2}{m\lambda}\right)^2\right\}.$$

Choosing $\lambda = 1/\sqrt{m}$, the derived learning rate is $\left(\frac{1}{m}\right)^{\frac{1}{2}}$ in expectation when $f_\rho \in \mathcal{H}_K$ and $\mathcal{E}(f_\rho) > 0$. By the Markov inequality, $\|f_{\mathbf{z},\lambda} - f_\rho\|_{L^2_{\rho_X}}^2 \leq \frac{C}{\delta}\left(\frac{1}{m}\right)^{\frac{1}{2}}$ with confidence $1 - \delta$.

In [24], a functional analysis approach was employed for the error analysis of the scheme (15). The main result asserts that for any $0 < \delta < 1$, with confidence $1 - \delta$,

$$\left|\mathcal{E}(f_{\mathbf{z},\lambda}) - \mathcal{E}(f_\lambda)\right| \leq \frac{M\kappa^2}{\sqrt{m}}\left(1 + \frac{\kappa}{\sqrt{\lambda}}\right)\left(1 + \sqrt{2\log(2/\delta)}\right).$$

The corresponding learning rate in [24] is the following: when f_ρ lies in the range of L_K, that is, $L_K^{-1} f_\rho \in L^2_{\rho_X}$, for any $0 < \delta < 1$, with confidence $1 - \delta$, there holds

$$\|f_{\mathbf{z},\lambda} - f_\rho\|_{L^2_{\rho_X}}^2 \leq C\left(\frac{\log(2/\delta)}{m}\right)^{\frac{2}{5}}, \qquad \text{if} \qquad \lambda = \left(\frac{\log(2/\delta)}{m}\right)^{\frac{1}{5}}.$$

Thus the confidence is improved from $1/\delta$ to $\log(2/\delta)$, while the rate is weakened to $\left(\frac{1}{m}\right)^{\frac{2}{5}}$.

In [66], a modified McDiarmid inequality [44,82] was used to improve the capacity independent error bounds. If f_ρ is in the range of L_K, then for any $0 < \delta < 1$, with confidence $1 - \delta$ there hold

$$\|f_{\mathbf{z},\lambda} - f_\rho\|_K^2 \leq \widetilde{C}\left(\frac{\left(\log(4/\delta)\right)^2}{m}\right)^{\frac{1}{3}} \qquad \text{by taking} \qquad \lambda = \left(\frac{\left(\log(4/\delta)\right)^2}{m}\right)^{\frac{1}{3}}$$

and

$$\|f_{\mathbf{z},\lambda} - f_\rho\|^2_{L^2_{\rho_X}} \leq \widetilde{C}\frac{\log(4/\delta)}{\sqrt{m}} \quad \text{by taking} \quad \lambda = \left(\frac{\left(\log(4/\delta)\right)^2}{m}\right)^{\frac{1}{4}}.$$

Thus the confidence for the learning rate $m^{-\frac{1}{2}}$ is improved. Moreover, the error in the $\mathcal{H}_K$-norm can be estimated.

In [67], a Bennett inequality for vector-valued random variables with values in Hilbert spaces is applied, which yields better error bounds: If f_ρ is in the range of L_K, then we have

$$\|f_{\mathbf{z},\lambda} - f_\rho\|^2_{L^2_{\rho_X}} \leq \widetilde{C}\left(\log(4/\delta)\right)^2 \left(\frac{1}{m}\right)^{2/3} \quad \text{by taking} \quad \lambda = \log(4/\delta)\left(\frac{1}{m}\right)^{1/3}.$$

All the above results are capacity independent error bounds. When $f_\rho \in \mathcal{H}_K$ and $s < 2$, the learning rate given by Theorem 27 with $r = 1/2$ is better than existing results.

4. Classification Algorithms via Regularization

Classification algorithms provide ways to separate data into different classes. Consider the case of two different classes, labeled by 1 or -1. A binary classifier $f : X \to Y = \{1, -1\}$ is a function from X to Y which divides the input space X into two classes. It makes a decision for each case (an input $x \in X$) measured in many quantitative ways. If X is a subset of $\mathbb{R}^n$, each point $x \in X$ is a vector with n components corresponding to values of n measurements. These measurements are used to make the decision or to predict some event. For weather prediction, we observe the current weather in n different ways and get a vector $x \in X \subset \mathbb{R}^n$, then a classifier (a method for weather prediction) gives an output value $y = f(x) \in Y$. It predicts whether it will rain (if $f(x) = 1$) or not (if $f(x) = -1$). For the diagnosis of a special disease, a patient is tested by n different instruments which produce a vector $x \in X \subset \mathbb{R}^n$. Then a doctor makes a decision whether the patient gets the disease based on the vector x, according to the values $f(x) \in Y$. A good doctor corresponds to a good classifier. A good classifier has small error, measured by misclassification error.

Definition 28. Let ρ be a probability distribution on $Z = X \times Y$. The *misclassification error* $\mathcal{R}(f)$ of a classifier $f : X \to Y$ is defined to be the probability of a wrong prediction, that is, of the event $\{f(x) \neq y\}$,

$$\mathcal{R}(f) = \text{Prob}_{\mathbf{z}\in Z}\{f(x) \neq y\} = \int_X P(y \neq f(x)|x)d\rho_X.$$

By discussing for each $x \in X$, we can easily see that the best classifier minimizing the misclassification error, called the *Bayes rule*, is given by $f_c(x) = \text{sgn}(f_\rho)$, where

$$\operatorname{sgn}(f)(x) = \begin{cases} 1, & \text{if } f(x) \geq 0, \\ -1, & \text{if } f(x) < 0. \end{cases}$$

The purpose of classification algorithms is to find good approximations $f_{\mathbf{z}}$ of the Bayes rule from the random sample $\mathbf{z} = \{(x_i, y_i)\}_{i=1}^m$ drawn according to the probability distribution ρ. We hope that the approximating classifier $f_{\mathbf{z}}$ will approach the Bayes rule as long as the number of samples is large enough, in the sense that $\mathcal{R}(f_{\mathbf{z}}) \to \mathcal{R}(f_c)$ as $m \to \infty$.

Here we discuss classification algorithms generated from regularization schemes associated with reproducing kernel Hilbert spaces. The classifier is $\operatorname{sgn}(f_{\mathbf{z},\lambda})$ where

$$f_{\mathbf{z},\lambda} := \arg \min_{f \in \mathcal{H}_K} \left\{ \frac{1}{m} \sum_{i=1}^m \phi(y_i f(x_i)) + \lambda \|f\|_K^2 \right\}. \tag{19}$$

Here $\phi : \mathbb{R} \to \mathbb{R}_+$ is a loss function for classification.

4.1. Support vector machine classification

Before we study the general classification scheme (19), we describe a classical learning algorithm, the support vector machine (SVM) classifier [13].

Let us start with the hard margin SVM by hyperplanes in the separable case. Suppose $X \subset \mathbb{R}^n$. The sample $\mathbf{z}$ consists of two classes: $I = \{i : y_i = 1\}$ and $II = \{j : y_j = -1\}$. We say that these two classes are separable by a hyperplane $w \cdot x = b$ with $w \in \mathbb{R}^n$, $|w| = 1$, and $b \in \mathbb{R}$ if

$$\begin{cases} w \cdot x_i > b, & \text{if } i \in I, \\ w \cdot x_j < b, & \text{if } j \in II. \end{cases}$$

That is, points from the different classes lie on different sides of the hyperplane. We always assume that I and II are both nonempty.

Note that the (signed) distance from a point $x^* \in \mathbb{R}^n$ to the plane $w \cdot x = 0$ is $w \cdot x^*$, where a negative distance means x^* is in the other side of w. For $b \in \mathbb{R}$, the hyperplane $w \cdot x = b$ is parallel to $w \cdot x = 0$ with distance b. Thus, the distance from $x^* \in \mathbb{R}^n$ to the hyperplane $w \cdot x = b$ is $w \cdot x^* - b$. Positive and negative signs of the distance stand for different sides of the hyperplane $w \cdot x = b$.

Now if the two classes of the sample $\mathbf{z}$ are separable by the hyperplane $w \cdot x = b$, we know that $b_1(w) := \min_{i \in I} \{w \cdot x_i - b\} > 0$ and $b_2(w) := \max_{j \in II} \{w \cdot x_j - b\} < 0$. To balance, we shift the hyperplane to $w \cdot x = c(w)$, where $c(w) = \frac{1}{2}\{b_1(w) + b_2(w)\}$. Then we see that with $\Delta(w) = \frac{1}{2}\{b_1(w) - b_2(w)\} > 0$, there holds

$$\begin{cases} w \cdot x_i - c(w) \geq \min\limits_{\ell \in I} w \cdot x_\ell - c(w) = \Delta(w), & \text{if } i \in I, \\ w \cdot x_j - c(w) \leq \max\limits_{\ell \in II} w \cdot x_\ell - c(w) = -\Delta(w), & \text{if } j \in I. \end{cases}$$

We call the hyperplanes $w \cdot x = c(w) \pm \Delta(w)$ the separating hyperplanes. We know that each separating hyperplane contains some points from that class. The distance between these two separating hyperplanes is $2\Delta(w)$. So the number $\Delta(w)$ is called the margin associated with the hyperplane $w \cdot x = c(w)$.

The SVM seeks the best hyperplane by requiring $\Delta(w)$ to be the largest. Given the data $\mathbf{z} = \{(x_i, y_i)\}_{i=1}^m$, we solve the optimization problem

$$\Delta(w^*) = \max_{|w|=1} \frac{1}{2}\left\{ \min_{y_i=1} w \cdot x_i - \max_{y_j=-1} w \cdot x_j \right\}. \tag{20}$$

If w^* is a solution, then $\mathbf{z}$ is separable by a hyperplane if and only if $\Delta(w^*) > 0$. In this case, the hyperplane, called the *optimal hyperplane*, is given by $w \cdot x = c(w^*)$, and the number $\Delta(w^*)$ is called the *maximal margin* of the data. The vectors x_i on the separating hyperplanes $w^* \cdot x = c(w^*) \pm \Delta(w^*)$ are called *support vectors*, because these are the vectors determining the separating hyperplanes. The classifier is then given by $f(x) = \operatorname{sgn}(w^* \cdot x - c(w^*))$.

The optimization problem (20) has a unique solution for separable data, which is stated in [75] as follows.

Theorem 29. *If $\mathbf{z}$ is separable by a hyperplane, then the optimization problem (20) has a unique solution w^*. Moreover $w^* = \frac{\widetilde{w}}{|\widetilde{w}|}$ where $\widetilde{w}$ solves the optimization problem*

$$\widetilde{w} := \arg\min_{w \in \mathbb{R}^n} \min_{b \in \mathbb{R}} |w|^2 \tag{21}$$

$$\text{subject to } y_i(w \cdot x_i - b) \geq 1, \quad i = 1, \ldots, m$$

Also, $\Delta(w^*) = \frac{1}{|\widetilde{w}|}$ is the maximal margin.

4.2. SVM soft margin classifier

The optimization problem (21) is called the SVM hard margin classifier. It works well for separable data. But most data are not separable by hyperplanes.

In the nonseparable case, there is no $(w, b) \in \mathbb{R}^n \times \mathbb{R}$ such that $\mathbf{z}$ is separated by the hyperplane $w \cdot x = b$. Then we use the *SVM soft margin classifier* by introducing slack variables $\xi = (\xi_i)_{i=1}^m$

$$(\widetilde{w}, \widetilde{b}) := \arg\min_{w \in \mathbb{R}^n, b \in \mathbb{R}} \min_{\xi \in \mathbb{R}^n} \left\{ |w|^2 + \frac{2C}{m} \sum_{i=1}^m \xi_i \right\}, \tag{22}$$

$$\text{subject to } y_i(w \cdot x_i - b) \geq 1 - \xi_i, \xi_i \geq 0, \quad i = 1, \ldots, m$$

Here $C > 0$ is a trade-off parameter. The SVM soft margin classifier is then defined as $\operatorname{sgn}(\widetilde{w} \cdot x - \widetilde{b})$.

The soft margin classifier (22) can be reduced to the hard margin classifier (21) in the separable case by taking $C = +\infty$ and $\xi_i = 0$.

Define the *hinge loss* ϕ by $\phi(t) = \max\{1-t, 0\}$, we see that the optimal ξ in (22) is given by $\xi_i = \max\{1 - y_i(w \cdot x_i - b), 0\} = \phi(y_i(w \cdot x_i - b))$. Hence (22) can be expressed by means of the hinge loss ϕ as a regularization scheme

$$\min_{w \in \mathbb{R}^n, b \in \mathbb{R}} \left\{ \frac{1}{m} \sum_{i=1}^{m} \phi(y_i(w \cdot x_i - b)) + \frac{1}{2C}|w|^2 \right\}.$$

If we choose a Mercer kernel $K(x,y) = x \cdot y$, then $|w|^2 = \|w \cdot x\|_K^2$ and the above scheme can be written [19] as

$$\min_{f \in \mathcal{H}_K, b \in \mathbb{R}} \left\{ \frac{1}{m} \sum_{i=1}^{m} \phi(y_i(f(x_i) - b)) + \frac{1}{2C}\|f\|_K^2 \right\}. \tag{23}$$

This is the regularization scheme (19) associated with the hinge loss ϕ, except the constant term b, called the *offset*. We shall not discuss the offset. For an approach, see [17].

4.3. Classifying loss functions

The goal of the regularization scheme (19) is to approximate the Bayes rule f_c by $\mathrm{sgn}(f_{\mathbf{z},\lambda})$. Note that f_c is a minimizer of the misclassification error $\mathcal{R}(f) = \int_Z \chi_{\{y \neq f(x)\}} d\rho = \int_Z \chi_{\{yf(x)<0\}} d\rho$. It would then be natural to discretize $\mathcal{R}(f)$ into $\frac{1}{m}\sum_{i=1}^{m} \chi_{\{y_i f(x_i)<0\}}$ and consider the regularization scheme (19) with the misclassification loss

$$\phi_0(t) = \chi_{\{t<0\}}.$$

However, the corresponding optimization problem is not convex when we write $f_{\mathbf{z},\lambda} = \sum_{i=1}^{m} c_i K_{x_i}$. Thus, we choose the loss function ϕ in (19) according to two criteria: (1) ϕ should be convex; (2) ϕ should be close to ϕ_0.

Definition 30. We say that $\phi : \mathbb{R} \to \mathbb{R}_+$ is a *classifying loss function* if it is convex, differentiable at 0 with $\phi'(0) < 0$, and $\inf_{t \in \mathbb{R}} \phi(t) = 0$.

For $(x, y) \in Z$, the value $\phi(yf(x))$ is the local error suffered from the use of f as a model for the process reproducing y at x. The condition $\phi'(0) < 0$ tells us that the error has a major drop as the sign of $f(x)$ changes from $-y$ to y (turning from the other class to the class labeled by y). The last requirement means that for some pairs (x, y) and function f, the error $\phi(yf(x))$ is small. This verifies the ability of "classifying".

Example 31.

(a) For $1 \leq q < \infty$, the SVM q-norm soft margin classifier corresponds to the loss function $\phi(t) = (\max\{1 - t, 0\})^q$.

(b) The least-square loss corresponds to $\phi(t) = (1 - t)^2$ since $y^2 = 1$ for $y \in Y$ implies $\phi(yf(x)) = (1 - yf(x))^2 = (y - f(x))^2$.

(c) The exponential loss is $\phi(t) = e^{-t}$, see [87,43].

(d) The logistic regression is $\phi(t) = \log(1 + e^{-t})$, see [87,43].

In particular, the hinge loss $\phi(t) = \max\{1 - t, 0\}$ is a good approximation of ϕ_0.

Define the *ϕ-error* or ϕ-risk associated with the loss ϕ as

$$\mathcal{E}^{\phi}(f) = \int_Z \phi(yf(x))d\rho,$$

and the *empirical ϕ-error* as $\mathcal{E}^{\phi}_{\mathbf{z}}(f) = \frac{1}{m}\sum_{i=1}^{m}\phi(y_i f(x_i))$. Then the classification algorithm (19) can be studied as the regression scheme (12). In particular, we can see that $\mathcal{E}^{\phi}(f_{\mathbf{z},\lambda}) \to \inf_{f\in\mathcal{H}_K}\mathcal{E}^{\phi}(f)$ with confidence (as $m \to \infty$ and $\lambda(m) \to 0$). However, for the classification algorithm, we are interested in the misclassification error, not the ϕ-error. In [81], we presented a counterexample, and showed that in general one cannot have $\mathcal{R}(\mathrm{sgn}(f_{\mathbf{z},\lambda})) \to \inf_{f\in\mathcal{H}_K}\mathcal{R}(\mathrm{sgn}(f))$ with confidence (as $m \to \infty$) for any $\lambda = \lambda(m)$.

Problem 32. If $\mathcal{H}_K$ is not dense in $C(X)$, for which distributions ρ and choice $\lambda = \lambda(m)$ can we have $\lim_{m\to\infty}\mathcal{R}(\mathrm{sgn}(f_{\mathbf{z},\lambda})) = \inf_{f\in\mathcal{H}_K}\mathcal{R}(\mathrm{sgn}(f))$ with confidence?

Fortunately, the excess misclassification error $\mathcal{R}(\mathrm{sgn}(f)) - \mathcal{R}(f_c)$ can be estimated by the excess ϕ-error $\mathcal{E}^{\phi}(f) - \mathcal{E}^{\phi}(f^{\phi}_{\rho})$ when $\mathcal{H}_K$ is dense in $C(X)$ [68]. Here f^{ϕ}_{ρ} is a minimizer of the ϕ-error:

$$f^{\phi}_{\rho}(x) = \arg\min_{t\in\mathbb{R}} \int_Y \phi(yt)d\rho(y|x) = \arg\min_{t\in\mathbb{R}} \Phi_x(t), \qquad x \in X, \tag{24}$$

and $\Phi_x(t) = \phi(t)P(y=1|x) + \phi(-t)P(y=-1|x)$. For the hinge loss, $f^{\phi}_{\rho} = f_c$. The properties of f^{ϕ}_{ρ} can be investigated as for f^{ψ}_{ρ} in Theorem 16, see [41,78].

Zhang [87] derived the following result for the hinge loss. See [11,43] for related works on boosting methods.

Theorem 33. *Let $\phi(t) = \max\{1-t, 0\}$. For any measurable function $f : X \to \mathbb{R}$,*

$$\mathcal{R}(sgn(f)) - \mathcal{R}(f_c) \le \mathcal{E}^{\phi}(f) - \mathcal{E}^{\phi}(f_c).$$

For general loss functions, the problem was studied in [7,17]. We get the following comparison theorem in [17].

Theorem 34. *If ϕ is a classifying loss function satisfying $\phi''(0) > 0$, then there is a constant c_{ϕ} such that for any measurable function $f : X \to \mathbb{R}$, there holds*

$$\mathcal{R}(sgn(f)) - \mathcal{R}(f_c) \le c_{\phi}\sqrt{\mathcal{E}^{\phi}(f) - \mathcal{E}^{\phi}(f^{\phi}_{\rho})}.$$

Thus, we can estimate the excess misclassification error by means of the excess ϕ-error. But the binary classification problem has a special feature. The classifiers take values in $\{1, -1\}$ and are often signs of real-valued functions. Therefore, it is natural for us [17] to introduce the concept of projection.

Definition 35. The *projection operator* π_M is defined for $M > 0$ on the space of measurable functions $f : X \to \mathbb{R}$ as

$$\pi_M(f)(x) = \begin{cases} M, & \text{if } f(x) > M, \\ -M, & \text{if } f(x) < -M, \\ f(x), & \text{if } -M \le f(x) \le M. \end{cases}$$

By the projection and Theorems 33 and 34, for any $M > 0$ we only need to estimate $\mathcal{E}^\phi(\pi_M(f_{\mathbf{z},\lambda})) - \mathcal{E}^\phi(f_\rho^\phi)$ for the classification algorithm (19).

4.4. Error analysis for regularized classifiers

There has been an extensive investigation on the error analysis of kernel related classification algorithms, especially for SVM with the hinge loss. ERM methods were analyzed in [75,62,5,20,41]. A model selection process was introduced in [10,43,11,69] for error bounds. Leave-one-out techniques including stability analysis were used in [14,86,7]. Modified ERM methods were employed in [70]. A regularization approach was used in [81,17].

Based on previous methods, the error analysis for the general classification algorithm (19) associated with a general classifying loss function ϕ was given in [78].

We need the ℓ^2-empirical covering number $\mathcal{N}_2(\mathcal{H}, \mathbf{x}, \eta)$. The advantage of using this empirical covering number for the RKHS $\mathcal{H}_K$ is that the following is always true for some $p \in (0, 2]$ and $c_p > 0$:

$$\sup_{\mathbf{x} \in X^m, m \in \mathbb{N}} \log \mathcal{N}_2(B_1, \mathbf{x}, \eta) \le c_p (1/\eta)^p, \qquad \forall 0 < \eta \le 1. \tag{25}$$

In fact, we have a sharper bound: $\int_0^1 \left\{\sup_{\mathbf{x} \in X^m, m \in \mathbb{N}} \log \mathcal{N}_2(B_1, \mathbf{x}, \eta)\right\}^{1/2} d\eta < \infty$. For more detailed discussions on empirical covering numbers, see [74,6,10].

As for regressing loss functions, we need an exponent concerning the dominance of the variance by expected values.

Definition 36. We say that the classifying loss ϕ has *variancing exponent* $\tau \in [0, 1]$ if for each $M > 0$ there is a constant C_M such that for any measurable function $f : X \to [-M, M]$, there holds

$$\int \left(\phi(yf(x)) - \phi(yf_\rho^\phi(x))\right)^2 d\rho \le C_M \left\{\mathcal{E}^\phi(f) - \mathcal{E}^\phi(f_\rho^\phi)\right\}^\tau. \tag{26}$$

The variancing exponent τ depends on the strong convexity [45,6,7] of ϕ. For example, for the q-norm SVM loss $\phi(t) = (\max\{1 - t, 0\})^q$, we have $\tau = 1$ if $1 < q \le 2$, and $\tau = 2/q$ if $q > 2$, $\tau = 0$ if $q = 1$.

Theorem 37. *Assume that K satisfies (25), the distribution ρ satisfies (26) and*

$$\mathcal{D}(\lambda) = \inf_{f \in \mathcal{H}_K} \left\{\mathcal{E}^\phi(f) - \mathcal{E}^\phi(f_\rho^\phi) + \lambda \|f\|_K^2\right\} \le c_\beta |\lambda|^\beta, \qquad \forall \lambda > 0 \tag{27}$$

for some $\beta \in (0,1]$ and $c_\beta > 0$. If ϕ has a minimal zero M and with some $q > 0$ it satisfies $\phi(t) \le c_q|t|^q$ for any $|t| \ge M$, then for any $\epsilon > 0$ and $0 < \delta < 1$, there is a constant $\widetilde{c}$ independent on m such that with $\lambda = \lambda(m) = m^{-\gamma}$ there holds

$$\mathcal{E}^\phi(\pi_M(f_{\mathbf{z},\lambda})) - \mathcal{E}^\phi(f_\rho^\phi) \le \widetilde{c}m^{-\theta}$$

with confidence $1-\delta$, where $\gamma := \min\left\{\frac{2}{\beta(4-2\tau+p\tau)+p(1-\beta)}, \frac{2}{2\beta+q-\beta q}\right\}$ and

$$\theta := \min\left\{\frac{2\beta}{\beta(4-2\tau+p\tau)+p(1-\beta)} - \epsilon, \frac{2\beta}{2\beta+q-\beta q}\right\}.$$

In Theorem 37, we assume that the loss ϕ has zeros on $\mathbb{R}$ and ϕ has at most polynomial increment. This excludes the exponential loss $\phi(t) = e^{-t}$or the logistic loss $\phi(t) = \log(1+e^{-t})$. Though the convergence with these loss functions were studied in [11,43] and some error analysis was done in [78], the convergence rates are in general unknown.

Problem 38. For the exponential loss $\phi(t) = e^{-t}$ or the logistic loss $\phi(t) = \log(1+e^{-t})$, what is the convergence rate in terms of the variancing exponent τ, the order β for the regularization error, and capacity index p?

For the hinge loss, $\tau = 0$ and $q = 1$. So Theorem 37 in connection with Theorem 33 yields the convergence rate $\mathcal{R}(\text{sgn}(f_{\mathbf{z},\lambda})) - \mathcal{R}(f_c) = O(m^{-\theta})$ with $\theta = \min\left\{\frac{2\beta}{4\beta+p(1-\beta)} - \epsilon, \frac{2\beta}{\beta+1}\right\}$. When K is C^∞, p can be arbitrarily small, and learning rate θ can be arbitrarily close to $\frac{1}{2}$ if $\beta \ge 1/3$.

In the same way, we can get learning rates for the least-square loss, which can be arbitrarily close to $\frac{1}{2}$ when enough approximation order is assumed.

When the distribution ρ satisfies some noise conditions, the learning rates can be further improved.

4.5. Noise conditions for classification algorithms

In this subsection we discuss some noise conditions. To demonstrate the main idea, we restrict our discussion to the hinge loss $\phi(t) = \max\{1-t, 0\}$.

The classical noise condition is for strictly separable data.

Definition 39. We say that the probability distribution ρ is ***strictly separable*** by $\mathcal{H}_K$ with margin $\Delta > 0$ if there is some $f_{\text{sp}} \in \mathcal{H}_K$ such that $\|f_{\text{sp}}\|_K = 1$ and $y f_{\text{sp}}(x) \ge \Delta$ almost surely.

The strictly separable condition tells us that $\mathcal{R}(f_c) = 0$ and $\mathcal{E}^\phi(f_c) = 0$. By taking the function f_{sp}/Δ in the definition (19), we see that $\mathcal{E}_{\mathbf{z}}^\phi(f_{\mathbf{z},\lambda}) + \lambda\|f_{\mathbf{z},\lambda}\|_K^2 \le \frac{\lambda}{\Delta^2}$, since $\phi(y f_{\text{sp}}(x)/\Delta) = 0$ for almost every $(x,y) \in Z$. It follows that $\|f_{\mathbf{z},\lambda}\|_K \le \frac{1}{\Delta}$

and $\mathcal{E}^{\phi}_{\mathbf{z}}(f_{\mathbf{z},\lambda}) \leq \frac{\lambda}{\Delta^2}$. Thus we can apply the ratio probability inequality, Proposition 26, with $\tau = 1$ to the function set $B_{1/\Delta}$ to estimate the first term of the following error decomposition

$$\mathcal{R}(\mathrm{sgn}(f_{\mathbf{z},\lambda})) \leq \mathcal{E}^{\phi}(f_{\mathbf{z},\lambda}) \leq \mathcal{E}^{\phi}(f_{\mathbf{z},\lambda}) - \mathcal{E}^{\phi}_{\mathbf{z}}(f_{\mathbf{z},\lambda}) + \frac{\lambda}{\Delta^2}.$$

For any $0 < \delta < 1$, with confidence $1 - \delta$ we have

$$\mathcal{R}(\mathrm{sgn}(f_{\mathbf{z},\lambda})) \leq 2\varepsilon^*(m, \delta) + \frac{2\lambda}{\Delta^2},$$

where $\varepsilon^*(m, \delta)$ is the unique positive solution to the equation

$$\log \mathcal{N}(B_1, \frac{\Delta\varepsilon}{4}) - \frac{3m\varepsilon}{128(1 + \kappa/\Delta)} = \log \delta.$$

In particular, if $\log \mathcal{N}(B_1, \eta) = O((\log(1/\eta))^p)$ for some $p > 0$, then

$$\mathcal{R}(\mathrm{sgn}(f_{\mathbf{z},\lambda})) \leq \widetilde{c}\left\{\frac{(\log m)^p}{m}\log(2/\delta)\left(1 + \frac{1}{\Delta}\right) + \frac{\lambda}{\Delta^2}\right\}.$$

A weaker separable condition was introduced in [17] as follows.

Definition 40. We say that the probability distribution ρ is *(weakly) separable* by $\mathcal{H}_K$ if there is some $f_{\mathrm{sp}} \in \mathcal{H}_K$ such that $\|f_{\mathrm{sp}}\|_K = 1$ and $y f_{\mathrm{sp}}(x) > 0$ almost surely. It has *separation exponent* $\theta \in (0, \infty]$ if there are positive constants Δ, c_θ such that

$$\rho_X\left\{x \in X : |f_{\mathrm{sp}}(x)| < \Delta t\right\} \leq c_\theta t^\theta, \qquad \forall t > 0.$$

When $\theta = \infty$, this concept is reduced to the strictly separable condition.

Take a regularizing function $f_\lambda = (c_\theta/\lambda)^{\frac{1}{2+\theta}} \Delta^{-\frac{\theta}{2+\theta}} f_{\mathrm{sp}}$ and consider the error decomposition

$$\mathcal{E}^{\phi}(f_{\mathbf{z},\lambda}) + \lambda\|f_{\mathbf{z},\lambda}\|_K^2 \leq \mathcal{E}^{\phi}(f_{\mathbf{z},\lambda}) - \mathcal{E}^{\phi}_{\mathbf{z}}(f_{\mathbf{z},\lambda}) + \mathcal{E}^{\phi}_{\mathbf{z}}(f_\lambda) - \mathcal{E}^{\phi}(f_\lambda) + \mathcal{E}^{\phi}(f_\lambda) + \lambda\|f_\lambda\|_K^2.$$

We can apply the ratio probability inequality, Proposition 26, with $\tau = 1$ again to the function set B_R with $R := 2(c_\theta/\lambda)^{\frac{1}{2+\theta}} \Delta^{-\frac{\theta}{2+\theta}} + \sqrt{\frac{7\log(2/\delta)}{6m\lambda}}$. Then we know that if $\log \mathcal{N}(B_1, \eta) = O((\log(1/\eta))^p)$ for some $p > 0$, then with confidence $1 - \delta$,

$$\mathcal{R}(\mathrm{sgn}(f_{\mathbf{z},\lambda})) \leq \widetilde{c}\left\{m^{-\frac{\theta}{1+\theta}}(\log m)^p + \frac{\log(2/\delta)}{m} + \lambda^{\frac{\theta}{2+\theta}}\right\}.$$

The previous two noise conditions are for deterministic distributions since $\mathcal{R}(f_c) = 0$. Another noise condition, defined in [73], is also for non-deterministic distributions.

Definition 41. We say that the probability distribution ρ has *Tsybakov's noise exponent* $\zeta \in [0, \infty]$ if for some constant $c_\zeta > 0$,

$$\rho_X\left\{x \in X : |f_\rho(x)| \leq c_\zeta t\right\} \leq t^\zeta, \qquad \forall t > 0.$$

For some detailed discussions, see [67]. The Tsybakov's noise condition helps improve the variancing exponent τ in (26). The following result can be found in [70]. We give a simple proof in [80].

Theorem 42. *If ρ has Tsybakov's noise exponent $\zeta \in [0, \infty]$, then for every function $f : X \to [-1, 1]$, (26) holds with $\tau = \frac{\zeta}{\zeta+1}$, $M = 1$ and $C_1 = 8\big(1/(2c_\zeta)\big)^{\zeta/(\zeta+1)}$.*

Proof. Since $f(x) \in [-1, 1]$, we have $\phi(yf(x)) - \phi(yf_c(x)) = y(f_c(x) - f(x))$. It follows that

$$\mathcal{E}^\phi(f) - \mathcal{E}^\phi(f_c) = \int_X (f_c(x) - f(x)) f_\rho(x) d\rho_X = \int_X |f_c(x) - f(x)|\, |f_\rho(x)| d\rho_X$$

and

$$\int \Big(\phi(yf(x)) - \phi(yf_c(x))\Big)^2 d\rho = \int_X |f_c(x) - f(x)|^2 d\rho_X.$$

Let $t > 0$ and separate the domain X into two sets: $X_t^+ = \{x \in X : |f_\rho(x)| > c_\zeta t\}$ and $X_t^- = \{x \in X : |f_\rho(x)| \leq c_\zeta t\}$. On X_t^+ we have $|f_c(x) - f(x)|^2 \leq 2|f_c(x) - f(x)| \frac{|f_\rho(x)|}{c_\zeta t}$. On X_t^- we have $|f_c(x) - f(x)|^2 \leq 4$. It follows from the Tsybakov's noise condition that

$$\int_X |f_c(x) - f(x)|^2 d\rho_X \leq \frac{2\big(\mathcal{E}^\phi(f) - \mathcal{E}^\phi(f_c)\big)}{c_\zeta t} + 4\rho_X(X_t^-) \leq \frac{2\big(\mathcal{E}^\phi(f) - \mathcal{E}^\phi(f_c)\big)}{c_\zeta t} + 4t^\zeta.$$

Choosing $t = \big\{(\mathcal{E}^\phi(f) - \mathcal{E}^\phi(f_c))/(2c_\zeta)\big\}^{1/(\zeta+1)}$ yields the desired bound. □

Theorem 42 together with Theorems 37 and 33 yields the following convergence rate.

Corollary 43. *Let $\phi(t) = \max\{1-t, 0\}$. Assume (25) and (27) for some $\beta \in (0, 1]$. If ρ has Tsybakov's noise exponent $\zeta \in [0, \infty]$, then for any $\epsilon > 0$ and $0 < \delta < 1$, there is a constant $\widetilde{c}$ independent on m such that with confidence $1 - \delta$*

$$\mathcal{R}(sgn(f_{\mathbf{z},\lambda})) - \mathcal{R}(f_c) \leq \widetilde{c} m^{-\theta},$$

where $\theta := \min\left\{\frac{(\zeta+1)\beta}{\beta(\zeta+2)+(\zeta+1-\beta)p/2} - \epsilon, \frac{2\beta}{\beta+1}\right\}$.

The rate derived in [87] is $\theta = \frac{\beta}{1+\beta}$, at most 1/2 when $\beta = 1$. The rate in [70] is $\theta = \frac{4\beta(1+\zeta)}{(1+\beta)(4+2\zeta+p\zeta)} - \epsilon$. Corollary 43 improves these results because of the projection operator. Another noise condition can be found in [70,61].

5. Further Topics

Learning theory is a rapidly growing field. Many people are working on it, theoretically or practically, from different points of view. But there are still a lot of open areas. For some topics, even fundamental questions remain unsolved. In this survey, we only discuss the regularization approach. It works well with reproducing kernel Hilbert spaces, for regression and classification. Let us briefly mention some further research directions involving regularization schemes.

(1) *Feature selection.* One purpose is to understand structures of high dimensional data. Topics include manifold learning (e.g. [8,53,27,18]) and dimensionality reduction (see [32] and references therein). Another purpose is to determine important features (variables) of huge dimensional functions. Two approaches are filter method and wrapper method [37]. There are some attempts to use regularization schemes like support vector machines for feature selection, e.g. [33,34]. But not so much is known about feature selection in general. See the special issue [32]. Recently, a least-square type algorithm for learning covariances via gradients was introduced in [50]. It involves solving linear systems only and provides promising results for variable selection in some applications of gene expression analysis.

(2) *Multi-kernel regularization schemes.* Let $K_\Sigma = \{K_\sigma : \sigma \in \Sigma\}$ be a set of Mercer kernels on X such as Gaussian kernels with variances σ^2 running over $(0, \infty)$. The *multi-kernel regularization scheme* associated with K_Σ is defined as

$$f_{\mathbf{z},\lambda} = \arg \inf_{\sigma \in \Sigma} \inf_{f \in \mathcal{H}_{K_\sigma}} \left\{ \frac{1}{m} \sum_{i=1}^{m} V\big(y_i, f(x_i)\big) + \lambda \|f\|_{K_\sigma}^2 \right\}.$$

Here $V : \mathbb{R}^2 \to \mathbb{R}_+$ is a general loss function. This multi-kernel scheme is motivated by recent work on learning algorithms with varying kernels. In [16] support vector machines with multiple parameters are investigated. In [40,57] mixture density estimation is considered and Gaussian kernels with variance σ^2 changing on an interval $[\sigma_1^2, \sigma_2^2]$ with $0 < \sigma_1 < \sigma_2 < +\infty$ are used for deriving bounds. Multi-task learning algorithms involve kernels from a convex hull of several Mercer kernels and spaces with changing norms, e.g. [30,35]. Learning of kernel functions is studied in [38,47]. The existence of minimizers of the multi-kernel scheme has been verified for convex hulls of several kernels in [38,47], and for the case when the set of kernels is compact with respect to a metric in Σ in [78], based on the dual formulation in [88]. The error analysis of multi-kernel regularization scheme for classification with has been done in [78].

Problem 44. It would be desirable to have good algorithms for the multi-kernel regularization scheme.

Another related class of multi-kernel regularization schemes are those generated by polynomial kernels $\{K_d(x, y) = (1 + x \cdot y)^d\}$ with $d \in \mathbb{N}$, or in general dot product kernels [54,42]. In [92] we provide some convergence rates in the univariate case $n = 1$ for the multi-kernel regularized classifiers generated by polynomial kernels. The method is Durrmeyer operators from approximation theory. But the situation in the multivariate case is not so clear due to the difficulty of bounding the RKHS norm of the multivariate Durrmeyer operators.

(3) *Regularization with penalty functionals other than* $\|f\|_K^2$. One interesting class deals with the KL-divergence or the entropy such as the maximum entropy discrimination [36]. The convergence and error analysis is not so clear due to the complicated penalty functional (which is not symmetric). Another class is the ν-SVM algorithms [60]. In [80] we investigate the linear programming SVM. For this algorithm, the hypothesis space depends on the sample $\mathbf{z}$, which makes the analysis difficult [77]. We get some error bounds and show that the convergence of the linear programming SVM is not slower than the 1-norm soft margin SVM [80]. But it is unknown whether the convergence of the linear programming SVM is actually faster. Other regularization schemes include one-class SVM [72] and learning of vector-valued functions [46].

(4) *Online learning algorithms.* These algorithms improve the efficiency of learning methods. The convergence is investigated in [75,15], while the error with respect to the step size has been analyzed for the least-square regularized regression in [63] and for regularized classification with a general classifying loss in [84]. But the analysis for online algorithms and related game theory problems needs further study.

(5) *Connections to sampling theory and signal processing.* In many fields like sampling theory, signal processing, and inverse problems [29], the nodes $\overline{x} = \{x_i\}_{i=1}^m$ are deterministic and one needs to recover the function relation from values at these nodes. So the regularization scheme has the same form, but only $\{y_i\}_{i=1}^m$ are (random) samples. Some error analysis is done in [65]. In particular, we consider the following scheme

$$f_{\mathbf{z},\lambda} := \arg \min_{f \in \mathcal{H}_{K,\bar{t}}} \left\{ \frac{1}{m} \sum_{i=1}^{m} (y_i - f(x_i))^2 + \lambda \|f\|_K^2 \right\}.$$

Here $\bar{t}$ is a discrete subset of X and $\mathcal{H}_{K,\bar{t}}$ is the closed subspace of $\mathcal{H}_K$ generated by $\{K_t\}_{t \in \bar{t}}$. The convergence of the above scheme for the case when $\bar{t} \subset \overline{x}$ is part of the random sample, by means of $\#\bar{t}/m \to 0$ has not been investigated so far.

Acknowledgments

This work is supported partially by the Research Grants Council of Hong Kong, Project No. CityU 1144/01P.

References

1. Alon, N., Ben-David, S., Cesa-Bianchi, N. and Haussler, D., Scale-sensitive dimensions, uniform convergence and learnability, *J. ACM* **44** (1997), 615–631.
2. Anthony, M. and Bartlett, P. L., *Neural Network Learning: Theoretical Foundations*, Cambridge University Press, Cambridge, 1999.

3. Aronszajn, N., Theory of reproducing kernels, *Trans. Amer. Math. Soc.* **68** (1950), 337–404.
4. Barron, A. R., Complexity regularization with applications to artificial neural networks, in: *Nonparametric Functional Estimation* (G. Roussa, Ed.), Kluwer, Dordrecht, 1990, pp. 561–576.
5. Bartlett, P. L., The sample complexity of pattern classification with neural networks: the size of the weights is more important than the size of the network, *IEEE Trans. Information Theory* **44** (1998), 525–536.
6. Bartlett, P. L., Bousquet, O. and Mendelson, S., Local Rademacher complexities, *Ann. Stat.*, to appear.
7. Bartlett, P. L., Jordan, M. I. and McAuliffe, J. D., Convexity, classification, and risk bounds, preprint, 2003.
8. Belkin, M. and Niyogi, P., Semisupervised learning on Riemannian manifolds, *Machine Learning* **56** (2004), 209–239.
9. Binev, P., Cohen, A., Dahmen, W., DeVore, R. and Temlyakov, V., Universal algorithms for learning theory, Part I: piecewise constant functions, preprint, 2004.
10. Blanchard, B., Bousquet, O. and Massart, P., Statistical performance of support vector machines, preprint, 2003.
11. Blanchard, B., Lugosi, G. and Vayatis, N., On the rate of convergence of regularized boosting classifiers, *J. Mach. Learning Res.* **4** (2003), 861–894.
12. Breiman, L., Arcing classifiers, *Ann. Stat.* **26** (1998), 801–824.
13. Boser, B. E., Guyon, I. and Vapnik, V., A training algorithm for optimal margin classifiers, in: *Proc. Fifth Annual Workshop of Computational Learning Theory* (D. Haussler, Ed.), ACM, Pittsburgh, 1992, pp. 144–152.
14. Bousquet, O. and Elisseeff, A., Stability and generalization, *J. Mach. Learning Res.* **2** (2002), 499–526.
15. Cesa-Bianchi, N., Long, P. M. and Warmuth, M. K., Worst-case quadratic loss bounds for prediction using linear functions and gradient descent, *IEEE Trans. Neural Networks* **7** (1996), 604–619.
16. Chapelle, O., Vapnik, V., Bousquet, O. and Mukherjee, S., Choosing multiple parameters for support vector machines, *Machine Learning* **46** (2002), 131–159.
17. Chen, D. R., Wu, Q., Ying, Y. and Zhou, D. X., Support vector machine soft margin classifiers: error analysis, *J. Machine Learning Res.* **5** (2004), 1143–1175.
18. Coifman, R. R., Lafon, S., Lee, A. B., Maggioni, M., Nadler, B., Warner, F. and Zucker, S. W., Geometric diffusions as a tool for harmonic analysis and structure definition of data: diffusion maps, *Proc. National Academy of Sciences of USA* **102** (2005), 7426–7431.
19. Cortes, C. and Vapnik, V., Support-vector networks, *Machine Learning* **20** (1995), 273–297.
20. Cristianini, C. and Shawe-Taylor, J., *An Introduction to Support Vector Machines*, Cambridge University Press, Cambridge, 2000.
21. Cucker, F. and Smale, S., On the mathematical foundations of learning, *Bull. Amer. Math. Soc.* **39** (2001), 1–49.

22. Cucker, F. and Smale, S., Best choices for regularization parameters in learning theory: On the bias-variance problem, *Foundations Comput. Math.* **2** (2002), 413–428.
23. Cucker, F. and Zhou, D. X., *Learning Theory: an Approximation Theory Viewpoint*, Cambridge University Press, in preparation.
24. De Vito, E., Caponnetto, A. and Rosasco, L., Model selection for regularized least-squares algorithm in learning theory, *Foundations Comput. Math.* **5** (2005), 59–85.
25. De Vito, E., Rosasco, L., Caponnetto, A., Piana, M. and Verri, A., Some properties of regularized kernel methods, *J. Machine Learning Res.* **5** (2004), 1363–1390.
26. Devroye, L., Györfi, L. and Lugosi, G., *A Probabilistic Theory of Pattern Recognition*, Springer, New York, 1997.
27. Donoho, D. L., For most large underdetermined systems of linear equations the minimal ℓ^1-norm solution is also the sparsest solution, preprint, 2004.
28. Dudley, R. M., Giné, E. and Zinn, J., Uniform and universal Glivenko-Cantelli classes, *J. Theor. Prob.* **4** (1991), 485–510.
29. Engl, H. W., Hanke, M. and Neubauer, A., *Regularization of Inverse Problems*, Mathematics and Its Applications, Vol. 375, Kluwer, Dordrecht, 1996.
30. Evgeniou T. and Pontil, M., Regularized multi-task learning, in: *Proc. 17th SIGKDD Conf. on Knowledge Discovery and Data Mining* (W. Kim, R. Kohavi, J. Gehrke and W. DuMouchel, Eds.), ACM, Seattle, 2004.
31. Evgeniou, T., Pontil, M. and Poggio, T., Regularization networks and support vector machines, *Adv. Comput. Math.* **13** (2000), 1–50.
32. Guyon, I. and Elisseeff, A., An introduction to variable and feature selection, *J. Machine Learning Res.* **3** (2003), 1157–1182.
33. Guyon, I., Weston, J., Barnhill, S. and Vapnik, V., Gene selection for cancer classification using support vector machines, *Machine Learning* **46** (2002), 389–422.
34. Hardin, D., Tsamardinos, I. and Aliferis, C. F., A theoretical characterization of linear SVM-based feature selection, in: *21st Int. Conf. on Machine Learning*, Proc. Banff 2004 (C. E. Brodley, Ed.), ACM, Seattle, 2004.
35. Herbster, M., Relative loss bounds and polynomial-time predictions for the K-LMS-NET algorithm, in: *Proc. 15th Int. Conf. on Algorithmic Learning Theory*, (J. Case and A. Maruoka, Eds.), Lecture Notes in Computer Science, Vol. 3244, Springer, Berlin, 2004.
36. Jaakkola, T., Meila, M. and Jebara, T., Maximum entropy discrimination, MIT AI-Lab Technical Report, Boston, 1999.
37. Kohavi, R. and John, G., Wrappers for feature selection, *Artificial Intelligence* **97** (1997), 273–324.
38. Lanckriet, G. R. G., Cristianini, N., Bartlett, P., El Ghaoui, L. and Jordan, M. I., Learning the kernel matrix with semidefinite programming, *J. Machine Learning Res.* **5** (2004), 27–72.
39. Lee, W. S., Bartlett, P. and Williamson, R., The importance of convexity in learning with least square loss, *IEEE Trans. Information Theory* **44** (1998),

1974–1980.

40. Li, J. and Barron, A., Mixture density estimation, in: *Advances in Neural Information Processing Systems, Vol. 12* (S. A. Solla, T. K. Leen and K. R. Muller, Eds.), Morgan Kaufman Publ., San Mateo, 1999, pp. 279–285.
41. Lin, Y., Support vector machines and the Bayes rule in classification, *Data Mining and Knowledge Discovery* **6** (2002), 259–275.
42. Lu, F. and Sun, H., Positive definite dot product kernels in learning theory, *Adv. Comput. Math.* **22** (2005), 181–198.
43. Lugosi, G. and Vayatis, N., On the Bayes-risk consistency of regularized boosting methods, *Ann. Stat.* **32** (2004), 30–55.
44. McDiarmid, C., Concentration, in: *Probabilistic Methods for Algorithmic Discrete Mathematics* (M. Habib, C. McDiarmid, J. Ramirez-Alfonsin and B. Reed, Eds.), Springer, Berlin, 1998, pp. 195–248.
45. Mendelson, S., Improving the sample complexity using global data., *IEEE Trans. Information Theory* **48** (2002), 1977–1991.
46. Micchelli, C. A. and Pontil, M., On learning vector-valued functions, *Neural Comp.* **17** (2005), 177–204.
47. Micchelli, C. A. and Pontil, M., Learning the kernel function via regularization, technical report, Dept. of Computer Science, University College London, 2004.
48. Micchelli, C. A., Pontil, M., Wu, Q. and Zhou, D. X., Error bounds for learning the kernel, preprint, 2005.
49. Mukherjee, S., Niyogi, P., Poggio, T. and Rifkin, R., Learning theory: stability is sufficient for generalization and necessary and sufficient for empirical risk minimization, *Adv. Comput. Math.*, to appear.
50. Mukherjee, S. and Zhou, D. X., Learning covariances via gradients, preprint, 2005.
51. Niyogi, P., *The Informational Complexity of Learning*, Kluwer, Dordrecht, 1998.
52. Niyogi, P. and Girosi, F., On the relationship between generalization error, hypothesis complexity, and sample complexity for radial basis functions, *Neural Comp.* **8** (1996), 819–842.
53. Niyogi, P., Smale, S. and Weinberger, S., Finding the homology of submanifolds with high confidence from random samples, preprint, 2004.
54. Pinkus, P., Strictly positive definite kernels on a real inner product space, *Adv. Comput. Math.* **20** (2004), 263–271.
55. Poggio, T., Rifkin, R., Mukherjee, S. and Niyogi, P., General conditions for predictivity in learning theory, *Nature* **428** (2004), 419–422.
56. Poggio, T. and Smale, S., The mathematics of learning: dealing with data, *Notices Amer. Math. Soc.* **50** (2003), 537–544.
57. Rakhlin, A., Panchenko, D. and Mukherjee, S., Risk bounds for mixture density estimation, *ESAIM: Probability and Statistics* **9** (2005), 220-229.
58. Schaback, R. and Werner, J., Linearly constrained reconstruction of functions by kernels, with applications to machine learning, *Adv. Comput. Math.*, to appear.

59. Schölkopf, B. and Smola, A. J., *Learning with Kernels*, MIT Press, Cambridge, 2002.
60. Schölkopf, B., Smola, A. J., Williamson, R. C. and Bartlett, P. L., New support vector algorithms, *Neural Comp.* **12** (2000), 1207–1245.
61. Scovel, C., Hush, D. and Steinwart, I., Learning rates for density level detection, *Anal. Appl.*, to appear.
62. Shawe-Taylor, J., Bartlett, P. L., Williamson, R. C. and Anthony, M., Structural risk minimization over data-dependent hierarchies, *IEEE Trans. Information Theory* **44** (1998), 1926–1940.
63. Smale, S. and Yao, Y., Online learning algorithms, *Found. Comput. Math.*, to appear.
64. Smale, S. and Zhou, D. X., Estimating the approximation error in learning theory, *Anal. Appl.* **1** (2003), 17–41.
65. Smale, S. and Zhou, D. X., Shannon sampling and function reconstruction from point values, *Bull. Amer. Math. Soc.* **41** (2004), 279–305.
66. Smale, S. and Zhou, D. X., Shannon sampling II. Connections to learning theory, *Appl. Comput. Harmonic Analysis*, to appear.
67. Smale, S. and Zhou, D. X., Learning theory estimates via integral operators and their approximations, *Constr. Approximation*, special issue on Learning Theory, submitted.
68. Steinwart, I., Support vector machines are universally consistent, *J. Complexity* **18** (2002), 768–791.
69. Steinwart, I., On the influence of the kernel on the consistency of support vector machines, *J. Machine Learning Res.* **2** (2001), 67–93.
70. Steinwart, I. and Scovel, C., Fast rates for support vector machines, in: *Proceedings of the Conference on Learning Theory, COLT-2005*, to appear.
71. Sun, H., Mercer theorem for RKHS on noncompact sets, *J. Complexity*, to appear.
72. Tax, D. M. J. and Duin, R. P. W., Support vector domain description, *Pattern Recognition Letters* **20** (1999), 1191–1199.
73. Tsybakov, A. B., Optimal aggregation of classifiers in statistical learning, *Ann. Stat.* **32** (2004), 135–166.
74. van der Vaart, A. W. and Wellner, J. A., *Weak Convergence and Empirical Processes*, Springer, New York, 1996.
75. Vapnik, V., *Statistical Learning Theory*, John Wiley & Sons, New York, 1998.
76. Wahba, G., *Spline Models for Observational Data*, CBMS-NSF Reg. Conf. Ser. in Applied Mathematics, Vol. 59, SIAM, Philadelphia, 1990.
77. Wu, Q., *Classification and Regularization in Learning Theory*, PhD thesis, City University of Hong Kong, 2005.
78. Wu, Q., Ying, Y. and Zhou, D. X., Multi-kernel regularized classifiers, submitted, 2004.
79. Wu, Q., Ying, Y. and Zhou, D. X., Learning rates of least-square regularized regression, *Found. Comput. Math.*, to appear.
80. Wu, Q. and Zhou, D. X., SVM soft margin classifiers: linear programming versus quadratic programming, *Neural Comp.* **17** (2005), 1160–1187.

81. Wu, Q. and Zhou, D. X., Analysis of support vector machine classification, submitted, 2004.
82. Ying, Y., McDiarmid inequalities of Bernstein and Bennett forms, Technical Report, City University of Hong Kong, 2004.
83. Ying, Y. and Zhou, D. X., Learnability of Gaussians with flexible variances, submitted, 2004.
84. Ying, Y. and Zhou, D. X., Online regularized classification algorithms, preprint, 2005.
85. Yurinsky, Y., *Sums and Gaussian Vectors*, Lecture Notes in Mathematics, Vol. 1617, Springer, Berlin, 1995.
86. Zhang, T., Leave-one-out bounds for kernel methods, *Neural Comp.* **15** (2003), 1397–1437.
87. Zhang, T., Statistical behavior and consistency of classification methods based on convex risk minimization, *Ann. Stat.* **32** (2004), 56–85.
88. Zhang, T., On the dual formulation of regularized linear systems with convex risks, *Machine Learning* **46** (2002), 91–129.
89. Zhou, D. X., The covering number in learning theory, *J. Complexity* **18** (2002), 739–767.
90. Zhou, D. X., Capacity of reproducing kernel spaces in learning theory, *IEEE Trans. Information Theory* **49** (2003), 1743–1752.
91. Zhou, D. X., Density problem and approximation error in learning theory, preprint, 2003.
92. Zhou, D. X. and Jetter, K., Approximation with polynomial kernels and SVM classifiers, *Adv. Comput. Math.*, to appear.

Topics in Multivariate Approximation and Interpolation
K. Jetter et al., Editors

Coherent States from Nonunitary Representations

Georg Zimmermann

Institut für Angewandte Mathematik und Statistik, Universität Hohenheim, D–70593 Stuttgart, Germany

For Ina
with thanks for her patience
and moral support.

Abstract

We try to obtain a transformation of wavelet-type with a reproducing property on the unit circle using the group of Möbius transformations. Since the natural unitary representations of this group are not square integrable, we use a nonunitary representation together with its contragredient representation. We also present conditions under which this approach works in general.

Key words: coherent states, nonunitary representation, Möbius transformation, reproducing property

2000 MSC: 22E45, 42C40, 43A65

Email address: Georg.Zimmermann@uni-hohenheim.de (Georg Zimmermann).

1. Introduction

1.1. Coherent states from unitary representations

In recent years, wavelets had enormous success in a broad variety of disciplines, practically in every area where large quantities of data need to be denoised or compressed. Wavelet bases turned out to be a versatile tool, and not only in applications, but also for mathematical problems. The central idea is the use of a single shape, described by a function $\varphi : \mathbb{R} \to \mathbb{R}$, and to vary its position and scale. In mathematical terms, we apply the affine group of translation and dilation operators, given by

$$\left(\mathrm{T}_b\,\mathrm{D}_a\,\varphi\right)(x) = \frac{1}{\sqrt{a}}\,\varphi\Big(\frac{x-b}{a}\Big), \qquad (a,b) \in \mathbb{R}^{+} \times \mathbb{R}.$$

This enables us to study given data at various positions (by translation), like using a microscope at various resolutions (by dilation).

The importance of the continuous wavelet transformation lies in the reproducing equation

$$\iint\limits_{\mathbb{R}^{+}\times\mathbb{R}} \Big\langle f(t), \frac{1}{\sqrt{a}}\,\varphi\Big(\frac{t-b}{a}\Big)\Big\rangle\, \frac{1}{\sqrt{a}}\,\varphi\Big(\frac{x-b}{a}\Big)\,\frac{da\,db}{a^2} = C_\varphi\, f(x)\,,$$

where C_φ is a constant that can be made equal one by simply replacing φ by an appropriate multiple, which we now want to assume to be done.

To get a better understanding of the principle behind this equation, we compose the left hand side in two steps. First, there is the ***analysis operator*** which maps a function $f \in \boldsymbol{L}^2(\mathbb{R})$ to its wavelet transform given by

$$\left(\mathcal{W}_\varphi f\right)(a,b) = \left\langle f, \mathrm{T}_b\,\mathrm{D}_a\,\varphi\right\rangle = \int\limits_{\mathbb{R}} f(x)\,\overline{\frac{1}{\sqrt{a}}\,\varphi\Big(\frac{x-b}{a}\Big)}\,dx\,.$$

This is followed by the ***synthesis operator***, the wavelet reconstruction

$$\iint\limits_{\mathbb{R}^{+}\times\mathbb{R}} \left(\mathcal{W}_\varphi f\right)(a,b)\,\frac{1}{\sqrt{a}}\,\varphi\Big(\frac{x-b}{a}\Big)\,\frac{da\,db}{a^2} = f(x)\,.$$

In a sense, we use the family $\left(\mathrm{T}_b\,\mathrm{D}_a\,\varphi\right)_{(a,b)\in\mathbb{R}^{+}\times\mathbb{R}}$ as building blocks to generate the (arbitrary) function f, and the wavelet transform $\mathcal{W}_\varphi f$ tells us the appropriate weight for each block. The important point is that this family of building blocks shows a lot of structure, since it is generated from a single function φ by applying a group of operators. This is usually referred to as a family of *coherent states.*

There is a general mathematical principle behind this phenomenon, of which we want to give a brief overview. Details can be found, *e.g.*, in [18], [21], [23], and [24]. Given a group G and a Hilbert space $\boldsymbol{H}$, a *representation* of G on $\boldsymbol{H}$ is a group homomorphism $\rho : G \to \boldsymbol{GL}(\boldsymbol{H})$. The representation is *unitary*, if $\rho(g)$ is unitary for each $g \in G$, *i.e.*, if we have $\rho : G \to \mathcal{U}(\boldsymbol{H})$. For such a unitary representation ρ, we can define the analysis operator $T_{\rho,\varphi}$ via

$$(T_{\rho,\varphi} f)(g) = \langle f, \rho(g)\varphi \rangle ,$$

for some appropriate φ. Note that $T_{\rho,\varphi}$ maps each vector $f \in \boldsymbol{H}$ to a function on G, the "ρ-transform" of f, also denoted a *matrix coefficient* of ρ. Unitarity of ρ implies that $\|\rho(g)\varphi\|_{\boldsymbol{H}} = \|\varphi\|_{\boldsymbol{H}}$ for all $g \in G$, and therefore $|\langle f, \rho(g)\varphi\rangle| \leq \|f\|_{\boldsymbol{H}} \|\varphi\|_{\boldsymbol{H}}$, so the function $T_{\rho,\varphi} f$ is bounded on G. If $T_{\rho,\varphi} f$ is even continuous for all φ and f, then we call ρ a *continuous representation.*

A subspace $\boldsymbol{H}_1$ of $\boldsymbol{H}$ is *ρ-invariant,* if $\rho(g)\,\boldsymbol{H}_1 \subseteq \boldsymbol{H}_1$ for all $g \in G$. The representation ρ is *irreducible,* if the only ρ-invariant closed subspaces of $\boldsymbol{H}$ are the trivial ones, *i.e.*, $\{0\}$ and $\boldsymbol{H}$ itself. Otherwise, ρ is *reducible.*

The (formal) adjoint of $T_{\rho,\varphi}$ is the synthesis operator $T^*_{\rho,\varphi}$, which maps an (integrable) function F on the group G back into our space $\boldsymbol{H}$ via

$$T^*_{\rho,\varphi} F = \int_G F(g)\, \rho(g)\varphi \, dm(g) .$$

For this integral to make sense at all, the group G has to be a topological group with a locally compact topology, since then, there exists a (left) Haar measure with respect to which we can integrate. Details on locally compact groups, left and right Haar measure, the modular function, and unimodularity can be found, *e.g.*, in [15], [19], or [25].

If we can choose φ in such a way that $T_{\rho,\varphi}$ maps $\boldsymbol{H}$ into $\boldsymbol{L}^2(G)$, *i.e.*, that $T_{\rho,\varphi} f$ is a square integrable function for each $f \in \boldsymbol{H}$, then we say that ρ is a *square integrable representation,* and φ is *admissible* for ρ. In that case, we have

$$T_{\rho,\varphi} : \boldsymbol{H} \to \boldsymbol{L}^2(G) \qquad \text{and thus} \qquad T^*_{\rho,\varphi} : \boldsymbol{L}^2(G) \to \boldsymbol{H} ,$$

so we can concatenate the two operators to obtain the linear map $T^*_{\rho,\varphi} T_{\rho,\varphi} : \boldsymbol{H} \to \boldsymbol{H}$. If, furthermore, the representation ρ is irreducible, then $T_{\rho,\varphi}$ can be shown to be an isometry (after proper normalization of φ), *i.e.*, we have

$$\|f\|^2_{\boldsymbol{H}} = \int_G |\langle f, \rho(g)\varphi\rangle|^2 \, dm(g) .$$

This implies that $T_{\rho,\varphi}$ is a unitary operator, and thus we have the reproducing property

$$T^*_{\rho,\varphi} T_{\rho,\varphi} = \mathrm{Id}_{\boldsymbol{H}} ,$$

or, explicitly,

$$\int_G \langle f, \rho(g)\varphi\rangle\, \rho(g)\varphi \, dm(g) = f .$$

This means that we can reconstruct any element of $\boldsymbol{H}$ using the building blocks $(\rho(g)\varphi)_{g\in G}$, a family of *coherent states from a unitary representation.*

It might help the intuitive understanding of what follows if we look at this construction as a generalization of an orthonormal basis. Consider a family of vectors $(v_i)_{i\in I}$ in a Hilbert space $\boldsymbol{H}$, and the associated analysis operator

$$T_v : \boldsymbol{H} \to \mathbb{C}^I , \quad x \mapsto \big(\langle x, v_i\rangle\big)_{i\in I} .$$

Its formal adjoint is the synthesis operator

$$T_v^* : \mathbb{C}^I \to \boldsymbol{H} , \quad (c_i)_{i\in I} \mapsto \textstyle\sum_{i\in I} c_i\, v_i ,$$

which a priori is only defined for finitely supported sequences. If $(v_i)_{i\in I}$ is a Bessel family, though, which means that T_v is a bounded map into $\ell^2(I)$, then T_v^* is well-defined on all of $\ell^2(I)$. In particular, this is the case if $(v_i)_{i\in I}$ is an orthonormal system. If, furthermore, this system is complete in $\boldsymbol{H}$, *i.e.*, an orthonormal basis, then T_v is an isometry between $\boldsymbol{H}$ and $\ell^2(I)$, since we have

$$\|x\|_{\boldsymbol{H}}^2 = \textstyle\sum_{i\in I} \big|\langle x, v_i\rangle\big|^2 .$$

Again, this implies that T_v is unitary, and we have the reproducing property

$$T_v^* \, T_v = \mathrm{Id}_{\boldsymbol{H}} ,$$

or, explicitly,

$$\textstyle\sum_{i\in I} \langle x, v_i\rangle\, v_i = x .$$

This simple setup exhibits a number of parallels to the one above: the Bessel property corresponds to ρ being square integrable; completeness somehow reflects irreducibility, since the latter ensures that the orbit of every nonzero element spans all of $\boldsymbol{H}$; and orthonormality corresponds to ρ being unitary, since together with the other properties, after an appropriate renormalization, this makes $T_{\rho,\varphi}$ a unitary operator.

1.2. Schur's Lemma and the reproducing property

The central principle behind the whole construction is given by Schur's Lemma, which in its basic form states the following.

Proposition 1 (Schur's Lemma). [23, Prop. 1.5] *A unitary representation $\rho : G \to \mathcal{U}(\boldsymbol{H})$ is irreducible if and only if the only bounded linear operators on $\boldsymbol{H}$ which commute with all $\rho(g)$, $g \in G$, are the multiples of the identity.*

A generalization of Schur's Lemma considers two unitary representations and their intertwining operators.

Proposition 2. [18, Prop. A.1] (i) *Consider an irreducible unitary representation $\rho_1 : G \to \mathcal{U}(\boldsymbol{H}_1)$ and another (not necessarily irreducible) unitary representation $\rho_2 : G \to \mathcal{U}(\boldsymbol{H}_2)$. Suppose that T is a closed linear operator from $\boldsymbol{H}_1$ to $\boldsymbol{H}_2$ with* *domain $\mathbb{D}$ dense in $\boldsymbol{H}_1$ and invariant under ρ_1, and assume that T intertwines ρ_1 with ρ_2. Then $\mathbb{D} = \boldsymbol{H}_1$, and T is a multiple of an isometry.*

(ii) *If, in addition, $\boldsymbol{H}_1 = \boldsymbol{H}_2$ and $\rho_1 = \rho_2$, then T is a multiple of the identity.*

For our purposes, the importance of Schur's Lemma lies in the so-called *orthogonality relations* for the matrix coefficients, since these result in the reproducing property we are interested in. The following results actually go back to [5], [8].

Lemma 3. *Let $\rho : G \to \mathcal{U}(\boldsymbol{H})$ be a square integrable, irreducible unitary representation. Then the following hold.*

(i) *If $\varphi \in \boldsymbol{H}$ is admissible, then $\langle f, \rho(g)\varphi\rangle \in \boldsymbol{L}^2(G)$ for all $f \in \boldsymbol{H}$, and the map*

$$T_{\rho,\varphi} : \boldsymbol{H} \to \boldsymbol{L}^2(G)\,, \quad f \mapsto \Big(g \mapsto \langle f, \rho(g)\varphi\rangle\Big)\,,$$

is a multiple of an isometry.

(ii) *The set of admissible vectors is invariant under $\rho(G)$ and hence dense in $\boldsymbol{H}$.*

Proposition 4. [18, Thm. 3.1] *Let $\rho : G \to \mathcal{U}(\boldsymbol{H})$ be a square integrable, irreducible unitary representation. Then there exists a unique self-adjoint positive operator C such that the following hold.*

(i) *The domain of C is the set of admissible vectors.*

(ii) *For any vectors f_1, $f_2 \in \boldsymbol{H}$ and any admissible vectors φ_1, $\varphi_2 \in \boldsymbol{H}$, we have*

$$\int_G \langle f_1, \rho(g)\varphi_1\rangle\, \overline{\langle f_2, \rho(g)\varphi_2\rangle}\, dm(g) = \langle f_1, f_2\rangle\, \overline{\langle C\varphi_1, C\varphi_2\rangle}\,. \tag{1}$$

(iii) *If G is unimodular, then C is a multiple of the identity, and thus all vectors are admissible.*

The Bessel property of a sequence of vectors in a Hilbert space has two aspects: on the one hand, the range of the coefficient map lies in $\boldsymbol{\ell}^2$, and on the other hand, the evaluation map (which is simply the adjoint of the former) is well defined from $\boldsymbol{\ell}^2$ into the Hilbert space.

Lemma 3.(i) states that for an admissible φ, the family $\big(\rho(g)\varphi\big)_{g\in G}$ has a generalized Bessel property, and thus, by considering the adjoint operator, obtain the following result.

Corollary 5. *Let $\rho : G \to \mathcal{U}(\boldsymbol{H})$ be a square integrable, irreducible unitary representation. Then the following hold.*

(i) *If $\varphi \in \boldsymbol{H}$ is admissible, then the map*

$$T^*_{\rho,\varphi} : F \mapsto \int_G F(g)\, \rho(g)\varphi\, dm(g)$$

is a bounded linear operator from $\boldsymbol{L}^2(G)$ onto $\boldsymbol{H}$, where the integral converges weakly in $\boldsymbol{H}$.

(ii) **(Reproducing Property)**
indexreproducing property If $\varphi_{1,2} \in \boldsymbol{H}$ are admissible, then $T^*_{\rho,\varphi_2} T_{\rho,\varphi_1} = \langle C\varphi_2, C\varphi_1 \rangle \operatorname{Id}_{\boldsymbol{H}}$, i.e.,

$$\int_G \langle f, \rho(g)\varphi_1 \rangle \, \rho(g)\varphi_2 \, dm(g) = \langle C\varphi_2, C\varphi_1 \rangle f \quad \text{for all } f \in \boldsymbol{H}, \tag{2}$$

where the integral converges weakly in $\boldsymbol{H}$, and C is the operator from Proposition 4.

1.3. The reducible case

In case the representation ρ is not irreducible, we call it *completely reducible*, if we can decompose the underlying Hilbert space $\boldsymbol{H}$ into irreducible, pairwise orthogonal closed subspaces

$$\boldsymbol{H} = \bigoplus_i \boldsymbol{H}^{(i)} . \tag{3}$$

Accordingly, the representation ρ decomposes into its irreducible components $\rho^{(i)} : G \to \boldsymbol{\mathcal{U}}(\boldsymbol{H}^{(i)})$ with $\rho^{(i)}(g) = \rho(g)\Big|_{\boldsymbol{H}^{(i)}}$.

In case the $\rho^{(i)}$ are pairwise nonequivalent, the orthogonality relation (1) becomes

$$\int_G \langle f_1, \rho(g)\varphi_1 \rangle \, \overline{\langle f_2, \rho(g)\varphi_2 \rangle} \, dm(g) = \sum_i \langle f_1^{(i)}, f_2^{(i)} \rangle \, \overline{\langle C^{(i)}\varphi_1^{(i)}, C^{(i)}\varphi_2^{(i)} \rangle}, \tag{1'}$$

and (2) changes accordingly. Nevertheless, the reproducing property does still hold if we can choose the $\varphi_{1,2}^{(i)}$ in such a way that the constants $\langle C^{(i)}\varphi_1^{(i)}, C^{(i)}\varphi_2^{(i)} \rangle$ do not depend on i. In case the decomposition (3) is finite, this is certainly possible.

1.4. Affine wavelets as an example

Let us have a closer look at affine wavelets from the aspect of the theory described above. Here, the group G is the affine group $\mathbb{R}^* \ltimes \mathbb{R}$ with $(a, b) \circ (a', b') = (a\,a', b + a\,b')$, also known as the $a\,x + b$-group. The left Haar measure on this group is given by

$$dm(a, b) = \frac{da\,db}{a^2} .$$

The Hilbert space $\boldsymbol{H}$ is $\boldsymbol{L}^2(\mathbb{R})$, and the representation is given by

$$\big(\rho(a, b)\varphi\big)(x) = \frac{1}{\sqrt{|a|}} \, \varphi\Big(\frac{x - b}{a}\Big) .$$

This representation is unitary, irreducible, and square integrable. The admissibility condition is given by

$$\int_{\widehat{\mathbb{R}}} \frac{\big|\widehat{\varphi}(\xi)\big|^2}{|\xi|} \, d\xi < \infty ,$$

and the operator C can be written as

$$C\varphi = \left(\frac{\widehat{\varphi}}{\sqrt{|\xi|}}\right)^{\vee}.$$

Consequently, we have the reproducing formula

$$\iint_{\mathbb{R}^2} \left\langle f(t), \frac{1}{\sqrt{|a|}}\varphi_1\left(\frac{t-b}{a}\right)\right\rangle \frac{1}{\sqrt{|a|}}\varphi_2\left(\frac{x-b}{a}\right)\frac{da\,db}{|a|^2} = C_{\varphi_2,\varphi_1} f(x),$$

where

$$C_{\varphi_2,\varphi_1} = \langle C\varphi_2, C\varphi_1\rangle = \int_{\widehat{\mathbb{R}}} \frac{\widehat{\varphi}_2(\xi)\,\overline{\widehat{\varphi}_1(\xi)}}{|\xi|}\,d\xi.$$

In applications, it is common to use for G the subgroup $\mathbb{R}^+ \ltimes \mathbb{R}$ of the affine group. For this G, the representation ρ given above is not irreducible anymore; instead, $\boldsymbol{H} = \boldsymbol{L}^2(\mathbb{R})$ is the orthogonal sum of the two irreducible subspaces $\boldsymbol{H}^+$ and $\boldsymbol{H}^-$, where

$$\boldsymbol{H}^+ = \{f \in \boldsymbol{L}^2(\mathbb{R}) : \operatorname{supp}\widehat{f} \subseteq [0,\infty[\}$$

is the Hardy space, and, analogously, $\boldsymbol{H}^-$ consists of the functions in $\boldsymbol{L}^2(\mathbb{R})$ whose Fourier transforms are supported on the left half axis. In this setup, the reproducing property still holds if

$$\langle C\varphi_2^-, C\varphi_1^-\rangle = \langle C\varphi_2^+, C\varphi_1^+\rangle,$$

which amounts to

$$\int_{-\infty}^{0} \frac{\widehat{\varphi}_2(\xi)\,\overline{\widehat{\varphi}_1(\xi)}}{|\xi|}\,d\xi = \int_{0}^{+\infty} \frac{\widehat{\varphi}_2(\xi)\,\overline{\widehat{\varphi}_1(\xi)}}{|\xi|}\,d\xi.$$

(The equality of the two integrals can be guaranteed, *e.g.*, by choosing $\varphi_{1,2}$ to be equal and real-valued.) If we normalize this constant to equal 1, we can reconstruct f in $\boldsymbol{L}^2(\mathbb{R})$ via

$$\int_0^{\infty}\int_{-\infty}^{+\infty} \left\langle f(t), \frac{1}{\sqrt{a}}\varphi_1\left(\frac{t-b}{a}\right)\right\rangle \frac{1}{\sqrt{a}}\varphi_2\left(\frac{x-b}{a}\right) db\,\frac{da}{a^2} = f(x). \tag{4}$$

1.5. Möbius wavelets on the unit circle

Since wavelets have been so successful, several approaches have been made to transfer the basic concept to other domains than just the real line or Euclidean space. We want to concentrate on the unit circle, where, as on any topological group, there exists a natural group of translation operators, which in this particular case are simply rotations of the circle. The problem is, however, as stated in [20], "On the circle it is difficult to define a good dilation operator."

Our approach to this problem is to use the group of Möbius transformations mapping the unit circle to itself while preserving the orientation, *i.e.*, mapping the unit disk onto itself. We shall see at the beginning of Section 2 that the construction described above does not work, since the natural unitary representations we would like to use are not square integrable. Therefore, we try a different approach and give up the unitarity condition on the representation. To make up for that, we have to use a different representation, the ***contragredient representation*** ρ^*, for the synthesis operator.

We show that these two representations are square integrable on the Korobov or Sobolev type spaces $\boldsymbol{H}_{-1/2}(\mathbb{T})$ and $\boldsymbol{H}_{1/2}(\mathbb{T})$, respectively (Proposition 7 and Corollary 8), and that their action is irreducible on the respective subspaces of holomorphic and antiholomorphic functions (Proposition 9 and Corollary 10), which is very much comparable to the situation in the case of affine wavelets on the real line. Finally, we show in Corollary 12 that, again after proper renormalization, we indeed have the reproducing property

$$\int_G \langle f, \rho^*(g)\varphi_1\rangle \, \rho(g)\varphi_2 \, dm(g) = f ,$$

where f, $\varphi_2 \in \boldsymbol{H}_{-1/2}(\mathbb{T})$ and $\varphi_1 \in \boldsymbol{H}_{1/2}(\mathbb{T})$.

It turns out, though, that behind the scenes, there nevertheless is a unitary representation at work. The strong advantage of our approach lies in the fact that we work with the standard inner product from $\boldsymbol{L}^2(\mathbb{T})$, which is certainly highly desirable for applications.

We finish Section 2 with a few graphical examples for the action of the group.

1.6. Reproducing properties from nonunitary representations

The approach that proves successful in this special example is again based on a general principle, which we describe in Section 3. To this end, we have to give up the Hilbert space setup and consider a Banach space $\boldsymbol{V}$ instead. In this case, we find that associated with a representation $\rho : G \to \boldsymbol{GL}(\boldsymbol{V})$, there exists a contragredient representation $\rho^* : G \to \boldsymbol{GL}(\boldsymbol{V}^*)$, given by

$$\rho^*(g) = \big(\rho(g^{-1})\big)^* = \big(\rho(g)^{-1}\big)^* .$$

Note that for a unitary representation on a Hilbert space, the definition of ρ^* yields $\rho^* = \rho$, and thus our construction will turn out to be a true generalization of the unitary approach described above.

In our analogy to bases discussed above, this approach corresponds to considering two families $(v_i)_{i\in I} \subset \boldsymbol{V}$ and $(w_i)_{i\in I} \subset \boldsymbol{V}^*$, and associated with these, the analysis and synthesis operators

$$T_w : \boldsymbol{V} \to \mathbb{C}^I , \; x \mapsto \big(\langle x, w_i\rangle\big)_{i\in I}, \quad \text{with} \quad T_w^* : \mathbb{C}^I \to \boldsymbol{V}^*, \; (c_i)_{i\in I} \to \textstyle\sum_{i\in I} c_i\, w_i ,$$

and

$$T_v : \boldsymbol{V}^* \to \mathbb{C}^I , \, y \mapsto \big(\langle v_i, y\rangle\big)_{i\in I}, \quad \text{with} \quad T_v^* : \mathbb{C}^I \to \boldsymbol{V}^{**} , (c_i)_{i\in I} \to \textstyle\sum_{i\in I} c_i\, v_i .$$

If $(v_i, w_i)_{i\in I}$ is a Schauder basis for $\boldsymbol{V}$ (and thus in particular, a complete biorthogonal system), then we have two reproducing properties for $\boldsymbol{V}$ and $\boldsymbol{V}^*$, respectively, given by

$$T_v^* T_w = \mathrm{Id}_{\boldsymbol{V}} \qquad \text{and} \qquad T_w^* T_v = \mathrm{Id}_{\boldsymbol{V}^*} .$$

This is due to the fact that $(v_i, w_i)_{i\in I}$ being a Schauder basis for $\boldsymbol{V}$ implies that $(w_i, v_i)_{i\in I}$ is a Schauder basis for $\boldsymbol{V}^*$. If we actually are considering Riesz bases, then the coefficient spaces are again $\ell^2(I)$, and $\boldsymbol{V}$ (and thus $\boldsymbol{V}^*$ also) is topologically isomorphic to a Hilbert space.

We can weaken these conditions somewhat by considering a pair of dual frames rather than bases, which in the above simply means that the operators T_v and T_w are injective, but not surjective, and thus have T_w^* and T_v^* as left inverses only.

Our results in Section 3 show that this analogy indeed can be carried over to a nonunitary representation ρ on a Banach space $\boldsymbol{V}$ together with its contragredient representation ρ^* on $\boldsymbol{V}^*$.

All in all, we show that in the standard method of generating reproducing identities from square integrable, irreducible unitary representations, it is possible to give up the unitarity condition, and to use a pair of mutually contragredient representations instead. As an example, we work with the group of Möbius transformations on the unit circle.

Ideas for further research include the application of the developed theory to a wavelet-type transformation on the sphere using the group of all Möbius transformations. These include all rotations of the sphere, which form the natural group of translations, and the maps $z \mapsto a\,z$ for $a > 0$ are a good family of generalized dilations. Again, we encounter the problem that the natural unitary representations are not square integrable. In [1], this problem has been approached differently, but is seems promising to try and apply our methods.

2. Möbius Wavelets on the Unit Circle

We want to study the following setup. For the Hilbert space, we want to use $\boldsymbol{H} = \boldsymbol{L}^2(\mathbb{T})$, where we use the unit circle $\mathbb{S}^1$ in $\mathbb{C}$ as model for $\mathbb{T}$, and for the group G the group of Möbius transformations of $\overline{\mathbb{C}} = \mathbb{C} \cup \{\infty\}$ which map $\mathbb{S}^1$ to itself, preserving its orientation.

2.1. The group and its Haar measure

Möbius transformations are complex functions of the form $z \mapsto \frac{a\,z+b}{c\,z+d}$, where a, b, c, $d \in \mathbb{C}$ with $a\,d - b\,c \neq 0$. They are the biholomorphic maps $\overline{\mathbb{C}} \to \overline{\mathbb{C}}$ and form a group under concatenation. A peculiarity of this group is the fact that the mapping

$$\begin{pmatrix} a & b \\ c & d \end{pmatrix} \mapsto \left(z \mapsto \frac{a\,z+b}{c\,z+d} \right)$$

is a group homomorphism from $\boldsymbol{GL}_2(\mathbb{C})$. Since $\frac{a\,z+b}{c\,z+d} \equiv \frac{a'\,z+b'}{c'\,z+d'}$ if and only if $\begin{pmatrix} a & b \\ c & d \end{pmatrix} = \lambda \begin{pmatrix} a' & b' \\ c' & d' \end{pmatrix}$ for some $\lambda \in \mathbb{C}^*$, the group of Möbius transformations is isomorphic to the quotient of $\boldsymbol{GL}_2(\mathbb{C})$ or $\boldsymbol{SL}_2(\mathbb{C})$ by their respective center, usually referred to as $\boldsymbol{PGL}_2(\mathbb{C})$ or $\boldsymbol{PSL}_2(\mathbb{C})$.

To determine the subgroup G which maps the unit circle to itself and preserves the orientation, note that since $|z| = 1$ is equivalent to $z = 1/\overline{z}$, an element $g \in G$ has to satisfy

$$g(z) = \frac{a\,z+b}{c\,z+d} = 1\big/\overline{g(z)} = \frac{\overline{c}\,\overline{z}+\overline{d}}{\overline{a}\,\overline{z}+\overline{b}} = \frac{\overline{c}+\overline{d}\,z}{\overline{a}+\overline{b}\,z} \qquad \text{for all } z \text{ with } z = 1/\overline{z}.$$

Since a holomorphic function is determined by its values on the set $\{|z| = 1\}$, we must have

$$\begin{pmatrix} a & b \\ c & d \end{pmatrix} = \lambda \begin{pmatrix} \overline{d} & \overline{c} \\ \overline{b} & \overline{a} \end{pmatrix}$$

which implies $\lambda\overline{\lambda} = 1$. Fixing a value for $\sqrt{\lambda}$, we define $\alpha = a/\sqrt{\lambda}$ and $\beta = b/\sqrt{\lambda}$ which yields $c/\sqrt{\lambda} = \sqrt{\lambda}\,\overline{b} = \overline{\beta}$ and $d/\sqrt{\lambda} = \sqrt{\lambda}\,\overline{a} = \overline{\alpha}$, *i.e.*, we have

$$g(z) = \frac{\alpha\,z+\beta}{\overline{\beta}\,z+\overline{\alpha}} \quad \text{for some } \alpha,\, \beta \in \mathbb{C} \text{ with } |\alpha|^2-|\beta|^2 \neq 0.$$

Since we want g to preserve the orientation of the unit circle, it has to map the unit disk onto itself, so $|g(0)| = |\beta/\overline{\alpha}| < 1$, *i.e.*, $|\alpha| > |\beta|$. Thus we may impose the normalization $|\alpha|^2-|\beta|^2 = 1$, which determines the pair (α, β) up to a factor of ± 1.

In other words, G is isomorphic to the quotient group of

$$\boldsymbol{SU}_{1,1} = \left\{ \begin{pmatrix} \alpha & \beta \\ \overline{\beta} & \overline{\alpha} \end{pmatrix} : |\alpha|^2-|\beta|^2 = 1 \right\}$$

by its center $\{\pm I_2\}$, usually referred to as $\boldsymbol{PSU}_{1,1}$, and we can use $\boldsymbol{SU}_{1,1}$ as a double cover for G.

In particular, Haar measure on G is equivalent to Haar measure on $\boldsymbol{SU}_{1,1}$, and to integrate over G we may integrate over $\boldsymbol{SU}_{1,1}$ (the extra factor of two simply means choosing a different normalization of the Haar measure). To be able to integrate over $\boldsymbol{SU}_{1,1}$, we use the map

$$\mathbb{R}^+ \times \mathbb{T} \times \mathbb{T} \to \boldsymbol{SU}_{1,1}\,, \quad (s, \phi, \theta) \mapsto \begin{pmatrix} e^{2\pi i\phi} \cosh(s) & e^{2\pi i\theta} \sinh(s) \\ e^{-2\pi i\theta} \sinh(s) & e^{-2\pi i\phi} \cosh(s) \end{pmatrix}.$$

This map covers $\boldsymbol{SU}_{1,1}$ except for the set $\left\{ \begin{pmatrix} e^{2\pi i\phi} & 0 \\ 0 & e^{-2\pi i\phi} \end{pmatrix} : \phi \in \mathbb{T} \right\}$ which is a submanifold of codimension 2 and as such a set of measure zero. We leave it as an exercise to the reader to show that both the left and the right Haar measure for $\boldsymbol{SU}_{1,1}$ on this map are given by

$$dm(s, \phi, \theta) = \sinh 2s \, ds \, d\phi \, d\theta \,,$$

so in particular, $\boldsymbol{SU}_{1,1}$ and thus G is unimodular. We will frequently use the substitution

$$(u,v) = (e^{2\pi i\phi}, e^{2\pi i\theta}) \quad \text{with} \quad d\phi\, d\theta = \frac{du}{2\pi i\, u}\,\frac{dv}{2\pi i\, v}\,.$$

To simplify notation, we shall write (α,β) for $\left(\begin{smallmatrix}\alpha & \beta\\ \overline{\beta} & \overline{\alpha}\end{smallmatrix}\right)$, and use the abbreviation

$$m_{(\alpha,\beta)}(z) := \frac{\alpha\, z+\beta}{\overline{\beta}\, z+\overline{\alpha}} \quad \text{for any } \alpha,\ \beta \in \mathbb{C} \text{ with } |\alpha|^2-|\beta|^2 \neq 0.$$

The following isomorphism should at least be mentioned here. We have $\boldsymbol{SL}_2(\mathbb{R}) \cong \boldsymbol{SU}_{1,1}$ via

$$\begin{pmatrix} a & b\\ c & d\end{pmatrix} \longleftrightarrow \frac{1}{2}\begin{pmatrix} 1 & -i\\ -i & 1\end{pmatrix}\begin{pmatrix} a & b\\ c & d\end{pmatrix}\begin{pmatrix} 1 & i\\ i & 1\end{pmatrix} = \frac{1}{2}\begin{pmatrix}(a+d)+i(b-c) & (b+c)+i(a-d)\\ (b+c)-i(a-d) & (a+d)-i(b-c)\end{pmatrix},$$

i.e., they are conjugate subgroups of $\boldsymbol{SL}_2(\mathbb{C})$. This carries over to $\boldsymbol{PSL}_2(\mathbb{R}) \cong \boldsymbol{PSU}_{1,1}$, which geometrically means the following. $\boldsymbol{PSL}_2(\mathbb{R})$ represents the subgroup of Möbius transformations mapping the real line to itself, preserving orientation. Since the Möbius transformation

$$z \mapsto \frac{z+i}{i\,z+1}$$

maps the unit circle to the real line and the unit disk to the upper half plane, it intertwines the two groups.

2.2. Unitary representation of G on $L^2(\mathbb{T})$

We want to work with a representation of G on $\boldsymbol{L}^2(\mathbb{T})$ of the form

$$\big(\rho(\alpha,\beta)\varphi\big)(z) = \omega(\alpha,\beta;z)\,\varphi\big(m_{(\alpha,-\overline{\beta})}(z)\big) = \omega(\alpha,\beta;z)\,\varphi\Big(\frac{\alpha\, z-\overline{\beta}}{-\beta\, z+\overline{\alpha}}\Big),$$

where ω is an appropriate weight function. The reason for this Ansatz instead of using $\varphi\big(m_{(\alpha,\beta)}(z)\big)$ is the fact that the latter yields

$$\rho\big((\alpha,\beta)\circ(\alpha',\beta')\big) = \rho(\alpha',\beta')\,\rho(\alpha,\beta)$$

instead of the correct group law (compare the case of affine wavelets, where for the very same reason we have to use $\varphi\big((x-b)/a\big)$ instead of $\varphi(a\,x+b)$). It might seem more natural to use $\varphi\big(m_{(\alpha,\beta)^{-1}}(z)\big) = \varphi\big(m_{(\overline{\alpha},-\beta)}(z)\big)$, but the above approach has become standard in the literature (*e.g.*, [23]).

To determine $\omega(\alpha,\beta;z)$, note that the substitution $z = \frac{\alpha w-\overline{\beta}}{-\beta\, w+\overline{\alpha}}$ with $dz = \frac{dw}{(-\beta\, w+\overline{\alpha})^2}$ yields

$$\oint\limits_{|z|=1} |\varphi(z)|^2\,\frac{dz}{2\pi i\, z} = \oint\limits_{|w|=1} \Big|\varphi\Big(\frac{\alpha\, w-\overline{\beta}}{-\beta\, w+\overline{\alpha}}\Big)\Big|^2\,\frac{w}{(\alpha\, w-\overline{\beta})(-\beta\, w+\overline{\alpha})}\,\frac{dw}{2\pi i\, w}\,.$$

So ρ being unitary is equivalent to

$$\left|\omega(\alpha,\beta;z)\right|^2 = \frac{z}{(\alpha\, z-\overline{\beta})(-\beta\, z+\overline{\alpha})} = \frac{1}{|-\beta\, z+\overline{\alpha}|^2}\,.$$

To ensure that $\rho(1,0)=I$, we must have $\omega(1,0;z)\equiv 1$, so natural candidates are $1/|-\beta\, z+\overline{\alpha}|$ and $1/(-\beta\, z+\overline{\alpha})$. These lead to the following two representations:

$$\big(\mathcal{P}^{+,0}(\alpha,\beta)\,\varphi\big)(z) = \frac{1}{|-\beta\, z+\overline{\alpha}|}\,\varphi\Big(\frac{\alpha\, z-\overline{\beta}}{-\beta\, z+\overline{\alpha}}\Big)\,,$$

which acts irreducibly on $\boldsymbol{L}^2(\mathbb{T})$, but is not square integrable; and

$$\big(\mathcal{P}^{-,0}(\alpha,\beta)\,\varphi\big)(z) = \frac{1}{(-\beta\, z+\overline{\alpha})}\,\varphi\Big(\frac{\alpha\, z-\overline{\beta}}{-\beta\, z+\overline{\alpha}}\Big)\,.$$

This representation is reducible, it acts irreducibly on the space of analytic functions on the disk and on the space of their conjugate complex functions, and it is not square integrable, either. Note, though, that $\mathcal{P}^{-,0}(-\alpha,-\beta) = -\mathcal{P}^{-,0}(\alpha,\beta)$, so $\mathcal{P}^{-,0}$ is not even a representation of G, but only of $\boldsymbol{SU}_{1,1}$; but that by itself would not make much difference for our purposes. For details on $\mathcal{P}^{\pm,0}$ and unitary representations of $\boldsymbol{SU}_{1,1}$ on other spaces, we refer to [23, § II.5&6].

The fact that these representations are not square integrable can be interpreted as saying that the constant C_{φ_1,φ_2} is always infinity. What can be done to nevertheless get useful results? One possibility is to restrict the representation to a subset of the group, where one has to ensure that the action is still irreducible. This has been used in [1] for the case $G = \boldsymbol{PSL}_2(\mathbb{C})$ and $\boldsymbol{H} = \boldsymbol{L}^2(\mathbb{S}^2)$. In this approach, technical problems arise from the loss of the group structure, unless there exists an appropriate subgroup.

We should also mention that in [4], the approach is totally different, since there, the authors consider the complete discrete series of representations of $\boldsymbol{SU}_{1,1}$.

2.3. A nonunitary representation of G

We want to take a different approach, namely, to give up the unitarity of the representation. This idea is similar to giving up the concept of an orthonormal basis and using a general basis instead; consequently, one has to work with a dual basis. In analogy, we should expect having to use something like a dual representation, called the *contragredient* representation, for the dual side to preserve the reproducing property. (This notion should not be confused with the *adjoint* representation.)

Instead of the unitary representation described above, we start from the $\boldsymbol{L}^1$-normalized version

$$\begin{aligned}\big(\rho(\alpha,\beta)\varphi\big)(z) &= \frac{z}{(\alpha\, z-\overline{\beta})(-\beta\, z+\overline{\alpha})}\,\varphi\Big(\frac{\alpha\, z-\overline{\beta}}{-\beta\, z+\overline{\alpha}}\Big)\\ &= \frac{1}{|-\beta\, z+\overline{\alpha}|^2}\,\varphi\Big(\frac{\alpha\, z-\overline{\beta}}{-\beta\, z+\overline{\alpha}}\Big)\,,\end{aligned}$$

which by the same calculation as in Section 2.2 can be shown to be isometric on $\boldsymbol{L}^1(\mathbb{T})$. Therefore, it seems natural to use on the dual side the $\boldsymbol{L}^\infty$-normalized version

$$\big(\rho^*(\alpha,\beta)\varphi\big)(z) = \varphi\Big(\frac{\alpha\, z - \overline{\beta}}{-\beta\, z + \overline{\alpha}}\Big)\,,$$

and we shall show that this yields the orthogonality relation

$$\int_G \big\langle f_1\,,\, \rho^*(\alpha,\beta)\varphi_1\big\rangle \big\langle \rho(\alpha,\beta)\varphi_2\,,\, f_2\big\rangle\, dm(\alpha,\beta) = \langle f_1, f_2\rangle\, \langle \varphi_2, \varphi_1\rangle$$

for appropriate $f_{1,2}$ and $\varphi_{1,2}$.

Note that by substituting $w = \frac{\alpha\, z - \overline{\beta}}{-\beta\, z + \overline{\alpha}}$, we have

$$\begin{aligned}\big\langle \varphi_1\,,\, \rho^*(\alpha,\beta)\varphi_2\big\rangle &= \oint_{|z|=1} \varphi_1(z)\,\overline{\varphi_2\Big(\frac{\alpha\, z - \overline{\beta}}{-\beta\, z + \overline{\alpha}}\Big)}\,\frac{dz}{2\pi i\, z} \\ &= \oint_{|w|=1} \varphi_1\Big(\frac{\overline{\alpha}\, w + \overline{\beta}}{\beta\, w + \alpha}\Big)\,\overline{\varphi_2(w)}\,\frac{w}{(\overline{\alpha}\, w + \overline{\beta})(\beta\, w + \alpha)}\,\frac{dw}{2\pi i\, w} \\ &= \big\langle \rho(\overline{\alpha}, -\beta)\varphi_1\,,\, \varphi_2\big\rangle = \big\langle \rho((\alpha,\beta)^{-1})\varphi_1\,,\, \varphi_2\big\rangle\,, \end{aligned} \tag{5}$$

i.e., $\rho^*(\alpha,\beta) = \big(\rho((\alpha,\beta)^{-1})\big)^*$.

2.4. The matrix coefficients

Let us calculate the matrix coefficients of ρ^* and ρ for $f(z) = z^n$ and $\varphi(z) = z^m$, *i.e.*,

$$\big\langle f\,,\, \rho^*(\alpha,\beta)\varphi\big\rangle = \oint_{|z|=1} z^n\, \overline{\Big(\frac{\alpha\, z - \overline{\beta}}{-\beta\, z + \overline{\alpha}}\Big)^m}\, \frac{dz}{2\pi i\, z} =: A^*_{n,m}(\alpha,\beta)$$

and

$$\big\langle \rho(\alpha,\beta)\varphi\,,\, f\big\rangle = \oint_{|z|=1} \frac{z}{(\alpha\, z - \overline{\beta})(-\beta\, z + \overline{\alpha})}\Big(\frac{\alpha\, z - \overline{\beta}}{-\beta\, z + \overline{\alpha}}\Big)^m\, \overline{z^n}\, \frac{dz}{2\pi i\, z} =: A_{m,n}(\alpha,\beta)\,.$$

Lemma 6. (i) *The matrix coefficients of ρ^* are given as follows.*

$$a)\; m > 0 : A^*_{n,m}(\alpha,\beta) = \begin{cases} \dfrac{1}{\overline{\alpha}^{n+m}} \displaystyle\sum_{j=1}^{\min\{n,m\}} \binom{m}{m-j}(-\beta)^{m-j}\binom{n-1}{n-j}\overline{\beta}^{n-j}\,, & n > 0\,, \\ \Big(-\dfrac{\beta}{\alpha}\Big)^m, & n = 0\,, \\ 0\,, & n < 0\,; \end{cases}$$

$b)\; m = 0 \,:\, A^*_{n,0}(\alpha,\beta) = \delta_{0,n}\;;$

$c)\; m < 0 \,:\, A^*_{n,m}(\alpha,\beta) = \overline{A^*_{-n,|m|}(\alpha,\beta)}\;.$

(ii) *The matrix coefficients of* ρ *satisfy*

$$A_{m,n}(\alpha,\beta) = A^*_{m,n}(\overline{\alpha},-\beta)\,.$$

Proof. Part (ii) follows from (5). In order to verify (i), we first consider the case $m > 0$ in three steps. For $n > 0$, we have

$$\begin{aligned} A^*_{n,m}(\alpha,\beta) &= \oint\limits_{|z|=1} \overline{z^n \left(\frac{\alpha z - \overline{\beta}}{-\beta z + \overline{\alpha}}\right)^m} \frac{dz}{2\pi i z} = \frac{1}{2\pi i} \oint\limits_{|z|=1} z^{n-1} \left(\frac{-\beta z + \overline{\alpha}}{\alpha z - \overline{\beta}}\right)^m dz \\ &= \frac{1}{2\pi i} \oint\limits_{|z|=1} H(z) \frac{1}{(z-\overline{\beta}/\alpha)^m}\, dz \\ &= \frac{1}{(m-1)!} H^{(m-1)}\left(\frac{\overline{\beta}}{\alpha}\right), \end{aligned}$$

where $H(z) = \left(-\frac{\beta}{\alpha}\right)^m z^{n-1} \left(z - \frac{\overline{\alpha}}{\beta}\right)^m$. By the Leibniz' product rule,

$$\begin{aligned} \frac{1}{(m-1)!} H^{(m-1)}(z) &= \\ &= \left(-\frac{\beta}{\alpha}\right)^m \frac{1}{(m-1)!} \sum_{\substack{\ell=0 \\ \ell \le n-1}}^{m-1} \binom{m-1}{\ell} \frac{(n-1)!}{(n-1-\ell)!} z^{n-1-\ell} \,\cdot \\ &\qquad \cdot \frac{m!}{(m-(m-1-\ell))!} \left(z - \frac{\overline{\alpha}}{\beta}\right)^{m-(m-1-\ell)} \\ &\underset{(j=\ell+1)}{=} \left(-\frac{\beta}{\alpha}\right)^m \sum_{j=1}^{\min\{n,m\}} \binom{n-1}{n-j} z^{n-j} \binom{m}{j} \left(z - \frac{\overline{\alpha}}{\beta}\right)^j \end{aligned}$$

and thus

$$\begin{aligned} A^*_{n,m}(\alpha,\beta) &= \left(-\frac{\beta}{\alpha}\right)^m \sum_{j=1}^{\min\{n,m\}} \binom{n-1}{n-j} \left(\frac{\overline{\beta}}{\alpha}\right)^{n-j} \binom{m}{j} \left(\frac{\overline{\beta}}{\alpha} - \frac{\overline{\alpha}}{\beta}\right)^j \\ &= \frac{1}{\alpha^{m+n}} \sum_{j=1}^{\min\{n,m\}} \binom{m}{m-j} (-\beta)^{m-j} \binom{n-1}{n-j} \overline{\beta}^{n-j}\,. \end{aligned}$$

For $n = 0$,

$$\begin{aligned} A^*_{0,m}(\alpha,\beta) &= \oint\limits_{|z|=1} \overline{1 \left(\frac{\alpha z - \overline{\beta}}{-\beta z + \overline{\alpha}}\right)^m} \frac{dz}{2\pi i z} = \frac{1}{2\pi i} \oint\limits_{|z|=1} \frac{1}{z} \left(\frac{-\beta z + \overline{\alpha}}{\alpha z - \overline{\beta}}\right)^m dz \\ &\qquad \text{(use } w = \tfrac{1}{z},\ z = \tfrac{1}{w},\ dz = -\tfrac{1}{w^2}\, dw\text{; this reverses the orientation:)} \\ &= -\frac{1}{2\pi i} \oint\limits_{|w|=1} w \left(\frac{-\beta/w + \overline{\alpha}}{\alpha/w - \overline{\beta}}\right)^m \frac{dw}{-w^2} \\ &= \frac{1}{2\pi i} \oint\limits_{|w|=1} \frac{1}{w} \left(\frac{-\beta + \overline{\alpha} w}{\alpha - \overline{\beta} w}\right)^m dw = \left(-\frac{\beta}{\alpha}\right)^m, \end{aligned}$$

and for $n < 0$,

$$A^*_{n,m}(\alpha,\beta) = \oint\limits_{|z|=1} z^n \overline{\left(\frac{\alpha\, z - \overline{\beta}}{-\beta\, z + \overline{\alpha}}\right)^m} \frac{dz}{2\pi i\, z} = \frac{1}{2\pi i} \oint\limits_{|z|=1} z^{n-1} \left(\frac{-\beta\, z + \overline{\alpha}}{\alpha\, z - \overline{\beta}}\right)^m dz$$

(use $w = \frac{1}{z}$, $z = \frac{1}{w}$, $dz = -\frac{1}{w^2}\, dw$; this reverses the orientation:)

$$= -\frac{1}{2\pi i} \oint\limits_{|w|=1} w^{1-n} \left(\frac{-\beta/w + \overline{\alpha}}{\alpha/w - \overline{\beta}}\right)^m \frac{dw}{-w^2}$$

$$= \frac{1}{2\pi i} \oint\limits_{|w|=1} w^{|n|-1} \left(\frac{-\beta + \overline{\alpha}\, w}{\alpha - \overline{\beta}\, w}\right)^m dw = 0\,.$$

For $m = 0$, we obtain

$$A^*_{n,0}(\alpha,\beta) = \oint\limits_{|z|=1} z^n\, 1\, \frac{dz}{2\pi i\, z} = \frac{1}{2\pi i} \oint\limits_{|z|=1} z^{n-1}\, dz = \delta_{0,n}\,.$$

Finally, for $m < 0$, we note that we have

$$\overline{\oint\limits_{|z|=1} f(z)\, \frac{dz}{2\pi i\, z}} = \overline{\int_0^1 f(e^{2\pi i\,\theta})\, d\theta}$$

$$= \int_0^1 \overline{f(e^{2\pi i\,\theta})}\, d\theta = \oint\limits_{|z|=1} \overline{f(z)}\, \frac{dz}{2\pi i\, z}\,.$$

Therefore,

$$A^*_{n,m}(\alpha,\beta) = \oint\limits_{|z|=1} z^n \overline{\left(\frac{\alpha\, z - \overline{\beta}}{-\beta\, z + \overline{\alpha}}\right)^m} \frac{dz}{2\pi i\, z} = \overline{\oint\limits_{|z|=1} \overline{z^n} \left(\frac{\alpha\, z - \overline{\beta}}{-\beta\, z + \overline{\alpha}}\right)^m \frac{dz}{2\pi i\, z}}$$

$$= \overline{\oint\limits_{|z|=1} z^{-n} \overline{\left(\frac{\alpha\, z - \overline{\beta}}{-\beta\, z + \overline{\alpha}}\right)^{-m}} \frac{dz}{2\pi i\, z}} = \overline{A^*_{-n,-m}(\alpha,\beta)}\,.$$

This completes the proof. □

2.5. Square integrability

We saw in Lemma 6 that the matrix coefficients of ρ^* for the constant function $f(z) \equiv 1$ are given by

$$A^*_{0,m}(\alpha,\beta) = \begin{cases} \left(-\frac{\beta}{\alpha}\right)^m, & m > 0\,, \\ 1\,, & m = 0\,, \\ \left(-\frac{\overline{\beta}}{\overline{\alpha}}\right)^{|m|}, & m < 0\,. \end{cases}$$

Since these are all not square integrable on G, it only makes sense to consider for f functions with zero mean.

Proposition 7. *For* $f(z) = \sum_{n\neq 0} a_n\, z^n$ *and* $\varphi(z) = \sum_{m\in\mathbb{Z}} b_m\, z^m$, *we have*

$$\int_G \left|\langle f\,,\, \rho^*(\alpha,\beta)\varphi\rangle\right|^2 dm(\alpha,\beta) =$$

$$= \Big(\sum_{n>0} |a_n|^2\, \tfrac{1}{n}\Big)\Big(\sum_{m>0} |b_m|^2\, m\Big) + \Big(\sum_{n<0} |a_n|^2\, \tfrac{1}{|n|}\Big)\Big(\sum_{m<0} |b_m|^2\, |m|\Big). \quad (6)$$

Proof. We begin by evaluating the integral

$$\int_G A^*_{n,m}(\alpha,\beta)\, \overline{A^*_{p,q}(\alpha,\beta)}\, dm(\alpha,\beta) =: B^*_{n,m,p,q}$$

for $n\,p \neq 0$. If $n\,m \leq 0$ or $p\,q \leq 0$, then $A^*_{n,m} \equiv 0$ or $A^*_{p,q} \equiv 0$ and thus $B^*_{n,m,p,q} = 0$.
For n, $m > 0$ and p, $q > 0$, we have by Lemma 6

$$\begin{aligned}
B^*_{n,m,p,q} &= \int_G \Big(\frac{1}{\alpha^{n+m}} \sum_{j=1}^{\min\{n,m\}} \binom{m}{m-j}(-\beta)^{m-j}\binom{n-1}{n-j}\overline{\beta}^{\,n-j}\Big) \\
&\qquad \Big(\frac{1}{\overline{\alpha}^{p+q}} \sum_{\ell=1}^{\min\{p,q\}} \binom{q}{q-\ell}(-\overline{\beta})^{q-\ell}\binom{p-1}{p-\ell}\beta^{p-\ell}\Big)\, dm(\alpha,\beta) \\
&= \sum_{j=1}^{\min\{n,m\}} \sum_{\ell=1}^{\min\{p,q\}} \binom{m}{m-j}\binom{n-1}{n-j}\binom{q}{q-\ell}\binom{p-1}{p-\ell}(-1)^{m-j+q-\ell} \\
&\qquad \int_0^\infty \oint_{|u|=1} \oint_{|v|=1} \frac{v^{m-n-q+p}}{u^{n+m-p-q}} \frac{(\sinh s)^{n+m-2j+p+q-2\ell}}{(\cosh s)^{n+m+p+q}} \frac{du}{2\pi i\, u} \frac{dv}{2\pi i\, v} \sinh 2s\, ds\,.
\end{aligned}$$

Now integration with respect to u and v yields 0 unless we have both $m+n-q-p = 0$ and $m-n-q+p = 0$, *i.e.*, $(n,m) = (p,q)$. In this case, we obtain

$$\begin{aligned}
B^*_{n,m,n,m} &= \sum_{j,\ell=1}^{\min\{n,m\}} \binom{m}{m-j}\binom{n-1}{n-j}\binom{m}{m-\ell}\binom{n-1}{n-\ell}(-1)^{j+\ell} \int_0^\infty \frac{(\sinh s)^{2m+2n-2j-2\ell+1}}{(\cosh s)^{2m+2n-1}}\, 2\, ds \\
&\underset{\text{(Lemma A.2)}}{=} \sum_{j,\ell=1}^{\min\{n,m\}} (-1)^{j+\ell}\binom{m}{m-j}\binom{n-1}{n-j}\binom{m}{m-\ell}\binom{n-1}{n-\ell} \frac{(m+n-j-\ell)!\,(j+\ell-2)!}{(m+n-1)!} \\
&= \frac{1}{n\binom{m+n-1}{m}} \sum_{j=1}^{\min\{n,m\}} (-1)^{n-j}\binom{m}{j} \sum_{\ell=1}^{\min\{n,m\}} (-1)^{n-\ell}\binom{n}{\ell}\binom{m+n-j-\ell}{n-j}\binom{j+\ell-2}{j-1} \\
&\underset{\text{(Lemma A.1)}}{=} \frac{1}{n\binom{m+n-1}{m}} (-1)^{n-1}\binom{m}{1}(-1)^{n+1}\binom{m+n-1}{n-1} = \frac{m}{n}\,.
\end{aligned}$$

For n, $m > 0$ and p, $q < 0$, we have

$$B^*_{n,m,p,q} = \int_G A^*_{n,m}(\alpha,\beta)\, \overline{A^*_{p,q}(\alpha,\beta)}\, dm(\alpha,\beta) = \int_G A^*_{n,m}(\alpha,\beta)\, A^*_{|p|,|q|}(\alpha,\beta)\, dm(\alpha,\beta)$$

$$= \int_G \left(\frac{1}{\alpha^{n+m}} \sum_{j=1}^{\min\{n,m\}} \binom{m}{m-j} (-\beta)^{m-j} \binom{n-1}{n-j} \overline{\beta}^{n-j} \right)$$

$$\left(\frac{1}{\alpha^{|p|+|q|}} \sum_{\ell=1}^{\min\{|p|,|q|\}} \binom{|q|}{|q|-\ell} (-\beta)^{|q|-\ell} \binom{|p|-1}{|p|-\ell} \overline{\beta}^{|p|-\ell} \right) dm(\alpha,\beta)$$

$$= \sum_{j=1}^{\min\{n,m\}} \sum_{\ell=1}^{\min\{|p|,|q|\}} \binom{m}{m-j} \binom{n-1}{n-j} \binom{|q|}{|q|-\ell} \binom{|p|-1}{|p|-\ell} (-1)^{m-j+|q|-\ell}$$

$$\int_0^\infty \oint_{|u|=1} \oint_{|v|=1} \frac{v^{m-n+|q|-|p|}}{u^{n+m+|p|+|q|}} \frac{(\sinh s)^{n+m-2j+|p|+|q|-2\ell}}{(\cosh s)^{n+m+|p|+|q|}} \frac{du}{2\pi i\, u} \frac{dv}{2\pi i\, v} \sinh 2s\, ds\,,$$

and integration with respect to u yields 0.

For n, $m < 0$, we have

$$B^*_{n,m,p,q} = \int_G A^*_{n,m}\, \overline{A^*_{p,q}}\, dm = \int_G \overline{A^*_{-n,-m}}\, A^*_{-p,-q}\, dm = \overline{B^*_{-n,-m,-p,-q}}\,.$$

All in all, we find for n, $p \neq 0$

$$B^*_{n,m,p,q} = \begin{cases} \delta_{(n,m),(p,q)}\, \dfrac{m}{n}\,, & n\,m > 0\,, \\ 0\,, & n\,m \leq 0\,. \end{cases}$$

From this, we obtain

$$\int_G \left| \langle f\,,\, \rho^*(\alpha,\beta)\varphi \rangle \right|^2 dm(\alpha,\beta) =$$

$$= \int_G \left\langle \sum_{n\neq 0} a_n\, z^n\,,\, \rho^*(\alpha,\beta) \sum_{m\in\mathbf{Z}} b_m\, z^m \right\rangle \overline{\left\langle \sum_{p\neq 0} a_p\, z^p\,,\, \rho^*(\alpha,\beta) \sum_{q\in\mathbf{Z}} b_q\, z^q \right\rangle}\, dm(\alpha,\beta)$$

$$= \sum_{n\neq 0} a_n \sum_{m\in\mathbf{Z}} \overline{b_m} \sum_{p\neq 0} \overline{a_p} \sum_{q\in\mathbf{Z}} b_q\, B^*_{n,m,p,q}$$

$$= \sum_{n,m>0} |a_n|^2\, |b_m|^2\, \frac{m}{n} + \sum_{n,m<0} |a_n|^2\, |b_m|^2\, \frac{m}{n}\,,$$

which yields the claim. □

Corollary 8. *For* $\varphi(z) = \sum_{p\neq 0} c_p\, z^p$ *and* $f(z) = \sum_{q\in\mathbb{Z}} d_q\, z^q$*, we have*

$$\int_G \left|\langle \rho(\alpha,\beta)\varphi\,,\, f\rangle\right|^2 dm(\alpha,\beta) =$$

$$= \Big(\sum_{p>0} |c_p|^2\, \tfrac{1}{p}\Big)\Big(\sum_{q>0} |d_q|^2\, q\Big) + \Big(\sum_{p<0} |c_p|^2\, \tfrac{1}{|p|}\Big)\Big(\sum_{q<0} |d_q|^2\, |q|\Big)\,. \tag{7}$$

Proof. We have

$$\int_G \left|\langle \rho(g)\varphi\,,\, f\rangle\right|^2 dm(g) \underset{(5)}{=} \int_G \left|\langle \varphi\,,\, \rho^*(g^{-1}) f\rangle\right|^2 dm(g) = \int_G \left|\langle \varphi\,,\, \rho^*(g) f\rangle\right|^2 dm(g)\,,$$

where the second equality holds because of the unimodularity of G, and the claim follows by Proposition 7. □

2.6. Irreducibility

Proposition 7 and Corollary 8 suggest the following approach. Consider the Korobov or Sobolev type spaces of periodic distributions with vanishing mean

$$\boldsymbol{H}_s(\mathbb{T}) = \left\{ f \in \mathcal{D}'(\mathbb{T}) : \widehat{f}[0] = 0\,,\, \|f\|_{\boldsymbol{H}_s} := \Big(\sum_{n\neq 0} |\widehat{f}[n]|^2\, |n|^{2s}\Big)^{1/2} < \infty \right\}.$$

These spaces are Hilbert spaces, but we can also interpret $\boldsymbol{H}_{-s}$ as the dual of $\boldsymbol{H}_s$ via the natural sesquilinear pairing

$$\langle f, \varphi\rangle = \sum_{n\neq 0} \widehat{f}[n]\,\overline{\widehat{\varphi}[n]}\,, \quad f \in \boldsymbol{H}_s\,,\, \varphi \in \boldsymbol{H}_{-s}\,.$$

These spaces decompose naturally into a holomorphic and an antiholomorphic part, *i.e.*, $\boldsymbol{H}_s = \boldsymbol{H}_s^+ \oplus \boldsymbol{H}_s^-$ where

$$\boldsymbol{H}_s^+(\mathbb{T}) = \{ f \in \boldsymbol{H}_s : \widehat{f}[n] = 0 \text{ for } n \leq 0 \}$$

and $\boldsymbol{H}_s^- = \overline{\boldsymbol{H}_s^+}$. Consequently, every element $f \in \boldsymbol{H}_s$ decomposes uniquely as

$$f = f^+ + f^- \quad \text{where } f^+ \in \boldsymbol{H}_s^+\,,\; f^- \in \boldsymbol{H}_s^-\,.$$

Thus we can rewrite Proposition 7 and Corollary 8 as

$$\|T_{\rho^*,\varphi} f\|^2_{\boldsymbol{L}^2(G)} = \|f^+\|^2_{\boldsymbol{H}_{-1/2}}\, \|\varphi^+\|^2_{\boldsymbol{H}_{1/2}} + \|f^-\|^2_{\boldsymbol{H}_{-1/2}}\, \|\varphi^-\|^2_{\boldsymbol{H}_{1/2}}$$

and

$$\|T_{\rho,\varphi} f\|^2_{\boldsymbol{L}^2(G)} = \|\varphi^+\|^2_{\boldsymbol{H}_{-1/2}}\, \|f^+\|^2_{\boldsymbol{H}_{1/2}} + \|\varphi^-\|^2_{\boldsymbol{H}_{-1/2}}\, \|f^-\|^2_{\boldsymbol{H}_{1/2}}\,.$$

Proposition 9. *$\rho(G)$ acts isometrically and irreducibly on $\boldsymbol{H}^{+}_{-1/2}(\mathbb{T})$ and on $\boldsymbol{H}^{-}_{-1/2}(\mathbb{T})$.*

Proof. (i) First, we need to show that $\rho(G)$ acts on $\boldsymbol{H}^{+}_{-1/2}$. To this end, note that $\varphi \in \boldsymbol{H}^{+}_{-1/2}$ implies that φ is holomorphic on the open unit disk $\mathbb{D}$ with $\varphi(0) = 0$, so we can write $\varphi(z) = z\,\psi(z)$ for some ψ holomorphic on $\mathbb{D}$. Then we have

$$\begin{aligned}\big(\rho(\alpha,\beta)\varphi\big)(z) &= \frac{z}{(\alpha z - \overline{\beta})(-\beta z + \overline{\alpha})}\,\varphi\Big(\frac{\alpha z - \overline{\beta}}{-\beta z + \overline{\alpha}}\Big)\\ &= \frac{z}{(-\beta z + \overline{\alpha})^2}\,\psi\Big(\frac{\alpha z - \overline{\beta}}{-\beta z + \overline{\alpha}}\Big)\\ &= z\left(\frac{1}{(-\beta z + \overline{\alpha})^2}\,\psi\Big(\frac{\alpha z - \overline{\beta}}{-\beta z + \overline{\alpha}}\Big)\right),\end{aligned}$$

i.e., we again have a function that is holomorphic on $\mathbb{D}$, multiplied by z. That $\rho(\alpha,\beta)\varphi$ is indeed in $\boldsymbol{H}^{+}_{-1/2}$ now follows from the isometry property.

To be able to show the isometry property, we first want to express the norm in $\boldsymbol{H}^{+}_{-1/2}$ in terms of $\varphi(z)$. To this end, note that for $\varphi(z) = \sum_{n\geq 1} a_n z^n$, we have

$$\int_0^z \frac{\varphi(\zeta)}{\zeta}\,d\zeta = \int_0^z \sum_{n=1}^{\infty} a_n \zeta^{n-1}\,d\zeta = \sum_{n=1}^{\infty} a_n \frac{\zeta^n}{n}\bigg|_0^z = \sum_{n=1}^{\infty} \frac{a_n}{n} z^n .$$

Therefore, we can write by invoking Parseval-Plancherel

$$\|\varphi\|^2_{\boldsymbol{H}^{+}_{-1/2}} = \sum_{n=1}^{\infty}\Big(\frac{a_n}{n}\Big)\overline{a_n} = \oint_{|z|=1}\Big(\int_0^z \frac{\varphi(\zeta)}{\zeta}\,d\zeta\Big)\overline{\varphi(z)}\,\frac{dz}{2\pi i\,z},$$

and thus

$$\begin{aligned}&\Big\|\rho(\alpha,\beta)\varphi\Big\|^2_{\boldsymbol{H}^{+}_{-1/2}} =\\ &= \oint_{|z|=1}\left(\int_0^z \frac{\varphi\Big(\frac{\alpha\zeta - \overline{\beta}}{-\beta\zeta + \overline{\alpha}}\Big)}{(\alpha\zeta - \overline{\beta})(-\beta\zeta + \overline{\alpha})}\,d\zeta\right)\frac{\overline{z}}{(\overline{\alpha}\overline{z} - \beta)(-\overline{\beta}\overline{z} + \alpha)}\,\overline{\varphi\Big(\frac{\alpha z - \overline{\beta}}{-\beta z + \overline{\alpha}}\Big)}\,\frac{dz}{2\pi i\,z}\\ &= \oint_{|z|=1}\left(\int_{-\frac{\overline{\beta}}{\overline{\alpha}}}^{\frac{\alpha z - \overline{\beta}}{-\beta z + \overline{\alpha}}} \frac{\varphi(\omega)}{\omega}\,d\omega\right)\frac{\overline{z}}{(\overline{\alpha}\overline{z} - \beta)(-\overline{\beta}\overline{z} + \alpha)}\,\overline{\varphi\Big(\frac{\alpha z - \overline{\beta}}{-\beta z + \overline{\alpha}}\Big)}\,\frac{dz}{2\pi i\,z}\\ &= \oint_{|w|=1}\left(\int_{-\overline{\beta}/\overline{\alpha}}^{w}\frac{\varphi(\omega)}{\omega}\,d\omega\right)\overline{\varphi(w)}\,\frac{dw}{2\pi i\,w}\\ &= \|\varphi\|^2_{\boldsymbol{H}^{+}_{-1/2}} + \oint_{|w|=1}\left(\int_{-\overline{\beta}/\overline{\alpha}}^{0}\frac{\varphi(\omega)}{\omega}\,d\omega\right)\overline{\varphi(w)}\,\frac{dw}{2\pi i\,w},\end{aligned}$$

where the extraneous term on the right vanishes, since we have

$$\oint_{|w|=1} \Big(\int_{-\overline{\beta}/\overline{\alpha}}^{0} \frac{\varphi(\omega)}{\omega}\, d\omega \Big) \overline{\varphi(w)}\, \frac{dw}{2\pi i\, w} = \Big(\int_{-\overline{\beta}/\overline{\alpha}}^{0} \frac{\varphi(\omega)}{\omega}\, d\omega \Big) \overline{\oint_{|w|=1} \varphi(w)\, \frac{dw}{2\pi i\, w}}$$

$$= \Big(\int_{-\overline{\beta}/\overline{\alpha}}^{0} \frac{\varphi(\omega)}{\omega}\, d\omega \Big) \overline{\varphi(0)}\,.$$

For the proof of the irreducibility, we follow [23]. Assume that $\boldsymbol{V} \subseteq \boldsymbol{H}^+_{-1/2}$ is a nonzero, closed, ρ-invariant subspace. Choose $\varphi \in \boldsymbol{V} \setminus \{0\}$. Since $\big(\rho(\alpha,\beta)\varphi\big)'(0) = -\varphi(-\overline{\beta}/\overline{\alpha})/(\overline{\alpha}\overline{\beta})$ whenever $\beta \neq 0$, we can choose $\lambda \in \mathbb{C}$ and (α,β) such that $\varphi_0 = \lambda\, \rho(\alpha,\beta)\varphi \in \boldsymbol{V}$ satisfies $\varphi_0'(0) = 1$. Now since $\big(\rho(e^{i\theta},0)\varphi\big)(z) = \varphi(e^{2i\theta} z)$ and $\boldsymbol{V}$ is closed, the function φ_1 given by

$$\varphi_1(z) = \frac{1}{2\pi} \int_0^{2\pi} e^{-2i\theta}\, \varphi_0(e^{2i\theta} z)\, d\theta = \sum_{n=1}^{\infty} \frac{\varphi_0^{(n)}(0)}{n!}\, \frac{1}{2\pi} \int_0^{2\pi} e^{-2i\theta}\, e^{2in\theta}\, z^n\, d\theta = z$$

is also in $\boldsymbol{V}$. For this function, we have

$$\big(\rho(\alpha,\beta)\varphi_1\big)(z) = \frac{z}{(-\beta\, z + \overline{\alpha})^2} = \sum_{n=1}^{\infty} \frac{n\, \beta^{n-1}}{\overline{\alpha}^{n+1}}\, z^n\,,$$

so fixing (α,β) with $\beta \neq 0$, we have $\varphi_2 = \rho(\alpha,\beta)\varphi_1 \in \boldsymbol{V}$ with $\varphi_2^{(n)}(0) \neq 0$ for all $n \in \mathbb{N}$. By the same reasoning as before, the function

$$\frac{1}{2\pi} \int_0^{2\pi} e^{-2in\theta}\, \varphi_2(e^{2i\theta} z)\, d\theta = \sum_{k=1}^{\infty} \frac{\varphi_2^{(k)}(0)}{k!}\, \frac{1}{2\pi} \int_0^{2\pi} e^{-2in\theta}\, e^{2ik\theta}\, z^k\, d\theta = \frac{\varphi_2^{(n)}(0)}{n!}\, z^n$$

is also in $\boldsymbol{V}$, so $z^n \in \boldsymbol{V}$ for all $n \in \mathbb{N}$, and thus $\boldsymbol{V} = \boldsymbol{H}^+_{-1/2}$ since $\boldsymbol{V}$ is closed.

(ii) To show the same for $\boldsymbol{H}^-_{-1/2}$, we consider the complex conjugation operator

$$CC : \boldsymbol{H}^+_{-1/2} \to \boldsymbol{H}^-_{-1/2}\,, \quad \varphi \mapsto \overline{\varphi}\,,$$

i.e., we have

$$\varphi(z) = \sum_{n=1}^{\infty} a_n\, z^n \quad \Longrightarrow \quad (CC\, \varphi)(z) = \sum_{n=1}^{\infty} \overline{a_n}\, z^{-n}\,.$$

Consequently,

$$\big(CC\, \rho(\alpha,\beta)\, \varphi\big)(z) = \overline{\frac{z}{(-\beta\, z + \overline{\alpha})(\alpha\, z - \overline{\beta})}\, \varphi\Big(\frac{\alpha\, z - \overline{\beta}}{-\beta\, z + \overline{\alpha}}\Big)}$$

$$= \frac{\overline{z}}{(-\overline{\beta}\, \overline{z} + \alpha)(\overline{\alpha}\, \overline{z} - \beta)} \sum_{n=1}^{\infty} \overline{a_n} \Big(\frac{\overline{\alpha}\, \overline{z} - \beta}{-\overline{\beta}\, \overline{z} + \alpha} \Big)^n$$

$$= \frac{z}{(-\overline{\beta}+\alpha z)(\overline{\alpha}-\beta z)} \sum_{n=1}^{\infty} \overline{a_n} \left(\frac{\overline{\alpha}-\beta z}{-\overline{\beta}+\alpha z}\right)^n$$

$$= \frac{z}{(-\beta z+\overline{\alpha})(\alpha z-\overline{\beta})} \sum_{n=1}^{\infty} \overline{a_n} \left(\frac{\alpha z-\overline{\beta}}{-\beta z+\overline{\alpha}}\right)^{-n}$$

$$= \Big(\rho(\alpha,\beta)\, CC\, \varphi\Big)(z),$$

i.e., $CC : \boldsymbol{H}^+_{-1/2} \to \boldsymbol{H}^-_{-1/2}$ is an antilinear isometry which commutes with all $\rho(\alpha,\beta)$. Therefore, $\rho(\alpha,\beta)$ is also an isometry on $\boldsymbol{H}^-_{-1/2}$, and the irreducibility of the action of ρ on $\boldsymbol{H}^+_{-1/2}$ is equivalent to that of its action on $\boldsymbol{H}^-_{-1/2}$. □

The contragredient representation ρ^*, on the other hand, creates a minor formal problem, since $\varphi^+(0) = \varphi^-(\infty) = 0$ does not necessarily imply $\big(\rho^*(\alpha,\beta)\varphi^+\big)(0) = \big(\rho^*(\alpha,\beta)\varphi^-\big)(\infty) = 0$. One possibility is to modify the definition of ρ^* by letting

$$\begin{aligned} &\big(\widetilde{\rho^*}(\alpha,\beta)\varphi^+\big)(z) = \varphi^+\Big(\frac{\alpha z-\overline{\beta}}{-\beta z+\overline{\alpha}}\Big) - \varphi^+\Big(\frac{-\overline{\beta}}{\overline{\alpha}}\Big) \\ \text{and}\quad &\big(\widetilde{\rho^*}(\alpha,\beta)\varphi^-\big)(z) = \varphi^-\Big(\frac{\alpha z-\overline{\beta}}{-\beta z+\overline{\alpha}}\Big) - \varphi^-\Big(\frac{\alpha}{-\beta}\Big), \end{aligned} \tag{8}$$

and we only have to check that $\widetilde{\rho^*}$ is still a group homomorphism.

A somewhat more elegant solution is the introduction of equivalence classes, as the norm on $\boldsymbol{H}_{1/2}$ already suggests. Consider the space

$$\boldsymbol{F} := \Big\{\sum_{n\in\mathbb{Z}} a_n z^n : \Big\|\sum_{n\in\mathbb{Z}} a_n z^n\Big\|_{\boldsymbol{F}} = \Big(\sum_{n\in\mathbb{Z}} |n|\,|a_n|^2\Big)^{1/2} < \infty\Big\}.$$

Obviously, $\|\cdot\|_{\boldsymbol{F}}$ is only a seminorm on $\boldsymbol{F}$ whose kernel $\boldsymbol{K}$ is just the space of constant functions. The constant functions, on the other hand, represent just those elements of $\boldsymbol{F}$ which, as functionals, vanish identically on $\boldsymbol{H}_{-1/2}$, and thus it is natural to define

$$\boldsymbol{H}_{1/2} := \boldsymbol{F}/\boldsymbol{K}$$

as dual space of $\boldsymbol{H}_{-1/2}$. We still have the natural decomposition

$$\boldsymbol{H}_{1/2} = \boldsymbol{H}^+_{1/2} \oplus \boldsymbol{H}^-_{1/2}\,.$$

Corollary 10. *$\rho^*(G)$ acts isometrically and irreducibly on the spaces $\boldsymbol{H}^+_{1/2}(\mathbb{T})$ and $\boldsymbol{H}^-_{1/2}(\mathbb{T})$.*

Proof. An element $\varphi \in \boldsymbol{H}^+_{1/2}$ represents an equivalence class of holomorphic functions in the unit disk, differing from each other by constants. Then

$$\big(\rho^*(\alpha,\beta)\varphi\big)(z) = \varphi\Big(\frac{\alpha z-\overline{\beta}}{-\beta z+\overline{\alpha}}\Big)$$

also represents holomorphic functions on the unit disk, since $z \mapsto \dfrac{\alpha z-\overline{\beta}}{-\beta z+\overline{\alpha}}$ maps the unit circle onto itself. That we indeed have $\rho^*(\alpha,\beta)\varphi \in \boldsymbol{H}^+_{1/2}$ follows once more from the isometry property, which again is an immediate consequence of $\rho^*(\alpha,\beta) = \left(\rho(\overline{\alpha},-\beta)\right)^*$ and the isometry property of ρ. Also, the irreducibility of the action of ρ^* follows from that of ρ by Lemma 18 and the reflexivity of our setup.

The same may be concluded for $\boldsymbol{H}^-_{1/2}$ since $CC : \boldsymbol{H}^+_{1/2} \to \boldsymbol{H}^-_{1/2}$ is again an antilinear isometry commuting with all $\rho^*(\alpha,\beta)$. □

Remark. The appearance of the space $\boldsymbol{H}^+_{1/2}(\mathbb{T})$ in this connection comes as no surprise. In [3], Möbius-invariant function spaces have been defined as spaces of analytic functions on the unit disk on which ρ^* acts isometrically, and in [2], it is shown that under mild technical assumptions, the space $\boldsymbol{H}^+_{1/2}(\mathbb{T})$, known as the *Dirichlet space*, is the unique Hilbert space with this property.

2.7. The orthogonality relation

Theorem 11. *For $f_1, \varphi_2 \in \boldsymbol{H}_{-1/2}$ and $f_2, \varphi_1 \in \boldsymbol{H}_{1/2}$, we have*

$$\int_G \langle f_1\,, \rho^*(\alpha,\beta)\varphi_1\rangle\,\langle \rho(\alpha,\beta)\varphi_2\,, f_2\rangle\, dm(\alpha,\beta) =$$

$$= \langle f_1^+, f_2^+\rangle\,\langle \varphi_2^+, \varphi_1^+\rangle + \langle f_1^-, f_2^-\rangle\,\langle \varphi_2^-, \varphi_1^-\rangle\,. \qquad (9)$$

Proof. In analogy to the proof of Proposition 7, we begin by evaluating the integral

$$\int_G A^*_{n,m}(\alpha,\beta)\, A_{p,q}(\alpha,\beta)\, dm(\alpha,\beta) =: C_{n,m,p,q}$$

for $n\,p \neq 0$. If $n\,m \leq 0$ or $p\,q \leq 0$, then $A^*_{n,m} \equiv 0$ or $A_{p,q} \equiv 0$ and thus $C_{n,m,p,q} = 0$.

For n, m, p, $q > 0$, we have by Lemma 6

$$\begin{aligned} C_{n,m,p,q} &= \int_G \left(\frac{1}{\alpha^{n+m}} \sum_{j=1}^{\min\{n,m\}} \binom{m}{m-j}(-\beta)^{m-j}\binom{n-1}{n-j}\overline{\beta}^{n-j}\right) \\ &\qquad \left(\frac{1}{\overline{\alpha}^{p+q}} \sum_{\ell=1}^{\min\{p,q\}} \binom{q}{q-\ell}\beta^{q-\ell}\binom{p-1}{p-\ell}(-\overline{\beta})^{p-\ell}\right) dm(\alpha,\beta) \\ &= \sum_{j=1}^{\min\{n,m\}} \sum_{\ell=1}^{\min\{p,q\}} \binom{m}{m-j}\binom{n-1}{n-j}\binom{q}{q-\ell}\binom{p-1}{p-\ell}(-1)^{m-j+p-\ell} \\ &\qquad \int_0^\infty \oint_{|u|=1} \oint_{|v|=1} \frac{v^{m-n+q-p}}{u^{n+m-p-q}} \frac{(\sinh s)^{m+n-2j+q+p-2\ell}}{(\cosh s)^{n+m+p+q}} \frac{du}{2\pi i\, u} \frac{dv}{2\pi i\, v} \sinh 2s\, ds\,. \end{aligned}$$

Now integration with respect to u and v yields 0 unless we have both $m-n+q-p=0$ and $n+m-p-q=0$, *i.e.*, $(n,m)=(q,p)$. In this case, we obtain

$$\begin{aligned}
C_{n,m,m,n} &= \sum_{j,\ell=1}^{\min\{n,m\}} \binom{m}{m-j}\binom{n-1}{n-j}\binom{n}{n-\ell}\binom{m-1}{m-\ell}(-1)^{j+\ell}\int_0^\infty \frac{(\sinh s)^{2m+2n-2j-2\ell+1}}{(\cosh s)^{2m+2n-1}}\,2\,ds \\
&\underset{\text{(Lemma A.2)}}{=} \sum_{j,\ell=1}^{\min\{n,m\}} (-1)^{j+\ell}\binom{m}{m-j}\binom{n-1}{n-j}\binom{n}{n-\ell}\binom{m-1}{m-\ell}\frac{(m+n-j-\ell)!\,(j+\ell-2)!}{(m+n-1)!} \\
&= \frac{1}{n\binom{m+n-1}{m-1}}\sum_{j=1}^{\min\{n,m\}}(-1)^{n-j}\binom{m}{j}\sum_{\ell=1}^{\min\{n,m\}}(-1)^{n-\ell}\binom{n}{\ell}\binom{m+n-j-\ell}{n-j}\binom{j+\ell-2}{j-1} \\
&\underset{\text{(Lemma A.1)}}{=} \frac{1}{n\binom{m+n-1}{m-1}}(-1)^{n-1}\binom{m}{1}(-1)^{n+1}\binom{m+n-1}{n-1} = 1\,.
\end{aligned}$$

For n, $m>0$ and p, $q<0$, we have

$$\begin{aligned}
C_{n,m,p,q} &= \int_G A^*_{n,m}(\alpha,\beta)\,A_{p,q}(\alpha,\beta)\,dm(\alpha,\beta) = \int_G A^*_{n,m}(\alpha,\beta)\,\overline{A_{|p|,|q|}(\alpha,\beta)}\,dm(\alpha,\beta) \\
&= \int_G \left(\frac{1}{\alpha^{n+m}}\sum_{j=1}^{\min\{n,m\}}\binom{m}{m-j}(-\beta)^{m-j}\binom{n-1}{n-j}\overline{\beta}^{n-j}\right) \\
&\qquad \left(\frac{1}{\overline{\alpha^{|p|+|q|}}}\sum_{\ell=1}^{\min\{|p|,|q|\}}\binom{|q|}{|q|-\ell}(-\overline{\beta})^{|q|-\ell}\binom{|p|-1}{|p|-\ell}\beta^{|p|-\ell}\right)dm(\alpha,\beta) \\
&= \sum_{j=1}^{\min\{n,m\}}\sum_{\ell=1}^{\min\{|p|,|q|\}}\binom{m}{m-j}\binom{n-1}{n-j}\binom{|q|}{|q|-\ell}\binom{|p|-1}{|p|-\ell}(-1)^{m-j+|q|-\ell} \\
&\qquad \int_0^\infty \oint_{|u|=1}\oint_{|v|=1} \frac{v^{m-n-|q|+|p|}}{u^{n+m+|p|+|q|}}\,\frac{(\sinh s)^{n+m-2j+|p|+|q|-2\ell}}{(\cosh s)^{n+m+|p|+|q|}}\,\frac{du}{2\pi i\,u}\,\frac{dv}{2\pi i\,v}\,\sinh 2s\,ds\,,
\end{aligned}$$

and integration with respect to u yields 0.

For n, $m<0$, we have

$$C_{n,m,p,q} = \int_G A^*_{n,m}\,A_{p,q}\,dm = \int_G \overline{A^*_{-n,-m}}\,\overline{A_{-p,-q}}\,dm = \overline{C_{-n,-m,-p,-q}}\,.$$

All in all, we find for n, $p\neq 0$

$$C_{n,m,p,q} = \begin{cases}\delta_{(n,m),(q,p)}\,, & nm>0\,,\\ 0\,, & nm\le 0\,.\end{cases}$$

From this, we obtain

$$\int_G \langle f_1\,,\,\rho^*(\alpha,\beta)\varphi_1\rangle\,\langle\rho(\alpha,\beta)\varphi_2\,,\,f_2\rangle\,dm(\alpha,\beta) =$$

$$= \int_G \Big\langle \sum_{n\neq 0} a_n\,z^n\,,\,\rho^*(\alpha,\beta)\sum_{m\in\mathbb{Z}} b_m\,z^m\Big\rangle \Big\langle \rho(\alpha,\beta)\sum_{p\neq 0} c_p\,z^p\,,\,\sum_{q\in\mathbb{Z}} d_q\,z^q\Big\rangle\,dm(\alpha,\beta)$$

$$= \sum_{n\neq 0} a_n \sum_{m\neq 0} \overline{b_m} \sum_{p\neq 0} c_p \sum_{q\neq 0} \overline{d_q}\; C_{n,m,p,q}$$

$$= \sum_{n,m>0} a_n\,\overline{b_m}\,c_m\,\overline{d_n} + \sum_{n,m<0} a_n\,\overline{b_m}\,c_m\,\overline{d_n}$$

$$= \Big(\sum_{n>0} a_n\,\overline{d_n}\Big)\Big(\sum_{m>0} c_m\,\overline{b_m}\Big) + \Big(\sum_{n<0} a_n\,\overline{d_n}\Big)\Big(\sum_{m<0} c_m\,\overline{b_m}\Big)\,,$$

which yields the claim. □

Corollary 12 (Reproducing Property). *Choose $\varphi_1 \in \boldsymbol{H}_{1/2}$ and $\varphi_2 \in \boldsymbol{H}_{-1/2}$ with*

$$\langle \varphi_2^+\,,\varphi_1^+\rangle = \langle \varphi_2^-\,,\varphi_1^-\rangle = 1\,.$$

Then we have for all $f \in \boldsymbol{H}_{-1/2}$ that

$$\int_G \langle f\,,\,\rho^*(\alpha,\beta)\varphi_1\rangle\,\rho(\alpha,\beta)\varphi_2\,dm(\alpha,\beta) = f \tag{10}$$

in the weak sense.

2.8. A different perspective

As we have shown in Proposition 9, $\rho(G)$ acts isometrically on $\boldsymbol{H}_{-1/2}(\mathbb{T})$. If we give up the idea of treating $\boldsymbol{H}_{1/2}$ as dual space of $\boldsymbol{H}_{-1/2}$ and switch back to the Hilbert space point of view instead, Proposition 9 says that ρ is a unitary representation on $\boldsymbol{H}_{-1/2}$. This leads to a different orthogonality relation and thus also to another reproducing formula. We want to show how these are related to the results above.

To this end, we consider the canonical isometry

$$S:\,\boldsymbol{H}_{1/2}\to\boldsymbol{H}_{-1/2}\,,\quad \sum_{n\neq 0} a_n\,z^n \mapsto \sum_{n\neq 0} |n|\,a_n\,z^n\,.$$

Obviously, we have

$$S:\,\boldsymbol{H}_{1/2}^+\to\boldsymbol{H}_{-1/2}^+ \quad\text{and}\quad S:\,\boldsymbol{H}_{1/2}^-\to\boldsymbol{H}_{-1/2}^-$$

and thus

$$(Sf)^+ = S(f^+) \quad\text{and}\quad (Sf)^- = S(f^-)\,.$$

Also, note that for $f = \sum a_n z^n \in \boldsymbol{H}_{-1/2}$ and $\varphi = \sum b_n z^n \in \boldsymbol{H}_{1/2}$, we have

$$\begin{aligned}\langle f, \varphi\rangle &= \sum_{n\neq 0} a_n \overline{b_n} \\ &= \sum_{n\neq 0} a_n \overline{|n|\, b_n}\, \frac{1}{|n|} = \langle f, S\varphi\rangle_{\boldsymbol{H}_{-1/2}}.\end{aligned}$$

Furthermore, S intertwines ρ and ρ^*, *i.e.*, we have the commutative diagram

$$\begin{array}{ccc} \boldsymbol{H}_{1/2} & \xrightarrow{\rho^*(g)} & \boldsymbol{H}_{1/2} \\ S\Big\downarrow & & \Big\downarrow S \\ \boldsymbol{H}_{-1/2} & \xrightarrow{\rho(g)} & \boldsymbol{H}_{-1/2} \end{array} .$$

To see this, let $f \in \boldsymbol{H}_{-1/2}$ and $\varphi \in \boldsymbol{H}_{1/2}$. Then we have for all $g \in G$ that

$$\begin{aligned}\langle f, S\rho^*(g)\varphi\rangle_{\boldsymbol{H}_{-1/2}} &= \langle f, \rho^*(g)\varphi\rangle \\ &= \langle \rho(g^{-1})f, \varphi\rangle \\ &= \langle \rho(g^{-1})f, S\varphi\rangle_{\boldsymbol{H}_{-1/2}} = \langle f, \rho(g)S\varphi\rangle_{\boldsymbol{H}_{-1/2}},\end{aligned}$$

where the last equality holds since ρ acts unitarily on $\boldsymbol{H}_{-1/2}$.

Consequently, we can prove Theorem 11 with the general methods described in Section 1 by the following argument.

Alternative proof of Theorem 11. With the aid of the map S, we want to deduce (9) from (1′).

We saw in Proposition 9 that ρ acts unitarily and irreducibly on $\boldsymbol{H}^+_{-1/2}$ and on $\boldsymbol{H}^-_{-1/2}$. Denoting the restrictions of ρ to these two spaces by ρ^+ and ρ^-, respectively, we need to show that ρ^+ and ρ^- are not unitarily equivalent. To this end, we shall show to which standard representations of $\boldsymbol{SU}_{1,1}$ they are equivalent. Defining the operator

$$(Z\psi)(z) := z\,\psi(z)$$

yields

$$\begin{aligned}\big(\rho(\alpha,\beta)Z\psi\big)(z) &= \frac{z}{(\alpha z - \overline{\beta})(-\beta z + \overline{\alpha})}\, \frac{\alpha z - \overline{\beta}}{-\beta z + \overline{\alpha}}\, \psi\Big(\frac{\alpha z - \overline{\beta}}{-\beta z + \overline{\alpha}}\Big) \\ &= z\, \frac{1}{(-\beta z + \overline{\alpha})^2}\, \psi\Big(\frac{\alpha z - \overline{\beta}}{-\beta z + \overline{\alpha}}\Big) = \big(Z\mathcal{D}_2^+(\alpha,\beta)\psi\big)(z),\end{aligned}$$

i.e., ρ^+ is equivalent to $\mathcal{D}_2^+$. Since ρ^- is obtained from ρ^+ by complex conjugation, it follows that ρ^- is equivalent to $\mathcal{D}_2^-$, and it is well known that the two representations $\mathcal{D}_2^+$ and $\mathcal{D}_2^-$ are not equivalent (*e.g.*, see [23]).

So everything is prepared for the application of Proposition 4 with (1′) instead of (1). Since G is unimodular, we know that $C^+ = \lambda^+ \operatorname{Id}_{\boldsymbol{H}^+_{-1/2}}$ and $C^- = \lambda^- \operatorname{Id}_{\boldsymbol{H}^-_{-1/2}}$. Finally, since ρ^+ and ρ^- are related via complex conjugation, we have

$\lambda^+ = \lambda^-$, and we may assume — after proper normalization of the Haar measure — that $\lambda^+ = \lambda^- = 1$.

Consequently, we have for $f_1, \varphi_2 \in \boldsymbol{H}_{-1/2}$ and $\varphi_1, f_2 \in \boldsymbol{H}_{1/2}$

$$\int_G \langle f_1 , \rho^*(\alpha,\beta)\varphi_1\rangle \langle \rho(\alpha,\beta)\varphi_2 , f_2\rangle \, dm(\alpha,\beta) =$$

$$\begin{aligned}
&= \int_G \langle f_1 , S\rho^*(\alpha,\beta)\varphi_1\rangle_{\boldsymbol{H}_{-1/2}} \langle \rho(\alpha,\beta)\varphi_2 , Sf_2\rangle_{\boldsymbol{H}_{-1/2}} \, dm(\alpha,\beta) \\
&= \int_G \langle f_1 , \rho(\alpha,\beta)S\varphi_1\rangle_{\boldsymbol{H}_{-1/2}} \overline{\langle Sf_2 , \rho(\alpha,\beta)\varphi_2\rangle_{\boldsymbol{H}_{-1/2}}} \, dm(\alpha,\beta) \\
&\underset{(1')}{=} \langle f_1^+ , (Sf_2)^+\rangle_{\boldsymbol{H}_{-1/2}} \overline{\langle (S\varphi_1)^+ , \varphi_2^+\rangle_{\boldsymbol{H}_{-1/2}}} \\
&\qquad + \langle f_1^- , (Sf_2)^-\rangle_{\boldsymbol{H}_{-1/2}} \overline{\langle (S\varphi_1)^- , \varphi_2^-\rangle_{\boldsymbol{H}_{-1/2}}} \\
&= \langle f_1^+ , f_2^+\rangle \langle \varphi_2^+ , \varphi_1^+\rangle + \langle f_1^- , f_2^-\rangle \langle \varphi_2^- , \varphi_1^-\rangle
\end{aligned}$$

as claimed. □

After realizing these equivalences, we should ask ourselves, why not work with the direct sum of $\mathcal{D}_2^+$ and $\mathcal{D}_2^-$ right away. This would cause technical problems since the two spaces on which these act are not disjoint, since both contain the constant functions. Also, the strong advantage of our approach is the usage of the standard inner product on the unit circle, which is certainly highly desirable for applications. Furthermore, we have found an example of a reproducing formula containing a pair of contragredient representations, and we shall study this phenomenon in general in Section 3. But first, we have a look at some graphical examples for the above setup.

2.9. Generators for $\rho(G)$

In order to get a visual impression of the action of the group under this representation, we describe two basic types of operators which generate $\rho(G)$. Note that for $\beta = 0$, we have $|\alpha| = 1$ and thus may write $\alpha = e^{-i\pi\omega}$, so we obtain

$$\big(\rho(\alpha,\beta)\varphi\big)(z) = \big(\rho(e^{-i\pi\omega},0)\varphi\big)(z) = \frac{z}{(e^{-i\pi\omega}\, z)(e^{i\pi\omega})}\, \varphi\Big(\frac{e^{-i\pi\omega}\, z}{e^{i\pi\omega}}\Big) = \varphi(e^{-2\pi i\,\omega}\, z)\,.$$

So $\rho(e^{-i\pi\omega},0)$ induces a rotation of $\mathbb{S}^1$ by the angle $2\pi\omega$; in other words, it is the translation operator T_ω on $\mathbb{T}$ by ω. On the other hand, if both α and β are nonnegative, the Möbius transformation

$$m_{(\alpha,-\overline{\beta})}(z) = m_{(\cosh(s),-\sinh(s))}(z) = \frac{\cosh(s)\, z - \sinh(s)}{-\sinh(s)\, z + \cosh(s)}$$

satisfies

$$m_{(\cosh(s),-\sinh(s))}(1) = 1 \quad \text{and} \quad m_{(\cosh(s),-\sinh(s))}(-1) = -1\,,$$

and locally around $\xi = 0$ behaves like a dilation on $\mathbb{T}$, which we want to show. To this end, let $t = e^{2s}$, so $\cosh(s) = \frac{t+1}{2\sqrt{t}}$ and $\sinh(s) = \frac{t-1}{2\sqrt{t}}$.

Lemma 13. *The mapping*

$$m_{(t+1,-(t-1))} : \mathbb{S}^1 \to \mathbb{S}^1\,, \quad z \mapsto \frac{(t+1)z-(t-1)}{-(t-1)z+(t+1)}$$

corresponds to the mapping

$$d_t : \mathbb{T} \to \mathbb{T}\,, \quad \xi \mapsto \frac{1}{\pi}\arctan\bigl(t\tan(\pi\xi)\bigr)\,.$$

Proof.

$$\begin{aligned} d_t(\xi) &= \frac{1}{2\pi}\arg\Bigl(\frac{(t+1)\,e^{2\pi i\xi}-(t-1)}{-(t-1)\,e^{2\pi i\xi}+(t+1)}\Bigr) \\ &= \frac{1}{2\pi}\arg\Bigl(\frac{(t+1)\,e^{i\pi\xi}-(t-1)\,e^{-i\pi\xi}}{-(t-1)\,e^{i\pi\xi}+(t+1)\,e^{-i\pi\xi}}\Bigr) \\ &= \frac{1}{2\pi}\arg\Bigl(\frac{2\cos(\pi\xi)+2\,i\,t\sin(\pi\xi)}{2\cos(\pi\xi)-2\,i\,t\sin(\pi\xi)}\Bigr) \\ &= \frac{1}{2\pi}\,2\arg\bigl(\cos(\pi\xi)+i\,t\sin(\pi\xi)\bigr) = \frac{1}{\pi}\arctan\bigl(t\tan(\pi\xi)\bigr)\,, \end{aligned}$$

where we used $\arg(w/\overline{w}) = 2\arg(w)$. □

In Figure 1, we show the graphs of d_t for $t = 2$, 4, and 8.

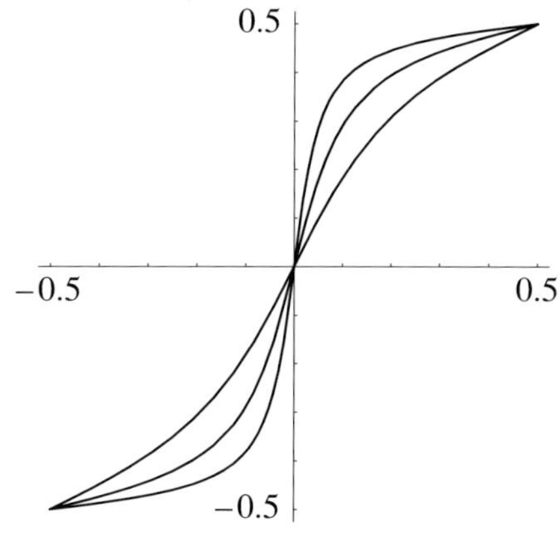

Fig. 1. Graphs of d_2, d_4, and d_8.

It is worth noting that if we transfer the stereographic projection $\mathbb{S}^1 \to \overline{\mathbb{R}}$ (with $1 \mapsto 0$) to $\mathbb{T}$, we obtain the mapping $\mathbb{T} \to \overline{\mathbb{R}}$, $\xi \mapsto \tan(\pi\xi)$. Thus the above shows that the transformations $d_t : \mathbb{T} \to \mathbb{T}$ and $m_{(t+1,-t+1)} : \mathbb{S}^1 \to \mathbb{S}^1$ correspond to the usual dilation $\mathbb{R} \to \mathbb{R}$, $x \mapsto t\,x$ in the sense that the stereographic projection intertwines $m_{(t+1,-t+1)}$ with this dilation on $\mathbb{R}$. Consequently, $\{d_t : t \in \mathbb{R}^+\}$ is a group with $d_{t_1}\, d_{t_2} = d_{t_1 t_2}$, which of course can also be seen directly from

$$\begin{pmatrix} \frac{t_1+1}{2\sqrt{t_1}} & \frac{t_1-1}{2\sqrt{t_1}} \\ \frac{t_1-1}{2\sqrt{t_1}} & \frac{t_1+1}{2\sqrt{t_1}} \end{pmatrix} \begin{pmatrix} \frac{t_2+1}{2\sqrt{t_2}} & \frac{t_2-1}{2\sqrt{t_2}} \\ \frac{t_2-1}{2\sqrt{t_2}} & \frac{t_2+1}{2\sqrt{t_2}} \end{pmatrix} = \begin{pmatrix} \frac{t_1 t_2+1}{2\sqrt{t_1 t_2}} & \frac{t_1 t_2-1}{2\sqrt{t_1 t_2}} \\ \frac{t_1 t_2-1}{2\sqrt{t_1 t_2}} & \frac{t_1 t_2+1}{2\sqrt{t_1 t_2}} \end{pmatrix}.$$

Therefore it makes sense to refer to

$$D_t := \rho\Big(\frac{t+1}{2\sqrt{t}}, \frac{t-1}{2\sqrt{t}}\Big)$$

as *(generalized) dilation operator* on $\mathbb{S}^1$ or $\mathbb{T}$.

From these two classes of operators, $\{T_\omega : \omega \in \mathbb{T}\}$ and $\{D_t : t \geq 1\}$, we can generate all $\rho(\alpha, \beta)$. Since

$$\begin{pmatrix} e^{2\pi i\phi}\frac{t+1}{2\sqrt{t}} & e^{2\pi i\theta}\frac{t-1}{2\sqrt{t}} \\ e^{-2\pi i\theta}\frac{t-1}{2\sqrt{t}} & e^{-2\pi i\phi}\frac{t+1}{2\sqrt{t}} \end{pmatrix} =$$

$$= \begin{pmatrix} e^{-i\pi(-\phi-\theta)} & 0 \\ 0 & e^{i\pi(-\phi-\theta)} \end{pmatrix} \begin{pmatrix} \frac{t+1}{2\sqrt{t}} & \frac{t-1}{2\sqrt{t}} \\ \frac{t-1}{2\sqrt{t}} & \frac{t+1}{2\sqrt{t}} \end{pmatrix} \begin{pmatrix} e^{-i\pi(\theta-\phi)} & 0 \\ 0 & e^{i\pi(\theta-\phi)} \end{pmatrix},$$

we have

$$\rho\Big(e^{2\pi i\phi}\frac{t+1}{2\sqrt{t}}, e^{2\pi i\theta}\frac{t-1}{2\sqrt{t}}\Big) = T_{-\phi-\theta}\, D_t\, T_{-\phi+\theta},$$

so each operator is a combination of a translation, a dilation, and another translation.

In order to allow the generalized dilation operator to be centered at some other point $\xi \in \mathbb{T}$, we define

$$D_{\xi,t} := T_\xi\, D_t\, T_{-\xi} = \rho\Big(\frac{t+1}{2\sqrt{t}}, e^{-2\pi i\xi}\frac{t-1}{2\sqrt{t}}\Big).$$

Then we can write

$$\rho\Big(e^{2\pi i\phi}\frac{t+1}{2\sqrt{t}}, e^{2\pi i\theta}\frac{t-1}{2\sqrt{t}}\Big) = T_{-2\phi}\, T_{\phi-\theta}\, D_t\, T_{-(\phi-\theta)} = T_{-2\phi}\, D_{\phi-\theta,t},$$

i.e., a dilation centered at the point $\xi = \phi - \theta$ followed by a translation.

This corresponds to a well-known parameterization of G given by

$$e^{i\gamma}\frac{z-a}{-\bar{a}z+1}, \quad \gamma \in [0, 2\pi[\,, a \in \mathbb{D},$$

where we have $\gamma = 4\pi\phi$ and $a = e^{2\pi i(\theta-\phi)}\tanh(s) = e^{2\pi i(\theta-\phi)}\frac{t-1}{t+1}$. Note that the Möbius transformation $m_{(1,-a)}(z) = \frac{z-a}{-\bar{a}z+1}$ satisfies

$$m_{(1,-a)}(a) = 0, \quad m_{(1,-a)}(1/\bar{a}) = \infty, \quad \text{and} \quad m_{(1,-a)}(\pm\,\text{sign}(a)) = \pm\,\text{sign}(a).$$

So $m_{(1,-a)}$ indeed corresponds to a generalized dilation on $\mathbb{S}^1$ centered at $\text{sign}(a)$.

2.10. Graphical examples for the action of ρ and ρ^*

To visualize the action of $\rho(G)$ and $\rho^*(G)$, we use

$$\varphi(z) = z + z^{-1} \quad \left(= 2\cos(2\pi\xi) \text{ for } z = e^{2\pi i\xi} \right).$$

In Figures 2 and 3, we show some examples for $\rho(\alpha,\beta)\varphi$ and $\rho^*(\alpha,\beta)\varphi$.

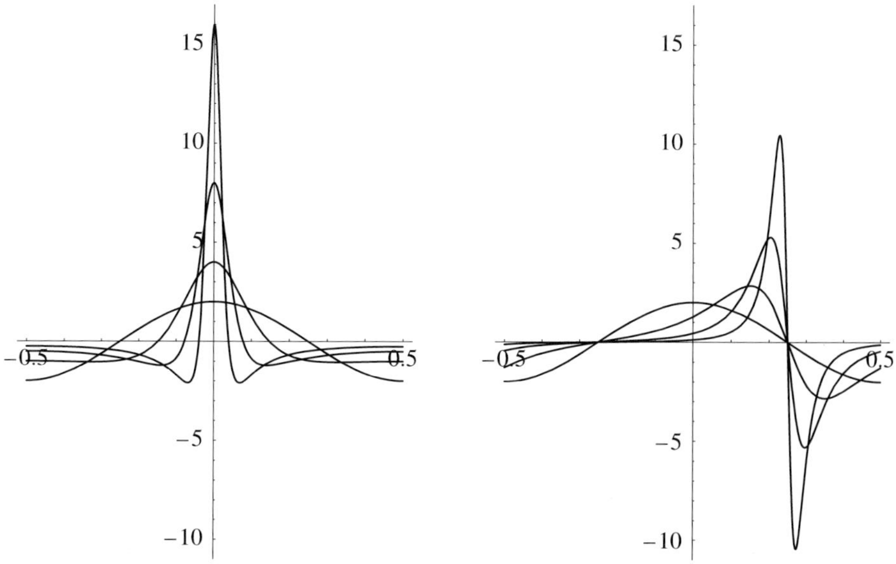

Fig. 2. Graphs of $\rho\left(\frac{t+1}{2\sqrt{t}}, \frac{t-1}{2\sqrt{t}}\right)\varphi$ and $\rho\left(\frac{t+1}{2\sqrt{t}}, e^{-2\pi i/4}\,\frac{t-1}{2\sqrt{t}}\right)\varphi$ for $t = 1$, 2, 4, and 8.

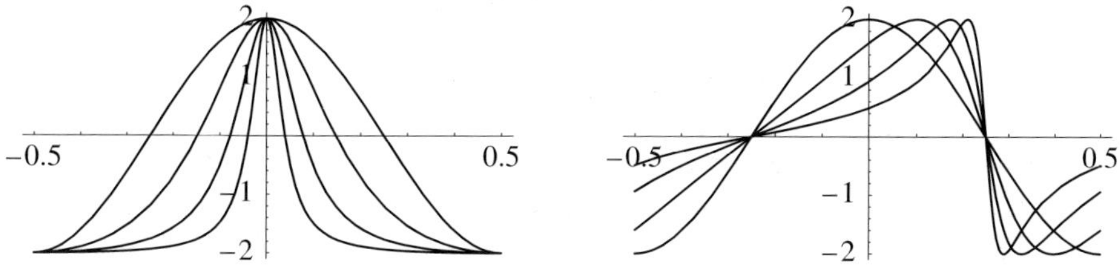

Fig. 3. Graphs of $\rho^*\left(\frac{t+1}{2\sqrt{t}}, \frac{t-1}{2\sqrt{t}}\right)\varphi$ and $\rho^*\left(\frac{t+1}{2\sqrt{t}}, e^{-2\pi i/4}\,\frac{t-1}{2\sqrt{t}}\right)\varphi$ for $t = 1$, 2, 4, and 8.

On the left, we apply D_t for various values of t, and on the right, we show the effect of choosing $\xi = \frac{1}{4}$ as center of dilation, *i.e.*, we apply $D_{\frac{1}{4},t}$. Note that since φ exhibits even symmetry about $\xi = 0$ and odd symmetry about $\xi = \frac{1}{4}$, we obtain functions of both symmetry types from the same generating function. This is highly desirable; in [6], it was shown how this can be achieved for the case of affine wavelets on $\mathbb{R}$. Using $k = 3$ instead of $k = 2$ as scaling factor in the multiresolution analysis leads to a wavelet basis generated from two basic functions, and the authors exhibit conditions making it possible to choose one of these to be even and the other one odd. Note, however, that this actually requires two generating functions, while due to the richness of operators in our setup, we obtain functions of both symmetry types from the same generator.

Allowing arbitrary points as center of dilation yields functions without any symmetry, though, as Figure 4 shows.

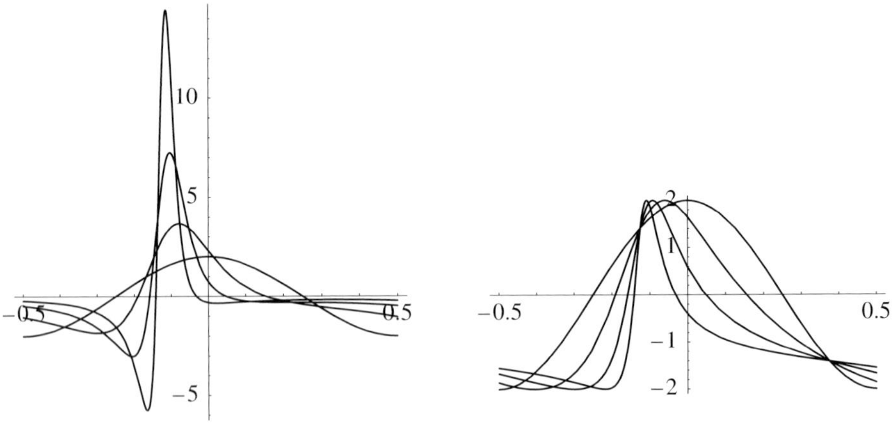

Fig. 4. Graphs of $\rho\left(\frac{t+1}{2\sqrt{t}}, e^{2\pi i/8}\frac{t-1}{2\sqrt{t}}\right)\varphi$ and $\rho^*\left(\frac{t+1}{2\sqrt{t}}, e^{2\pi i/8}\frac{t-1}{2\sqrt{t}}\right)\varphi$ for $t = 1, 2, 4,$ and 8.

3. Reproducing Properties from Nonunitary Representations

In the previous section, we presented a pair of representations which together gave rise to a reproducing property. Now we want to try and exhibit a general principle behind this phenomenon. To this end, we have to give up the Hilbert space setup and consider a Banach space $\boldsymbol{V}$ instead. Since we want to maintain the square integrability property, we have to discuss operators $\boldsymbol{V} \to \boldsymbol{L}^2(G)$ together with their adjoints. This turns out to be formally much easier if we do not work with the space $\boldsymbol{V}'$ of bounded linear functionals, but instead use $\boldsymbol{V}^*$, the space of bounded antilinear functionals, as dual space for $\boldsymbol{V}$. The two spaces $\boldsymbol{V}'$ and $\boldsymbol{V}^*$ are easily seen to be (antilinearly!) isometrically isomorphic, so essentially nothing changes; but the pairing between $\boldsymbol{V}$ and $\boldsymbol{V}^*$ is (not bilinear, but) sesquilinear, and thus a generalization of the inner product in a Hilbert space. Consequently, we use Hilbert space notation and define

$$\langle v, w\rangle_{(\boldsymbol{V},\boldsymbol{V}^*)} := \overline{w(v)}$$

for $v \in \boldsymbol{V}$ and $w \in \boldsymbol{V}^*$. We include some more details on $\boldsymbol{V}^*$ in Section A.1.

3.1. Representations and continuity

Definition 14. Let G be a locally compact group and $\boldsymbol{V}$ a Banach space. A *representation* of G on $\boldsymbol{V}$ is a group homomorphism $G \to \boldsymbol{GL}(\boldsymbol{V})$. The representation ρ is *bounded*, if

$$\|\rho\| := \sup_{g \in G} \|\rho(g)\|_{\mathcal{L}(\boldsymbol{V})} < \infty .$$

The representation ρ is *continuous*, if the mapping

$$G \times \boldsymbol{V} \to \boldsymbol{V}, \quad (g, v) \mapsto \rho(g)\, v$$

is continuous. ρ is *separately continuous*, if it is continuous with the strong operator topology on $\mathcal{L}(\boldsymbol{V})$, *i.e.*, if for each $v \in \boldsymbol{V}$, the mapping

$$G \to \boldsymbol{V}, \quad g \mapsto \rho(g)\, v$$

is continuous. (Note that for fixed $g \in G$, the mapping $v \mapsto \rho(g)\, v$ is continuous by definition.)

Furthermore, we say that ρ is *weakly continuous*, if it is continuous with the weak operator topology on $\mathcal{L}(\boldsymbol{V})$, *i.e.*, if the mapping

$$G \to \mathbb{C}, \quad g \mapsto \langle \rho(g)v, w \rangle$$

is continuous for all $v \in \boldsymbol{V}$ and $w \in \boldsymbol{V}^*$. Analogously, ρ is *weak*-continuous*, if $\boldsymbol{V}$ is a dual space and ρ is continuous with the weak* operator topology on $\mathcal{L}(\boldsymbol{V})$.

The following result shows that there is no need to distinguish between continuity and separate continuity. For related statements, see also [23] and [24].

Proposition 15. [22] *A separately continuous representation ρ of a locally compact group G on a Banach space $\boldsymbol{V}$ is continuous.*

Proof. Let $g \in G$ and $v \in \boldsymbol{V}$, and $\varepsilon > 0$ be given. By separate continuity of ρ, there is a neighborhood U of g such that $\|\rho(h)v - \rho(g)v\|_{\boldsymbol{V}} < \varepsilon/2$ for all $h \in U$, and since G is locally compact, we may choose U to be compact. For each $w \in \boldsymbol{V}$, the mapping $h \mapsto \rho(h)w$ is continuous and thus bounded on U. Therefore, by the uniform boundedness principle, there exists $C > 0$ such that $\|\rho(h)\|_{\mathcal{L}(\boldsymbol{V})} \leq C$ for all $h \in U$. Letting $\delta = \varepsilon/(2\,C) > 0$, we have for all $(h, w) \in U \times U_\delta(v)$ that

$$\begin{aligned}
\|\rho(h)w - \rho(g)v\|_{\boldsymbol{V}} &\leq \|\rho(h)w - \rho(h)v\|_{\boldsymbol{V}} + \|\rho(h)v - \rho(g)v\|_{\boldsymbol{V}} \\
&\leq C\,\|w - v\|_{\boldsymbol{V}} + \|\rho(h)v - \rho(g)v\|_{\boldsymbol{V}} \\
&< C\,\frac{\varepsilon}{2\,C} + \frac{\varepsilon}{2} = \varepsilon ,
\end{aligned}$$

so ρ is continuous. □

3.2. The contragredient representation

Lemma 16. *Let $\rho : G \to \boldsymbol{GL}(\boldsymbol{V})$ be a representation of a locally compact group G on a Banach space $\boldsymbol{V}$. Then the map $\rho^* : G \to \boldsymbol{GL}(\boldsymbol{V}^*)$ given by*

$$\rho^*(g) := \big(\rho(g^{-1})\big)^*$$

is a representation of G on the dual space $\boldsymbol{V}^*$, the contragredient representation *of ρ.*

ρ^* is bounded if and only if ρ is bounded.

If ρ is continuous, then ρ^* is weak-continuous.*

Proof. The definition of ρ^* immediately implies that $\rho^* : G \to \mathcal{L}(\boldsymbol{V}^*)$, and we have

$$\begin{aligned}\rho^*(g_1\, g_2) &= \big(\rho((g_1\, g_2)^{-1})\big)^* \\ &= \big(\rho(g_2^{-1} g_1^{-1})\big)^* \\ &= \big(\rho(g_2^{-1})\rho(g_1^{-1})\big)^* \\ &= \big(\rho(g_1^{-1})\big)^* \big(\rho(g_2^{-1})\big)^* = \rho^*(g_1)\, \rho^*(g_2)\,,\end{aligned}$$

so ρ^* is a group homomorphism, and thus $\rho^* : G \to \boldsymbol{GL}(\boldsymbol{V}^*)$.

Since $\|T^*\|_{\mathcal{L}(\boldsymbol{V}^*)} = \|T\|_{\mathcal{L}(\boldsymbol{V})}$, boundedness of ρ is equivalent to that of ρ^*.

ρ being continuous means that for each $v \in \boldsymbol{V}$, the mapping

$$G \to \boldsymbol{V}\,, \quad g \mapsto \rho(g)\, v\,,$$

is continuous. Consequently, for each $v \in \boldsymbol{V}$ and $w \in \boldsymbol{V}^*$, the mapping

$$G \to \mathbb{C}\,, \quad g \mapsto \langle v, \rho^*(g) w\rangle = \langle \rho(g^{-1})\, v, w\rangle\,,$$

is continuous, since $g \mapsto g^{-1}$ is continuous on the topological group G. So ρ^* is weak*-continuous as claimed. □

Remark. It is worth noting that for a unitary representation on a Hilbert space, we have $\rho^* = \rho$, since $\rho^*(g) = \big(\rho(g^{-1})\big)^* = \big(\rho(g^{-1})\big)^{-1} = \rho(g)$. Unitarity is imitated by the pair ρ and ρ^* in the sense that for all $g \in G$, we have

$$\langle \rho(g) v, \rho^*(g) w\rangle = \langle v, w\rangle \tag{11}$$

for $v \in \boldsymbol{V}$ and $w \in \boldsymbol{V}^*$.

In case $\boldsymbol{V}$ is a reflexive Banach space, we can strengthen the above statement somewhat (compare [15] and [19]).

Proposition 17. *Let $\rho : G \to \boldsymbol{GL}(\boldsymbol{V})$ be a continuous, bounded representation of a locally compact group G on a reflexive Banach space $\boldsymbol{V}$. Then the contragredient representation ρ^* is continuous.*

Proof. We know from Lemma 16 that ρ^* is weak*-continuous, which in a reflexive Banach space is the same as weakly continuous. Furthermore, under the assumption that $\boldsymbol{V}$ is reflexive, we may apply [19], Thm. (22.8) which states that in that case, a bounded, weakly continuous representation is strongly continuous. □

This result can also be found in [15] for unitary representations in a Hilbert space. We would like to extend it to the non-reflexive case, and to this end, we need a few preliminary results. We begin with comparing irreducibility of ρ and ρ^*.

Lemma 18. *Let ρ be a representation on a Banach space $\boldsymbol{V}$ and ρ^* its contragredient representation. If ρ^* is irreducible, then so is ρ.*

Proof. Assume that ρ is reducible, *i.e.*, that there exists a closed ρ-invariant subspace $\boldsymbol{W}$ of $\boldsymbol{V}$ with $\{0\} \neq \boldsymbol{W} \neq \boldsymbol{V}$. Then its orthogonal complement

$$\boldsymbol{W}^{\perp} = \{w \in \boldsymbol{V}^* : \langle v, w\rangle = 0 \text{ for all } v \in \boldsymbol{W}\}$$

is a closed subspace of $\boldsymbol{V}^*$ with $\{0\} \neq \boldsymbol{W}^{\perp} \neq \boldsymbol{V}^*$, which follows from the Hahn-Banach-extension theorem. Namely, for $v \in \boldsymbol{W} \setminus \{0\}$, there exists $w \in \boldsymbol{V}^*$ with $\langle v, w\rangle \neq 0$, so $\boldsymbol{W}^{\perp} \neq \boldsymbol{V}^*$; and on the other hand, for $v \in \boldsymbol{V} \setminus \boldsymbol{W}$, there exists $w \in \boldsymbol{W}^{\perp}$ with $\langle v, w\rangle \neq 0$, so $\boldsymbol{W}^{\perp} \neq \{0\}$. Furthermore, $\boldsymbol{W}^{\perp}$ is ρ^*-invariant. To see this, consider $g \in G$ and $w \in \boldsymbol{W}^{\perp}$. Then we have for all $v \in \boldsymbol{W}$ that

$$\langle v, \rho^*(g)w\rangle = \langle \rho(g^{-1})v, w\rangle = 0$$

by the ρ-invariance of $\boldsymbol{W}$, so $\rho^*(g)w \in \boldsymbol{W}^{\perp}$. Thus ρ^* is reducible also. □

Remark. Lemma 18 immediately implies that on a reflexive Banach space, irreducibility of ρ is equivalent to that of ρ^*, since $\rho^{**} = \rho$. If $\boldsymbol{V}$ is non-reflexive, though, we have that $\boldsymbol{V}$ is a nontrivial, closed, ρ^{**}-invariant subspace of $\boldsymbol{V}^{**}$, since $\rho^{**}(g)\big|_{\boldsymbol{V}} = \rho(g)$ for all $g \in G$. So even if ρ is irreducible, ρ^{**} is reducible, and the converse of Lemma 18 becomes false for at least one of the pairs (ρ, ρ^*) and (ρ^*, ρ^{**}).

The essential tool both in [15] and in [19] for the results corresponding to Proposition 17 is the algebra representation of $\boldsymbol{L}^1(G)$ induced from a bounded representation of G, which we now define.

Theorem 19. *Let ρ be a bounded, continuous representation of a locally compact group G in a Banach space $\boldsymbol{V}$, with contragredient representation ρ^*. Then*

$$\sigma^* : \boldsymbol{L}^1(G) \to \mathcal{L}(\boldsymbol{V}^*), \quad f \mapsto \int_G f(g)\, \rho^*(g)\, dm(g),$$

is a bounded algebra representation.

Proof. For $f \in \boldsymbol{L}^1(G)$ and $w \in \boldsymbol{V}^*$, we define $\sigma^*(f)\, w \in \boldsymbol{V}^*$ via

$$\langle v, \sigma^*(f)\, w\rangle = \int_G \overline{f(g)}\, \langle v, \rho^*(g)\, w\rangle\, dm(g).$$

Since, as we saw before, the map $g \mapsto \langle v, \rho^*(g)\, w\rangle$ is continuous, the integrand is measurable. Boundedness of ρ^* implies

$$\left|\overline{f(g)}\,\langle v, \rho^*(g)\, w\rangle\right| \le |f(g)|\, \|v\|_{\boldsymbol{V}}\, \|\rho^*\|\, \|w\|_{\boldsymbol{V}^*},$$

and thus the integral is well-defined. $\sigma^*(f)\, w$ is antilinear in v (recall that $w(v) = \langle v, w\rangle$) and satisfies

$$\left|\langle v, \sigma^*(f)\, w\rangle\right| \le \|f\|_{\boldsymbol{L}^1(G)}\, \|v\|_{\boldsymbol{V}}\, \|\rho^*\|\, \|w\|_{\boldsymbol{V}^*},$$

i.e., $\sigma^*(f)\, w \in \boldsymbol{V}^*$ with

$$\left\|\sigma^*(f)\, w\right\|_{\boldsymbol{V}^*} \le \|f\|_{\boldsymbol{L}^1(G)}\, \|\rho^*\|\, \|w\|_{\boldsymbol{V}^*}.$$

Furthermore, the mapping

$$w \mapsto \sigma^*(f)\, w = \int_G f(g)\, \rho^*(g)\, w\, dm(g)$$

is linear in w, and thus $\sigma^*(f) \in \mathcal{L}(\boldsymbol{V}^*)$ with $\left\|\sigma^*(f)\right\|_{\mathcal{L}(\boldsymbol{V}^*)} \le \|f\|_{\boldsymbol{L}^1(G)}\, \|\rho^*\|$, so σ^* is bounded. Obviously, σ^* is linear in f, and it remains to show that $\sigma^*(f * k) = \sigma^*(f)\, \sigma^*(k)$. Indeed, we have for $v \in \boldsymbol{V}$ and $w \in \boldsymbol{V}^*$ that

$$\begin{aligned}
\langle v, \sigma^*(f * k)\, w\rangle &= \int_G \int_G \overline{f(h)}\,\overline{k(h^{-1}g)}\, dm(h)\, \langle v, \rho^*(g)\, w\rangle\, dm(g) \\
&= \int_G \int_G \overline{f(h)}\,\overline{k(g)}\, \langle v, \rho^*(h\, g)\, w\rangle\, dm(g)\, dm(h) \\
&= \int_G \int_G \overline{f(h)}\,\overline{k(g)}\, \langle \rho(h^{-1})\, v, \rho^*(g)\, w\rangle\, dm(g)\, dm(h) \\
&= \int_G \overline{f(h)}\, \langle \rho(h^{-1})\, v, \sigma^*(k)\, w\rangle\, dm(h) \\
&= \int_G \overline{f(h)}\, \langle v, \rho^*(h)\, \sigma^*(k)\, w\rangle\, dm(h) = \langle v, \sigma^*(f)\, \sigma^*(k)\, w\rangle,
\end{aligned}$$

where we may apply Fubini's theorem several times, since all integrals are absolutely convergent. □

Remark. Similar statements can be found in [15], Thm. V.5.2 and also in [19], Thm. (22.3) for the reflexive Banach space case. We should note that in [19], we find a footnote stating that $\boldsymbol{V}$ be reflexive is needed in the computation corresponding to the above, since otherwise the adjoint of an operator on $\boldsymbol{V}^*$ could not be regarded as an operator on $\boldsymbol{V}$. Here, however, we do not encounter this problem since we are dealing with adjoints of operators to begin with.

The following lemma states the interaction between ρ^* and σ^*.

Lemma 20. *For $g \in G$ and $f \in \boldsymbol{L}^1(G)$, we have*

$$\rho^*(g)\,\sigma^*(f) = \sigma^*(L_g f)\,, \tag{12}$$

where L_g is left translation, i.e., $(L_g f)(h) = f(g^{-1}h)$.

Proof. The result follows from

$$\begin{aligned}\rho^*(g)\,\sigma^*(f) &= \int_G f(h)\,\rho^*(g\,h)\,dm(h)\\ &= \int_G f(g^{-1}h)\,\rho^*(h)\,dm(h) = \sigma^*(L_g f)\,.\end{aligned}$$

□

This property is the main ingredient for the proof of the following result, since it enables us to essentially derive continuity of ρ^* from the continuity of the left regular representation L on $\boldsymbol{L}^1(G)$.

Theorem 21. *Let $\rho : G \to \boldsymbol{GL}(\boldsymbol{V})$ be a bounded, continuous representation of a locally compact group G on a Banach space $\boldsymbol{V}$, with contragredient representation ρ^*. If ρ^* is irreducible, then it is continuous.*

Proof. We have to show that for all $w \in \boldsymbol{V}^*$ and $h \in G$, we have

$$\left\|\rho^*(g)w - \rho^*(h)w\right\|_{\boldsymbol{V}^*} \to 0 \quad (g \to h)\,.$$

If $w = 0$, there is nothing to prove. So assume $w \in \boldsymbol{V}^* \setminus \{0\}$, and let $h \in G$ and $\varepsilon > 0$ be given. Consider the space

$$\boldsymbol{W} = \overline{\left\{\sigma^*(f)w : f \in \boldsymbol{L}^1(G)\right\}}\,.$$

$\boldsymbol{W}$ is a closed subspace of $\boldsymbol{V}^*$, and we want to show that $\boldsymbol{W} = \boldsymbol{V}^*$. Because of (12), $\boldsymbol{W}$ is ρ^*-invariant, since $L_g\big(\boldsymbol{L}^1(G)\big) = \boldsymbol{L}^1(G)$ for all $g \in G$. So by the irreducibility of ρ^*, it suffices to show that $\boldsymbol{W}$ contains a nonzero element.

To this end, note that $w \neq 0$ means that there is $v \in \boldsymbol{V}$ with $\langle v, w\rangle \neq 0$, and after replacing v by an appropriate multiple we have

$$1 = \langle v, w\rangle = \big\langle v, \rho^*(e)w\big\rangle\,.$$

Since by Lemma 16, ρ^* is weakly continuous, there exists a neighborhood U of e in G such that

$$\mathrm{Re}\Big(\big\langle v, \rho^*(g)w\big\rangle\Big) > \tfrac{1}{2} \quad \text{for } g \in U,$$

and we may assume $0 < m(U) < \infty$, since G is locally compact. Therefore, the characteristic function χ_U of U is an element of $\boldsymbol{L}^1(G)$, and it satisfies

$$\mathrm{Re}\Big(\big\langle v, \sigma^*(\chi_U)w\big\rangle\Big) = \mathrm{Re}\Big(\int_U \big\langle v, \rho^*(g)w\big\rangle\,dm(g)\Big) > \tfrac{1}{2}\,m(U)\,,$$

hence we have $0 \neq \sigma^*(\chi_U)w \in \boldsymbol{W}$. So we may conclude that indeed $\boldsymbol{W} = \boldsymbol{V}^*$.

Consequently, we can choose $f \in \boldsymbol{L}^1(G)$ such that

$$\left\|\sigma^*(f)w - w\right\|_{\boldsymbol{V}^*} < \frac{\varepsilon}{2\|\rho^*\|\big(\|\rho^*\|+1\big)}.$$

Then we have for arbitrary $g \in G$ that

$$\begin{aligned}\left\|\rho^*(g)\sigma^*(f)w - \sigma^*(f)w\right\| &= \left\|\sigma^*(L_g f - f)w\right\| \\ &\leq \|\rho^*\| \left\|L_g f - f\right\|_{\boldsymbol{L}^1(G)} \|w\|_{\boldsymbol{V}^*}.\end{aligned}$$

By continuity of the left regular representation L, we have $\|L_g f - f\|_{\boldsymbol{L}^1(G)} \to 0$ as $g \to e$, so there exists a neighborhood U of e in G with

$$\left\|L_g f - f\right\|_{\boldsymbol{L}^1(G)} < \frac{\varepsilon}{2\|\rho^*\|^2 \|w\|_{\boldsymbol{V}^*}} \quad \text{for } g \in U.$$

Then hU is a neighborhood of h, and we have for all $g \in hU$ that $h^{-1}g \in U$ and thus

$$\begin{aligned}&\left\|\rho^*(g)w - \rho^*(h)w\right\|_{\boldsymbol{V}^*} \leq \\ &\quad \leq \|\rho^*(h)\|_{\mathcal{L}(\boldsymbol{V}^*)} \left\|\rho^*(h^{-1}g)w - w\right\|_{\boldsymbol{V}^*} \\ &\quad \leq \|\rho^*\| \Big(\left\|\rho^*(h^{-1}g)w - \rho^*(h^{-1}g)\sigma^*(f)w\right\|_{\boldsymbol{V}^*} + \\ &\qquad\qquad + \left\|\rho^*(h^{-1}g)\sigma^*(f)w - \sigma^*(f)w\right\|_{\boldsymbol{V}^*} + \left\|\sigma^*(f)w - w\right\|_{\boldsymbol{V}^*} \Big) \\ &\quad \leq \|\rho^*\| \Big(\big(\|\rho^*(h^{-1}g)\|_{\mathcal{L}(\boldsymbol{V}^*)} + 1\big) \left\|\sigma^*(f)w - w\right\|_{\boldsymbol{V}^*} + \\ &\qquad\qquad + \|\rho^*\| \left\|L_{h^{-1}g} f - f\right\|_{\boldsymbol{L}^1(G)} \|w\|_{\boldsymbol{V}^*} \Big) \\ &\quad < \|\rho^*\| \Big(\big(\|\rho^*\| + 1\big) \frac{\varepsilon}{2\|\rho^*\|\big(\|\rho^*\|+1\big)} + \|\rho^*\| \frac{\varepsilon}{2\|\rho^*\|^2 \|w\|_{\boldsymbol{V}^*}} \|w\|_{\boldsymbol{V}^*} \Big) = \varepsilon,\end{aligned}$$

which proves our claim. □

3.3. Square integrability and reproducing properties

Recall that for an irreducible, unitary representation ρ on a Hilbert space, we defined square integrability by the existence of a nonzero vector φ with $\langle \varphi, \rho(\cdot)\varphi \rangle \in \boldsymbol{L}^2(G)$, and called such a φ admissible. The reason for this approach was stated in Lemma 3.(i), namely, that as a consequence, we have $\langle v, \rho(\cdot)\varphi \rangle \in \boldsymbol{L}^2(G)$ for all $v \in \boldsymbol{H}$. In a Banach space, this approach does not make sense, and we have to modify the definitions accordingly.

Definition 22. A representation ρ on a Banach space $\boldsymbol{V}$ is *square integrable*, if there exists a nonzero vector $\varphi \in \boldsymbol{V}$ such that $\langle \rho(\cdot)\varphi, w \rangle \in \boldsymbol{L}^2(G)$ for all $w \in \boldsymbol{V}^*$, *i.e.*,

$$\int_G \left|\langle \rho(g)\,\varphi, w \rangle\right|^2 dm(g) < \infty \quad \text{for all } w \in \boldsymbol{V}^*. \tag{13}$$

A vector $\varphi \in \boldsymbol{V} \setminus \{0\}$ is *admissible* for ρ, if it satisfies (13).

A contragredient representation ρ^* on $\boldsymbol{V}^*$ is **-square integrable*, if there exists a nonzero vector $\psi \in \boldsymbol{V}^*$ such that $\langle v, \rho^*(\,\cdot\,)\psi\rangle \in \boldsymbol{L}^2(G)$ for all $v \in \boldsymbol{V}$, *i.e.*,

$$\int_G \left|\langle v, \rho^*(g)\,\psi\rangle\right|^2 dm(g) < \infty \quad \text{for all } v \in \boldsymbol{V}. \tag{14}$$

A vector $\psi \in \boldsymbol{V}^* \setminus \{0\}$ is **-admissible* for ρ^*, if it satisfies (14).

Lemma 23. *For a square integrable representation ρ, the set of admissible vectors is ρ-invariant.*

*For a *-square integrable contragredient representation ρ^*, the set of *-admissible vectors is ρ^*-invariant.*

Proof. Let φ be admissible for ρ. Then we have for $h \in G$ and for all $w \in \boldsymbol{V}^*$

$$\begin{aligned}\int_G \left|\langle \rho(g)\,(\rho(h)\varphi), w\rangle\right|^2 dm(g) &= \int_G \left|\langle \rho(g\,h)\varphi, w\rangle\right|^2 dm(g)\\ &= \int_G \left|\langle \rho(g)\varphi, w)\rangle\right|^2 \Delta(h^{-1})\, dm(g) < \infty\,,\end{aligned}$$

so $\rho(h)\varphi$ is admissible also.

The proof for ρ^* is analogous. □

Now we are ready to define the analysis and synthesis operators induced by the representations, and then we can discuss the reproducing property.

Lemma 24.

(i) *Let ρ be a square integrable representation of a locally compact group G on a Banach space $\boldsymbol{V}$, and $\varphi \in \boldsymbol{V}$ an admissible vector for ρ. Then the* analysis operator *associated with ρ and φ is the bounded linear transformation*

$$T_{\rho,\varphi} : \boldsymbol{V}^* \to \boldsymbol{L}^2(G)\,, \qquad (T_{\rho,\varphi}w)(g) = \overline{\langle \rho(g)\varphi, w\rangle}\,. \tag{15}$$

*We refer to $T_{\rho,\varphi}w$ as the ρ-*transform *of w with window φ. The* synthesis operator *is the adjoint transformation*

$$T^*_{\rho,\varphi} : \boldsymbol{L}^2(G) \to \boldsymbol{V}^{**}\,, \qquad k \mapsto \int_G k(g)\,\rho(g)\varphi\, dm(g)\,,$$

where the integral is defined in the weak sense.

(ii) *For $g \in G$, we have*

$$T_{\rho,\varphi}\,\rho^*(g) = L_g\,T_{\rho,\varphi} \quad \textit{and thus} \quad \rho^{**}(g)\,T^*_{\rho,\varphi} = T^*_{\rho,\varphi}\,L_g\,.$$

(iii) *If ρ^* is *-square integrable with *-admissible vector $\psi \in \boldsymbol{V}^*$, then the analysis operator associated with ρ^* and ψ is the bounded linear transformation*

$$T_{\rho^*,\psi} : \boldsymbol{V} \to \boldsymbol{L}^2(G)\,, \qquad (T_{\rho^*,\psi}v)(g) = \langle v, \rho^*(g)\psi\rangle\,.$$

We refer to $T_{\rho^*,\psi}v$ *as the* ρ^*-transform *of* v *with window* ψ. *The synthesis operator is the adjoint transformation*

$$T^*_{\rho^*,\psi} : \boldsymbol{L}^2(G) \to \boldsymbol{V}^*, \qquad f \mapsto \int_G f(g)\,\rho^*(g)\psi\,dm(g),$$

where the integral is defined in the weak sense.*

(iv) *For* $g \in G$, *we have*

$$T_{\rho^*,\psi}\,\rho(g) = L_g\,T_{\rho^*,\psi} \quad \textit{and thus} \quad \rho^*(g)\,T^*_{\rho^*,\psi} = T^*_{\rho^*,\psi}\,L_g\,.$$

Proof. (i) Under the assumptions made, $T_{\rho,\varphi}$ obviously is a well-defined linear map from $\boldsymbol{V}^*$ to $\boldsymbol{L}^2(G)$. In order to show that $T_{\rho,\varphi}$ is bounded, it suffices to show that it is closed, since then boundedness follows by the closed graph theorem. So assume that $w_n \to w$ in $\boldsymbol{V}^*$ with $T_{\rho,\varphi}w_n \to f$ in $\boldsymbol{L}^2(G)$. Then $(T_{\rho,\varphi}w_n)(g) = \overline{\langle \rho(g)\varphi, w_n\rangle} \to \overline{\langle \rho(g)\varphi, w\rangle} = (T_{\rho,\varphi}w)(g)$ for all $g \in G$, *i.e.*, $T_{\rho,\varphi}w_n \to T_{\rho,\varphi}w$ pointwise, and thus $T_{\rho,\varphi}w = f$.

Given $k \in \boldsymbol{L}^2(G)$, we have for all $w \in \boldsymbol{V}^*$ that

$$\begin{aligned}\langle T_{\rho,\varphi}w, k\rangle_{\boldsymbol{L}^2(G)} &= \int_G \overline{\langle \rho(g)\varphi, w\rangle_{(\boldsymbol{V},\boldsymbol{V}^*)}}\,\overline{k(g)}\,dm(g)\\ &= \int_G \langle w, k(g)\,\rho(g)\varphi\rangle_{(\boldsymbol{V}^*,\boldsymbol{V}^{**})}\,dm(g)\\ &= \left\langle w, \int_G k(g)\,\rho(g)\varphi\,dm(g)\right\rangle_{(\boldsymbol{V}^*,\boldsymbol{V}^{**})},\end{aligned}$$

so $T^*_{\rho,\varphi}$ does indeed have the form stated above.

(ii) For $h \in G$ and $w \in \boldsymbol{W}$, we have

$$\begin{aligned}\big(T_{\rho,\varphi}\,\rho^*(g)w\big)(h) &= \overline{\langle \rho(h)\varphi, \rho^*(g)w\rangle}\\ &= \overline{\langle \rho(g^{-1}h)\varphi, w\rangle}\\ &= \big(T_{\rho,\varphi}w\big)(g^{-1}h) = \big(L_g\,T_{\rho,\varphi}w\big)(h)\,.\end{aligned}$$

The second identity follows from $(L_g)^* = L_{g^{-1}}$ and $(\rho^*(g))^* = \rho^{**}(g^{-1})$.

(iii), (iv) The corresponding properties of $T_{\rho^*,\psi}$ and $T^*_{\rho^*,\psi}$ are shown analogously. □

Remark. It is worth noting that $T^*_{\rho^*,\psi}$ and σ^* (*cf.*, Thm. 19) are related via

$$T^*_{\rho^*,\psi}f = \sigma^*(f)\psi \quad \text{for } f \in \boldsymbol{L}^1 \cap \boldsymbol{L}^2(G).$$

Lemma 25. *If, in addition to the assumptions of Lemma 24,* ρ *is continuous, then we have*

$$T^*_{\rho,\varphi} : \boldsymbol{L}^2(G) \to \boldsymbol{V}\,.$$

Proof. For $k \in \boldsymbol{C}_c \subseteq \boldsymbol{L}^2(G)$, we know that

$$T^*_{\rho,\varphi} k = \int_G k(g)\, \rho(g)\varphi \, dm(g) \in \boldsymbol{V} ,$$

since the right hand side is a convergent Bochner integral. Furthermore, since $\boldsymbol{C}_c(G)$ is dense in $\boldsymbol{L}^2(G)$ and $\boldsymbol{V}$ is norm-closed in $\boldsymbol{V}^{**}$, the claim follows from the boundedness of the operator $T^*_{\rho,\varphi}$. □

The important assumption for Schur's Lemma is the commuting property, which we may conclude directly from Lemma 24.

Corollary 26.
(i) *Under the assumptions of Lemmata 24 and 25, the operators* $T^*_{\rho,\varphi} T_{\rho^*,\psi}$ *and* $T^*_{\rho^*,\psi} T_{\rho,\varphi}$ *commute with* $\rho(G)$ *and* $\rho^*(G)$*, respectively.*
(ii) *If furthermore* ρ^* *is irreducible, then the two operators either are both injective, or both vanish identically.*

Proof. (i) By Lemma 24, we know that $T^*_{\rho^*,\psi} T_{\rho,\varphi}$ is a bounded linear operator from $\boldsymbol{V}^*$ into itself, and it satisfies

$$T^*_{\rho^*,\psi}\, T_{\rho,\varphi}\, \rho^*(g) = T^*_{\rho^*,\psi}\, L_g\, T_{\rho,\varphi} = \rho^*(g)\, T^*_{\rho^*,\psi}\, T_{\rho,\varphi} \tag{16}$$

for all $g \in G$. Also, by Lemmas 24 and 25, we know that $T^*_{\rho,\varphi} T_{\rho^*,\psi}$ is a bounded linear operator from $\boldsymbol{V}$ into itself, and it satisfies

$$T^*_{\rho,\varphi}\, T_{\rho^*,\psi}\, \rho(g) = T^*_{\rho,\varphi}\, L_g\, T_{\rho^*,\psi} = \rho(g)\, T^*_{\rho,\varphi}\, T_{\rho^*,\psi} \tag{17}$$

for all $g \in G$.

(ii) Because of (16), the kernel of $T^*_{\rho^*,\psi} T_{\rho,\varphi}$ is invariant under $\rho^*(G)$ and hence can only be $\{0\}$ or all of $\boldsymbol{V}^*$, *i.e.*, the operator is injective or identically zero. By Lemma 18, ρ is also irreducible, and we may apply the analogous argument to $T^*_{\rho,\varphi} T_{\rho^*,\psi}$. Since $T^*_{\rho^*,\psi} T_{\rho,\varphi} = (T^*_{\rho,\varphi} T_{\rho^*,\psi})^*$, the two operators can only vanish simultaneously. □

In case $\boldsymbol{V}$ actually is (topologically isomorphic to) a Hilbert space, we can apply Schur's Lemma to obtain the desired reproducing property. In order to allow for the situation from Section 2, where we treated the Hilbert space $\boldsymbol{H}_{-1/2}$ as a Banach space with dual $\boldsymbol{H}_{1/2}$, we formulate the following theorem for this somewhat more general case.

Theorem 27 (Reproducing Property). *Let* ρ *be a representation of a locally compact group* G *on a Banach space* $\boldsymbol{V}$*, with contragredient representation* ρ^* *on* $\boldsymbol{V}^*$*. Assume that* ρ *is square integrable with admissible vector* φ*, and that* ρ^* *is irreducible and *-square integrable with *-admissible vector* ψ*. If* $\boldsymbol{V}$ *is topologically isomorphic to a Hilbert space, the following hold.*

(i) *The operator*

$$T^*_{\rho,\varphi}T_{\rho^*,\psi} : v \mapsto \int_G \langle v, \rho^*(g)\psi\rangle\, \rho(g)\varphi\, dm(g)$$

is a multiple of the identity on $\boldsymbol{V}$.

(ii) *The operator*

$$T^*_{\rho^*,\psi}T_{\rho,\varphi} : w \mapsto \int_G \overline{\langle \rho(g)\varphi, w\rangle}\, \rho^*(g)\psi\, dm(g)$$

is a multiple of the identity on $\boldsymbol{V}^*$.

Proof. (ii) If $\boldsymbol{V}$ is topologically isomorphic to a Hilbert space, then so is $\boldsymbol{V}^*$. So because of (16) and the irreducibility of ρ^*, the claim follows by Schur's Lemma.

(i) $\boldsymbol{V}$ being isomorphic to a Hilbert space implies that it is reflexive. Consequently, we have $T^*_{\rho,\varphi}T_{\rho^*,\psi} = (T^*_{\rho^*,\psi}T_{\rho,\varphi})^*$, and the claim follows from (ii). □

The multiplicative factor of the above operators can be determined as follows. Letting

$$T^*_{\rho,\varphi}T_{\rho^*,\psi} =: C_{\varphi,\psi}\, I_{\boldsymbol{V}}\,,$$

we have

$$\langle C_{\varphi,\psi}\varphi, \psi\rangle = \langle T^*_{\rho,\varphi}T_{\rho^*,\psi}\varphi, \psi\rangle = \langle T_{\rho^*,\psi}\varphi, T_{\rho,\varphi}\psi\rangle_{\boldsymbol{L}^2(G)}$$

and thus

$$C_{\varphi,\psi} = \frac{\langle T_{\rho^*,\psi}\varphi, T_{\rho,\varphi}\psi\rangle_{\boldsymbol{L}^2(G)}}{\langle \varphi, \psi\rangle} \quad \text{if } \langle \varphi, \psi\rangle \neq 0\,.$$

Consequently,

$$T^*_{\rho^*,\psi}T_{\rho,\varphi} = (T^*_{\rho,\varphi}T_{\rho^*,\psi})^* = \overline{C_{\varphi,\psi}}\, I_{\boldsymbol{V}^*}\,.$$

In the general case, however, we encounter technical difficulties. The proof of Schur's Lemma relies heavily on the spectral decomposition for operators in Hilbert spaces and thus can not be transferred to the Banach space case. On the other hand, the assumption of square integrability implies a strong interrelation between the spaces $\boldsymbol{V}$ and $\boldsymbol{V}^*$ on the one hand and an appropriate subspace of $\boldsymbol{L}^2(G)$ on the other. Indeed, the converse of Theorem 27 holds in the sense that the reproducing property implies that $\boldsymbol{V}$ (and thus $\boldsymbol{V}^*$) is isomorphic to a Hilbert space.

Theorem 28. *Let ρ be a continuous representation of a locally compact group G on a Banach space $\boldsymbol{V}$, with contragredient representation ρ^* on $\boldsymbol{V}^*$. Assume that ρ is square integrable with admissible vector φ, and that ρ^* is irreducible and *-square integrable with *-admissible vector ψ. If the operator $T^*_{\rho,\varphi}T_{\rho^*,\psi}$ does not vanish identically, then the following are equivalent.*

(i) $T^*_{\rho,\varphi}T_{\rho^*,\psi}$ *is a multiple of the identity on* $\boldsymbol{V}$.

(ii) $T^*_{\rho^*,\psi}T_{\rho,\varphi}$ *is a multiple of the identity on* $\boldsymbol{V}^*$.

(iii) $T_{\rho^*,\psi}(\boldsymbol{V})$ *is closed in* $\boldsymbol{L}^2(G)$.

(iv) *$T_{\rho,\varphi}(\boldsymbol{V}^*)$ is closed in $\boldsymbol{L}^2(G)$.*
(v) *$\boldsymbol{V}$ is topologically isomorphic to a Hilbert space.*
(vi) *$\boldsymbol{V}^*$ is topologically isomorphic to a Hilbert space.*
(vii) *The left regular representation acts irreducibly on $\overline{T_{\rho^*,\psi}(\boldsymbol{V})} \subseteq \boldsymbol{L}^2(G)$.*
(viii) *The left regular representation acts irreducibly on $\overline{T_{\rho,\varphi}(\boldsymbol{V}^*)} \subseteq \boldsymbol{L}^2(G)$.*

Proof. (i)⇒(ii). If $T^*_{\rho,\varphi}T_{\rho^*,\psi} = C\,I_{\boldsymbol{V}}$, then $T^*_{\rho^*,\psi}T_{\rho,\varphi} = (T^*_{\rho,\varphi}T_{\rho^*,\psi})^* = \overline{C}\,I_{\boldsymbol{V}^*}$.

(ii)⇒(i). As usual, consider $\boldsymbol{V}$ to be embedded in $\boldsymbol{V}^{**}$. Then $T^*_{\rho^*,\psi}T_{\rho,\varphi} = C\,I_{\boldsymbol{V}^*}$ implies $T^*_{\rho,\varphi}T^{**}_{\rho^*,\psi} = (T^*_{\rho^*,\psi}T_{\rho,\varphi})^* = \overline{C}\,I_{\boldsymbol{V}^{**}}$. Replacing $T^{**}_{\rho^*,\psi}$ by $T_{\rho^*,\psi}$ simply restricts this operator to $\boldsymbol{V}$ which yields the claim.

(i)⇒(iii). $T^*_{\rho,\varphi}T_{\rho^*,\psi} = C\,I_{\boldsymbol{V}}$ with, by assumption, $C \neq 0$ means $T^{-1}_{\rho^*,\psi} = C^{-1}\,T^*_{\rho,\varphi}$, so $T_{\rho^*,\psi}$ is boundedly invertible and thus a topological isomorphism between $\boldsymbol{V}$ and its image. Hence $T_{\rho^*,\psi}(\boldsymbol{V})$ is complete and therefore closed in $\boldsymbol{L}^2(G)$.

(ii)⇒(iv). Analogously, since $T^*_{\rho,\varphi}T_{\rho^*,\psi} \not\equiv 0$ implies $T^*_{\rho^*,\psi}T_{\rho,\varphi} = (T^*_{\rho,\varphi}T_{\rho^*,\psi})^* \not\equiv 0$.

(iii)⇒(v). As a closed subspace of $\boldsymbol{L}^2(G)$, the space $T_{\rho^*,\psi}(\boldsymbol{V})$ is a Hilbert space itself. From Corollary 26(ii), we may conclude that $T_{\rho^*,\psi}$ is injective. So by the open mapping theorem, the continuous bijection $T_{\rho^*,\psi}$ is a topological isomorphism between $\boldsymbol{V}$ and its range.

(iv)⇒(vi). Analogously.

(v)⇒(i). See Theorem 27.

(vi)⇒(v). Let $F : \boldsymbol{V}^* \to \boldsymbol{H}$ be a topological isomorphism between $\boldsymbol{V}^*$ and some Hilbert space $\boldsymbol{H}$. Then $\boldsymbol{V}^*$ is reflexive, and thus so is $\boldsymbol{V}$ (*e.g.*, see [7]). Consequently, we have the topological isomorpism $F^* : \boldsymbol{H} \to \boldsymbol{V}$.

(iii)⇒(vii). Since $T_{\rho^*,\psi}$ intertwines the actions of ρ and L, irreducibility of L on $T_{\rho^*,\psi}(\boldsymbol{V}) = \overline{T_{\rho^*,\psi}(\boldsymbol{V})}$ is equivalent to that of ρ on $\boldsymbol{V}$.

(iv)⇒(viii). Analogously.

(vii)⇒(i). Let $\boldsymbol{H}$ be the closure of $T_{\rho^*,\psi}(\boldsymbol{V})$ in $\boldsymbol{L}^2(G)$. Then we can treat the operator

$$T_{\rho^*,\psi}T^*_{\rho,\varphi} : \boldsymbol{L}^2(G) \to T_{\rho^*,\psi}(\boldsymbol{V})$$

as an element of $\boldsymbol{\mathcal{L}}(\boldsymbol{H})$. By Lemma 24, this operator commutes with the left regular representation L of G, so by the assumed irreducibility and Schur's Lemma, we have $T_{\rho^*,\psi}T^*_{\rho,\varphi} = C\,I_{\boldsymbol{H}}$ and thus

$$T^*_{\rho,\varphi}T_{\rho^*,\psi}T^*_{\rho,\varphi}T_{\rho^*,\psi} = C\,T^*_{\rho,\varphi}T_{\rho^*,\psi}\,,$$

so $T^*_{\rho,\varphi}T_{\rho^*,\psi}$ is a multiple of a projection. But from Corollary 26(ii), we know that $T^*_{\rho,\varphi}T_{\rho^*,\psi}$ is injective, and therefore $T^*_{\rho,\varphi}T_{\rho^*,\psi} = C\,I_{\boldsymbol{V}}$.

(viii)⇒(ii). Analogously. □

It is important to realize that in the proof of Theorem 28, we have explicitly described the Hilbert spaces for part (v) and (vi): namely, closed subspaces of $\boldsymbol{L}^2(G)$. Thus, if we do have the reproducing properties (i) and (ii), we can replace the norms on $\boldsymbol{V}$ and $\boldsymbol{V}^*$ by norms such that $T_{\rho^*,\psi}$ and $T_{\rho,\varphi}$ become isometries. Then ρ and ρ^* act isometrically, and thus we end up with unitary representations. This explains our observations from Section 2.8.

3.4. Other integrability properties

The only possibility to avoid this lies in giving up square integrability. What we do need for the explicit definition of the synthesis operators is the existence of the integral

$$\int_G \langle v, \rho^*(g)\psi\rangle \langle \rho(g)\varphi, w\rangle \, dm(g),$$

which we can also achieve by ensuring that the two factors are p- and q-integrable, respectively, for conjugate exponents p and q. This motivates the following definitions.

Definition 29. A representation ρ on a Banach space $\boldsymbol{V}$ is *p-integrable* for some $p \in [1, \infty[$, if there exists a nonzero vector $\varphi \in \boldsymbol{V}$ such that $\langle \rho(\,\cdot\,)\varphi, w\rangle \in \boldsymbol{L}^p(G)$ for all $w \in \boldsymbol{V}^*$, *i.e.*,

$$\int_G \left|\langle \rho(g)\varphi, w\rangle\right|^p dm(g) < \infty \quad \text{for all } w \in \boldsymbol{V}^*. \tag{18}$$

A vector $\varphi \in \boldsymbol{V} \setminus \{0\}$ is *p-admissible* for ρ, if it satisfies (18).

For the term 1-integrable, we also use *integrable* for short.

A contragredient representation ρ^* on $\boldsymbol{V}^*$ is **-q-integrable* for some $q \in [1, \infty[$, if there exists a nonzero vector $\psi \in \boldsymbol{V}^*$ such that $\langle v, \rho^*(\,\cdot\,)\psi\rangle \in \boldsymbol{L}^q(G)$ for all $v \in \boldsymbol{V}$, *i.e.*,

$$\int_G \left|\langle v, \rho^*(g)\psi\rangle\right|^q dm(g) < \infty \quad \text{for all } v \in \boldsymbol{V}. \tag{19}$$

A vector $\psi \in \boldsymbol{V}^* \setminus \{0\}$ is **-q-admissible* for ρ^*, if it satisfies (19).

Straightforward modifications of the proofs of Lemma 23–25 and Corollary 26 yield the following analogous statements.

Corollary 30. *For a p-integrable representation ρ, the set of p-admissible vectors is ρ-invariant.*

*For a *-q-integrable contragredient representation ρ^*, the set of *-q-admissible vectors is ρ^*-invariant.*

Corollary 31. *Let $p \in [1, \infty[$ and $\frac{1}{p} + \frac{1}{q} = 1$. Consider a continuous p-integrable representation ρ of a locally compact group G on a Banach space $\boldsymbol{V}$, and let $\varphi \in \boldsymbol{V}$ be a p-admissible vector for ρ. Then the analysis operator $T_{\rho,\varphi} : \boldsymbol{V}^* \to \boldsymbol{L}^p(G)$ and the synthesis operator $T^*_{\rho,\varphi} : \boldsymbol{L}^q(G) \to \boldsymbol{V}$ are bounded linear transformations. They satisfy*

$$T_{\rho,\varphi}\, \rho^*(g) = L_g\, T_{\rho,\varphi} \quad \text{and thus} \quad \rho(g)\, T^*_{\rho,\varphi} = T^*_{\rho,\varphi}\, L_g$$

for all $g \in G$, where ρ^* is the contragredient representation of ρ, and L_g is left translation in $\boldsymbol{L}^p(G)$ and $\boldsymbol{L}^q(G)$, respectively. Analogous statements hold for $T_{\rho^*,\psi}$ and $T^*_{\rho^*,\psi}$.

Corollary 32. *Let $p \in]1,\infty[$ and $\frac{1}{p}+\frac{1}{q}=1$. Let ρ be a p-integrable representation of a locally compact group G on a Banach space $\boldsymbol{V}$ with p-admissible vector φ, and assume that ρ^* is *-q-integrable with *-q-admissible vector ψ. Then the operators $T^*_{\rho,\varphi}T_{\rho^*,\psi}$ and $T^*_{\rho^*,\psi}T_{\rho,\varphi}$ commute with $\rho(G)$ and $\rho^*(G)$, respectively.*

The same holds if ρ is integrable and ρ^* is bounded, or vice versa.

We encounter the problem, though, that we can not apply Schur's Lemma to prove a reproducing property. Assuming Hilbert space structure for $\boldsymbol{V}$ and $\boldsymbol{V}^*$ does not make sense in this situation, since then a reproducing property would impose Hilbert space structure to closed subspaces of $\boldsymbol{L}^p(G)$ and $\boldsymbol{L}^q(G)$.

Integrable representations and their discretizations have been used for so-called atomic decompositions of Banach spaces in [10] and [11], where the authors also introduce and discuss the concept of *coorbit spaces*. They start from an integrable, irreducible, unitary, continuous representation ρ and define the coorbit of $\boldsymbol{Y}$ under ρ to be the space

$$\mathcal{C}o(\boldsymbol{Y}) = \{v : T_{\rho,\varphi}v \in \boldsymbol{Y}\}$$

for an appropriate φ, where v is allowed to range in some space of generalized distributions. In case $\boldsymbol{Y} = \boldsymbol{L}^p(G)$, this can be seen as a somehow reverse approach compared to the idea of a p-integrable representation.

For the special case of the wavelet transformation, this leads to characterizations, *e.g.*, of Besov spaces ([14], [27]). Applied to the Gabor transformation, this approach gives rise to the so-called modulation spaces (*e.g.*, see [9], [12], [13], [17], [26]).

Appendix

A.1. Antilinear functionals

As dual space of a Banach space $\boldsymbol{V}$, we usually consider the space of bounded linear functionals $\boldsymbol{V}'$. This space has vector space structure, since for $w_1, w_2 \in \boldsymbol{V}'$ and $\lambda \in \mathbb{C}$, we can define

$$(w_1 + w_2)(v) := w_1(v) + w_2(v) \quad \text{and} \quad (\lambda w_1)(v) := \lambda\, w_1(v) \quad \text{for all } v \in \boldsymbol{V}.$$

Also, for a bounded linear operator $T : \boldsymbol{V}_1 \to \boldsymbol{V}_2$ between two Banach spaces, we usually define the adjoint operator $T^* : \boldsymbol{V}_2' \to \boldsymbol{V}_1'$ via

$$(T^*w)(v) := w(T\,v)\,,$$

and the mapping $\mathcal{L}(\boldsymbol{V}_1, \boldsymbol{V}_2) \to \mathcal{L}(\boldsymbol{V}_2', \boldsymbol{V}_1')$, $T \mapsto T^*$, is linear.

When dealing with a Hilbert space $\boldsymbol{H}$, though, we encounter a formal problem. $\boldsymbol{H}$ is self-dual in the sense that for each bounded linear functional w on $\boldsymbol{H}$, there exists $v' \in \boldsymbol{H}$ such that $w(v) = \langle v, v' \rangle$ for all $v \in \boldsymbol{H}$, but the correspondence between $\boldsymbol{H}'$ and $\boldsymbol{H}$ given by $w \longleftrightarrow v'$ is antilinear in the sense that $\lambda w \longleftrightarrow \overline{\lambda} v'$.

Furthermore, for a bounded linear operator $T : \boldsymbol{H}_1 \to \boldsymbol{H}_2$ between Hilbert spaces, we define the Hilbert space adjoint or Hermitian adjoint operator $T^* : \boldsymbol{H}_2 \to \boldsymbol{H}_1$ via

$$\langle v, T^* w \rangle_{\boldsymbol{H}_1} := \langle T\, v, w \rangle_{\boldsymbol{H}_2}, \tag{A.1}$$

and here, the mapping $\boldsymbol{\mathcal{L}}(\boldsymbol{H}_1, \boldsymbol{H}_2) \to \boldsymbol{\mathcal{L}}(\boldsymbol{H}_2, \boldsymbol{H}_1)$, $T \mapsto T^*$, is antilinear.

As natural as these conventions are, they will create havoc as soon as we consider adjoints of operators between a Banach space and a Hilbert space. To avoid this problem, we follow the convention in [19] and replace $\boldsymbol{V}'$, the space of bounded linear functionals, by $\boldsymbol{V}^*$, the space of bounded antilinear functionals on $\boldsymbol{V}$, which satisfy

$$w(\lambda_1\, v_1 + \lambda_2\, v_2) = \overline{\lambda_1}\, w(v_1) + \overline{\lambda_2}\, w(v_2)\,.$$

This space also has vector space structure over $\mathbb{C}$, if we define

$$(w_1 + w_2)(v) := w_1(v) + w_2(v) \quad \text{and} \quad (\lambda\, w_1)(v) := \lambda\, w_1(v) \quad \text{for all } v \in \boldsymbol{V},$$

as before. Then we can introduce a sesquilinear pairing between $\boldsymbol{V}$ and $\boldsymbol{V}^*$ via

$$\langle v, w \rangle_{(\boldsymbol{V},\boldsymbol{V}^*)} := \overline{w(v)}\,.$$

(Here we deviate somewhat from [19], where the authors define $\langle w, v \rangle_{(\boldsymbol{V}^*,\boldsymbol{V})} = w(v)$.) This pairing is linear in the first and antilinear in the second argument, since we have

$$\begin{aligned}
\langle \lambda_1\, v_1 + \lambda_2\, v_2, w \rangle &= \overline{w(\lambda_1\, v_1 + \lambda_2\, v_2)} \\
&= \lambda_1\, \overline{w(v_1)} + \lambda_2\, \overline{w(v_2)} = \lambda_1\, \langle v_1, w \rangle + \lambda_2\, \langle v_2, w \rangle \\
\text{and} \qquad \langle v, \mu_1\, w_1 + \mu_2\, w_2 \rangle &= \overline{(\mu_1\, w_1 + \mu_2\, w_2)(v)} \\
&= \overline{\mu_1\, w_1(v) + \mu_2\, w_2(v)} = \overline{\mu_1}\, \langle v, w_1 \rangle + \overline{\mu_2}\, \langle v, w_2 \rangle\,.
\end{aligned}$$

The space $\boldsymbol{V}^*$ is antilinearly isomorphic to $\boldsymbol{V}'$ via

$$w : v \mapsto w(v) \qquad \longleftrightarrow \qquad \widetilde{w} : v \mapsto \overline{w(v)}\,,$$

since we have

$$\begin{aligned}
\widetilde{w}(\lambda_1\, v_1 + \lambda_2\, v_2) &= \overline{w(\lambda_1\, v_1 + \lambda_2\, v_2)} \\
&= \overline{\overline{\lambda_1}\, w(v_1) + \overline{\lambda_2}\, w(v_2)} = \lambda_1\, \widetilde{w}(v_1) + \lambda_2\, \widetilde{w}(v_2)\,,
\end{aligned}$$

i.e., $\widetilde{w}$ is a linear functional, and this correspondence is obviously isometric, so $\widetilde{w} \in \boldsymbol{V}'$.

We still have the canonical embedding $\phi : \boldsymbol{V} \hookrightarrow \boldsymbol{V}^{**}$ via

$$\langle w, \phi(v) \rangle_{(\boldsymbol{V}^*,\boldsymbol{V}^{**})} = \overline{\langle v, w \rangle_{(\boldsymbol{V},\boldsymbol{V}^*)}}\,.$$

This embedding is again linear, and $\boldsymbol{V}$ being reflexive in the usual sense is equivalent to ϕ being an isomorphism.

Consequently, we define for a bounded linear operator $T : \boldsymbol{V}_1 \to \boldsymbol{V}_2$ the adjoint operator $T^* : \boldsymbol{V}_2^* \to \boldsymbol{V}_1^*$ by

$$\langle v, T^* w\rangle_{(\boldsymbol{V}_1,\boldsymbol{V}_1^*)} := \langle T v, w\rangle_{(\boldsymbol{V}_2,\boldsymbol{V}_2^*)}$$

which includes (A.1) as a special case; in particular, the mapping $\mathcal{L}(\boldsymbol{V}_1, \boldsymbol{V}_2) \to \mathcal{L}(\boldsymbol{V}_2', \boldsymbol{V}_1')$, $T \mapsto T^*$, is antilinear. This definition still makes T^* a linear (and not antilinear) operator, since

$$\langle v, T^* \lambda w\rangle = \langle T v, \lambda w\rangle = \overline{\lambda}\,\langle T v, w\rangle = \overline{\lambda}\,\langle v, T^* w\rangle = \langle v, \lambda T^* w\rangle\,.$$

The advantage of this approach is that we actually have the natural *linear* isomorphism $\boldsymbol{H} \cong \boldsymbol{H}^*$ (instead of the *antilinear* isomorphism $\boldsymbol{H} \cong \boldsymbol{H}'$ mentioned above), and this avoids the ambiguity in the definition of T^*.

A.2. Two technical lemmata

Lemma A.1. For m, $n \in \mathbb{N}$ and $1 \le j \le \min\{m, n\}$, we have

$$\sum_{\substack{\ell=0\\ \ell\ge m-n}}^{m-1} (-1)^{m-1-\ell} \binom{m-1}{\ell} \binom{n+\ell-j}{m-j} = \delta_{1,j}\,,$$

and

$$\sum_{\ell=1}^{\min\{m,n\}} (-1)^{n-\ell} \binom{n}{\ell} \binom{m+n-j-\ell}{n-j} \binom{j+\ell-2}{j-1} = (-1)^{n+1} \binom{m+n-1}{n-1} \delta_{1,j}\,.$$

Proof. For a function f on $\mathbb{R}$, consider iterated forward differences on knots with integer distances:

$$\begin{aligned} f^{[0]}(x) &= f(x)\,,\\ f^{[1]}(x) &= f(x+1) - f(x)\,,\\ f^{[k]}(x) &= f^{[k-1]}(x+k) - f^{[k-1]}(x)\,. \end{aligned}$$

This yields

$$f^{[k]}(x) = \sum_{\ell=0}^{k} (-1)^{k-\ell} \binom{k}{\ell} f(x+\ell)\,,$$

which we show by induction on k.

For $k = 0$, we have $f^{[0]}(x) = f(x)$ by definition. For the induction step, consider

$$\begin{aligned} f^{[k+1]}(x) &= f^{[k]}(x+k+1) - f^{[k]}(x)\\ &\underset{\text{(ind. hyp.)}}{=} \sum_{\ell=0}^{k}(-1)^{k-\ell}\binom{k}{\ell} f(x+1+\ell) - \sum_{\ell=0}^{k}(-1)^{k-\ell}\binom{k}{\ell} f(x+\ell) \end{aligned}$$

$$\underset{(j=\ell+1)}{=} f(x+1+k) + \sum_{j=1}^{k}(-1)^{k-(j-1)} \binom{k}{j-1} f(x+j) +$$

$$+ \sum_{\ell=1}^{k}(-1)^{k-\ell+1} \binom{k}{\ell} f(x+\ell) + (-1)^{k+1} f(x)$$

$$= f(x+k+1) + \sum_{\ell=1}^{k}(-1)^{k+1-\ell} \left(\binom{k}{\ell-1} + \binom{k}{\ell}\right) f(x+\ell) + (-1)^{k+1} f(x)$$

$$= \sum_{\ell=0}^{k+1}(-1)^{k+1-\ell} \binom{k+1}{\ell} f(x+\ell),$$

which completes the induction.

In case the function under consideration is a polynomial $p(x)$ of degree r with leading coefficient a_r, it is easy to see that for $k \leq r$, the iterated difference $p^{[k]}(x)$ is a polynomial of degree $r-k$ with leading coefficient $\frac{r!}{(r-k)!} a_r$, and $p^{[k]}(x) = 0$ for $k > r$.

To prove our first claim, let $p(x) = \binom{n+x-j}{m-j} = \dfrac{(n+x-j)(n+x-j-1)\cdots(n+x-m+1)}{(m-j)!}$, which is a polynomial of degree $m-j$ with leading coefficient $a_{m-j} = \dfrac{1}{(m-j)!}$. Note that if $m > n$, we have for $\ell \in \mathbb{N}_0$ with $0 \leq \ell < m-n$ that $p(\ell) = 0$. Therefore,

$$\sum_{\substack{\ell=0\\ \ell\geq m-n}}^{m-1} (-1)^{m-1-\ell} \binom{m-1}{\ell} \binom{n+\ell-j}{m-j} = \sum_{\ell=0}^{m-1}(-1)^{m-1-\ell} \binom{m-1}{\ell} p(\ell) = p^{[m-1]}(0),$$

which equals 0 for $m-1 > m-j$, *i.e.*, $j > 1$, and $(m-1)!\, a_{m-1} = 1$ for $j = 1$, as claimed.

For the second claim, let $p(x) = \binom{m+n-j-x}{n-j}\binom{j+x-2}{j-1}$, which is a polynomial in x of degree $(n-j)+(j-1) = n-1$, so $p^{[n]}(x) \equiv 0$. Note that $p(\ell) = 0$ for $\ell > m$, and for $j > 1$, also $p(0) = 0$, hence

$$\sum_{\ell=1}^{\min\{m,n\}} (-1)^{n-\ell} \binom{n}{\ell} \binom{m+n-j-\ell}{n-j} \binom{j+\ell-2}{j-1} = \sum_{\ell=0}^{n}(-1)^{n-\ell} \binom{n}{\ell} p(\ell) = p^{[n]}(0) = 0.$$

For $j = 1$, though, we have $p(0) = \binom{m+n-1}{n-1}$, so

$$\sum_{\ell=1}^{\min\{m,n\}} (-1)^{n-\ell} \binom{n}{\ell} \binom{m+n-j-\ell}{n-j} \binom{j+\ell-2}{j-1} = p^{[n]}(0) - (-1)^n p(0) = (-1)^{n+1} \binom{m+n-1}{n-1},$$

which yields the claim. □

Lemma A.2. For $p, q \in \mathbb{N}$ with $p < q$, we have

$$\int_0^\infty \frac{(\sinh s)^{2p+1}}{(\cosh s)^{2q+1}}\, ds = \frac{1}{2}\, \frac{p!\,(q-p-1)!}{q!}.$$

Proof.

$$\int_0^\infty \frac{(\sinh s)^{2p+1}}{(\cosh s)^{2q+1}}\, ds \underset{(t=\cosh s)}{=} \int_1^\infty \frac{(t^2-1)^p}{t^{2q+1}}\, dt$$
$$= \sum_{r=0}^{p} \binom{p}{r} (-1)^r \int_1^\infty t^{2(p-r)-(2q+1)}\, dt$$
$$= \sum_{r=0}^{p} \binom{p}{r} (-1)^r \left[\frac{t^{2(p-r-q)}}{2(p-r-q)}\right]_1^\infty$$
$$= \frac{1}{2} \sum_{r=0}^{p} \binom{p}{r} \frac{(-1)^r}{q-p+r}\,.$$

So it suffices to show that

$$\sum_{r=0}^{p} \binom{p}{r} \frac{(-1)^r}{q-p+r} = \frac{p!\,(q-p-1)!}{q!} \qquad \text{for } q > p, \tag{A.2}$$

for which we use induction on p.

For $p = 0$, (A.2) becomes

$$\frac{1}{q} = \frac{0!\,(q-1)!}{q!} \qquad \text{for } q \in \mathbb{N},$$

which is obviously true.

For the induction step, we find that for $q > p+1$,

$$\sum_{r=0}^{p+1} \binom{p+1}{r} \frac{(-1)^r}{q-(p+1)+r} = \frac{1}{q-p-1} + \sum_{r=1}^{p} \left(\binom{p}{r} + \binom{p}{r-1}\right) \frac{(-1)^r}{q-(p+1)+r} + \frac{(-1)^{p+1}}{q}$$
$$= \frac{1}{(q-1)-p} + \sum_{r=1}^{p} \binom{p}{r} \frac{(-1)^r}{(q-1)-p+r} + \sum_{s=0}^{p-1} \binom{p}{s} \frac{(-1)^{s+1}}{q-p+s} + \frac{(-1)^{p+1}}{q}$$
$$= \sum_{r=0}^{p} \binom{p}{r} \frac{(-1)^r}{(q-1)-p+r} - \sum_{s=0}^{p} \binom{p}{s} \frac{(-1)^s}{q-p+s}$$
$$\underset{(\text{ind. hyp.})}{=} \frac{p!\,(q-1-p-1)!}{(q-1)!} - \frac{p!\,(q-p-1)!}{q!}$$
$$= \frac{p!\,(q-p-2)!}{q!} \big(q - (q-p-1)\big) = \frac{(p+1)!\,(q-(p+1)-1)!}{q!}\,,$$

which proves the claim. □

It is worth noting that the left hand side of (A.2) is just the pth order forward difference $f^{[p]}(q-p)$ of the function $f(x) = \frac{1}{x}$ at the point $x = q-p$, and this is what we actually make use of in the proof. For more details on iterated differences, *e.g.*, see [16].

Acknowledgements

The author wishes to express his sincere thanks to Elena Berdysheva, Lars Schneider, and Markus Stroppel for fruitful discussions, to the referees for their time and efforts, and most of all to Kurt Jetter for his support and lasting patience.

References

1. Ali, S. T., Antoine, J.-P. and Gazeau, J.-P., *Coherent States, Wavelets and Their Generalizations*, Graduate Texts in Contemporary Physics, Springer, New York, 2000.
2. Arazy, J. and Fisher, S. D., The uniqueness of the Dirichlet space among Möbius-invariant Hilbert spaces, *Illinois J. Math.* **29** (1985), 449–462.
3. Arazy, J., Fisher, S. D. and Peetre, J., Möbius-invariant function spaces, *J. Reine Angew. Math.* **363** (1985), 110–145.
4. Barut, A. O. and Girardello, L., New "coherent" states associated with non-compact groups, *Comm. Math. Phys.* **21** (1971), 41–55.
5. Carey, A. L., Square-integrable representations of non-unimodular groups, *Bull. Austral. Math. Soc.* **15** (1976), 1–12.
6. Chui, C. K. and Lian, J. A., A study of orthonormal wavelets, *Appl. Numer. Math.* **20** (1996), 273–298.
7. Day, M. M., *Normed Linear Spaces*, Ergebnisse der Mathematik und ihrer Grenzgebiete, Vol. 21, Springer, Berlin, 1962.
8. Duflo, M. and Moore, C. C., On the regular representation of a nonunimodular locally compact group, *J. Funct. Anal.* **21** (1976), 209–243.
9. Feichtinger, H. G., Minimal Banach spaces and atomic representations, *Publ. Math. Debrecen* **34** (1987), 231–240.
10. Feichtinger, H. G. and Gröchenig, K., A unified approach to atomic characterizations via integrable group representations, in: *Function Spaces and Applications*, Proc. US-Swed. Semin. Lund/Swed. (M. Cwikel, J. Peetre, Y. Sagher and H. Wallin, Eds.), Lecture Notes Math., Vol. 1302, Springer, Berlin, 1988, pp. 52–73.
11. Feichtinger, H. G. and Gröchenig, K., Banach spaces related to integrable group representations and their atomic decompositions, I, *J. Funct. Anal.* **86** (1989), 307–340.
12. Feichtinger, H. G. and Gröchenig, K., Gabor wavelets and the Heisenberg group: Gabor expansions and short time Fourier transform from the group theoretical point of view, in: *Wavelets: A Tutorial in Theory and Applications* (C. K. Chui, Ed.), Academic Press, Boston, 1992, pp. 359–397.
13. Feichtinger, H. G. and Zimmermann, G., A Banach space of test functions for Gabor analysis, in: *Gabor Analysis and Algorithms: Theory and Applications* (H. G. Feichtinger and T. Strohmer, Eds.), Birkhäuser, Boston, 1998, pp. 123–170.

14. Frazier, M. and Jawerth, B., Decomposition of Besov spaces, *Indiana Univ. Math. J.* **34** (1985), 777–799.
15. Gaal, S. A., *Linear Analysis and Representation Theory*, Grundlehren Math. Wiss., Vol 198, Springer, Berlin, 1973.
16. Gould, H. W., *Combinatorial Identities*, Gould, Morgantown, 1972.
17. Gröchenig, K., *Foundation of Time–Frequency Analysis*, Birkhäuser, Boston, 2001.
18. Grossmann, A., Morlet, J. and Paul, T., Transforms associated to square integrable group representations. I. General results, *J. Math. Phys.* **26** (1985), 2473–2479.
19. Hewitt, E. and Ross, K. A., *Abstract Harmonic Analysis I*, Grundlehren Math. Wiss., Vol. 115, Springer, New York, 1979.
20. Holschneider, M., Wavelet analysis on the circle, *J. Math. Phys.* **31** (1990), 39–44.
21. Holschneider, M., *Wavelets: An Analysis Tool*, Oxford Univ. Press, New York, 1995.
22. Kirillov, A. A., *Elements of the Theory of Representations*, Grundlehren Math. Wiss., Vol. 220, Springer, Berlin, 1976 (Translated from the Russian by E. Hewitt).
23. Knapp, A., *Representation Theory of Semisimple Groups. An Overview Based on Examples*, Princeton Math. Ser., Vol. 36, Princeton University Press, Princeton, 1986.
24. Lang, S., $SL_2(\mathbb{R})$, Grad. Texts in Math., Vol. 105, Springer, New York, 1985.
25. Loomis, L. H., *An Introduction to Abstract Harmonic Analysis*, Univ. Ser. Higher Math., Van Nostrand, Princeton, 1953.
26. Triebel, H., Modulation spaces on the Euclidean n-space, *Z. Anal. Anw.* **2** (1983), 443–457.
27. Triebel, H., *Theory of Function Spaces*, Birkhäuser, Basel, 1983.

INDEX